PROPAGATION OF VISIBLE AND INFRARED RADIATION IN THE ATMOSPHERE

PROPAGATION OF VISIBLE AND INFRARED RADIATION IN THE ATMOSPHERE

V. E. Zuev

Translated from Russian by D. Lederman
Translation edited by P. Greenberg

A HALSTED PRESS BOOK

JOHN WILEY & SONS
New York · Toronto

ISRAEL PROGRAM FOR SCIENTIFIC TRANSLATIONS
Jerusalem · London

Sole distributors for the Western Hemisphere and Japan

HALSTED PRESS, a division of
JOHN WILEY & SONS, INC., NEW YORK

Library of Congress Cataloging in Publication Data

Zuev, Vladimir Evseevich.
Propagation of visible and infrared radiation in the atmosphere.

Translation of *Rasprostranenie vidimykh i infrakrasnykh voln v atmosfere.*
"A Halsted Press book."
Bibliography: p.
1. Atmosphere. 2. Infra-red rays. 3. Meteorological optics. 4. Ionospheric radio wave propagation.
I. Title.
QC880.Z8413 551.56 73-13827
ISBN 0-470-98460-0

Distributors for the U.K., Europe, Africa and the Middle East

JOHN WILEY & SONS, LTD., CHICHESTER

Distributed in the rest of the world by

KETER PUBLISHING HOUSE JERUSALEM LTD.

ISBN 0 7065 1312 6
IPST cat. no. 22058

This book is a translation from Russian of
RASPROSTRANENIE VIDIMYKH I INFRAKRASNYKH
VOLN V ATMOSFERE
"Sovetskoe Radio"
Moscow, 1970

Printed in Israel

PREFACE

Recent years have seen increased interest on the part of engineers and scientists in the behavior of electromagnetic radiation of optical wavelengths propagating in the earth's atmosphere. This is due primarily to the promise held by lasers in systems for communications, information transmission, telemetry, radar and other devices which operate through the atmosphere. In addition, a wide range of meteorological, geophysical and astrophysical problems require ever-increasing accuracy of quantitative data on the characteristics of atmospheric propagation of visible and infrared radiation. Such problems include: interpretation of satellite radiation data and the related inverse problems of satellite meteorology; calculation of radiative heat transfer in the atmosphere and determination of the thermal budget of the earth as a planet; derivation of the radiation term in weather-forecasting equations; reception of astrophysical observation data free from the effect of the earth's atmosphere.

The solution to the above problems involves knowledge of a large volume of quantitative data on the atmospheric propagation of optical radiation for large variations in both the radiation characteristics and propagation conditions. In this connection the radiation characteristics comprise: width of spectral interval or degree of monochromaticity; power and its time-variation (continuous and discrete radiation); coherence; directivity and geometric dimensions (from infinite waves to tight beams characterized by initial diameter and angle of divergence); spatial homogeneity of the field. Propagation conditions depend on instantaneous and average temperature values, total and partial pressures of the gaseous components of the atmosphere, chemical composition, concentration, spectra of dimensions and forms of atmospheric aerosols; all these quantities affect the parameters of radiation propagating through the atmosphere.

The author's first monograph /1/ and the recently published monographs of Kondrat'ev /2/ and Goody /3/ considered some aspects of the problem of interaction between optical radiation and the atmosphere. In /1/ we summarized data on absorption, scattering and attenuation of thermal radiation in the visible and the infrared by the atmosphere, under various meteorological conditions. The computation of heat transfer in the earth's atmosphere was discussed systematically in /2/. Goody's monograph /3/ is devoted primarily to the theory of absorption of thermal radiation by atmospheric gases.

This monograph gives results of theoretical and experimental studies of various aspects of atmospheric propagation of visible and infrared radiation, obtained both in the USSR and elsewhere.

The book consists of three parts. In the first we consider questions of absorption of visible and infrared radiation from thermal sources and lasers by atmospheric gases, under different propagation conditions. The possible variations of these conditions and average vertical profiles of temperature, pressure, humidity and concentration of absorbing gases in the atmosphere are described in Chapter 1. Chapter 2 is devoted to the origin of the absorption spectra of atmospheric gases (water vapor, carbon dioxide, ozone, methane, nitrogen oxides, carbon monoxide, oxygen). Chapters 3 and 4 analyze theoretical and experimental research on the absorption of optical radiation propagating in horizontal and oblique directions under various atmospheric conditions.

In the second part of the monograph we study problems of aerosol and molecular scattering of visible and infrared radiation from thermal sources and optical lasers, under different atmospheric conditions. These conditions are described in Chapter 5. The behavior of optical radiation scattered by a single particle is examined in Chapter 6. Chapter 7 summarizes theoretical and experimental investigations of the scattering of visible and infrared radiation under different conditions in the earth's atmosphere.

The third part of the book comprises a systematic examination of various aspects of the propagation of tight radiation beams from lasers and thermal sources in the atmosphere. The basic radiation parameters of lasers of various types are considered in Chapter 8. In Chapter 9 we analyze results concerning the structure of the light field in the cross-section of a beam propagating through various scattering media. Nonlinear effects accompanying the propagation of high-power light radiation in the atmosphere are discussed in Chapter 10. In the eleventh and last chapter we consider fluctuations in the parameters of tight radiation beams due to turbulent motions in the atmosphere.

It is the author's pleasant duty to express his heartfelt gratitude to his colleagues at the infrared radiation laboratory of the Siberian Physicotechnical Institute of Tomsk University: B. A. Antipov, P. A. Bokhan, A. G. Borovyi, E. A. Vedernikova, O.K. Voitsekhovskaya, V. N. Genin, L. I. Gerasimova, I. I. Ippolitov, M. V. Kabanov, G. M. Krekov, P. N. Kokhanenko, M. Ya. Krylova, V. P. Lopasov, E. V. Lugin, Yu. S. Makushkin, A. A. Nalivaiko, L. I. Nesmelova, B. A. Savel'ev, I. V. Samokhvalov, V. A. Sapozhnikova, V. K. Sonchik, V. V. Fomin, S. S. Khmelevtsov, N. A. Chernyavskaya; all these assisted in various ways in the preparation of the manuscript. The author is particularly indebted to Stanislav Dmitrievich Tvorogov, who collaborated in the writing of Chapters 4, 6 and 10, and Sec. 2 of

Chapter 7. Thanks are also due to G. P. Puchkov and S. I. Sokolov for their great help in technical matters.

The author is grateful to Professors K. Ya. Kondrat'ev and A. A. Semenov for their valuable critical remarks and advice offered during their examination of the manuscript.

CONTENTS

Preface v

Part One

ABSORPTION OF VISIBLE AND INFRARED RADIATION IN THE EARTH'S ATMOSPHERE

1. The earth's atmosphere as an absorbing medium 1
1. Composition of the atmosphere, pressure and temperature variations . 1
2. Water vapor distribution in the atmosphere 4
3. Distribution of atmospheric ozone 21
4. Distribution of CO_2, O_2, CH_4, N_2O and CO in the atmosphere . . . 22

2. The origin and general description of absorption spectra of atmospheric gases 25
1. Molecular energy and spectra 25
2. Rotational energy and rotational spectra of molecules 27
3. Vibrational energy and vibration-rotation spectra of molecules 35
4. Electronic energy and electronic molecular spectra 42
5. Shape of spectral lines 43
6. Spectral line strength 49
7. General characteristics of absorption spectra of atmospheric gases . . 50

3. Absorption of visible and infrared radiation in a homogeneous atmosphere . 66
1. Basic definitions 66
2. Absorption of a single line 68
3. Absorption of groups of overlapping lines. 73
4. Limits of applicability of the Bouguer-Beer law 82
5. Fine structure of the vibration-rotation and pure rotational spectrum of water vapor 86
6. Absorption function in narrow regions of the vibration-rotation and pure rotational water vapor spectrum 104
7. Fine structure of the absorption spectrum of carbon dioxide. 114
8. Absorption functions in different regions of the CO_2 spectrum. . . . 117
9. Fine structure and absorption functions in the ozone spectrum. . . . 120
10. Fine structure and absorption functions of the oxygen spectrum . . . 122
11. Fine structure and absorption functions of spectra of minor atmospheric constituents 122

12. Absorption of overlapping lines and bands 126
13. Absorption of radiation from selective sources 128
14. Absorption of laser radiation in the atmosphere 130

4. Absorption of visible and infrared radiation in an inhomogeneous atmosphere **146**
1. Spectral transmission (absorption). Transmission (absorption) function 146
2. Direct computation of spectral transmission and the transmission function . . . 147
3. Appropriate computation of absorption function H 150
4. Approximate allowance for temperature in computation of function H 154
5. Approximate determination of the spectral transmission of an inhomogeneous atmosphere 155
6. Computation of the absorption function with allowance for the earth's curvature 156
7. Experimental study of absorption function H 159
8. Absorption of laser radiation in an inhomogeneous atmosphere . . . 165

Part Two

SCATTERING OF VISIBLE AND INFRARED RADIATION IN THE EARTH'S ATMOSPHERE

5. The earth's atmosphere as a scattering medium 168
1. Cloud and fog 168
2. Haze 177
3. Precipitation 187

6. Scattering of visible and infrared radiation by a single particle 189
1. Representation of polarized light 189
2. Scattering of a plane monochromatic wave 191
3. Coefficients of scattering, absorption and extinction 193
4. Scalar approximation in the light-scattering problem 195
5. Scattering of an inhomogeneous field 199
6. Scattering of visible and infrared radiation by a single spherical particle 204
7. Scattering of light by nonspherical particles 221

7. Scattering of visible and infrared radiation in the atmosphere 223
1. Molecular scattering of visible and infrared radiation 224
2. Scattering of waves by a system of particles 225
3. Investigation of the scattering matrix 228
4. Energy characteristics of scattering of visible and infrared radiation by clouds and fog 229
5. Energy characteristics of scattering of visible and infrared radiation by haze 246

6. Energy characteristics of scattering of visible and infrared radiation by precipitation 262
7. Scattering diagrams of polydisperse aerosols 263
8. Scattering of laser radiation in the atmosphere 266
9. Effect of measuring apparatus on experimental results in scattering of visible and infrared radiation in the atmosphere 269

Part Three

PROPAGATION OF TIGHT LIGHT BEAMS IN THE ATMOSPHERE

8. Radiation from various types of lasers 276
Introduction 276
1. Solid crystalline lasers 276
2. Glass lasers 279
3. Gas lasers 280
4. Semiconductor lasers 282
5. Variation in laser radiation parameters due to various causes . . . 283

9. Propagation of tight light beams in atmospheric aerosols 286
1. Brightness of forward-scattered radiation. 286
2. Extinction of direct radiation in scattering media. Limits of applicability of the Bouguer law 301
3. Brightness of back-scattered radiation 304
4. Polarization characteristics of scattered radiation 307
5. Propagation of a pulse of directed radiation in a scattering and absorbing medium 310

10. Nonlinear effects in the propagation of high-power light radiation in the atmosphere 320
1. Probability of multiple-photon processes 320
2. Computation of the transition probability by successive approximations 323
3. Estimates for the probability of multiple-photon photoionization and photodissociation 325
4. Spectroscopic saturation effect 332
5. Absorption of light with allowance for saturation effects 335
6. Self-focusing of a laser beam 336
7. Action of a high-power radiation field on atmospheric aerosol particles 338

11. Propagation of tight light beams in a turbulent atmosphere 348
1. General remarks 349
2. Theoretical studies of fluctuations in the parameters of tight beams propagating in the atmosphere 351

3. Experimental studies of fluctuations in the parameters of tight beams propagating in a turbulent atmosphere 356

Appendix 1 368

Appendix 2 371

Bibliography 376

Index 403

Part One

ABSORPTION OF VISIBLE AND INFRARED RADIATION IN THE EARTH'S ATMOSPHERE

1. THE EARTH'S ATMOSPHERE AS AN ABSORBING MEDIUM

The earth's atmosphere is an absorbing medium possessing very high space-time variability. The absorption of visible and infrared radiation in the atmosphere depends essentially on the pressure, temperature and concentration of absorbing gases. Each of these quantities varies widely with altitude and time, and also as a function of geography. Without knowing how these elements vary we cannot solve a single problem requiring quantitative data on the absorption of the visible and the infrared in the earth's atmosphere.

This chapter deals with the behavior of the pressure, temperature and concentration of absorbing atmospheric gases. Particular attention is paid to the water vapor distribution in the atmosphere, since it is the principal absorbing component; this component is also more variable than the others.

Variations considered in greatest detail will be those of all the elements of interest in the lower 30-km atmospheric layer, which contains the bulk of all the gases absorbing radiation in the visible and infrared spectral regions.

1. COMPOSITION OF THE ATMOSPHERE, PRESSURE AND TEMPERATURE VARIATIONS

Pressure is the only element affecting absorption of visible and infrared radiation. The time-variation in pressure can almost always be neglected throughout the lower 30-km layer of the atmosphere. Vertical pressure profiles for different regions of the globe can also be treated as virtually identical.

The atmosphere consists primarily of nitrogen, oxygen and argon, the percent content of which is tabulated in Table 1.1. The table also presents data on other components of the atmosphere, with the exception of water vapor. The atmosphere also contains trace amounts of other gases which, however, have no perceptible effect on the absorption of visible and infrared radiation. In individual local regions one may encounter gases which absorb visible and infrared radiation, but are not listed in Table 1.1.

TABLE 1.1. Composition of dry air /3/

Constituents of the atmosphere	Average volume concentration in the troposphere, %	Remark
N_2	78.084	Dissociates in the upper ionosphere, mixed at lower levels.
O_2	20.946	Dissociates above 95 km, mixed at lower levels.
Ar	$9.34\cdot10^{-1}$	Mixed up to 110 km, diffusion separation above this level.
CO_2	$3.3\cdot10^{-2}$	Undergoes insignificant changes, mixed up to 100 km, dissociates above this level.
Ne	$1.818\cdot10^{-3}$	Mixed up to 100 km, diffusion separation above this level.
He	$5.24\cdot10^{-4}$	Mixed up to 110 km, diffusion separation above this level.
CH_4	$1.6\cdot10^{-4}$	Mixed in the troposphere, oxidized in the stratosphere, dissociates in the mesosphere.
Kr	$1.14\cdot10^{-4}$	Mixed up to 110 km, diffusion separation above this level.
H_2	$5\cdot10^{-5}$	Mixed in the troposphere and stratosphere, dissociates above this level.
N_2O	$3.5\cdot10^{-5}$	Changes slightly at the ground, gradually dissociates in the stratosphere and mesosphere.
CO	7.10^{-6}	Oxidation product, variable content.
O_3	10^{-6}	Changes quite markedly, photochemical origin.
NO_2 NO	from 0 to $2\cdot10^{-6}$	Industrial origin in the troposphere, photochemical nature in the mesosphere and ionosphere.

The sea-level pressure averaged over the globe is 1013 mb. This quantity ranges between 980 and 1040 mb /4/. Deviations from mean pressure thus lie within approximately ±3%. The reduction in pressure with altitude (vertical profile) is described quite satisfactorily by an exponential law, which follows from the barometric height equation. The numerical values of pressure at various altitudes in the bottom 30-km layer, according to standard models of the atmosphere, are listed in Appendix 1. Data on pressure at altitudes up to 100 km are also tabulated there.

The average vertical temperature profile points to the presence of a number of layers in the atmosphere. The lowest layer, in which the average temperature decreases with altitude, is conventionally termed the troposphere. The vertical thickness of this layer depends on latitude and season; in tropical regions it is 16—18 km, in polar regions it is 7—10 km. The next layer of the atmosphere is the stratosphere, which extends to approximately 50 km. A characteristic feature of this layer is an initially (up to approximately 20 km) slow and then rapid rise in temperature. The boundary layer between the troposphere and stratosphere, in which the vertical temperature gradient changes sign, is termed the tropopause.

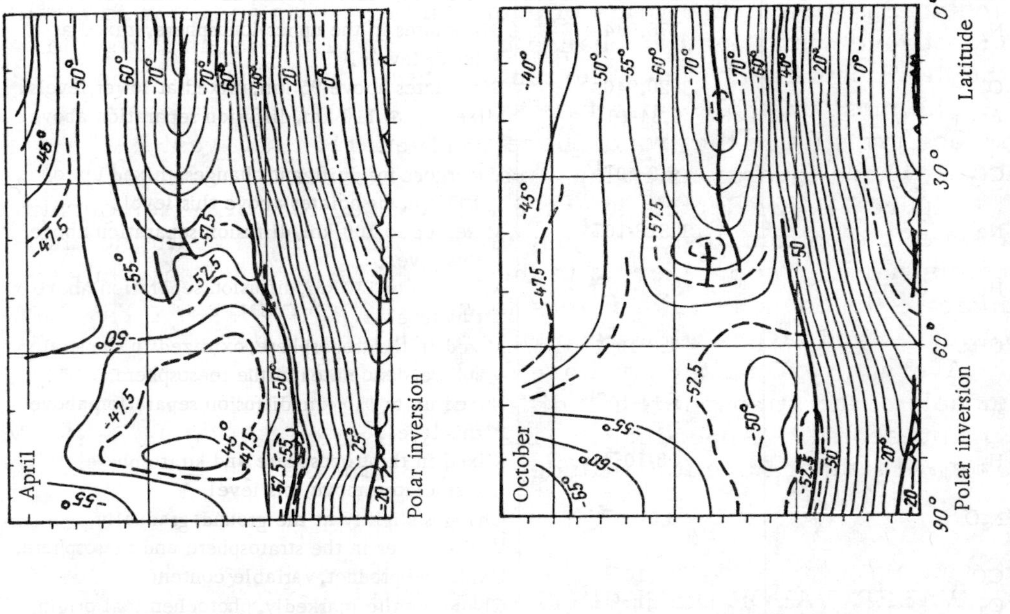

FIGURE 1.1. Annual variation in average atmospheric isotherms along 80°W at altitudes from 0 to 30 km.

The temperature attains its maximum at the upper boundary of the stratosphere. The latter is followed by the stratopause, in which the lapse rate changes sign again. The atmospheric layer situated above 50 km is called the mesosphere and extends to an altitude of 80—90 km. Here the temperature decreases with altitude and attains an average value of about 200°K. This is followed by the mesopause with a further change in the sign of the lapse rate, and the thermosphere in which the temperature steadily increases.

The most unstable lapse rate within the above atmospheric layers prevails in the troposphere, particularly its lowest 2—3-km layer, where inversions in lapse rate are frequently observed.

Figure 1.1 /3/ depicts average vertical sections of the temperature field of the bottom 30-km atmospheric layer at 80°W in the northern hemisphere. The isotherms in the figure are in °C. Heavy curves indicate the boundaries of layers in which temperature inversion is observed and layers with steep changes in lapse rate.

Average vertical temperature profiles assumed in standard models of the atmosphere for different altitudes from 0 to 30 km are presented in Appendix 1. This appendix also lists data on the average temperature for altitudes between 25 and 100 km.

2. WATER VAPOR DISTRIBUTION IN THE ATMOSPHERE

The humidity of air is usually specified by several means: 1) absolute humidity a, expressed in g/m^3; 2) partial water vapor pressure e, expressed in mm Hg or mb; 3) relative humidity r, which is the ratio of the actual partial pressure of the water vapor in the atmosphere to the saturation vapor pressure E at the same temperature, expressed as a percentage (%); 4) specific humidity q, which is the mass of water vapor per unit mass of moist air; 5) mixing ratio δ, which is the mass of water vapor per unit mass of dry air; 6) precipitable water layer ω, which is the thickness of the water layer obtained by fully condensing the water vapor in the given atmospheric layer; 7) deficit moisture content d, which is the difference between the maximum and actual partial water vapor pressure at given temperature and pressure; 8) dew point τ, which is the temperature at which the saturation vapor pressure is equal to the partial vapor pressure in the atmosphere.

The humidity characteristics are interrelated by the following expressions /4, 5/:

$$a = 217\frac{e}{T}(\text{mb}) = 289.4\frac{e}{T}\ (\text{mm Hg}); \tag{1.1}$$

$$q=\frac{\delta}{1+\delta}=0.622\,\frac{e}{P-0.378e}\,;$$

$$\delta=\frac{q}{1-q}=0.622\,\frac{e}{P-e}\,; \qquad (1.2)$$

$$r=\frac{e}{E}\cdot 100\%;\quad d=E-e;$$

$$e=E_0\cdot 10^{b\tau/c+\tau}, \qquad (1.3)$$

where T is absolute temperature, P is air pressure, measured in the same units as e and $E_0=4.58$ mm Hg $=6.11$ mb; τ is measured in °C; b and c for water are 7.5 and 237.3, and for ice 9.5 and 265.5, respectively.

Since in the atmosphere $e \ll P$; $q \ll 1$ and $\delta \ll 1$, it follows from (1.2) that

$$q \simeq \delta \simeq 622\,\frac{e}{P}\,.$$

The precipitable water layer ω' per 1 km is related to the absolute humidity (expressed in g/m³) by the simple relationship

$$\omega' = 10e. \qquad (1.4)$$

The concept of precipitable water layer is used extensively in the study of radiation absorption by water vapor.

Table 1.2 lists values of E, a, q and δ for $P=1000$ mb (surface layer) and different temperatures.

TABLE 1.2. Saturation vapor pressure E over a flat pure-water surface, absolute humidity a, specific humidity q, and mixing ratio δ for pressure $P=1000$ mb and different temperatures

t, °C	E, mb	a, g/m³	q, g/kg	δ, g/kg
−20	1.254	1.073	0.780	0.781
−10	2.863	2.357	1.782	1.786
0	6.108	4.845	3.808	3.822
10	12.270	9.39	7.67	7.73
20	23.37	17.27	14.67	14.89
30	42.43	30.33	26.82	27.56
40	73.78	65.30	47.20	49.54
50	123.40	82.76	80.51	87.56

Water vapor concentration is highly variable in time and space, so much so that each specific measurement of the vertical humidity profile is nonrepeatable. Nevertheless average statistical horizontal and vertical profiles in the troposphere have been studied quite satisfactorily.

The average vertical distribution of water vapor concentration in the troposphere satisfies the exponential law

$$e(z) = e(0)\,e^{-\alpha z}, \qquad (1.5)$$

where $e(z)$ and $e(0)$ are water vapor pressures at altitude z and at the surface, respectively; α is an empirical constant. The water vapor concentration in the ground layer can vary as a function of time and location between approximately 0 and 7%.

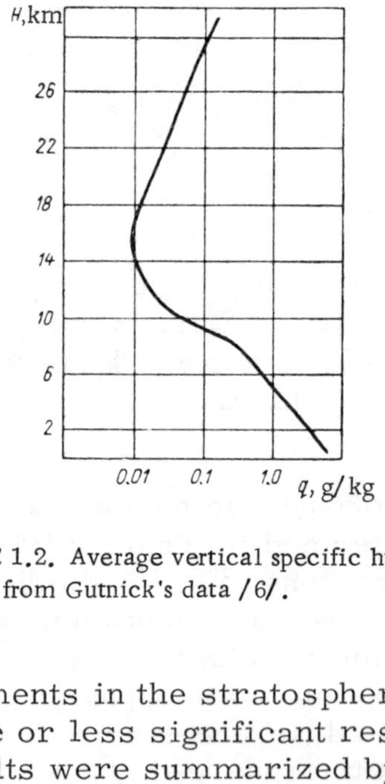

FIGURE 1.2. Average vertical specific humidity profile from Gutnick's data /6/.

Humidity measurements in the stratosphere encounter difficulties, for which reason more or less significant results were only recently obtained. These results were summarized by Gutnick /6/, who constructed the average vertical profile of specific humidity (Figure 1.2) cited elsewhere /7, 8, 13/. Malkevich et al. /7/ is of the opinion that two types of vertical water vapor distribution exist in the stratosphere. The first type possesses a steep drop in water vapor concentration down to very low specific humidity values (of the order of 10^{-3}—10^{-2} g/kg), in the lower layers of the stratosphere (to 16 km). Above this layer the concentration increases, attaining at the maximum (for which data are available) altitude of 31 km a value which is 1—2 orders higher than at the 16-km level. The second type of distribution possesses constant specific humidity, starting at some altitude in the stratosphere.

The altitudinal dependence of specific humidity in both cases assumes the form

$$q(z)=\begin{cases} q_0 e^{-\alpha_0 z}, & 0\leqslant z\leqslant z_1, \\ q_1 e^{\alpha_1(z-z_1)}, & z_1\leqslant z\leqslant z_{max}. \end{cases} \qquad (1.6)$$

It is seen from this equation that in describing the second type of vertical humidity distribution one should set $\alpha_1 = 0$. Table 1.3 lists values of the parameters appearing in equation (1.6), obtained by analyzing various measurements.

TABLE 1.3. Specific humidity q_0 and q_1 and parameters α_0 and α_1 in equation (1.6) from data of Malkevich et al. /7/

q_0, g/kg	α_0, km^{-1}	q_1, g/kg	α_1, km^{-1}	z_1, km
12.2	0.52	0.012	0.08	12
11.5	0.49	0.006	0.23	14
1.6	0.43	0.05	0	8
5.1	0.42	0.01	0.27	15
17.5	0.56	0.04	0	10
3.6	0.51	0.002	0.18	14
5.7	0.48	0.002	0.24	16
2.6	0.55	0.002	0.31	16

The above two types of specific humidity distribution in the stratosphere have been termed in Western literature the "wet" and "dry" models of the stratosphere /10/. Both models assume an identical vertical water vapor distribution up to the tropopause level. The difference between the models is insignificant when calculating infrared radiation absorption in atmospheric layers, including the lower layers. Perceptible differences arise when calculating radiation absorption in the upper atmosphere.

The controversy between the adherents of the dry and wet stratosphere continues to this day, since later publications support sometimes one and sometimes the other hypothesis.

It is this author's opinion that this controversy cannot be resolved solely by studying individual measurements. It requires statistical processing of systematic measurements of the vertical water vapor concentration distribution. This approach was made /9, 11–13/, the principal results of which are subsequently presented.

1. Average vertical profile of specific humidity. The average vertical profile of specific humidity for the winter and summer seasons in moderate latitudes was obtained from data on the average temperature and relative humidity at the principal isobaric surfaces, obtained at 95 aerological stations of the Soviet Union by means of the A-22 radiosonde. Average values of these meteorological elements were calculated from 280—370 observations over 48 hours in the 400—100-mb layer and from 20—100 observations at the 50-, 30-, 20- and 10-mb layers. In the latter case use was made of IQSY data. The average values of specific humidity were corrected for the systematic error of measurements due to inertia of the radiosonde sensor.

Figure 1.3 shows average vertical specific humidity profiles for January and July, together with Gutnick's profile. A sharp reduction in specific humidity with altitude occurs in the upper troposphere in all cases. Qualitatively the distributions in Figure 1.3 are approximately similar to that of Gutnick. However, the absolute values of specific humidity at different altitudes differ quite markedly. The January- and July- average specific humidity profiles for moderate latitudes of the USSR lie to the right of the Gutnick profile. On the other hand, profiles of individual regions lie to both the right and left of the Gutnick profile.

Table 1.4 lists the monthly average values of specific humidity at the ground and at different isobaric surfaces for January and July in the Baltic Sea region, Central Asia and Eastern Siberia, determined from single observations at the Riga, Tashkent and Irkutsk stations.

TABLE 1.4. Average specific humidity in January and July for the Riga, Tashkent and Irkutsk stations from data of Komarov /9/, g/kg

Isobaric surface, mb	Riga station		Tashkent station		Irkutsk station	
	January	July	January	July	January	July
Ground	2.30	8.70	2.78	7.80	1.62	9.40
900	2.00	6.50	2.30	5.50	1.10	8.40
850	1.75	5.50	2.10	4.70	0.90	6.90
700	0.88	3.20	1.42	3.60	0.60	4.50
600	0.53	2.00	0.86	2.50	0.35	3.00
500	0.33	1.15	0.53	1.60	0.25	1.90
400	0.10	0.28	0.16	0.60	0.07	0.68
300	0.03	0.11	0.04	0.17	0.031	0.20
200	**0.020**	**0.07**	**0.03**	0.06	**0.024**	**0.048**
150	0.029	0.08	0.04	0.024	0.041	0.061
100	0.036	0.09	0.05	**0.011**	0.068	0.088
50	0.048	0.16	0.06	0.037	0.130	0.14
30	0.053	0.31	0.12	0.12	0.245	0.30
20	0.070	0.59	0.20	0.25	0.36	0.49
10	0.180	1.50	0.65	1.20	0.85	1.50

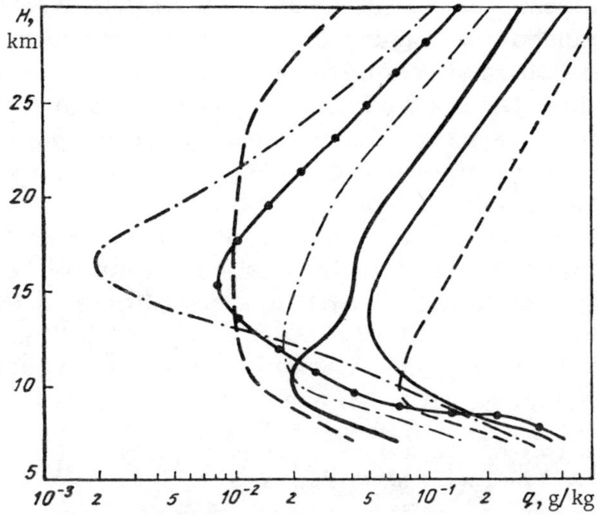

FIGURE 1.3. Average vertical specific humidity profile for moderate latitudes of the USSR (——), for the Barentsburg (— — —) and Ashkhabad (— · —) stations, and Gutnick's profile (—•—).

Heavy and fine curves denote January and July profiles, respectively.

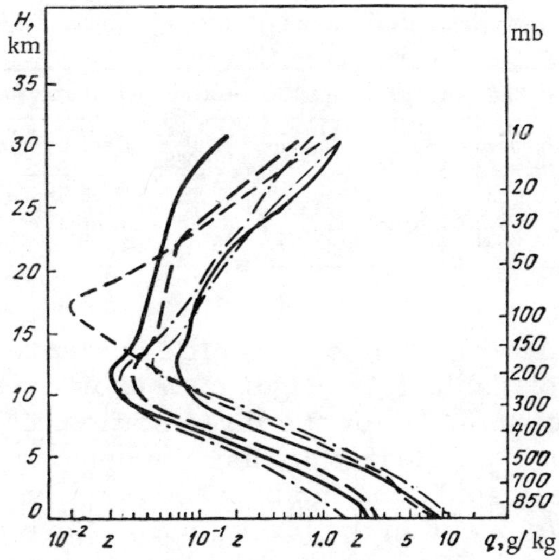

Figure 1.4. Monthly average vertical specific humidity profiles for the Riga (—), Tashkent (— — —), and Irkutsk (– · –) stations from Komarov's data.

Heavy and fine curves denote January and July profiles, respectively.

The statistical processing was based on data of four daily radiosoundings of the atmosphere at the ground surface and at isobaric surfaces from 900 to 100 mb for January and July from 1961 to 1966. Data used for the 50- and 10-mb level were obtained during the IQSY (1964—1965). At levels between the ground and 100 mb, 560—740 observations were carried out at each station in January and July; 30—150 observations were conducted in January and 50—180 in July at the 50- to 10-mb levels.

The minimum value in each column of Table 1.4 has been emphasized. Note that the data in Table 1.4 were not corrected for the systematic measurement error due to radiosonde inertia.

TABLE 1.5. Average vertical specific humidity gradient (g/kg/100 m) at the Riga, Tashkent and Irkutsk stations

Layer, mb	Riga station		Tashkent station		Irkutsk station	
	January	July	January	July	January	July
0—900	0.03	0.22	0.05	0.23	0.05	0.10
900—850	0.05	0.20	0.04	0.16	0.04	0.30
850—700	0.07	0.15	0.05	0.07	0.02	0.25
700—600	0.04	0.12	0.06	0.11	0.03	0.15
600—500	0.02	0.09	0.03	0.09	0.01	0.11
500—400	0.02	0.04	0.02	0.05	0.01	0.06
400—300	0.004	0.009	0.006	0.021	0.002	0.024
300—200	0.0003	0.0013	0.0003	0.0036	0.0002	0.005
200—150	—0.0005	—0.0005	0.0005	0.0018	—0.0008	—0.0006
150—100	—0.0002	—0.0003	—0.0003	0.0004	—0.0009	—0.0009
100—50	—0.0003	—0.0018	—0.0002	—0.0006	—0.0013	—0.0010
50—30	—0.0002	—0.0050	—0.0020	—0.0028	—0.0038	—0.0053
30—20	—0.0008	—0.0140	—0.0040	—0.0065	—0.0058	—0.0095
20—10	—0.0027	—0.0180	—0.0110	—0.0190	—0.0120	—0.0202

Table 1.5 lists average vertical specific humidity gradients calculated from data of Table 1.4. Figure 1.4 shows vertical profiles of average specific humidity for the three stations in question. Table 1.4 and Figure 1.4 clearly display the behavior of the individual profiles. The most rapid drop in specific humidity with altitude is shown (Table 1.4) to occur in the lower troposphere to the 600-mb level (0— 4 km range). For all stations the minimum monthly average specific humidity is observed near the tropopause, approximately 0.5—1 km above the minimum-temperature level.

Analysis of individual observations at any of the stations discloses a high scatter of the vertical specific humidity profiles, which can be subdivided into three types. The first consists of profiles with a perceptible rise in specific humidity with altitude above the tropopause level for sufficiently high moisture content of the middle stratosphere (of the order of 0.1—0.5 g/kg). The second type possesses constant specific humidity at different altitudes in the stratosphere for low moisture content throughout the stratosphere (of the order of 0.01—0.02 g/kg). Finally, the third type comprises profiles with a slow decrease in specific humidity with altitude in the stratosphere for low moisture content of the middle stratosphere (of the order of 0.01—0.02 g/kg). Figure 1.5 gives examples of these three types of profile.

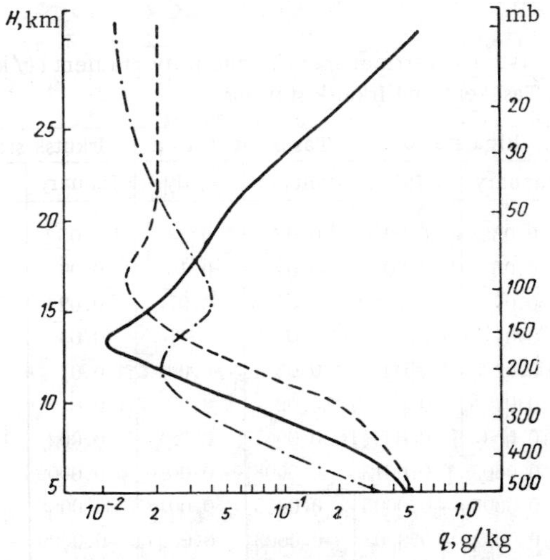

FIGURE 1.5. Examples of different types of vertical specific humidity profile obtained at Irkutsk on 28 July 1964 (—), Tashkent on 7 July 1963 (— — —), and Riga on 5 January 1966 (— · —).

Hence during each specific observation the stratosphere can be both "wet" and "dry." However, the average statistical vertical specific humidity profile points to a quite perceptible increase in this quantity with altitude in the stratosphere for all the stations, in both January and July. This gives rise to the question: where does so much moisture in the upper layers of the stratosphere come from?

2. Transport of water vapor from the troposphere to the stratosphere. One answer to this question can be derived from a study conducted by Komarov /13/ of the vertical transport of water vapor from the troposphere to the stratosphere. Komarov processed statistical observation data at 12 USSR stations distributed more or less uniformly over the country, and only examined radiosonde data on humidity and wind speed and direction pertaining to the 400-, 300-, 200-, 150-, 100-, 50-, 30-, 20- and 10-mb isobaric surfaces for the duration of the IQSY (1964—1965). Information was thus obtained on the frequency of vertical motion, which characterizes the frequency of tropopause penetration by ascending airflows.

The frequency of vertical air motion in the tropopause region in January for 9 of the 12 stations is 52—67%, and 52—64% in July for 10 of the 12 stations. For the remaining three stations in winter and two in summer, descending air motion prevails with frequency from 52 to 60%.

Table 1.6 lists average frequencies of upward and downward air motion for the middle latitudes of the USSR in January and July for different atmospheric layers on clear and cloudy days, for $n \geqslant 30$ cases. The frequency of ascending and descending flows depends quite perceptibly on atmospheric cloudiness. Under cloudy conditions the frequency of upward air motion predominates in both winter and summer over the entire troposphere and up to the 100-mb level in the stratosphere. In summer cloudy conditions ascending airflow predominates in the troposphere as well as the stratosphere up to the maximum radiosonde ceiling. These flows supply moisture to the atmosphere. The stratosphere should therefore be drier on clear days and more moist on cloudy days. This conclusion is in agreement with corresponding average vertical specific humidity profiles of the stratosphere.

3. Variability of the humidity field at different altitudes in the atmosphere. The humidity field in the troposphere and stratosphere is highly variable as a function of time and physicogeographical location. Hence, alongside knowledge of the mean-statistical vertical humidity profiles, it is important to have data on the corresponding confidence intervals. The necessary calculations were carried out for four stations lying in different climatic regions of the USSR. The calculations were based on radiosonde data processing over an extended period of time at levels up to 100 mb, and over the IQSY period at levels from 50 to 10 mb.

Figures 1.6 through 1.13 depict monthly-average vertical specific humidity profiles for January and July and the confidence intervals. The latter characterize the number of cases in which the profiles lay within these intervals.

The specific humidity can change during very short periods of time owing to fluctuations in water vapor content, which bring about corresponding fluctuations in the beam propagating through the atmosphere /14/. The question of humidity fluctuations is only at the initial stage of study.

TABLE 1.6. Average frequency of ascending (asc.) and descending (desc.) vertical air motion in different atmospheric layers for middle latitudes of the USSR /13/

Layer, mb	January						July							
	clear		Ci, Cs and Ce clouds		average		clear		Ci, Cs and Ce clouds		Cb clouds		average	
	asc.	desc.	asc.	desc.	asc.	desc.	asc.	desc.	asc.	desc.	asc.	desc.	asc.	desc.
400—300	49	51	63	37	56	44	51	49	57	43	56	44	55	45
300—200	51	49	58	42	56	44	44	56	52	48	54	46	52	48
200—150	43	57	58	42	58	48	48	52	52	48	53	47	52	48
150—100	40	60	51	49	51	49	46	54	62	38	56	44	53	47
100—50	32	68	50	50	40	60	57	43	60	40	56	44	55	45
50—30	31	69	37	63	40	60	60	40	60	40	50	50	54	46
30—20	32	68	—	—	44	56	51	49	58	42	42	58	55	45

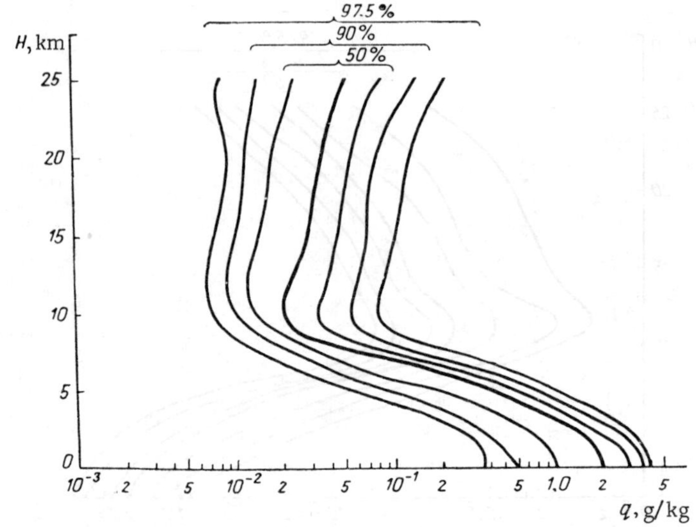

FIGURE 1.6. Average vertical specific humidity profile in January for the Voeikovo station and corresponding confidence intervals (50, 90 and 97.5%).

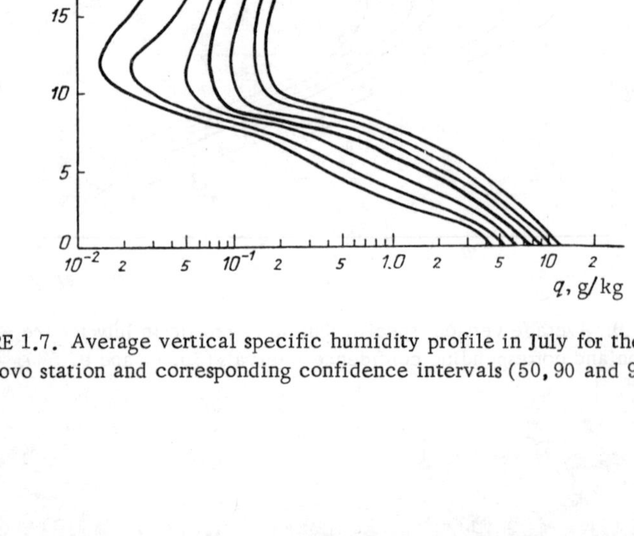

FIGURE 1.7. Average vertical specific humidity profile in July for the Voeikovo station and corresponding confidence intervals (50, 90 and 97.5%).

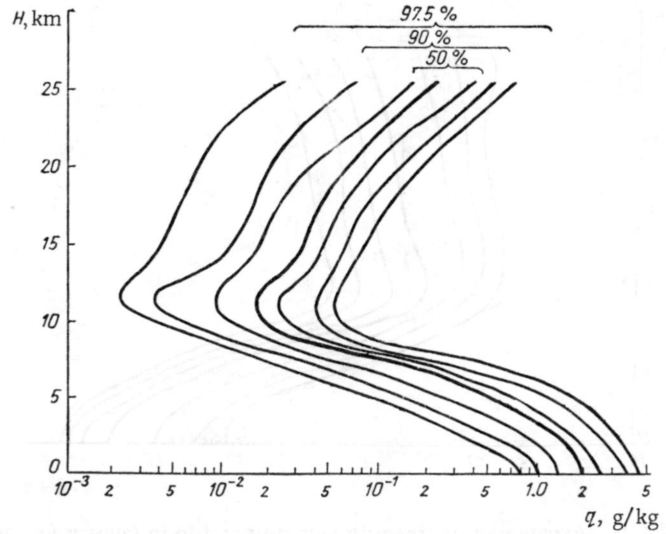

FIGURE 1.8. Average vertical specific humidity profile in January for the Alma-Ata station and corresponding confidence intervals (50, 90, and 97.5%).

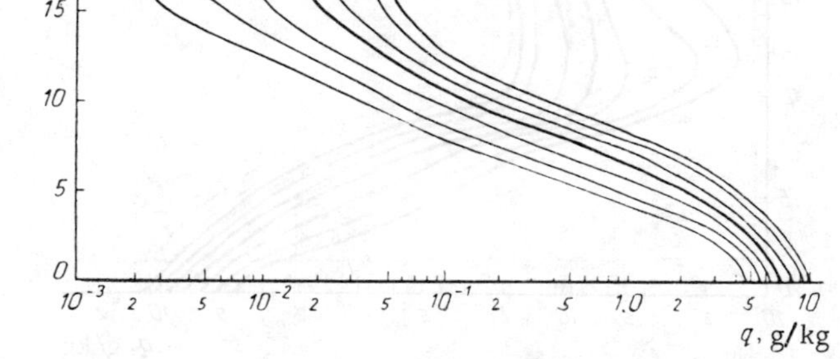

FIGURE 1.9. Average vertical specific humidity profile in July for the Alma-Ata station and corresponding confidence intervals (50, 90, and 97.5%).

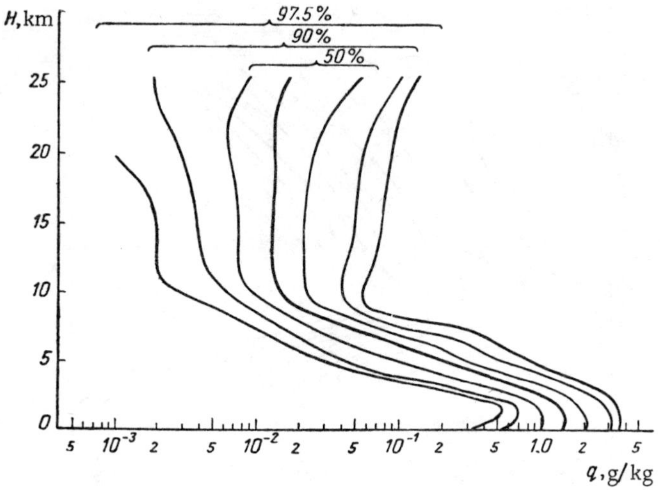

FIGURE 1.10. Average vertical specific humidity profile in January for the Murmansk station and corresponding confidence intervals (50, 90 and 97.5%).

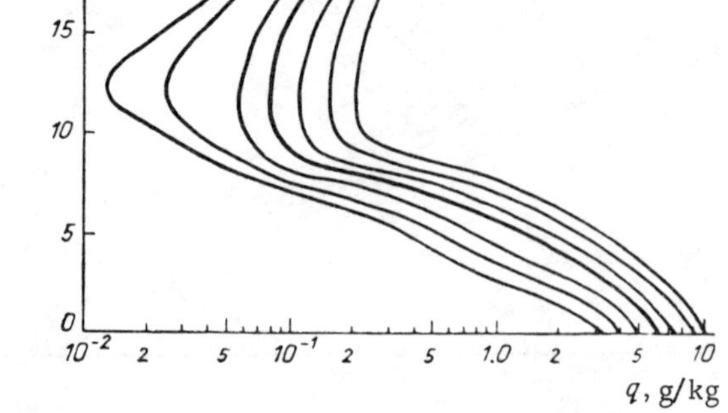

FIGURE 1.11. Average vertical specific humidity profile in July for the Murmansk station and corresponding confidence intervals (50, 90, and 97.5%).

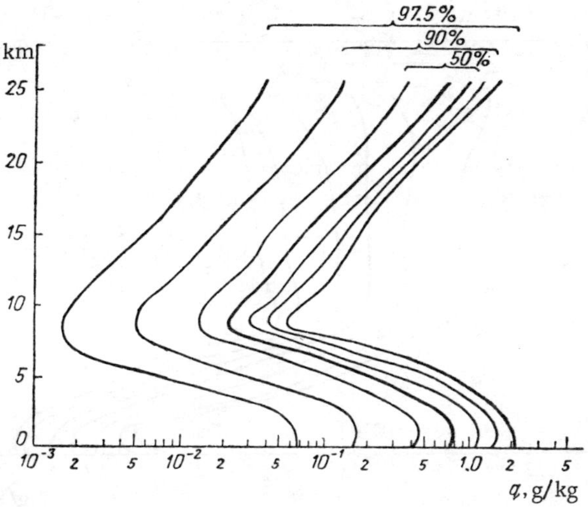

FIGURE 1.12. Average vertical specific humidity profile in January for the Nagaevo station and corresponding confidence intervals (50, 90 and 97.5%).

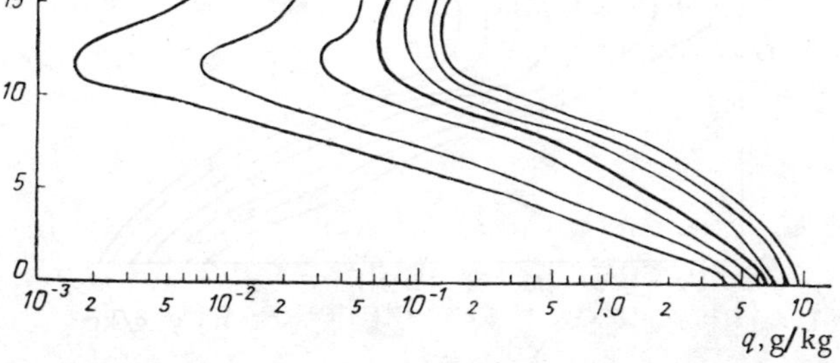

FIGURE 1.13. Average vertical specific humidity profile in July for the Nagaevo station and corresponding confidence intervals (50, 90 and 97.5%).

4. Total water vapor content of different atmospheric layers. Data on average vertical specific humidity profiles can yield the total water vapor content in different layers of the atmosphere in both vertical and horizontal directions (Tables 1.7 and 1.8).

TABLE 1.7. Monthly-average values of precipitable water (cm/km in layers at different altitudes for three stations in January and July /13/

Layer, mb	Riga station		Tashkent station		Irkutsk station	
	January	July	January	July	January	July
Ground–850	0.2070	0.7240	0.2490	0.6380	0.1290	0.8340
850—700	0.1340	0.4440	0.1790	0.4230	0.0770	0.5800
700—500	0.0620	0.2230	0.1000	0.1300	0.0430	0.3270
500—400	0.0110	0.0370	0.0180	0.0560	0.0560	0.0660
400—300	0.0030	0.0102	0.0051	0.0200	0.0024	0.0250
300—200	0.0013	0.0046	0.0018	0.0060	0.0017	0.0063
200—150	0.0007	0.0020	0.0009	0.0010	0.0010	0.0014
150—100	0.0006	0.0014	0.0008	0.0003	0.0013	0.0013
100—50	0.0001	0.0012	0.0001	0.0003	0.0019	0.0013
50—30	0.0004	0.0016	0.0006	0.0005	0.0017	0.0015
30—20	0.0001	0.0015	0.0005	0.0006	0.0011	0.0014
20—10	0.0003	0.0027	0.0011	0.0010	0.0016	0.0026

Knowledge of the precipitable water layer between two points in the atmosphere is very important for solving problems associated with calculation of visible and infrared atmospheric radiation absorption. In this connection we note a method for determining this quantity between two points in the atmosphere with allowance for the earth's curvature and refraction, when the vertical gas-concentration distribution is assumed known /15/. The mass of absorbing gas is obtained /15/ from the expression

$$\omega = \omega_{z_1} f(z_1) - \omega_{z_2} f(z_2), \qquad (1.7)$$

where ω_{z_1} and ω_{z_2} are the masses of absorbing gas in the vertical layer of the atmosphere from altitudes z_1 and z_2 to ∞. Function $f(z)$ is given in /16/ by the approximate expression

$$f(z) = (\cos z + k e^{-m \cos z})^{-1}, \qquad (1.8)$$

where m and k are constants.

5. Comparison of available data on specific humidity in the stratosphere. Available data on specific humidity in the stratosphere point to the presence of a marked scatter of quantitative values of this quantity. Both the average yearly profile of Gutnick /6/ and the average monthly profiles of Komarov /12/ differ quite markedly from many other profiles based on individual measurements, including those obtained during recent years. Houghton /8/ examined a large number of stratospheric humidity measurements, on the basis of which he concluded that the mixing ratio is constant within the limits $1-4\cdot10^{-3}$ g/kg, so supporting the dry model of the stratosphere. The minimum specific humidity in the Gutnick model at the tropopause level is $9\cdot10^{-3}$ g/kg; according to the January-average and July-average profiles of Komarov it is even higher. Above the tropopause level the specific humidity increases with altitude quite perceptibly according to both Gutnick and Komarov. A recent humidity measurement in the mesosphere at an altitude of 70—80 km points to a perceptible moisture content at these altitudes /17/. Fedynskii et al. /117/ obtained a specific humidity of 1.7 g/kg. A wet stratosphere is also supported by the presence of nacreous and luminescent clouds.

TABLE 1.8. Monthly-average values of precipitable water in various layers of the atmosphere in the vertical direction /13/ for three stations

Layer, mb	Riga station				Tashkent station				Irkutsk station			
	January		July		January		July		January		July	
	cm	%	cm	%	cm	%	cm	%	cm	%	cm	%
Ground — 10*	0.673		2.334		0.903		2.026		0.442		3.022	
Ground—200*	0.666	99.0	2.300	98.5	0.893	98.8	2.014	99.5	0.414	93.5	2.962	98.0
Ground—500*	0.635	94.0	2.197	94.0	0.842	94.3	1.851	91.4	0.394	89.3	2.772	91.8
200 — 10	0.008	1.0	0.034	0.034	0.010	1.2	0.012	0.5	0.028	6.5	0.061	2.0

* These quantities were computed for precipitable water in the earth — 10 mb layer.

Nevertheless, it is as yet impossible to resolve the substantial discrepancies in specific humidity data obtained by the different investigators. It appears to this author that it is first necessary to determine very accurately the errors in measurements carried out by the various instruments and techniques. It is above all necessary to resolve the problem of errors in determining humidity by means of very bulky radiosondes, because this is the only method which could yield sufficient data for statistically reliable analysis.

3. DISTRIBUTION OF ATMOSPHERIC OZONE

The substantial effect of ozone on radiation processes in the atmosphere has given rise to much work in this field /18—21, 557, 560, 562, 865/.

Some idea as to the vertical profile of ozone concentration can be obtained from Figure 1.14, which indicates that the average monthly ozone concentration at altitudes corresponding to the location of the maximum can vary approximately sixfold. Above and below the maximum the variation of average monthly concentrations is smaller, the further away the given altitude from the layer maximum. This pattern is followed in various geographic regions, although the absolute concentration also undergoes perceptible variations as a function of time of day and season of year, and latitude and longitude.

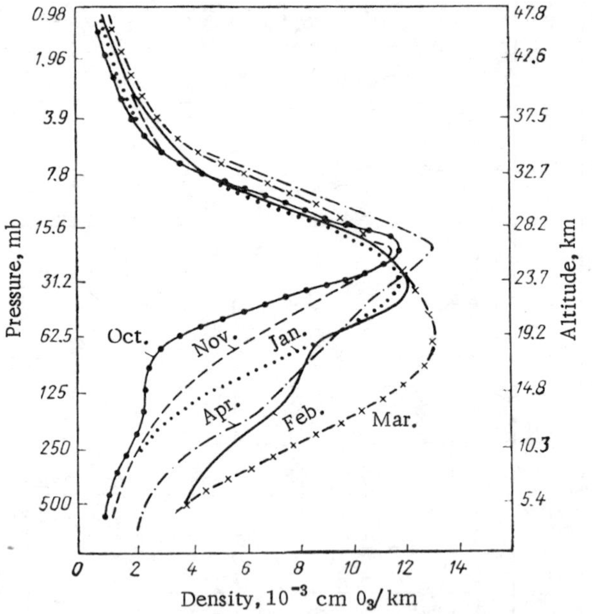

FIGURE 1.14. Average monthly vertical ozone profiles for the Arosa station obtained by Dütsch /22/ in 1956—1957.

Table 1.9 shows that the total ozone content varies from one point to another and from month to month by a factor of 3 (0.172—0.536 cm). The average yearly total O_3 content varies within narrower limits (0.394—0.204 cm, or approximately by a factor of 2).

In spite of the many ozone-concentration measurements in the atmosphere, an average statistical vertical O_3 profile has not yet been constructed. Various sources list averaged data pertaining to series of measurements (for example, Table 1.10).

TABLE 1.9. Monthly-average total ozone content (in cm) for some stations participating in the 1958 IGY program /21/

Serial No.	Station	Location		I	II	III	IV	V	VI	VII	VIII	IX	X	XI	XII	For the year
		latitude	longitude													
1	Alert	82°30′ N	62°20′ E	0.259	0.480	0.536	0.470	0.429	0.355	0.331	0.292	0.299	0.320	0.366	0.370	0.376
2	Resolute	74°43′	94°59′W	0.376	0.502	0.499	0.491	0.419	0.373	0.344	0.306	0.323	0.339	0.427	0.321	0.394
3	Larvik	60°08′	01°11′W	0.335	0.397	0.408	0.385	0.415	0.358	0.346	0.321	0.289	0.290	0.258	0.266	0.339
4	Oxford	51°46′	01°16′W	0.342	0.353	0.396	0.392	0.398	0.381	0.360	0.337	0.300	0.292	0.284	0.317	0.346
5	Arosa	46°47′	09°41′E	0.347	0.340	0.394	0.401	0.358	0.352	0.338	0.314	0.292	0.292	0.288	0.308	0.335
6	Elbrus	43°17′	42°25′E	0.299	0.320	0.362	0.402	0.345	0.304	0.256	0.228	0.222	0.250	0.241	0.274	0.295
7	Washington	38°51′	77°02′W	0.330	0.373	0.359	0.367	0.364	0.333	0.313	0.315	0.289	0.295	0.285	0.315	0.328
8	Torishima	30°29′	140°18′E	0.279	0.292	0.337	0.339	0.329	0.329	0.321	0.300	0.285	0.269	0.246	0.256	0.298
9	Tamanrasset	22°48′	05°31′E	0.172	0.196	0.227	0.258	0.214	0.185	0.173	0.212	0.212	0.213	0.199	0.192	0.204
10	Maunaloa	19°29′	155°36′W	0.256	0.251	0.270	0.290	0.289	0.284	0.271	0.271	0.268	0.264	0.257	0.253	0.269
11	Kodaikanal	10°14′	77°28′W	0.227	0.224	0.233	0.247	0.240	0.249	0.246	0.244	0.241	0.239	0.237	0.217	0.237
12	Leopoldville	4°30′ S	15°00′W	0.257	0.253	0.252	0.245	0.239	0.244	0.245	0.249	0.254	0.245	0.251	0.244	0.248
13	Brisbane	27°28′	153°02 W	0.285	0.276	0.282	0.289	0.292	0.303	0.314	0.318	0.348	0.329	0.320	0.337	0.308
14	Wellington	41°17′	174°46′W	0.294	0.271	0.261	0.269	0.281	0.319	0.340	0.339	0.370	0.364	0.331	0.289	0.311
15	Argentine Islands	65°15′	64°16′W	0.295	0.259	0.234	0.251	0.289	0.301	0.278	0.298	0.324	0.300	0.350	0.309	0.291
16	Little America	78°12′	162°15′W	0.314	0.310	0.334	0.266	0.380	0.480	0.368	0.536	0.334	0.373	0.420	0.316	0.360

TABLE 1.10. Ozone concentration C in various atmospheric layers /23/, $C = C_i \cdot 10^{-3}$

H, km	0—5	5—10	10—13	13—16	16—18	18—20	20—21	21—23
C_1, cm/km	1.9	4.1	5.8	7.6	11.1	15.6	16.8	17.5
C_2, cm/km	1.6	2.8	3.9	6.1	10.3	14.1	14.3	15.2

H, km	23—24	26—28	28—40.5	40.5—45.5	45.5—50.5	50.5—55.5
C_1, cm/km	10.8	6.5	1.6	0.3	0.1	0.1
C_2, cm/km	11.3	5.4	1.4	0.3	0.09	0.04

In exact calculations of the absorption of visible and infrared radiation in the atmosphere consideration must be given to variations both in total ozone content and in the vertical distribution of its concentration. This is especially important when computing spectral intervals in which radiation is absorbed primarily by O_3. This applies mainly to the 9.0—10.3 μ wavelength range, which is occupied by the most intensive ozone absorption band, located at the center of the long-wave atmospheric window.

4. DISTRIBUTION OF CO_2, O_2, CH_4, N_2O AND CO IN THE ATMOSPHERE

A feature common to all these gases is the approximate constancy of the specific concentration at all altitudes, at least for the bottom 30-km atmospheric layer. The absolute concentration of oxygen is virtually independent of time or location, up to approximately 90 km.

The concentration of CO_2 has been investigated quite extensively. Measurements have been conducted over the most characteristic locations of the globe: oceans, land, the Arctic and Antarctic, villages and industrial towns, mountains, and the free atmosphere. Comprehensive data, with specification of measurement location and of average minimum and maximum CO_2 concentrations in the atmosphere /24—38/, enabled the problem to be studied in depth. On the basis of measurements of average CO_2 concentrations at 64 points of the globe in ppm of the air volume, these values in 60 cases lie between 280 and 340. In the remaining 4 cases they were: 253 (England, at altitude 4—10 km), 256 (near Horn Head, 56°N), 354 (Pacific), and 650 (London). It is of interest that not far from London the average concentration is approximately 300.

The maximum deviation of absolute concentration from its average value was 46% in only one case (London, maximum concentration 900, minimum 425). However, the CO_2 concentration in London was measured as long ago as 1936 /24/ and refinement is certainly necessary. Voitikova /41/ noted that references /39, 40/

list minimum and maximum CO_2 concentrations of 0.015 and 0.078% by volume, respectively. In line with these data the absolute minimum and maximum CO_2 concentration in the atmosphere should be taken as 0.015 and 0.09%, respectively.

It follows that the difference between the absolute minimum and maximum concentration is quite high. However, in calculating the absorption of infrared radiation by CO_2 one may assume its concentration to be constant and equal to 0.03% by volume, since large departures from this value are observed very infrequently.

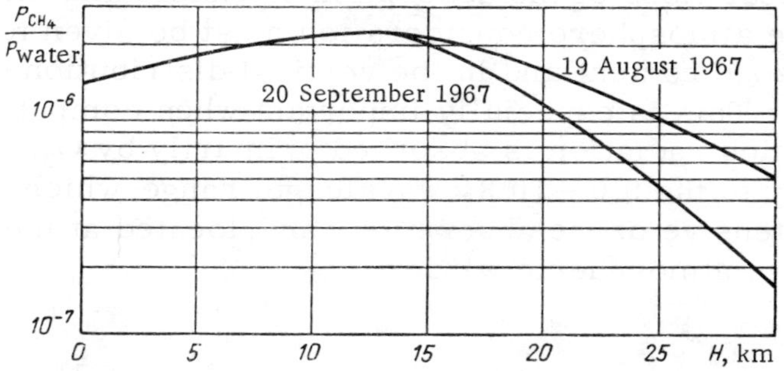

FIGURE 1.15. Vertical profile of the ratio of the pressure of methane to the pressure of air from two measurements by means of an automatic aerostat-borne spectrometer /866/.

Data on concentrations of minor atmospheric admixtures, such as methane (CH_4), nitrous oxide (N_2O) and carbon monoxide (CO), were obtained in only a few studies. Seeley and Houghton /42/ found mixing ratios of CH_4, CO and N_2O equal to $1.05 \cdot 10^{-6}$, $0.12 \cdot 10^{-6}$ and $0.12 \cdot 10^{-6}$ g/g, respectively. For these values the total content of these gases in the entire vertical thickness of the atmosphere is 0.84, 0.10 and 0.10 atm-cm.

Seeley and Houghton /42/ note that these data are in satisfactory agreement with the measurements of the Migeotte group /43/. Bowman and Shaw /44/ obtained for ground-level air two concentrations of these gases in ppm: 1.7 ± 0.03 (CH_4), 1.3 (CO), 0.28 ± 0.04 (N_2O). Considerable variation is noted in the abundance of CO. The same conclusion was reached elsewhere /45/, where the CO content of the entire atmospheric layer was found to vary between 0.04 and 0.13 atm-cm. The abundance of CH_4 in the earth's atmosphere was determined most precisely by Fink et al. /46/ and was found to be 1.11 ± 0.16 atm-cm.

Vasilevskii et al /866/ employed a high-resolution automatic aerostat-borne spectrometer to study the transmission spectra of the atmosphere in two narrow regions in the vicinity of the 3.31 and 3.39μ wavelengths at altitudes from 0 to 30 km. Processing of their measured data yielded data on the vertical profile of the abundance of methane in the atmosphere (Figure 1.15).

The mixing ratio for N_2O values listed in /47—49/ was found to be respectively $0.27 \cdot 10^{-6}$, $0.27 \cdot 10^{-6}$ and $0.05 \cdot 10^{-6}$ g/g. Discrepancies are due to difficulties in interpreting spectrographic data of /42/.

Available data thus show that CH_4, CO and N_2O concentrations in ppm lie, respectively, within the ranges 1.05—1.7, 0.04—1.3, and 0.12—0.5. Refinement of these limits (such as determination of the statistically average concentrations of CH_4, N_2O and CO in the atmosphere) requires appropriate systematic measurements, which are not presently being carried out anywhere.

2. THE ORIGIN AND GENERAL DESCRIPTION OF ABSORPTION SPECTRA OF ATMOSPHERIC GASES

1. MOLECULAR ENERGY AND SPECTRA

In the first approximation the energy of a molecule can be expressed in the form

$$E = E_{\text{trans}} + E_{\text{el}} + E_{\text{vib}} + E_{\text{rot}}, \tag{2.1}$$

where E_{trans} is the energy of translational motion, which is a function of the velocity of this motion and can assume any value; E_{el} is the electronic energy; E_{vib} and E_{rot} are the vibrational and rotational energies. Quantities E_{el}, E_{vib} and E_{rot} assume only discrete values and therefore only vary jumpwise, the variation being accompanied by either absorption or emission of a photon of some frequency.

Formula (2.1) applies only when interaction between the different types of molecular motion can be neglected. In general the energy of a molecule is given by

$$E = E_{\text{trans}} + E_{\text{el}} + E_{\text{vib}} + E_{\text{rot}} + E_{\text{el vib}} + E_{\text{el rot}} + E_{\text{vib rot}}, \tag{2.2}$$

where the last three terms are due to interactions between the different types of molecular motion.

The magnitudes of the electronic, vibrational and rotational molecular energies are of different orders. The energy of electronic transitions is of the order of several eV, the vibrational energy amounts to tenths and hundredths of eV, while the rotational energy is equal to thousandths and ten-thousandths of an eV. Depending on the energy, so the molecular electronic spectra occupy the ultra-violet and visible parts of the spectrum, the vibrational spectra occupy the near infrared, and the rotational spectra occupy the far infrared and microwave regions.

Since a molecule possesses electronic, vibrational and rotational energies which generally vary simultaneously during emission or absorption, pure electronic and vibrational spectra do not appear. Thus we deal with electronic-vibration-rotation, vibration-rotation, or pure rotational spectra, depending on the wavelength range. These are frequently termed for brevity electronic, vibrational and rotational spectra, respectively.

For each electron state there exist vibrational energy levels, characterized by a specific value of the vibrational quantum number v (Figure 2.1). In its turn, for each vibrational level there exist rotational levels, characterized by a specific value of the rotational quantum number j.

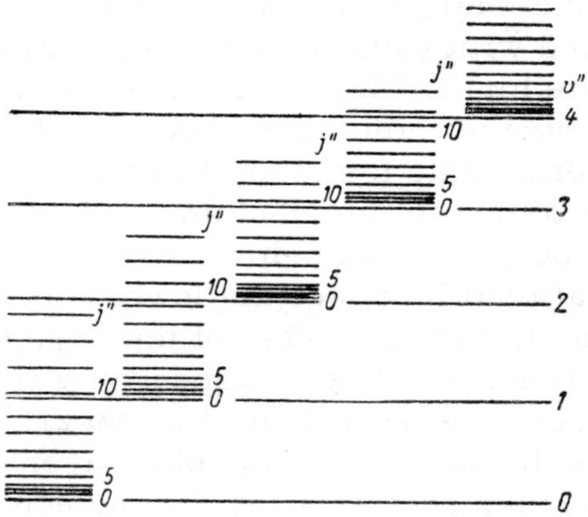

a

b

FIGURE 2.1. Schematic of vibrational and rotational levels of two electronic states (a and b) of a molecule (single and double primes of v and j denote that the corresponding levels belong to higher and lower electronic states of the molecule).

Figure 2.1 can be used to explain why pure electronic and vibrational transitions in a molecule cannot be observed. In a group of molecules we are dealing with their statistical distribution by electronic, vibrational and rotational energies. During each electronic transition a molecule simultaneously undergoes a change in its vibrational and rotational energies. The entire group of electronic transitions forms the electronic-vibration-rotation spectrum, the line intensity of which is determined by the energy-level distribution of the molecules and the probabilities of corresponding transitions.

The electronic-vibration-rotation spectrum consists of a system of bands. For each group of transitions of a molecule between two electronic states there exists an electronic band consisting of a set of vibration-rotation bands.

The number of different vibrational transitions accompanying a variation in electronic energy and the number of different rotational transitions associated with variations in the vibrational energy of a molecule can be quite high. This circumstance is responsible for the quite complex structure of molecular spectra. This structure depends on the selection rule for transitions between energy levels with specific values of the set of quantum numbers. More details on selection rules will be presented below.

The highest vibrational levels of the lower electronic state of the molecule can overlap the lower levels of the adjacent, higher electronic state. The rotational levels of adjacent vibrational states can overlap similarly (this is evident from Figure 2.1). This overlapping of levels results in overlapping of adjacent electronic and vibrational bands, further complicating the molecular band spectrum.

The rotational, vibrational and electronic energies and their interaction should be treated separately in order to properly understand the structure of different molecular spectra.

2. ROTATIONAL ENERGY AND ROTATIONAL SPECTRA OF MOLECULES

A rotating molecule can be regarded approximately as a rigid rotator. The rotation of molecules and their rotational spectra depend on the relationship between the three principal moments of inertia $\mathcal{J}_a$, $\mathcal{J}_b$ and $\mathcal{J}_c$ relative to the three principal molecular axes of rotation. Molecules with rotational properties are subdivided into four groups (Table 2.1).

1. **Linear molecules**. The rotational energy of a linear molecule is /51/

$$E=\frac{\vec{M}_p^2}{2\mathcal{J}}, \tag{2.3}$$

where $\vec{M}_p^2$ is the square of the angular momentum, and $\mathcal{J}$ is the moment of inertia about the axis of rotation perpendicular to the axis of the molecule passing through its center of mass.

TABLE 2.1. Classification of molecules by their rotational properties

Serial number	Moments of inertia	Type of molecule	Atmospheric gases belonging to the corresponding type of molecule
1	$\mathcal{J}_a=0\ \mathcal{J}_b=\mathcal{J}_c\neq 0$	Linear	CO_2, N_2O, NO, CO, O_2, N_2
2	$\mathcal{J}_a\neq 0\ \mathcal{J}_b=\mathcal{J}_c\neq 0$	Symmetric rotator	Nonexistent among atmospheric gases on a planetary scale
3	$\mathcal{J}_a=\mathcal{J}_b=\mathcal{J}_c$	Spherical rotator	CH_4
4	$\mathcal{J}_a\neq\mathcal{J}_b\neq\mathcal{J}_c$	Asymmetric rotator	H_2O, O_3, HDO

Quantity $\vec{M}_p^2$ is quantized by the expression

$$\vec{M}_p^2=\hbar^2 j\,(j+1), \tag{2.4}$$

where $\hbar=\frac{h}{2\pi}$, h is Planck's constant, and j is the rotational quantum number which assumes integral values:

$$j=0,\ 1,\ 2,\ 3,\ \ldots \tag{2.5}$$

Relationships (2.4) and (2.3) yield the quantized rotational energy of the molecule in the form

$$E_j=\frac{\hbar^2}{2\mathcal{J}}\,j\,(j+1)=Bj\,(j+1), \tag{2.6}$$

where

$$B=\frac{\hbar^2}{2\mathcal{J}}=\frac{2.80\cdot 10^{-39}}{\mathcal{J}}\,(\mathrm{cm}^{-1})=$$
$$=\frac{0.83\cdot 10^{-28}}{\mathcal{J}}\,(\mathrm{sec}^{-1})=\frac{0.83\cdot 10^{-34}}{\mathcal{J}}\,(\mathrm{MH_Z}) \tag{2.7}$$

is termed the rotational constant. Table 2.2 lists values of E_j for various j.

TABLE 2.2. Rotational energy E_j for various rotational quantum numbers j

j	0	1	2	3	4	5	6	7	8	9	10
E_j	0	$2B$	$6B$	$12B$	$20B$	$30B$	$42B$	$56B$	$72B$	$90B$	$110B$

The spacing between successive energy levels

$$E_{j+1}-E_j=2B\,(j+1) \tag{2.8}$$

is proportional to j.

It is seen from (2.8) and (2.7) that the absolute spacings between rotational energy levels are inversely proportional to the molecular moment of inertia. The heavier and larger the molecule, the longer the waves comprising its rotational spectrum.

Simultaneously with quantization of the square of angular momentum $\vec{M}_p^2$ one quantizes its z component

$$M_{pz}=\hbar m_j, \tag{2.9}$$

where m_j is the magnetic quantum number which assumes $2j+1$ values:

$$m_j=j,\ j-1,\ \dots,\ -j. \tag{2.10}$$

The energy of any free molecule is determined by its angular momentum and does not depend on a component of this momentum, and consequently on quantum number m_j. However, the ensemble of quantum numbers j and m_j fully defines the rotational motion of a linear molecule possessing two degrees of freedom of rotational motion.

As mentioned above, for given j, m_j assumes $2j+1$ values. A linear molecule thus has $2j+1$ degenerate states for given j. The statistical weight of a rotational level with given j is $g_j=2j+1$ ($j=0, 1, 2, \dots$).

For thermal equilibrium the number of molecules possessing a given value of j is

$$n_j=(2j+1)\,n_0 e^{-\frac{\varepsilon_j}{kT}}=(2j+1)\,n_0 e^{-\frac{Bj(j+1)}{kT}}, \tag{2.11}$$

where n_0 denotes the number of molecules in the state $j=0$.

Only molecules with nonzero constant dipole moment can possess pure rotational spectra. Linear molecules can possess a constant dipole moment when the centers of mass of the positive and negative charges do not coincide.

The selection rules allow pure rotational transitions when

$$\Delta j=\pm 1. \tag{2.12}$$

It is seen from (2.12) and (2.8) that the frequencies of possible successive rotational transitions differ by $2B$. The pure rotational spectrum of a linear molecule in the rigid-rotator approximation thus consists of equally spaced lines.

This approximation assumes invariability of the distance between the nuclei of the rotating molecule. Actually these distances increase due to centrifugal stretching (prolateness) [arising from the nonrigidity of the rotating molecule], which is the greater, the higher the rotational energy, and hence the larger the rotational quantum number j. Rotational constant B decreases with increasing nuclear separation (see (2.7)). This results in more closely spaced lines in the spectrum.

The quantum-mechanical perturbation theory accounting for moderate prolateness of the molecule leads to the following approximate formula for the molecular rotational energy:

$$E_j = Bj(j+1) - D[j(j+1)]^2. \tag{2.13}$$

Constant D does not exceed $10^{-4}B$ and must be taken into account only for high j.

2. **Spherical rotator molecules.** These molecules ($\mathcal{I}_a = \mathcal{I}_b = \mathcal{I}_c \neq 0$) are the simplest type of nonlinear molecule and their rotational energy levels are defined by the same expression as the energy levels of linear molecules (see equation (2.6)). However, the properties of the rotational levels of these two types of molecule differ markedly. Spherical rotator molecules have three degrees of rotational freedom, and so the complete description of rotational levels requires specification of three quantum numbers. In addition to quantum numbers j and m_j, which characterize the rotational motion, we introduce quantum number k, which determines the angular momentum component along an arbitrary axis rigidly fixed to the molecule.

The angular momentum (component) along an axis fixed to the molecule is quantized in the same manner as the momentum component along a stationary axis. Hence quantum number k assumes $2j+1$ values for given j, similar to quantum number m_j.

The energy of a molecule depends only on quantum number j. Hence the general degree of degeneracy of rotational levels of a spherical rotator molecule with respect to m_j and k is

$$g_j = (2j+1)^2. \tag{2.14}$$

For each value of j the molecule can have $(2j+1)$ orientations relative to a stationary axis not fixed to the molecule and $(2j+1)$ orientations relative to an axis moving together with the molecule.

The molecular distribution over rotational states now has the form

$$n_j = (2j+1)^2 n_0 e^{-\frac{E_j}{kT}} = (2j+1)^2 n_0 e^{-\frac{Bj(j+1)}{kT}} \tag{2.15}$$

Comparison of expressions (2.15) and (2.11) shows that the number of spherical-rotator molecules at high rotational levels (higher than j) is greater than the number of linear molecules at the same B and .

The effect of prolateness during rotation of spherical-rotator molecules is taken into account in the same manner as for linear molecules and is determined by (the same) formula (2.13).

3. Symmetric-rotator molecules. No molecules of this type are present among atmospheric gases prevalent on the planetary scale. Nevertheless, this problem must be considered.

Let a, b and c be axes relative to which the principal molecular moments of inertia satisfy the inequality

$$\mathcal{J}_a \leqslant \mathcal{J}_b \leqslant \mathcal{J}_c. \tag{2.16}$$

The symmetric rotator can be both prolate and oblate. For a prolate rotator

$$\mathcal{J}_a < \mathcal{J}_b = \mathcal{J}_c. \tag{2.17}$$

For an oblate rotator

$$\mathcal{J}_a = \mathcal{J}_b < \mathcal{J}_c.$$

Each moment of inertia has a corresponding rotational constant (A, B or C):

$$A = \frac{h^2}{8\pi^2 \mathcal{J}_a}; \quad B = \frac{h^2}{8\pi^2 \mathcal{J}_b}; \quad C = \frac{h^2}{8\pi^2 \mathcal{J}_c}. \tag{2.18}$$

The rotational energy of a rigid prolate rotator is

$$E_{jk} = Bj(j+1) + (A - B)k^2; \quad (A > B), \quad j = 0, 1, 2, \ldots; \quad k = 0; \pm 1; \pm 2; \ldots; \pm j. \tag{2.19}$$

For an oblate rotator the rotational energy is

$$E_{jk} = Bj(j+1) + (C - B)k^2; \quad (C < B); \quad j = 0, 1, 2, \ldots; \quad k = 0; \pm 1; \pm 2; \ldots; \pm j. \tag{2.20}$$

The first term in the latter two expressions has the same form as that for the rotational energy of a linear molecule and of a spherical-rotator molecule (see 2.7)). The second term depends on the absolute value of quantum number k, which defines the angular momentum component along the molecular axis of symmetry.

Whereas for a spherical rotator each level with given j possesses degree of degeneracy $2j+1$ relative to quantum number k, for symmetric-rotator molecules it splits into $j+1$ levels, corresponding to $|k|=0, 1, 2. \ldots, j$. Hence the sign of k need not be indicated.

Molecular levels with given j and k have degree of degeneracy $2(2j+1)$, with the exception of the level with $k=0$. In the latter case the degree of degeneracy is $(2j+1)$.

The selection rules for pure rotational transitions in symmetric-rotator molecules allow transitions satisfying

$$\Delta j = \pm 1; \quad \Delta k = 0. \tag{2.21}$$

Relationships (2.18), (2.19) and (2.20) yield the energy of molecular transition between levels with quantum numbers $j'k$ and $j''k$ in the form

$$\Delta E = E_{j'k} - E_{j''k} = B[j'(j'+1) - j''(j''+1)]. \tag{2.22}$$

Hence the transition energy between adjacent levels $(\Delta j = 1)$ is

$$\Delta E = E_{j+1,\,k} - E_{jk} = 2B(j+1). \tag{2.23}$$

It is thus seen that, as in the case of symmetric-rotator molecules, the pure rotational molecular spectrum consists of equally spaced lines. Here, rotational absorption and emission spectra are allowed only for those symmetric-rotator molecules with nonzero constant dipole moments.

The prolateness of molecules is taken into account by the following formula for the rotational energy of symmetric-rotator molecules:

$$E_{jk} = E_{jk}^{(0)} - D_j j^2 (j+1)^2 - D_{jk} j (j+1) k^2 - D_k k^4, \tag{2.24}$$

where $E_{jk}^{(0)}$ is the rotational energy of a rigid symmetric rotator given by (2.19) and (2.20); D_j, D_{jk} and D_k are constants, small relative to B, A and C.

For the allowed rotational molecular transitions expressions (2.21) and (2.24) give

$$\Delta E = E_{j+1,k} - E_{jk} = 2B(j+1) - \\ - 4D_j (j+1)^3 - 2D_{jk} (j+1) k^2. \tag{2.25}$$

Hence the rotational spectrum of a real molecule consists of convergent lines, and each value of j has correspondingly $j+1$ levels for $k = 0, 1, \ldots, j$.

4. **Asymmetric-rotator molecule.** In this case

$$\mathcal{J}_a < \mathcal{J}_b < \mathcal{J}_c; \quad A > B > C. \tag{2.26}$$

The asymmetric rotator (unlike those considered above) cannot be described by simple analytical expressions. Determination of the rotational energy with $j = 2$ and $j = 3$ involves the solution of quadratic equations. Energy levels with $j > 3$ are obtained by solving equations of third, fourth, etc., degrees. Corresponding calculations have been conducted up to $j = 40$ /52/.

The qualitative picture of the rotational-energy level of an asymmetric-rotator molecule can be obtained by treating this molecule as intermediate between the limiting cases of prolate and oblate rotators.

It is evident that the less $\mathcal{J}_b$ differs from $\mathcal{J}_a$ or $\mathcal{J}_c$, the closer the asymmetric rotator is to the corresponding symmetric rotator. The rotator asymmetry characteristics are expressed by the parameter

$$\varkappa = \frac{2B - A - C}{A - C}. \tag{2.27}$$

For a prolate rotator $\varkappa = -1$, while for an oblate rotator $\varkappa = 1$. When $B = \frac{A+C}{2}$ (most asymmetric rotators), $\varkappa = 0$. For water molecules, $\varkappa = -0.430$.

For an asymmetric-rotator molecule the level with given j splits into $2j + 1$ levels, designated by subscript τ which assumes the values

$$\tau = -j, \ -j+1, \ \ldots, \ j-1, \ j. \tag{2.28}$$

The levels are situated in order of increasing energies from j_{-j} to j_{+j} (see, for example, Table 2.3).

TABLE 2.3. Energies of rotational molecular levels for water (cm^{-1}) /51/

$j = 0$	$j = 1$		$j = 2$		$j = 3$		$j = 4$	
0.0	1_{-1}	23.78	2_{-2}	70.09	3_{-3}	136.76	4_{-4}	222.07
	1_0	37.09	2_{-1}	79.43	3_{-2}	142.23	4_{-3}	224.81
	1_1	42.31	2_0	95.15	3_{-1}	173.31	4_{-2}	275.50
			2_1	134.98	3_0	206.35	4_{-1}	300.44
			2_2	136.24	3_1	212.24	4_0	315.83
					3_2	286.18	4_1	384.03
					3_3	286.93	4_2	385.44
							4_3	490.64
							4_4	490.79

The prolateness of asymmetric-rotator molecules results only in shifting energy levels, which can be substantial for light molecules and high j. Incorporation of prolateness at rotational levels is quite a complicated problem. Solutions of this problem for various cases are presented elsewhere /53/.

The spectra of asymmetric-rotator molecules are very complex. The selection rules allow transitions with changes of rotational quantum number given by

$$\Delta j = 0, \pm 1. \tag{2.29}$$

Here transitions with $\Delta j = 0$, $+1$ and -1 correspond to the Q, R and P branches.

In addition to selection rules (2.29) asymmetric-rotator molecules are subjected to transition limitations due to the symmetry properties of the ellipsoid of inertia. In the first case reference is had to an even function (designated by e), and in the second to an odd function (designated by o). Rotation of the ellipsoid through 180° relative to any two axes is equivalent to a rotation through the same angle relative to the third axis. It is thus sufficient to consider ellipsoid rotations relative to two axes. Here the following types of molecular symmetry are possible: ee, eo, oe and oo, or $++$, $+-$, $-+$ and $--$.

Depending on the principal axis along which the dipole moment of the molecule is directed, a distinction is made between three types of selection rule: A, B and C. In the first case the dipole moment is directed along the a axis, in the second along the b axis, and in the third along the c axis.

The dipole moment of a water molecule is directed along the b axis, in which case the selection rules are of type B:

$$++ \leftrightarrow -- \quad \text{and} \quad +- \leftrightarrow -+. \tag{2.30}$$

For types A and C we have, respectively,

$$++ \leftrightarrow -+ \quad \text{and} \quad +- \leftrightarrow -- \tag{2.31}$$

and

$$++ \leftrightarrow +- \quad \text{and} \quad -+ \leftrightarrow --. \tag{2.32}$$

If the direction of the dipole moment does not coincide with that of any of the principal axes of revolution, then all three types of selection rule apply.

As above, pure rotational spectra can be associated only with asymmetric-rotator molecules which possess nonzero constant dipole moment. Among atmospheric gases these are the molecules of the principal absorbing constituents of the atmosphere: water vapor and ozone. Pure rotational spectra of H_2O and O_3 are highly complicated and have a large number of intensive absorption lines, extending over a wide range of wavelengths.

3. VIBRATIONAL ENERGY AND VIBRATION-ROTATION SPECTRA OF MOLECULES

1. **Vibrational energy of molecules.** A molecule consisting of N atoms has $3N$ degrees of freedom, of which 3 pertain to the translational motion of the center of mass of the molecule as a single entity, while 3 or 2 pertain to the rotational motion. Linear molecules have 2 degrees of rotational freedom, and nonlinear molecules have 3 degrees of freedom. A nonlinear molecule thus has $3N-6$ vibrational degrees of freedom and, in general, $3N-6$ different vibrational frequencies. The linear molecule has $3N-5$ degrees of rotational freedom and $3N-5$ different frequencies. In symmetric molecules some frequencies coincide as a result of degeneration.

The simplest model of vibrational molecular motion is that of the harmonic oscillator, for which the vibrational energy is

$$E_v = h\nu\left(v+\frac{1}{2}\right), \tag{2.33}$$

where

$$\nu = \frac{1}{2\pi}\sqrt{\frac{k}{M}} \tag{2.34}$$

is the classical vibrational frequency, v is the vibrational quantum number assuming integral values $v = 0, 1, 2, 3, \ldots$, k is a quasi-elastic force constant, and M is the oscillator mass.

Expression (2.33) indicates that the energy levels of the harmonic oscillator are equally spaced. The energy difference of adjacent levels is

$$E_{v+1} - E_v = h\nu. \tag{2.35}$$

For a harmonic oscillator the frequency of quantum transition between adjacent vibrational levels coincides with the classical vibrational frequency.

If the molecular vibrations are treated as an ensemble of vibrations of harmonic oscillators with frequencies $\nu_1, \nu_2, \nu_3, \ldots$, then according to quantum mechanics the molecular vibrational energy has the form

$$E_v = \left(v_1 + \frac{1}{2}\right) h\nu_1 + \left(v_2 + \frac{1}{2}\right) h\nu_2 + \\ + \left(v_3 + \frac{1}{2}\right) h\nu_3 + \ldots, \qquad (2.36)$$

where v_1, v_2, v_3, . . . are vibrational quantum numbers.

The vibrational molecular states are designated by a set of values of these numbers. For example, the (001) state corresponds to the value $v_1 = 0$, $v_2 = 0$ and $v_3 = 1$. Frequencies ν_1, ν_2, ν_3, etc. are termed fundamental vibrational frequencies.

The selection rules allow transitions between the energy levels of a harmonic oscillator, satisfying the condition

$$\Delta v = \pm 1. \qquad (2.37)$$

If the molecule is here represented by an ensemble of harmonic oscillators, then the quantum number of only one of the vibrations should change. The selection rules thus allow transitions only between adjacent energy levels with the same frequency ν (see (2.35)).

The harmonic oscillator model is an oversimplification of the true vibrations. Consideration of anharmonicity of molecular vibrations leads to the appearance in energy expression (2.36) of terms of second and higher orders. For a diatomic molecule one can derive the following approximate expression for E_v:

$$E_v = \left(v + \frac{1}{2}\right) h\nu - \left(v + \frac{1}{2}\right)^2 xh\nu, \qquad (2.38)$$

where x is the anharmonic constant.

For an N-atomic molecule, the vibrational energy with allowance for anharmonicity is /54/

$$E(v_1, v_2, \ldots) = \\ = \sum_j h\nu_i \left(v_i + \frac{1}{2}\right) + \sum_j \sum_k x_{ik} \left(v_i + \frac{1}{2}\right)\left(v_k + \frac{1}{2}\right) + \ldots, \qquad (2.39)$$

where x_{ik} are anharmonicity constants and coupling factors of individual vibrations.

Anharmonicity of vibrations is responsible for reducing the spacing between adjacent levels of molecular vibrational energy with an increase in the vibrational quantum number. Instead of one frequency there appears a series of frequencies decreasing in the series $v = 0 \leftrightarrow v = 1$, $v = 1 \leftrightarrow v = 2$, etc. The distance between the levels decreases gradually with an increase in vibrational quantum numbers v_1, v_2, and v_3 (Figure 2.2).

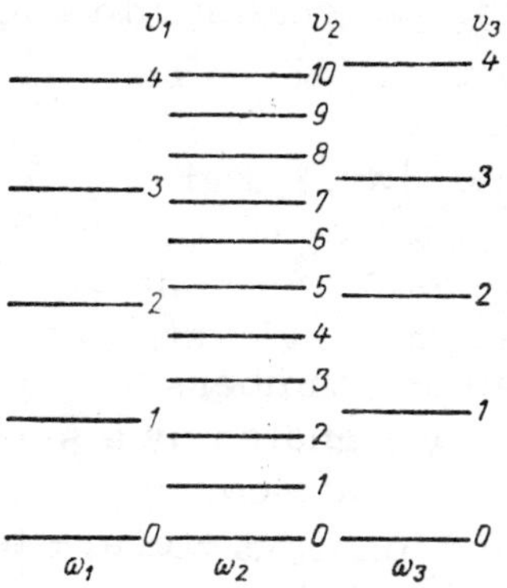

FIGURE 2.2. Vibrational-energy levels of the H_2O molecule.

Allowance for anharmonicity essentially alters the selection rules. The permitted transitions now satisfy the condition

$$\Delta v = 2,\ 3,\ 4 \text{ etc.} \tag{2.40}$$

In addition, transitions during which two quantum numbers change simultaneously are also permitted.

Transitions characterized by changes of 2, 3, etc., in Δv result in the appearance of corresponding overtones of the fundamental frequencies. Simultaneous variation of two or more vibrational quantum numbers give rise to combination frequencies (sum and difference frequencies). The number of frequencies of a real molecule thus exceeds markedly the number of its fundamental frequencies.

The anharmonic nature of vibrations can bring about strong interaction between vibrations in the presence of random resonance. Thus, for example, the CO_2 molecule has almost coinciding energy levels $\nu_1 = 1337$ cm^{-1} ($v_1 = 1,\ v_2 = 0,\ v_3 = 0$) and $2\nu_2$ ($v_1 = 0,\ v_2 = 2,\ v_3 = 0$). Such a random coincidence of energy levels belonging to the same symmetry class (Fermi-type resonance) brings about mutual disturbance of vibrational level energies, exhibited by their shifting in opposite directions.

2. **Vibration-rotation molecular spectra.** The energy of vibrational and rotational molecular photons is of the order of $100-1000^{-1}$ cm and $1-10^{-1}$ cm, respectively. Therefore whereas pure rotational spectra can be observed, pure vibrational spectra can be observed only in special, extremely infrequent cases. A change in the vibrational state of a molecule, associated with emission or absorption of energy, nearly always causes a change in the rotational state too. The smallness of rotational, as compared with vibrational, energy is responsible for the fact that the rotation does not affect the vibrational structure of the molecular spectrum. Since each vibrational energy level has a corresponding series of rotational levels, each vibrational transition in the molecular ensemble forms

a series of lines or a spectral band. Each of these lines corresponds to the same variation in vibrational energy and to a quite definite variation in rotational energy.

The spectrum of pure vibrational molecular transitions should thus be treated as a skeleton, around the elements of which vibration-rotation bands cluster. If the interaction between the vibrations and rotations is neglected, then the molecular vibration-rotation spectrum can be easily represented as an additive superposition of vibrational and rotational transitions, permitted by the corresponding selection rules.

Consider first the nature of the vibrational structure of spectra, which is the basis of vibration-rotation bands. This structure is simplest in diatomic molecules. The condition for appearance of vibrational bands in the spectrum is nonzero variation of the dipole moment during the corresponding quantum transition. Diatomic molecules, consisting of identical atoms and consequently possessing zero dipole moment which does not change during vibrations, cannot have vibration-rotation spectra. The frequencies of such vibrations are termed optically inactive.

The vibrational-transition spectrum of a diatomic molecule consisting of nonidentical atoms can be derived from expression (2.38), in which case the vibrational frequencies ν_k are given by

$$\nu_k = \nu(1-x)(v''-v') - \nu x(v''^2 - v'^2), \qquad (2.41)$$

where v'' and v' are vibrational quantum numbers of the higher and lower state of the molecule.

It is seen from (2.41) that the vibrational spectrum of a diatomic molecule consists of a series of bands. Corresponding to each band there exists an ensemble of transitions of the molecule from a given vibrational level to the adjacent level. We recall that the selection rules for the anharmonic oscillator, for which formula (2.38) was obtained, allow transitions with any variations of the vibrational quantum number. For example, for the series starting with the zero vibrational level [ground vibrational state] ($v'=0$) expression (2.41) yields

$$\nu_k(0, v'') = \nu(1-x)v'' - \nu x v''^2. \qquad (2.42)$$

This leads to the following ensemble of frequencies of vibrational transitions, and consequently of vibration-rotation bands:

$$\nu_1 = \nu(1-2x); \quad \nu_2 = 2\nu(1-3x); \quad \nu_3 = 3\nu(1-4x); \ \ldots \qquad (2.43)$$

It is seen from (2.43) that the spacing between successive lines decreases with increasing v'' and with increase in frequency. The lines of the series converge to some limit, corresponding to molecular dissociation. All the other series have a similar structure.

This pattern of vibration-rotation bands also occurs in the case of polyatomic molecules. However, as a result of the complex scheme of vibrational levels and the possibility of simultaneous excitation of a large number of vibrations (fundamental frequencies, harmonics, combination frequencies) similar structures of individual vibrations overlap and the resulting spectrum is quite complex.

Consider now the fine structure of the spectrum of individual vibration-rotation bands. It is expedient to start with diatomic molecules. The vibration-rotation energy for two levels characterized by vibrational and rotational quantum numbers v'', j'' and v', j' can be expressed in the form

$$E'' = E_{v''} + B_{v''} j'' (j'' + 1), \tag{2.44}$$

$$E' = E_{v'} + B_{v'} j' (j' + 1), \tag{2.45}$$

where $E_{v''}$ and $E_{v'}$, are defined by expressions such as (2.38), while the rotational energy is given for simplicity in the form $Bj\,(j+1)$.

The interaction of molecular vibrations and rotations in expressions (2.44) and (2.45) is taken into account by means of the rotational constant, which is a function of the vibrational quantum number v:

$$B_v = B_e - \alpha \left(v + \frac{1}{2} \right), \tag{2.46}$$

where B_e is the rotational constant of a nonvibrating molecule; α is a constant, the ratio of which to B_e does not exceed several hundredths.

Relationship (2.46) indicates that B_v decreases linearly with increase in v. This is due to the fact that the distance between the atoms in a vibrating molecule increases, and consequently the moment of inertia increases too, while constant B_v decreases (see (2.7)). Expressions (2.44) and (2.45) yield

$$E'' - E' = E_{v''} - E_{v'} + B_{v''} j'' (j'' + 1) - B_{v'} j' (j' + 1). \tag{2.47}$$

Quantity $E_{v''} - E_{v'} = \nu_{00}$ defines the frequency of a pure vibrational transition which would have been obtained for $j'' = j' = 0$. The transition corresponding to the zero line is prohibited by selection rules. The location of this line is determined by calculation. Because the first difference in (2.47) is much greater than the second, the location of the vibration-rotation band in the spectrum is determined by the zero line.

Unlike pure rotational spectra, transitions defined by (2.47) must involve consideration of not only the values of $\Delta j = j'' - j' > 0$, but also of $\Delta j<0$, and in individual cases also $\Delta j=0$. Then:

$$\left.\begin{array}{l}\text{when } \Delta j=j''-j'=+1 \text{ the transitions form the } R \text{ branch,}\\ \text{when } \Delta j=j''-j'=0 \text{ the transitions form the } Q \text{ branch,}\\ \text{when } \Delta j=j''-j'=-1 \text{ the transitions form the } P \text{ branch.}\end{array}\right\} \quad (2.48)$$

Transitions with $\Delta j=+1$ correspond to frequencies exceeding ν_{00}, those with $\Delta j=-1$ correspond to frequencies smaller than ν_{00} (Figure 2.3). Hence the R and P branches are also termed positive and negative, respectively.

Transitions with $\Delta j=0$ (Q branch) in diatomic molecules are permitted when the electronic angular momentum is nonzero (for example, in the NO molecule). For the Q branch, expression (2.47) for transition frequency ν_j gives

$$\nu_j=\nu_{00}+(B_{v''}-B_{v'})\,j\,(j+1). \qquad (2.49)$$

Since the difference $B_{v''}-B_{v'}$ is very small, the lines in the Q branch are closely spaced.

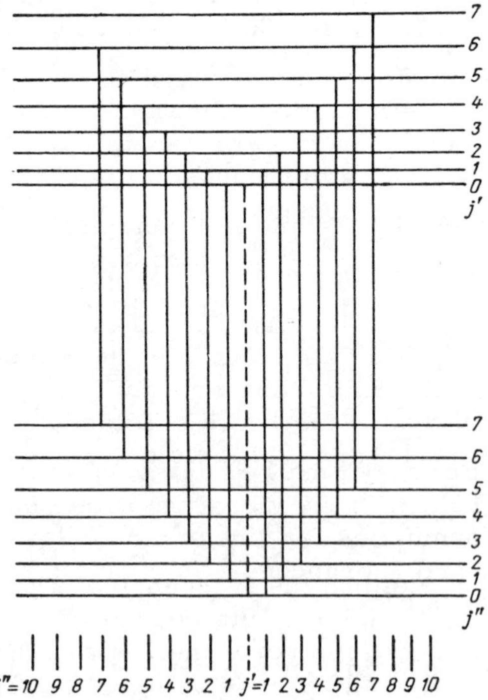

FIGURE 2.3. Schematic of energy transitions of a vibration-rotation band.

It is common practice to subdivide the infrared absorption bands into parallel (||) and perpendicular ($\perp$). Parallel bands arise when the direction of the dipole moment coincides with that of the pertinent axis of symmetry of the molecule. Perpendicular bands are defined similarly.

Consider the fine structure of vibration-rotation bands of molecules, differentiated by their rotational properties (linear, spherical-, symmetric- and asymmetric-rotator molecules).

1. Linear polyatomic molecules. The location of rotational lines for a linear polyatomic molecule is given by an expression similar to (2.47), except that now the rotational constants depend on all the vibrational quantum numbers of the molecule:

$$B_{v_1, v_2, \dots, v_k} = B_e - \sum_i \alpha_i \left(v_i + \frac{1}{2} g_i \right), \tag{2.50}$$

where g_i is the order of the degeneracy of the i-th frequency.

For perpendicular bands the selection rules in the case at hand allow transitions in all three (R, Q and P) branches. For parallel bands there is no Q branch, and the bands themselves have a structure similar to those of a diatomic molecule with no electronic angular momentum. The fundamental vibration-rotation bands of linear molecules include both parallel and perpendicular bands. For bands corresponding to vibrations in the direction of valent inter-atomic couplings the dipole moment is directed along the molecule axis (parallel bands). Deformation vibrations are responsible for the onset of perpendicular bands. For example, in the case of CO_2 the fundamental bands (001) with frequency $\nu_3 = 2349.3$ cm^{-1} and (010) with $\nu_2 = 667.3$ cm^{-1} are respectively parallel and perpendicular. Both parallel and perpendicular bands are encountered among the overtones and combination frequencies. We note that the nature of the band (parallel or perpendicular) can be determined from the external form of the band spectrum. A parallel band has a minimum due to the absence of the Q branch. A perpendicular band in the region of the zero line possesses a maximum owing its origin to the Q branch.

2. Spherical-rotator molecules. In this case the location of rotational lines in the vibration-rotation band is determined by the same formula as for linear molecules. The absorption bands consist of R, Q and P branches. Interaction between the vibrational and rotational molecular motions causes splitting of threefold degenerate vibrations.

3. Symmetric-rotator molecules. As in the case of linear molecules, these have both parallel and perpendicular bands for which different selection rules exist. The nature of the vibration-rotation spectrum is found to be more complex than for linear and spherical-rotator molecules, in particular when the direction of the dipole moment does not coincide with the molecule

axis. There are no symmetric-rotator molecules among atmospheric gases, and so the fine structure of the spectrum of this molecule will not be treated in detail.

4. Asymmetric-rotator molecules. Molecules of this type have the most complex rotational, and correspondingly vibration-rotation spectra. The dependence of rotational constants A, B and C on the vibrational quantum number is similar in form to (2.50). In the particular case of a triatomic molecule

$$\left.\begin{aligned} A &= A_e - \alpha_{A_1}\left(v_1+\frac{1}{2}\right) - \alpha_{A_2}\left(v_2+\frac{1}{2}\right) - \alpha_{A_3}\left(v_3+\frac{1}{2}\right), \\ B &= B_e - \alpha_{B_1}\left(v_1+\frac{1}{2}\right) - \alpha_{B_2}\left(v_2+\frac{1}{2}\right) - \alpha_{B_3}\left(v_3+\frac{1}{2}\right), \\ C &= C_e - \alpha_{C_1}\left(v_1+\frac{1}{2}\right) - \alpha_{C_2}\left(v_2+\frac{1}{2}\right) - \alpha_{C_3}\left(v_3+\frac{1}{2}\right). \end{aligned}\right\} \quad (2.51)$$

Constants α can be expressed in terms of anharmonicity constants.

The selection rules for transitions in these molecules depend on the axis along which the dipole moment of the transition is directed. In general one obtains three types of bands: type A, type B and type C (the dipole moment is directed respectively along the a, b or c axis). In the case of a symmetric triatomic molecule XY_2 the dipole moment is directed either along the axis of symmetry of the molecule, or along an axis perpendicular to it and lying in the plane of the molecule. Both perpendicular and parallel bands are possible here. For the particular case of the water molecule the molecular axis coincides with the axis of the average moment of inertia, parallel bands belonging to type B and perpendicular bands to type A. The majority of H_2O bands are perpendicular.

The nature of the vibration-rotation structure of bands of molecules of all types, the location of individual lines, their shape and intensity are substantially affected by interaction between molecular vibrations and rotations. An exact solution to the problem of the location of fine-structure lines involves allowance for anharmonicity of the vibrations, centrifugal stretching of the molecules, resonance excitation of energy levels, and the effect of the Coriolis force. Proper incorporation of all these effects is a complicated and difficult problem.

4. ELECTRONIC ENERGY AND ELECTRONIC MOLECULAR SPECTRA

The systematics of molecular electronic-energy levels and the selection rules for electronic transitions are quite complex. Further, all atmospheric gases (with the exception of oxygen) in the visible and infrared regions of electromagnetic waves of interest do not possess electronic spectra. Consequently it does not appear expedient to examine in detail the problem of electronic energy and electronic molecular spectra. The visible-range absorption spectrum of oxygen will be described below.

In spite of the great complexity involved in quantitative calculation of the electronic-vibration-rotation spectrum, it is precisely the solution to this problem which must be regarded as the most promising direction in the study of vibration-rotation spectra. This problem will be treated in more detail in the next chapter.

5. SHAPE OF SPECTRAL LINES

It was assumed in examining transitions between discrete molecular energy levels that absorption or emission processes associated with these transitions can be characterized by fully defined frequencies. The spectra resulting from these transitions were assumed to consist of individual monochromatic lines. However, one never deals with strictly monochromatic lines. Actual lines have fully defined profiles, the shape of which depends on the conditions under which light absorption or emission by molecules occurs. In the earth's atmosphere the spectral line profile is determined primarily by the following three effects: 1) radiation damping, 2) Doppler effect, and 3) molecular collision effects.

1. **Natural broadening of spectral lines.** Transition of a molecule from one quantum energy state into another occurs over a finite time interval. It follows from classical radiation theory that the vibrational process accompanying this transition represents an infinite set of frequencies over a specific spectral range. The width of this interval depends on the time during which the molecule is in the excited state. If molecules absorbing or emitting a certain frequency are not subjected to any external disturbances, then the emission (or absorption) intensity frequency distributions will be the natural profiles of the spectral lines. Quantum mechanics provides the following expression for this profile (Figure 2.4, curve 2):

$$k(\nu)=\frac{1}{\pi}\,\frac{\gamma_N}{(\nu-\nu_0)^2+\gamma_N^2}, \tag{2.52}$$

where

$$\gamma_N=\frac{1}{2\pi\tau} \tag{2.53}$$

is the natural line width, ν_0 is the frequency of the center of the line or the transition frequency for a hypothetical case corresponding to molecular excitation time $\tau=\infty$.

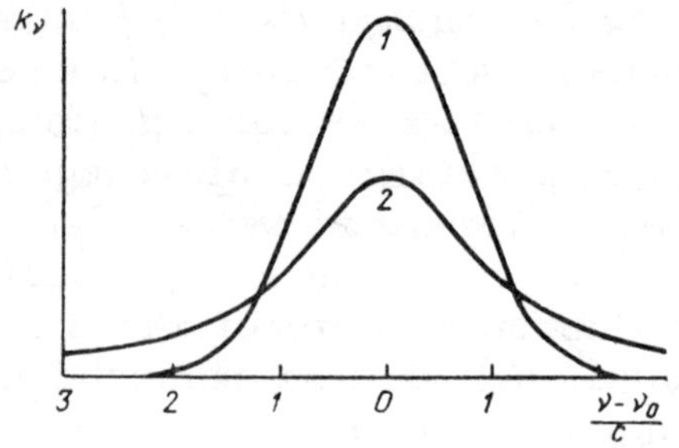

FIGURE 2.4. Profile of lines broadened due to the Doppler effect (1) and due to the collision or radiation damping effect (2) for similar intensities and line widths.

The line width is defined as the spectral interval between points ν_1 and ν_2 of the profile satisfying the condition

$$k(\nu_1) = k(\nu_2) = \frac{1}{2} k(\nu_0). \tag{2.54}$$

Equation (2.52) is the dispersion or Lorentz expression. When reference is usually had to a disperse spectrum one refers to the shape of line profiles obtained by considering the effect of collisions.

The value of γ_N for atmospheric gases is very small. The most probable value of τ for vibrational and rotational molecular transitions in atmospheric gases is of the order of 0.1 sec. For this value of τ, $\gamma_N = 3 \cdot 10^{-11}$ cm^{-1} /3/. This value is negligible compared to Doppler or collision line broadening. Hence in problems of visible and infrared wave propagation in the atmosphere natural broadening of spectral lines is disregarded.

2. Doppler broadening. On the assumption that it is possible to postulate thermodynamic equilibrium for the translational degrees of molecular freedom, associated with a Maxwellian molecular velocity distribution, the following expression is obtained for the spectral-line profile (Figure 2.4, curve 1) which is broadened due to the Doppler effect alone:

$$k(\nu) = \frac{S}{\sqrt{\pi}\alpha_D} \exp\left[-\left(\frac{\nu - \nu_0}{\alpha_D}\right)^2\right], \tag{2.55}$$

where S is the line strength (defined in the next section),

$$\alpha_D = \frac{\nu_0}{c} \sqrt{\frac{2kT \ln 2}{m}} \tag{2.56}$$

is the Doppler line half-width, T is the absolute temperature, m is the mass of a molecule, c is the speed of light, and k is the Boltzmann constant.

Hence α_D is a function of ν_0, m and T. The temperature in the lower atmosphere (troposphere and stratosphere) changes on the average by not more than 50%. Thus, for a given spectral line (ν_0 = const, m = const), temperature variations give rise to values of α_D within ±15% from the average value corresponding to the average temperature.

Goody /3/ gives two extreme values of α_D for atmospheric conditions when $T = 300°$K: 1) for the forbidden line of atomic oxygen with wavelength about 5577 Å,

$$\alpha_D = 3.3 \cdot 10^{-2}\ \text{cm}^{-1};$$

2) for the line of the rotational H_2O band about 200 cm^{-1},

$$\alpha_D = 3.5 \cdot 10^{-4}\ \text{cm}^{-1}.$$

3. Collision broadening. The interaction between an absorbing or emitting molecule with neighboring molecules results in deformation of the energy levels and its associated spectral line broadening. It is clear that this interaction should depend on pressure. Hence this effect is frequently termed pressure or collision broadening. If absorption or emission occurs in a pure gas, one sometimes refers to natural line broadening.

The effect of molecular collisions on broadening of spectral lines with consideration of all the inherent features of collisions has not yet been elaborated in full. In all cases one considers only double collisions, while the complex forces of the interacting molecules are described by interaction models such as: dipole-dipole, dipole-quadrupole, dipole-induced dipole.

A distinction is made between strong and weak collisions, depending on the magnitude of the variation in the phase of molecular absorption or emission induced by intermolecular collision. The collisions considered in kinetic theory are strong. The first theories of collision spectral-line broadening treated only strong collisions. The first of all these theories is the Lorentz theory /55/, developed in 1906. The results of this theory are still in use.

According to the theory of strong collisions the center of spectral lines does not shift during the interaction. During weak collisions, however, the line center does shift. Both types of collisions occur under real conditions. Under atmospheric conditions (pressure $P < 1$ atm) in the visible and infrared regions the line profiles obtained from the theory of strong collisions describe satisfactorily the central parts of spectral lines, the centers of which do not shift with variations in pressure.

From the standpoint of quantum mechanics, energy transitions during interaction between colliding molecules can occur in both molecules. Collisions accompanied by energy transitions are conventionally termed nonadiabatic. Adiabatic collisions are those in which only the energy of the levels is altered. In this classification weak collisions can be regarded as adiabatic.

It follows from qualitative considerations that the theories of nonadiabatic collisions provide a better description of cases with high transition probabilities. Conversely, theories of adiabatic collisions are more suitable for describing low-probability transitions. For atmospheric conditions in the infrared region the probabilities of molecular transition are high, so theories of nonadiabatic collisions are preferable here. An example of this type of theory, used extensively in calculations, is the Anderson theory.

In addition to the above classifications of pressure broadening theories, one should note the subdivision into impact and statistical theories. The latter assumes that the duration of collisions between molecules is negligible compared with the interval between the collisions. In the statistical theory one considers the distribution of the shift in energy level of a molecule at rest induced by its interaction with all the other molecules.

Since the theories used most extensively to describe spectral line broadening by collisions of molecules of atmospheric gases in the visible and infrared regions are those of Lorentz and Anderson, only these two theories will be examined here.

The Lorentz theory for the absorption spectral line profile yields

$$k(\nu) = \frac{S}{\pi} \frac{\gamma_L}{(\nu - \nu_0)^2 + \gamma_L^2}, \tag{2.57}$$

frequently termed the dispersion or Lorentz profile. Here S is the line strength, ν_0 is the location (frequency) of the line center, and

$$\gamma_L = \frac{1}{4\pi\tau(v)} \tag{2.58}$$

is the line half-width. Quantity $\tau(v)$ is the excited lifetime of the absorbing or emitting molecule and is a function of its velocity v.

Determination of γ_L involves consideration of $\tau(v)$. Usually γ_L is found as the line width corresponding to the average value of $\tau(v) = \overline{\tau}$. In this case the kinetic theory of gases yields the expression

$$\gamma_L = \frac{1}{2\pi\tau} = \sum_i n_i \sigma_i^2 \left[\frac{2kT}{\pi}\left(\frac{1}{m} + \frac{1}{m_i}\right)\right]^{1/2}, \tag{2.59}$$

where n_i is the molecular concentration of the i-th class, σ_i is the effective distance between the emitting or absorbing molecule and the molecule of the i-th class, m and m_i are the masses of these molecules. The term effective denotes the distance at which collisions result in line broadening.

In any gas mixture the molecular concentration n_i is proportional to pressure. It thus follows from (2.59) that the line width is proportional to pressure. Since the pressure under atmospheric conditions varies quite widely, the dependence of γ_L on P is very important. The dependence of γ_L on σ is also very important, since the value of σ depends not only on the class of colliding molecules, but also on the energy levels of the molecules, or on the rotational quantum number j. The temperature dependence of γ_L is quite weak, and can be neglected in many problems of visible and infrared wave propagation in the atmosphere. Subsequently we shall frequently return to the dependence of γ_L on various quantities.

For a two-component mixture γ_L is given by the simpler expression

$$\gamma_L = \frac{1}{4}\left\{2\pi kT\left[N_a D_{aa}^2\left(\frac{2}{m_a}\right)^{1/2} + N_b D_{ab}^2\left(\frac{1}{m_a}+\frac{1}{m_b}\right)^{1/2}\right]\right\}^{1/2}, \qquad (2.60)$$

where subscripts a and b pertain to absorbing and nonabsorbing molecules, N_a and N_b are the numbers of a and b molecules per unit volume, D_{aa} and D_{ab} are the sums of the optical collision diameters between molecules $a \leftrightarrow a$ and $a \leftrightarrow b$; m_a and m_b are the molecular masses.

Quantity γ_L for a two-component mixture frequently assumes the even simpler form

$$\gamma_L = \gamma_L^0 \frac{P}{P_0}\sqrt{\frac{T_0}{T}}, \qquad (2.61)$$

where γ_L^0 is the line half-width under standard conditions ($P_0 = 1$ atm, $T_0 = 273\,^\circ\mathrm{K}$).

The Anderson theory of pressure broadening of spectral lines, developed in 1949 /56/, is a quantum-mechanical nonadiabatic theory for molecular spectra. According to this theory it is assumed that transitions may occur in both colliding molecules. The starting formula for γ_L is then formula (2.59) of collision theory. However, the effective distances or optical collision diameters are determined with allowance for possible energy transitions in the colliding molecules, leading to the following expression for γ_L:

$$\gamma_L^{ij} = \sum_{j_2}\left[\frac{2kT}{\pi}\left(\frac{1}{m}+\frac{1}{m_0}\right)\right]^{1/2} n(j_2)\,\sigma^2(i, j, j_2), \qquad (2.62)$$

where $n(j_2)$ is the concentration of perturbing molecules:

$$\sigma(i, j, j_2) = \sum_{j_2', i'} \sigma(i, i'; j_2, j_2'; j) + \sum_{j_2', j'} \sigma(i; j_2, j_2'; j, j'); \qquad (2.63)$$

i, i', and j, j' are indices characterizing the states of the absorbing molecule for which transition $i \rightarrow j$ occurs without interaction with the perturbing molecule, j_2 and j_2' are indices characterizing the state of the perturbing molecule. Molecular collision results in the appearance of $i \rightarrow i'$ and $j \rightarrow j'$ transitions in the absorbing molecule and $j_2 \rightarrow j_2'$ in the perturbing molecule.

Benedict and Kaplan /57, 58/ used the Anderson theory to analyze in detail the line half-widths of the pure rotational spectrum of water vapor during collision between its molecules and those of nitrogen and oxygen. The results showed that the value of γ_L for $H_2O - N_2$ collisions changes from line to line within the limits 0.03—0.1 cm^{-1} at $P = 1$ atm. The results of Benedict and Kaplan /57, 58/ are examined in the next chapter.

4. Combined Doppler and collision broadening. Under real conditions Doppler and collision broadening occur simultaneously. However, the role of these two effects differs substantially at different altitudes, since Doppler broadening is not a function of pressure. Collision broadening attains a maximum in the ground layer and decreases with altitude in proportion to pressure reduction. In the ground layer the Doppler line width is negligible compared with the half-width due to collisions. The width of each collision-broadened line at some altitude equals the Doppler width. On further rise in altitude the Doppler width becomes increasingly larger compared with the dispersion width.

If one assumes that $\gamma_L = 0.08$ cm^{-1}, which is the most probable value under standard conditions, then altitudes at which $\gamma_L = \gamma_D$ for limiting values $\gamma_D = 3.3 \cdot 10^{-2}$ cm^{-1} and $3.5 \cdot 10^{-3}$ cm^{-1} (see part 2 of this section) are found to be 7 and 37 km. For narrower dispersion lines these altitudes will be smaller, and vice versa. These estimates imply that combined consideration of spectral-line broadening effects can be mandatory in some cases, starting as low as in the upper troposphere.

Combined consideration of Doppler and collision broadening yields the following expression for the absorption coefficient:

$$k(\nu) = \frac{k_0 y}{\pi} \int_{-\infty}^{\infty} \frac{e^{-t^2}}{y^2 + (x - t)^2} dt, \tag{2.64}$$

where

$$k_0 = \frac{S}{\gamma_D} \left(\frac{\ln 2}{\pi} \right)^{1/2};$$

$$y = \frac{\gamma_L}{\gamma_D} (\ln 2)^{1/2}; \qquad x = \frac{\nu - \nu_0}{\gamma_D} (\ln 2)^{1/2}, \tag{2.65}$$

S is the line strength and ν_0 is the center of the line.

5. Shape of the far wings of the line. It is seen from Figure 2.4 that the Doppler line profile drops in the wings much more rapidly than the dispersion profile. This means that the far wings of the line for a simultaneous Doppler and collision effect should be disperse.

However, the question as to whether the disperse profile describes correctly the far wings of lines has not yet been resolved. Qualitative considerations underlying the assumptions of various pressure broadening theories indicate that the shape of the far wings of a line should be better described by statistical than by impact theories. At present there is insufficient experimental data with which to support one or another line-wing theory. Available results show that the shape of the far wings varies for different molecules.

6. SPECTRAL LINE STRENGTH

Spectral line strength S is expressed in terms of the absorption coefficient $k(\nu)$:

$$S = \int_0^\infty k(\nu)\, d\nu. \tag{2.66}$$

Quantum mechanics provides the following expression for the strength of a single molecular transition from state j to state i /3/:

$$S_{i,j} = \frac{n_j}{g_i n} \frac{8\pi^3 \nu_{i,j} |R_{i,j}|^2}{3hc} (1 - e^{-h\nu_{i,j}/kT}), \tag{2.67}$$

where n_j is the molecular concentration in the lower state, n is the total molecular concentration, g_i is the statistical weight, $\nu_{i,j}$ is the transition frequency, $|R_{i,j}|^2$ is the square of the matrix component of the dipole moment, T is the absolute temperature, h, k and c are the Planck and Boltzmann constants and the speed of light.

Matrix component $\vec{M}$ of the dipole moment is expressed in terms of the wave functions of molecular states ψ_i and ψ_j:

$$R_{i,j} = \int \psi_i^* \vec{M} \psi_j \, dv. \tag{2.68}$$

Integration is here carried out over the entire configuration state, while dv is an element of this space. The wave functions are obtained by solving the Schrödinger equation.

For wave functions we have the orthogonality condition

$$\int \psi_i^* \psi_j \, dv = 0 \; (i \neq j). \tag{2.69}$$

For $\vec{M}$ = const, equations (2.69) and (2.68) yield

$$\vec{M} \int \psi_i^* \psi_j \, dv = 0. \tag{2.70}$$

The last expression justifies the previous assumption, that a molecular transition from one energy level to another during dipole emission occurs only when the dipole moment changes. Note that emission or absorption of light can occur not only during interaction between the electromagnetic field and a substance. The variation of the magnetic dipole or the electrical quadrupole of a molecule, accompanying the corresponding energy transition, also involves the appearance of spectral lines. However, the strength of these lines is very low. Transitions associated with variation in the electrical dipole moment of the molecule yield a 10^5 to 10^8 greater line strength than variations in the magnetic dipole and electrical quadrupole moments.

Matrix components $R_{i,j}$ are related to known Einstein coefficients $A_{i,j}$ characterizing the probability of induced emission and absorption between molecular levels i and j:

$$A_{i,j} = \frac{64\pi^4 \nu_{i,j}^3}{3hg_i c^3} |R_{i,j}|^2. \qquad (2.71)$$

The value of $A_{i,j}$ for electrical dipole emission is of the order of 10^8, 10 and 1 $\sec^{-1}$ for electronic, vibrational and rotational transitions, respectively.

The most difficult problem in determining line strengths is to find $|R_{i,j}|^2$.

7. GENERAL CHARACTERISTICS OF ABSORPTION SPECTRA OF ATMOSPHERIC GASES

Of interest here are absorption spectra of atmospheric gases in the visible and infrared regions. We shall consider spectra of the principal absorbing gases (water vapor, carbon dioxide, ozone), as well as the spectra of gases whose contribution to the integral absorption in the visible and infrared in the atmosphere is less marked, sometimes due to the relatively low absorption coefficient (oxygen) and sometimes due to their low concentration (methane, carbon monoxide, nitrous oxide). The principal constituent of the atmosphere — nitrogen — possesses absorption bands only in the far ultraviolet. The general picture of the absorption spectra of the earth's atmosphere is shown in Figure 2.5 /3/. We now consider the absorption spectra of individual atmospheric constituents.

1. Water vapor. The water molecule is an equilateral triangle with the oxygen atom at the vertex (Figure 2.6). Angle ϑ between the O—H bonds is 104°30'. The distance between the oxygen and hydrogen atoms is 0.958 Å.

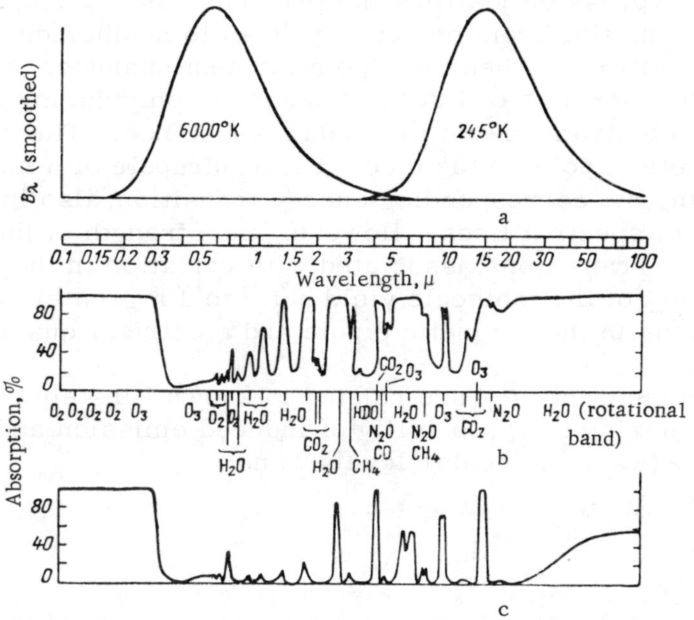

FIGURE 2.5. General picture of the absorption spectrum of atmospheric gases:

a – blackbody emission for 6000 and 245°K; b – absorption spectrum for solar radiation received at the earth's surface; c – absorption spectrum for solar radiation reaching an altitude of 11 km. (Spectra b and c were obtained for middle latitudes at 40° solar zenith angle under the most probable conditions.)

Electronic transitions in the H_2O molecule occur in a region with wavelength shorter than 1860 Å (far ultraviolet).

The vibration-rotation absorption spectrum of water vapor is very complex, due primarily to the fact that the molecular moments of inertia relative to the three principal axes of the ellipsoid of inertia differ greatly (they are $\mathcal{I}_a = 1.004 \cdot 10^{-40}$, $\mathcal{I}_b = 1.929 \cdot 10^{-40}$ and $\mathcal{I}_c = 3.104 \cdot 10^{-40}$ g·cm²). In this case the rotational structure of spectra of asymmetric-rotator molecules is random and complex. Figure 2.6 depicts the *a*, *b* and *c* axes relative to which the moments of inertia $\mathcal{I}_a$, $\mathcal{I}_b$ and $\mathcal{I}_c$ of the H_2O molecule are determined. The absolute values of the moments of inertia are small, since the molecule contains two hydrogen atoms. Hence the distance between the rotational lines is found to be large.

The high values of the dipole moments of the H_2O molecule and of its isotopes are the reason for an intensive rotational spectrum. Dipole moments M of the H_2O molecule and of its isotopes are: $M_b = 1.94\,D$ for H_2O, $M_b = 1.87\,D$ for D_2O, $M_a = 0.64\,D$ and $M_b = 1.70\,D$ for HDO.

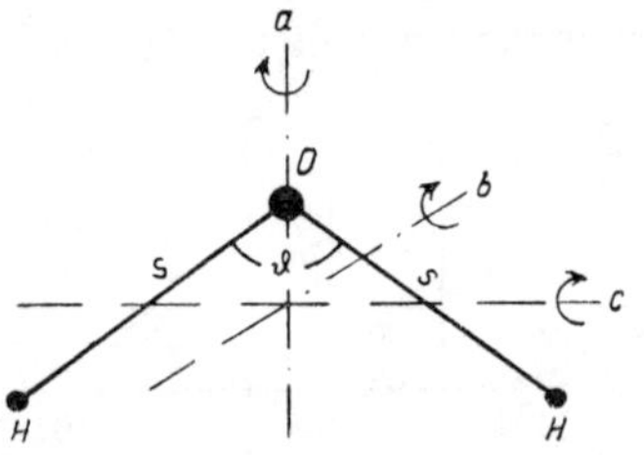

FIGURE 2.6. Structure of the H_2O molecule.

The solar spectrum was found to contain lines of the following four isotopes of H_2O: H_2O^{16}, H_2O^{18}, H_2O^{17} and HDO^{16}, which are present in the atmosphere in proportions 99.73, 0.2039, 0.0373 and 0.0298. The rotational and vibrational constants of the first three isotopes differ little. The lines of H_2O^{18} molecules are shifted relative to the corresponding lines of H_2O^{16} by from 1 to 11 cm^{-1}. For the H_2O^{17} molecule this shift is twice as great /3/. The rotational constants for the H_2O^{16} and HDO^{16} molecules are listed in Table 2.4.

TABLE 2.4. Rotational constants of HHO^{16} and HDO^{16} molecules in the ground state (cm^{-1}) /3/

Axis	HHO^{16}	HDO^{16}
a	27.79	23.38
b	14.51	9.06
c	9.29	6.38

The pure rotational absorption spectrum of water vapor extends over quite a wide region, approximately from 8μ to several centimeters in terms of wavelength.

In addition to the three fundamental vibration-rotation bands, which owe their origin to the fundamental vibrations with frequencies ν_1, ν_2 and ν_3 (Table 2.5), the vibration-rotation spectrum of H_2O also contains many harmonic bands and bands of combination frequencies in the visible and infrared (Tables 2.6 and 2.7).

TABLE 2.5. Frequencies of fundamental bands of the H_2O molecule /3/

Band	Transition	Band center, cm^{-1}	
		HHO^{16}	HDO^{16}
ν_1	000 — 100	3657.05	2723.66
ν_2	000 — 010	1594.78	1403.3
ν_3	000 — 001	3755.92	3707.47

TABLE 2.6. H_2O bands in the visible spectral region /3/

Transition	Band center		Type	Strength, cm
	cm^{-1}	μ		
000—411	18 394	0.54	*A*	$2\cdot10^{-23}$
000—203	17 495	0.57	*A*	$1\cdot10^{-22}$
000—401	16 899	0.59	*A*	$3\cdot10^{-22}$
000—302	16 898	0.59	*B*	$3\cdot10^{-23}$
000—321	16 822	0.59	*A*	$2\cdot10^{-22}$
000—113	15 832	0.63	*A*	$2\cdot10^{-23}$
000—311	15 348	0.66	*A*	$2\cdot10^{-22}$
000—103	14 319	0.69	*A*	$1\cdot10^{-21}$
000—400	14 221	0.70	*B*	$1\cdot10^{-22}$
000—301	13 831	0.72	*A*	$3\cdot10^{-21}$
000—202	13 828	0.72	*B*	$2\cdot10^{-23}$
000—221	13 653	0.73	*A*	$6\cdot10^{-21}$
000—013	12 565	0.79	*A*	$1\cdot10^{-22}$
000—112	12 408	0.81	*B*	$6\cdot10^{-21}$
000—211	12 151	0.82	*A*	$6\cdot10^{-23}$
000—210	12 140	0.82	*B*	$1\cdot10^{-22}$
000—131	11 813	0.85	*A*	$2\cdot10^{-21}$

Figure 2.7 shows the fundamental vibrations of the molecule, responsible for the appearance of frequencies ν_1, ν_2 and ν_3. The deformation frequency ν_2 is the lowest. Molecular transitions associated with frequencies ν_1 and ν_3 produce overlapping bands, the region of which also contains the first overtone of the ν_2 band. Overlapping of ν_1, ν_3 and $2\nu_2$ bands results in complex interaction between the corresponding energy levels.

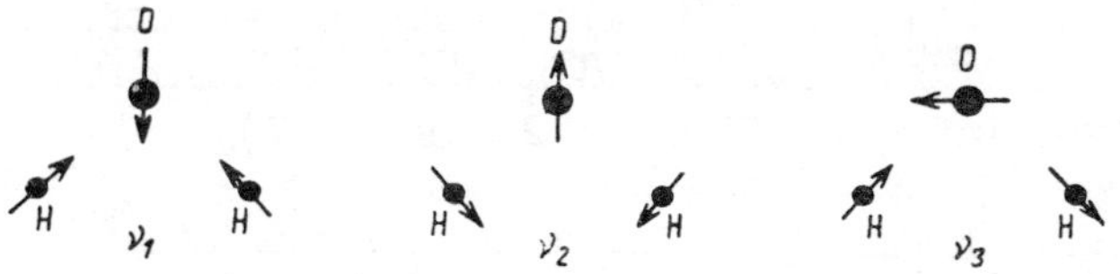

FIGURE 2.7. Fundamental vibrations of the H_2O molecule.

In addition to the fundamental bands, overtones and combination frequencies, the solar spectrum also contains bands of higher states, in which vibrational transition does not occur from the ground state (Table 2.8). The strength of higher-state bands depends greatly on the temperature, since it has a predominant effect on the population of the corresponding vibrational levels of the H_2O molecule. On the other hand, the population of the ground vibrational level can be regarded as being virtually independent of temperature.

TABLE 2.7. H_2O bands in the near-infrared spectral region /3/

Band	Transition	Band center		Type	Strength, cm
		cm^{-1}	μ		
ρ	000—003	11 032	0.91	*A*	$2\cdot10^{-21}$
	000—102	10 869	0.92	*B*	$4\cdot10^{-22}$
σ	000—201	10 613	0.94	*A*	$1\cdot10^{-20}$
	000—300	10 600	0.94	*B*	$6\cdot10^{-22}$
τ	000—121	10 329	0.97	*A*	$2\cdot10^{-21}$
	000—220	10 284	0.97	*B*	$<4\cdot10^{-23}$
	000—041	9 834	1.01	*A*	$6\cdot10^{-23}$
φ	000—012	9 000	1.11	*B*	$3\cdot10^{-22}$
	000—121	8 807	1.13	*A*	$8\cdot10^{-21}$
	000—210	8 762	1.14	*B*	$1\cdot10^{-23}$
	000—130	8 274	1.20	*B*	$7\cdot10^{-24}$
	000—031	8 374	1.19	*A*	$3\cdot10^{-23}$
ψ	000—002	7 445	1.34	*B*	$1\cdot10^{-21}$
	000—101	7 250	1.37	*A*	$1.5\cdot10^{-19}$
	000—200	7 201	1.38	*B*	$1.5\cdot10^{-20}$
	000—021	6 871	1.45	*A*	$1\cdot10^{-20}$
	000—120	6 775	1.47	*B*	$2\cdot10^{-22}$
Ω	000—011	5 331	1.87	*A*	$2.2\cdot10^{-19}$
	000—110	5 235	1.91	*B*	$7\cdot10^{-21}$
	000—030	4 667	2.14	*B*	$3\cdot10^{-22}$

The vibration-rotation spectra of the water isotopes HHO^{18} and HHO^{17} are similar to the spectrum of HHO^{16}, except that they are somewhat shifted. The ratios of line strengths of these isotopes to the HHO^{16} lines correspond to the ratios of their concentrations. The HDO^{16} bands (Table 2.9) differ markedly from those of HHO^{16}.

TABLE 2.8. Bands of higher states of H_2O /3/

Transition	Frequency		Type	Strength (at 259°K), cm
	cm^{-1}	μ		
010—001	2161.14	4.62	*A*	$2\cdot10^{-22}$
010—100	2062.27	4.84	*B*	$8\cdot10^{-23}$
010—020	1556.82	6.42	*B*	$9\cdot10^{-22}$
010—010	0—500	∞ — 20	*B*	$4.6.10^{-20}$

Analysis of the vibration-rotation spectrum of water vapor shows that the most intensive and wide absorption band is the fundamental band ν_2, corresponding to deformation vibration of the molecule. The center of this band lies near the 6.25μ wavelength. In the literature this band is frequently termed the 6.3μ band. In the vertical column of the atmosphere this band fully absorbs the solar radiation in the wavelength range of approximately 5.5 to 7.5μ /59/. The center of band ν_3, which has the next highest strength, lies at about 2.66μ. The centers of the ν_1 and $2\nu_2$ bands lie near wavelengths 2.74 and 3.17μ. Bands ν_1, ν_3 and $2\nu_2$ provide for complete absorption of solar radiation in the vertical column of the atmosphere in the spectral region of approximately 2.6 to 3.3μ. Other vibration-rotation bands of water vapor cluster and form (in a spectrum obtained with low-resolution) absorption bands with centers near wavelengths 1.87, 1.38, 1.1, 0.94, 0.81 and 0.72μ. Several weak bands exist in the visible region.

TABLE 2.9. Bands of HDO^{16} /3/

Transition	Frequency		Strength, cm	
	cm^{-1}	μ	component *A*	component *B*
000—010	1403.3	7.12	$5\cdot10^{-23}$	$2.3\cdot10^{-22}$
000—100	2723.66	3.67	$5\cdot10^{-22}$	$<5\cdot10^{-24}$
000—020	2782.16	3.59	$1.4\cdot10^{-23}$	$9\cdot10^{-24}$
000—001	3707.47	2.69	$3\cdot10^{-22}$	$1.4\cdot10^{-22}$
000—110	4100.05	2.43	$2\cdot10^{-23}$	$<1\cdot10^{-24}$
000—030	4145.59	2.41	$2\cdot10^{-23}$	$<1\cdot10^{-24}$
000—011	5089.59	1.96	$3\cdot10^{-23}$	$1.6\cdot10^{-23}$
000—200	5363.59	1.86	$3\cdot10^{-23}$	$<2\cdot10^{-24}$
000—101	6415.64	1.55	$1\cdot10^{-24}$	$9\cdot10^{-24}$
000—021	6452.05	1.54	$2\cdot10^{-24}$	$<5\cdot10^{-25}$
000—012	8611.22	1.16	$5\cdot10^{-24}$	$<5\cdot10^{-25}$

The fine structure of the vibration-rotation spectrum of water vapor is very complex and confusing. Each band of this spectrum consists of hundreds and even thousands of individual lines, the identification of which is a highly complicated task. The utilization of spectral instruments with even the highest resolution does not always result in complete resolution of all lines of the fine structure of the spectrum. As an illustration Figure 2.8 shows a part of the spectrum of the vibration-rotation band of H_2O which is only 0.07μ wide in the region of the 1.38μ band, recorded by a high-resolution instrument /60/. Theoretical calculation of the location of centers, strengths and half-widths of individual lines requires proper incorporation of the complex interaction between the vibrational and rotational levels of the molecule.

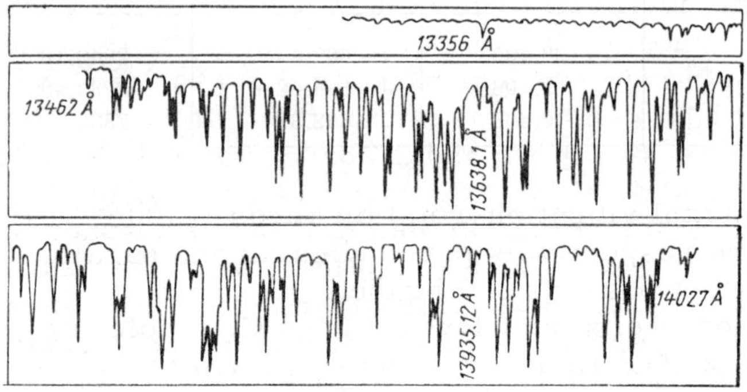

FIGURE 2.8. Absorption spectra of a part of the vibration-rotation band of water vapor near 1.38 μ /60/.

2. Carbon dioxide. The linearly symmetric CO_2 molecule, in which the length of the C—O bond in the fundamental vibrational state is 1.1632 Å, has three fundamental vibrational frequencies ν_1, ν_2 and ν_3, which are depicted schematically in Figure 2.9. The rotational constant of the molecule in the fundamental vibrational state is 0.3895 cm^{-1}. Deformation vibration ν_2 is twofold degenerate. In the latter case, in addition to the selection rules considered in section 3 of this chapter, it is necessary to consider selection rules for the vibrational quantum numbers of the vibrational angular momentum:

$$\Delta l = 0, \ \pm 1. \tag{2.72}$$

As a result of molecule symmetry the dipole moment for ν_1 varies, and hence this vibration does not appear in the infrared spectrum. The value of frequency ν_1 is approximately twice that of ν_2.

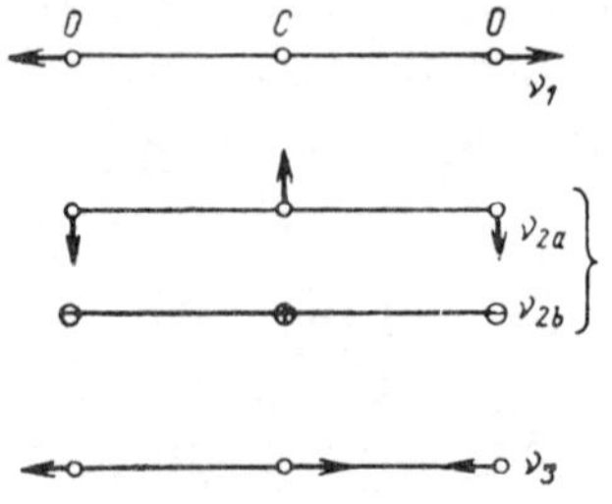

FIGURE 2.9. Fundamental vibrations of the CO_2 molecule.

TABLE 2.10. Fundamental ν_2 and ν_3 bands of the CO_2 molecule /3/

Isotope	Content, %	Band center, cm^{-1}	
		ν_2	ν_3
$C^{12}O^{16}O^{16}$	98.420	667.40	2349.16
$C^{13}O^{16}O^{16}$	1.108	648.52	2283.48
$C^{12}O^{16}O^{18}$	0.408	662.39	2333

The fundamental vibration-rotation bands ν_2 $(00^\circ0-01^11)$ and ν_3 $(00^\circ0-00^\circ1)$ are respectively perpendicular and parallel. The ν_3 band thus does not possess a Q branch.

The statistical weights of levels of the CO_2 molecule with even values of j are zero. Hence the fine structure of the vibration-rotation bands contains only lines with odd j.

The CO_2 molecule has the following isotopes: $C^{12}O_2^{16}$, $C^{13}O_2^{16}$, $C^{12}O^{16}O^{17}$, and $C^{12}O^{16}O^{18}$, the content of which in percents of the total abundance of CO_2 in the atmosphere is respectively 98.420, 1.108, 0.0646 and 0.4078%. In addition, two very rare isotopes of CO_2 are also present. Table 2.10 lists the location of the centers of the ν_2 and ν_3 bands of the three CO_2 isotopes most prevalent in the atmosphere /3/.

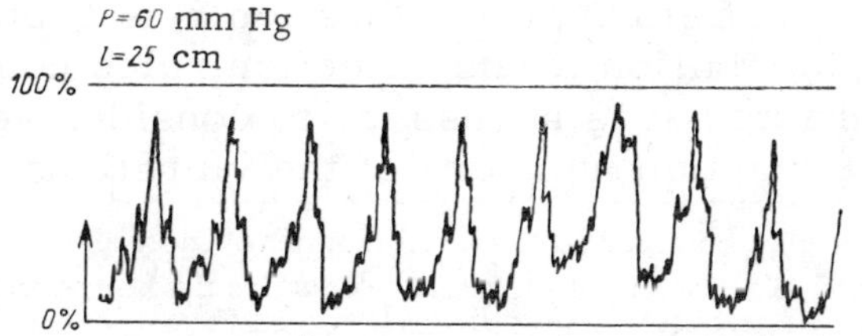

FIGURE 2.10. A part of the absorption spectrum of the 15.0 μ CO_2 band recorded with high resolution /61/.

The fundamental vibration-rotation band ν_2 with center about the 15μ wavelength together with 14 bands of higher states occupies quite a wide spectral interval, from approximately 12 to 20μ. In the vicinity of the central part of this band (~13.5—16.5μ) the vertical column of the atmosphere completely absorbs the solar radiation. This entire set of bands is frequently termed the 15.0μ band (Table 2.11). Figure 2.10 depicts a part of the spectrum of the 15.0μ band of CO_2, from which it is seen that not all the lines of this spectral interval have been resolved by the instrument. The region of the 15μ band contains a large number of fine-structure lines. For example, Drayson /62/ in calculating the absorption coefficients of CO_2 in the vicinity of this band (more precisely, of this band series) took into account about 2000 lines.

TABLE 2.11. CO_2 bands in the 15 μ region /3/

Isotope	Transition	Band center		Molecular strength of band*, cm	E'', cm^{-1}
		cm^{-1}	μ		
$C^{12}O^{16}O^{16}$	00^00—01^10	667.40	15.0	$7.89\cdot10^{-18}$	0.0
$C^{13}O^{16}O^{16}$	00^00—01^10	648.52	15.4	$7.9\cdot10^{-20}$	0.0
$C^{12}O^{18}O^{16}$	00^00—01^10	662.39	15.1	$3.7\cdot10^{-20}$	0.0
$C^{12}O^{16}O^{16}$	01^10—02^00	618.03	16.2	$1.75\cdot10^{-19}$	667.4
—	01^10—10^00	720.83	13.9	$2.3\cdot10^{-19}$	667.4
—	01^10—02^20	667.76	15.0	$6.2\cdot10^{-19}$	667.4
—	10^00—03^10	647.02	15.5	$4.2\cdot10^{-20}$	1285.43
—	10^00—11^10	791.48	12.6	$8.2\cdot10^{-22}$	1285.43
—	02^20—03^10	597.29	16.7	$5.83\cdot10^{-21}$	1335.16
—	02^20—11^10	741.75	13.5	$5.2\cdot10^{-21}$	1335.16
—	02^20—03^30	668.3	15.0	$3.2\cdot10^{-20}$	1335.16
—	02^00—03^10	544.26	18.4	$1.64\cdot10^{-22}$	1388.19
—	03^30—04^20	581.2	17.2	$1.56\cdot10^{-22}$	2003.28
—	03^30—12^20	756.75	13.2	$2.2\cdot10^{-22}$	2003.28
—	03^10—12^20	828.18	12.1	$1.8\cdot10^{-23}$	1932.45
—	03^10—12^00	740.5	13.5	$5.2\cdot10^{-22}$	1932.45
—	04^40—13^30	769.5	13.0	$1.5\cdot10^{-23}$	2674.76

* These data were obtained for the total number of molecules of all the isotopes. Each value of the strength contains the Boltzmann factor of the ground state.

The fundamental vibration-rotation absorption band of the $C^{12}O_2^{16}$ molecule ($\nu_3 = 2349.16$ cm^{-1}) is responsible for quite high radiation absorption in the atmosphere in the 4.3μ region. This band is overlapped by two others: the fundamental band of the $C^{13}O_2^{16}$ molecule ($\nu_3 = 2283.48$ cm^{-1}) and the combination band $\nu_1+\nu_3-2\nu_2$ ($02°0-10°1$) with center near 2429.37 cm^{-1}. All these bands are parallel and consequently have no Q branches. The 4.3μ band is quite narrow; its strength is so high that up to altitudes of the order of 20 km the solar radiation over the range of approximately 4.2—4.4μ is totally absorbed in the vertical layer of the atmosphere.

In addition to the above 15.0 and 4.3μ bands, in which absorption is primarily due to the fundamental vibration-rotation bands ν_2 and ν_3 of the $C^{12}O_2^{16}$ molecule CO_2 has bands whose centers lie near wavelengths 10.4, 9.4, 5.2, 4.8, 2.7, 2.0, 1.6 and 1.4μ and a number of weak bands in the range 1.24 to 0.78μ. All these bands are relatively narrow (with a width of the order of 0.1μ).

TABLE 2.12. 5.2 and 4.8μ absorption bands of CO_2 /3/

Band, μ	Isotope	Transition	Band center		E'', cm^{-1}	Molecular strength of band at 259°K, cm
			cm^{-1}	μ		
5.2	$C^{12}O_2^{16}$	$000-03^10$	1932.45	5.17	0	$2.4 \cdot 10^{-21}$
4.8	$C^{13}O_2^{16}$	$000-11^10$	2037.08	4.90	0	$4.0 \cdot 10^{-22}$
	$C^{12}O_2^{16}$	$000-11^10$	2076.86	4.81	0	$3.2 \cdot 10^{-20}$
	$C^{12}O_2^{16}$	01^10-12^20	2093.35	4.77	667.40	$2.0 \cdot 10^{-21}$
	$C^{12}O_2^{16}$	01^10-200	2129.79	4.69	667.40	$8.0 \cdot 10^{-22}$

The two parallel bands in the 10μ region have frequencies of 1063.8 cm^{-1} (020—001 transition) and 961.0 cm^{-1} (100—001 transition). Absorption in the 1.4, 1.6, 2.0, 2.7, 5.2 and 4.8μ region is due to bands, data on which are listed in Tables 2.12, 2.13 and 2.14. In these tables the band strengths are defined for a single CO_2 molecule with incorporation of all the isotopes of the molecule.

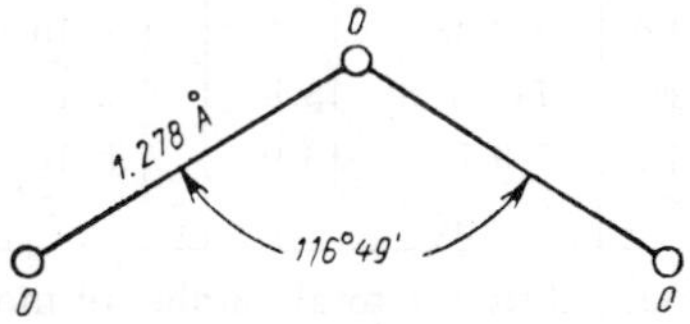

FIGURE 2.11. Structure of the ozone molecule.

3. Ozone. The dipole moment of the O_3 molecule (Figure 2.11) is 0.58*D*. Electronic transitions in the molecule produce Hartley and Huggins bands, situated in the ultraviolet spectral region (wavelengths shorter than 3400 Å), and Chappuis bands in the 4500—7400 Å region (visible region). The maximum value of the absorption coefficient in the Chappuis bands is $5 \cdot 10^{-21}$ cm^2, which is responsible for the 7% absorption of solar radiation for an atmospheric mass of two.

All three fundamental frequencies of the O_3 molecule are active in infrared absorption. They form vibration-rotation bands with centers $\nu_1 = 1110$ cm^{-1} (9.1μ), $\nu_2 = 710$ cm^{-1} (14.1μ) and $\nu_3 = 1043$ cm^{-1} (9.6μ). The ν_1 band is very weak and overlaps the ν_3 band. This results in strong interaction between the corresponding molecular energy levels.

TABLE 2.13. CO_2 absorption band near 2.7 μ /3/

Isotope	Transition	Band center		E'', cm^{-1}	Molecular strength of band at 259°K, cm
		cm^{-1}	μ		
$C^{12}O_2^{16}$	000—02^{0}1	3613.03	2.76	0	$1.0\cdot10^{-18}$ $(1.4\times\times10^{-18})$
—	000—101	3714.56	2.69	0	$1.3\cdot10^{-18}$ $(2\cdot10^{-18})$
—	01^{1}0—03^{1}0	3580.81	2.79	667.40	—
—	01^{1}0—11^{1}1	3723.05	2.68	667.40	—
$C^{13}O_2^{16}$	000—02^{0}1	3527.70	2.83	0	$1.1\cdot10^{-20}$
—	000—101	3632.92	2.75	0	$1.4\cdot10^{-20}$
—	01^{1}0—03^{1}1	3498.72	2.85	667.40	—

TABLE 2.14. CO_2 absorption band near 2.0, 1.6 and 1.4 μ /3/

Band, μ	Isotope	Transition	Band center		E'', cm^{-1}	Molecular strength of band at 259°K, cm
			cm^{-1}	μ		
2.0	$C^{12}O_2^{16}$	00^{0}0—04^{0}1	5100	1.96	0	$1.6\cdot10^{-20}$
	$C^{12}O_2^{16}$	00^{0}0—12^{0}1	4978	2.00	0	$3.7\cdot10^{-20}$
	$C^{12}O_2^{16}$	00^{0}0—20^{0}1	4853	2.06	0	$1.0\cdot10^{-20}$
	$C^{12}O_2^{16}$	01^{1}0—05^{1}0	5132	1.94	667.40	—
	$C^{12}O_2^{16}$	01^{1}0—13^{1}1	4965	2.01	667.40	—
	$C^{12}O_2^{16}$	00^{1}0—21^{1}1	4808	2.07	667.40	—
	$C^{13}O_2^{16}$	00^{0}0—04^{0}1	5046	1.98	0	$1.8\cdot10^{-22}$
	$C^{13}O_2^{16}$	00^{0}0—12^{0}1	4887	2.04	0	$4.1\cdot10^{-22}$
	$C^{13}O_2^{16}$	00^{0}0—20^{0}1	4748	2.10	0	$1.1\cdot10^{-22}$
	$C^{12}O^{16}O^{18}$	00^{0}0—04^{0}1	5042	1.98	0	$7\cdot10^{-23}$
	$C^{12}O^{16}O^{18}$	00^{0}0—12^{0}1	4905	2.03	0	$1.5\cdot10^{-23}$
	$C^{12}O^{16}O^{18}$	00^{0}0—20^{0}1	4791	2.08	0	$4\cdot10^{-23}$
1.6	$C^{12}O_2^{16}$	00^{0}0—06^{0}1	6503	1.53	0	—
	$C^{12}O_2^{16}$	00^{0}0—14^{0}1	6350	1.57	0	—
	$C^{12}O_2^{16}$	00^{0}0—22^{0}1	6228	1.60	0	$2.9\cdot10^{-22}$
	$C^{12}O_2^{16}$	00^{0}0—20^{0}1	6076	1.64	0	—
1.4	$C^{12}O_2^{16}$	00^{0}0—00^{0}3	6973	1.43	0	$8.6\cdot10^{-22}$

The atmosphere is known to contain the following three isotopes of the ozone molecule: O_3^{16}, $O^{16}O^{18}O^{16}$ and $O^{16}O^{16}O^{18}$ (Table 2.15).

TABLE 2.15. Location of the fundamental bands of the O_3 molecule /3/

Band	O_3^{16}, cm^{-1}	$O^{16}O^{18}O^{16}$, cm^{-1}	$O^{16}O^{16}O^{18}$, cm^{-1}
ν_1	1110 (?)	1080	1095
ν_2	701.42	697	688
ν_3	1045.16	1008	1029
Constant, %	99.4	0.21	0.41

The overtones and combination frequencies of O_3 vibrations produce vibration-rotation bands of this molecule in the 2.7, 3.27, 3.59, 4.75 and 5.75 μ regions, the most intensive of which is the 4.75μ band. The widths of all these bands are of the order of 0.1 cm^{-1}.

In considering the absorption of solar radiation by the earth's atmosphere one usually considers only the absorption of the 9.6μ band of O_3, situated at the center of the long-wave "window" of the atmosphere (8—13μ). Its central part is about 1.0μ wide and absorbs, in the vertical column of the atmosphere, approximately one-half of the solar radiation. Other bands either overlap the stronger absorption bands of H_2O and CO_2, or possess low intensity. The 9.6μ band contains a large number of fine-structure lines which have not yet been resolved experimentally. However, it was possible to obtain experimentally the molecular constants for excitation of the rotational state. These constants, together with those of the ground state, obtained from the pure rotational spectrum of O_3 in the microwave region, make it possible to calculate the fine structure of the band. The O_3 molecule has an intensive, pure rotational spectrum in the microwave range.

4. Oxygen. Atmospheric oxygen exists in molecular and atomic states. Atomic oxygen appears as a result of dissociation of molecular oxygen by ultraviolet solar radiation. This dissociation results in atoms from which ozone forms.

Atomic oxygen in the visible and infrared regions does not possess permitted electronic transitions of dipole emission. The series of forbidden transitions yields a single line with wavelength 5571 Å and a multiplet in the 6300—6364 Å region. The strengths of these lines are low and should be taken into account only when considering airglow in the upper atmosphere.

TABLE 2.16. Red bands of the O_2^{16} molecule /3/

Transition	Band center, μ	Band strength,* cm	Characteristic band strength in the solar spectrum**
0.0	0.7621	$4.3\cdot10^{-22}$	s
0.1	0.6884	$2.0\cdot10^{-23}$	s
0.2	0.6288	$5.1\cdot10^{-25}$	ms
0.3	0.5796	$2.0\cdot10^{-26}$	ms
0.4	0.5384	$8.1\cdot10^{-28}$	—
1.1	0.7710	$2.7\cdot10^{-27}$	ls
1.2	0.6970	$2.3\cdot10^{-23}$	ls
1.3	0.6379	—	ms

* The band strengths were referred to the number of molecules in the initial state.

** s denotes a strong band, ms denotes a band of medium strength, ls denotes a band of low strength.

Molecular oxygen has strong electronic bands in the ultraviolet region and relatively weak bands in the red and near-infrared regions (Table 2.16). In the near-infrared region the O_2^{16} molecule has two perceptible bands, the centers of which lie near wavelengths 1.2683μ (0—0 transition) (Figure 2.12) and 1.0674μ (0—1 transition). Owing to the symmetry of the O_2^{16} molecule the spectra of its bands exhibit only lines with alternate values of the rotational quantum number *j*.

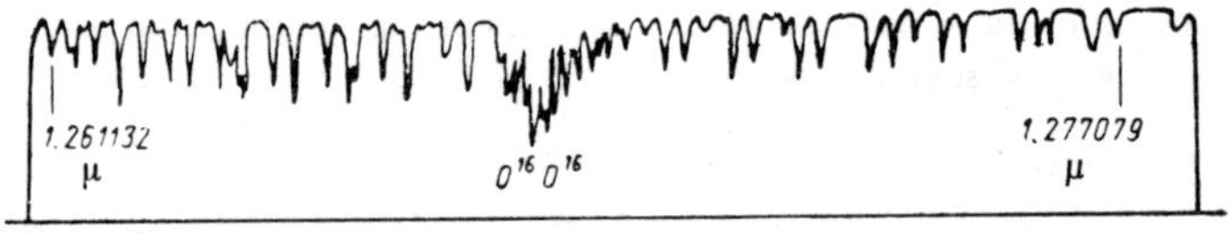

FIGURE 2.12. Solar spectrum in the vicinity of the 1.2683μ absorption band of O_2^{16} /63/.

The $O^{16}O^{18}$ molecule has bands lying in the red region near wavelengths 0.7620μ (0—0 transition), 0.6901μ (0—1 transition) and 0.6317 μ (0—2 transition). The bands contain lines with all values of *j*, due to a reduction in the symmetry of the $O^{16}O^{18}$ molecule compared with the O_2^{16} molecule.

In addition to the molecular electronic bands considered above, oxygen has in the ultraviolet, visible and near-infrared regions diffuse bands, which owe their origin to the complex molecules $[O_2]_2$. The spectra of these molecules were obtained in /64, 65/. The bands of these complex molecules possess low strength. In addition, they overlap the more intensive absorption bands of ozone and water vapor. Absorption in the $[O_2]_2$ bands apparently must be considered only in special cases (e.g., for slant radiation paths close to the horizon).

5. Nitrous oxide. The N_2O molecule is a linear asymmetric molecule, the dipole moment and rotational constant of which are respectively 0.166*D* and 0.4182 cm^{-1} for the principal isotope $N_2^{14}O^{16}$. Strong electronic bands of N_2O are located in the far ultraviolet region. All the fundamental frequencies, i.e., $\nu_1 = 1285.6$ cm^{-1} (7.8μ), $\nu_2 = 588.8$ cm^{-1} (17.0μ) and $\nu_3 = 2223.5$ cm^{-1} (4.6μ) (Figure 2.13), are active in the infrared spectra.

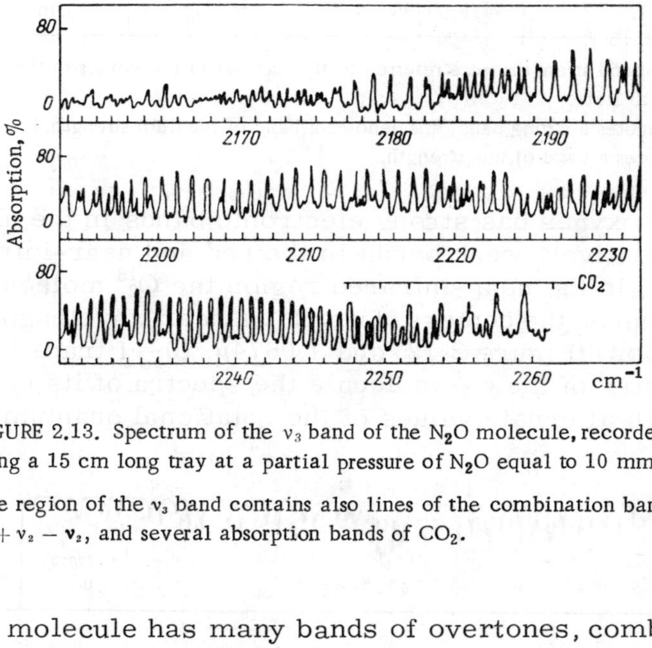

FIGURE 2.13. Spectrum of the ν_3 band of the N_2O molecule, recorded using a 15 cm long tray at a partial pressure of N_2O equal to 10 mm Hg /3/.

The region of the ν_3 band contains also lines of the combination band $\nu_3 + \nu_2 - \nu_2$, and several absorption bands of CO_2.

The N_2O molecule has many bands of overtones, combination frequencies and higher-state bands. Thus, the region with frequencies in excess of 8000 cm^{-1} contains 23 bands, that between 4417 and 1880 cm^{-1} contains 18 bands, that between 4062 and 7431 cm^{-1} contains 14 bands, and the region between 2209.53 and 2798.30 cm^{-1} contains 7 bands, excluding higher-state bands /3/. Most of these bands are very weak. The solar spectrum exhibits 9 bands, some data on which are listed in Table 2.17.

The N_2O molecule has 12 stable isotopes, formed by combinations of the N^{14}, N^{15}, O^{16}, O^{17} and O^{18} atoms. The spectrum which has been more or less thoroughly investigated is the vibration-rotation spectrum of $N_2^{14}O^{16}$.

TABLE 2.17. N_2O bands in the solar spectrum /3/

Band center		Transition	Type	Molecular band strength under standard conditions, cm
cm^{-1}	μ			
4730.86	2.11	$00^{0}0$—$20^{0}1$	Parallel	—
4630.31	2.16	$00^{0}0$—$12^{0}1$	"	—
4417.51	2.27	$00^{0}0$—$00^{0}2$	"	—
4389.06	2.28	$01^{1}0$—$01^{1}2$	"	—
3481.2	2.87	$00^{0}0$—$10^{0}1$	"	$1.29 \cdot 10^{-18}$
3365.6	2.97	$00^{0}0$—$02^{0}1$	"	—
2798.6	3.57	$00^{0}0$—$01^{1}1$	Perpendicular	$9 \cdot 10^{-20}$
2577	3.88	$01^{1}0$—$21^{1}0$	Parallel	—
2563.5	3.90	$00^{0}0$—$20^{0}0$	"	$1.6 \cdot 10^{-18}$
2461.5	4.06	$00^{0}0$—$12^{0}0$	"	$4.3 \cdot 10^{-19}$
2223.5	4.50	$00^{0}0$—$00^{0}1$	"	$6.88 \cdot 10^{-17}$
2210	4.52	$01^{1}0$—$01^{1}1$	"	—
1285.0	7.78	$00^{0}0$—$10^{0}0$	"	$9.78 \cdot 10^{-18}$
1167.0	9.56	$00^{0}0$—$02^{0}0$	"	$4.08 \cdot 10^{-19}$
588.8	17.0	$00^{0}0$—$01^{1}0$	Perpendicular	$7.75 \cdot 10^{-19}$

6. Methane. The methane (CH_4) molecule is a tetrahedron. The electronic bands of CH_4 lie in the far-ultraviolet region (wavelengths less than 1450 Å). The high degree of molecular symmetry is responsible for strong degeneration of vibrational energy levels. Of the nine fundamental frequencies, one is twofold degenerate and two are threefold degenerate. Thus the molecule has only 4 fundamental vibrations with different frequencies. The ν_1 band is fully symmetric, the ν_2 band is twofold degenerate, while ν_3 and ν_4 are threefold degenerate. In the infrared region only the $\nu_3 = 3020.3\ cm^{-1}$ (3.3μ) and $\nu_4 = 1306.2\ cm^{-1}$ (7.7μ) bands appear. The frequencies of the optically inactive bands ν_1 and ν_2 are 2914.2 and 1526 cm^{-1}, respectively. The frequencies of the first harmonic of the ν_2 band and of the ν_1 and ν_3 bands are close. Overlapping of these three bands results in strong interaction between vibration-rotation transitions. This interaction is responsible for removing the degeneracy of the fundamental ν_3 and ν_4 bands and for substantial complication of the fine structure of the spectrum.

In addition to the two fundamental vibration-rotation bands, methane possesses many bands of harmonics and combination frequencies (Table 2.18), nine of which were detected in the solar spectrum. Methane has no rotational spectrum.

TABLE 2.18. Methane bands in the solar spectrum /3/

Band center		Transition	Molecular strength of band under standard conditions, cm
cm^{-1}	μ		
6005	1.67	0000—0020	$3.6 \cdot 10^{-20}$
5861	1.71	0000—1101	—
5775	1.73	0000—0111	—
4420	2.20	0000—0110	—
4313	2.32	0000—0011	—
4216	2.37	0000—1001	—
4123	2.43	0000—0102	—
3019	3.31	0000—0010	$1.26 \cdot 10^{-17}$
3823	3.55	0000—0101	—
2600	3.85	0000—0002	$1.04 \cdot 10^{-17}$
1306	7.66	0000—0001	$6.9 \cdot 10^{-18}$

7. Carbon monoxide. Due to its simplicity, the spectrum of carbon monoxide has been studied in great detail. The dipole moment of this molecule is 0.1*D*. The pure rotational spectrum lies in the far-infrared and microwave regions. The fundamental rotation-vibration band of the $C^{12}O^{16}$ molecule lies near 2143.2 cm^{-1} (4.67 μ). The first overtone of this band has its center near 4260 cm^{-1}. Table 2.19 lists the locations of the centers of different CO bands /66/.

The carbon monoxide molecule has two isotopes: $C^{12}O^{16}$ and $C^{13}O^{16}$.

TABLE 2.19. Location of centers of CO bands /66/

Band	1—0	2—0	3—1	4—2
Calculated, cm^{-1}	2143,2740	4260,0646	4207,1664	4154,4056
Observed, cm^{-1}	2143,2740	4260,0646	4207,1680	4154,4040

Band	5—3	6—4	7—5	3—0	4—0
Calculated, cm^{-1}	4101,7820	4049,2958	3996,9466	6350,4404	8414,4702
Observed, cm^{-1}	—	—	—	6350,436	8414,458

The fine structure of many vibration-rotation bands of CO has been quite well studied. The results of these investigations will be considered in the next chapter.

3. *ABSORPTION OF VISIBLE AND INFRARED RADIATION IN A HOMOGENEOUS ATMOSPHERE*

1. BASIC DEFINITIONS

1. Absorption coefficient. Optical thickness. Spectral transmission (absorption). The term absorption coefficient $k(\nu)$ of a medium for radiation of frequency ν is the constant of proportionality in the Bouguer law expressed in differential form:

$$dI(\nu) = -k(\nu)\, I(\nu)\, dl, \tag{3.1}$$

where $dI(\nu)$ is the attenuation of radiation of intensity $I(\nu)$ upon passing through a layer of the medium with thickness dl.

Integration of (3.1) over the layer thickness l for a homogeneous medium yields the Bouguer law

$$I(\nu) = I_0(\nu)\, e^{-k(\nu)l}, \tag{3.2}$$

where $I_0(\nu)$ and $I(\nu)$ are the radiation intensities entering and leaving the layer of thickness l.

The term optical thickness τ of layer l of the medium denotes the exponent in equation (3.2):

$$\tau = k(\nu)\, l. \tag{3.3}$$

The limits of applicability of (3.1) and (3.2) will be considered in Sec. 4.

Quantities

$$f(\nu) = \frac{I(\nu)}{I_0(\nu)}, \quad A(\nu) = \frac{I_0(\nu) - I(\nu)}{I_0(\nu)} \tag{3.4}$$

will be termed spectral transmission and spectral absorption, respectively.

2. Absorption and transmission functions. Integral absorption. Absorption function A of radiation in the spectral interval $\Delta\nu = \nu_2 - \nu_1$ is the ratio of radiation absorbed by the layer of the medium to that incident on the layer. Transmission function T is defined similarly.

Hence

$$A=\frac{\int\limits_{\nu_1}^{\nu_2}[I_0(\nu)-I(\nu)]\,d\nu}{\int\limits_{\nu_1}^{\nu_2}I_0(\nu)\,d\nu},\quad T=\frac{\int\limits_{\nu_1}^{\nu_2}I(\nu)\,d\nu}{\int\limits_{\nu_1}^{\nu_2}I_0(\nu)\,d\nu}. \tag{3.5}$$

These expressions for a homogeneous layer can also assume the form

$$A=\frac{\int\limits_{\nu_1}^{\nu_2}I_0(\nu)A(\nu)\,d\nu}{\int\limits_{\nu_1}^{\nu_2}I_0(\nu)\,d\nu},\quad T=\frac{\int\limits_{\nu_1}^{\nu_2}I_0(\nu)f(\nu)\,d\nu}{\int\limits_{\nu_1}^{\nu_2}I_0(\nu)\,d\nu}. \tag{3.6}$$

It follows that the absorption and transmission functions in the spectral interval $[\nu_1, \nu_2]$ characterize the average spectral absorption and spectral transmission in this interval.

In the particular case $I(\nu)=\mathrm{const}\in[\nu_1,\nu_2]$,

$$A=\frac{\int\limits_{\nu_1}^{\nu_2}A(\nu)\,d\nu}{\nu_2-\nu_1},\quad T=\frac{\int\limits_{\nu_1}^{\nu_2}f(\nu)\,d\nu}{\nu_2-\nu_1}, \tag{3.7}$$

or, with the aid of (3.4) and (3.2),

$$\begin{gathered}A=\frac{1}{\nu_2-\nu_1}\int\limits_{\nu_1}^{\nu_2}(1-e^{-k(\nu)l})\,d\nu,\\ T=\frac{1}{\nu_2-\nu_1}\int\limits_{\nu_1}^{\nu_2}e^{-k(\nu)l}\,d\nu.\end{gathered} \tag{3.7a}$$

In another particular case $k(\nu)=\mathrm{const}\in[\nu_1,\nu_2]$,

$$A=A(\nu),\quad T=f(\nu). \tag{3.8}$$

Functions A and T depend (see (3.5) and (3.6) not only on the absorption properties of the medium, expressed by $A(\nu)$ or $f(\nu)$, but also on the spectral intensity of the source radiation density $I_0(\nu)$. Therefore to describe the absorption properties of the medium (which do not depend on $I_0(\nu)$) over a given spectral interval $[\nu_2, \nu_1]$, we introduce the integral absorption B, where

$$B=\int\limits_{\nu_1}^{\nu_2}A(\nu)\,d\nu. \tag{3.9}$$

The integral absorption has a maximum when $A(\nu)=1$ over the entire spectral region $\Delta\nu=\nu_2-\nu_1$. The integral absorption is expressed in units of frequency and can be treated as the width of some spectral interval in which for $A(\nu) = 1$ the integral absorption is equal to the given value. In this sense integral absorption is frequently termed the equivalent width of the given interval (line, band, etc.).

Expressions (3.1)—(3.9) were obtained for the case when frequency ν is taken as the spectral coordinate. Formally identical expressions are obtained when the wavelength λ is taken as the spectral coordinate.

Quantities ν and λ are related in the form

$$\nu\,(\mathrm{cm}^{-1}) = \frac{10^4}{\lambda\,(\mu)}; \tag{3.10}$$

it can therefore be shown that (3.5) and (3.6), expressed in terms of ν and λ, yield the same numerical values of A and T. At the same time formulas (3.7) expressed in terms of ν and λ are not equivalent:

$$\left.\begin{aligned} \frac{\int_{\nu_1}^{\nu_2} A(\nu)\,d\nu}{\nu_2-\nu_1} &\neq \frac{\int_{\lambda_1}^{\lambda_2} A(\lambda)\,d\lambda}{\lambda_2-\lambda_1};\\ \frac{\int_{\nu_1}^{\nu_2} f(\nu)\,d\nu}{\nu_2-\nu_1} &\neq \frac{\int_{\lambda_1}^{\lambda_2} f(\lambda)\,d\lambda}{\lambda_2-\lambda_1}. \end{aligned}\right\} \tag{3.11}$$

The smaller the spectral interval $[\nu_1, \nu_2]$ or its equivalent $[\lambda_1, \lambda_2]$, the weaker the inequality in (3.11). In the limit $\nu_2 - \nu_1 \to 0$ or $\lambda_2 - \lambda_1 \to 0$, $A(\nu) = A(\lambda)$ and $T(\nu) = f(\lambda)$.

Note that relationship (3.10) is strictly true in vacuum. When radiation propagates through a medium with refractive index $n \neq 1$, it is necessary to take into account that $\lambda = \lambda_0 n^{-1}$, where λ and λ_0 respectively are the wavelength in the material and in vacuum.

It should be emphasized that expressions (3.7) and (3.7a) for the absorption and transmission functions are frequently cited in the literature without stipulating that they are valid only when the frequency dependence of the intensity of incident frequency can be neglected in the spectral interval $\Delta\nu = \nu_2 - \nu_1$. In practice these expressions can be used for calculating absorption functions for thermal radiation over narrow spectral sections (of the order of 0.1μ).

2. ABSORPTION OF A SINGLE LINE

The absorption spectrum of any individual molecule consists of an ensemble of spectral lines, the position, strength and shape of which are uniquely defined by the corresponding quantum variations of the electronic, vibrational and rotational molecular energies. The spectral line is thus the principal component of the molecular spectrum. Hence the absorption of radiation by an individual spectral line is the subject matter of many studies, the results of which are surveyed elsewhere /67—71/. Absorption by a single line was considered to some extent in the second chapter of this book. Here we shall examine only several problems which are relevant to an understanding of the subsequent material.

1. **Line profile shape.** The shape of the absorption line in the earth's atmosphere is determined by the finiteness of molecular life in the excited state, the Doppler effect, and the effect of molecular collisions of the absorbing gas with one another and with molecules of nonabsorbing or foreign gases. The line shape functions due to corresponding effects or to their combined action were presented above (see (2.55), (2.57) and (2.64)).

Strictly speaking, none of these expressions describes exactly the line profile under actual atmospheric conditions. It is hence very important to known in which cases and with what degree of accuracy one or another expression can be used to describe a real line profile.

The requirements as to the description of the shape of an actual line profile depends essentially on the nature of the problem at hand. Suppose, for example, we wish to determine the integral absorption function over a wide spectral interval for black or gray radiation. In this case the requirements as to the accuracy of the shapes of individual lines are the least rigid, since the central part of the line and its wings both contribute to the integral absorption. If, however, we require the absorption function over a relatively narrow interval containing only the central parts of several lines, or containing also the wings of far lines, then the distribution of energy absorbed by the line in the central part and in the wings assumes great importance. Finally, if the absorption of monochromatic radiation in the atmosphere is to be determined, the requirements as to the accuracy of the line profile description are the most strict.

In the overwhelming majority of studies on the absorption of visible and infrared light in the atmosphere one uses the dispersion or Lorentz line profile. This is because the dispersion profile is very simple and describes quite satisfactorily the central part of lines at altitudes up to 30 km on the average, where the bulk of gases absorbing in the visible and infrared are concentrated /1/.

The past few years have seen new problems in whose solution the dispersion profile either does not ensure the required degree of accuracy, or cannot be used at all. These are primarily problems involving the use of lasers in various systems intended for operation through the atmosphere, inverse problems of satellite meteorology, and the more exact calculation of radiant fluxes in the atmosphere.

These problems involve data on the profile of the central part of the line for altitudes at which the Doppler effect can no longer be neglected compared with collision line broadening. The dispersion profile is a priori inapplicable for describing the true absorption by the far wings of the line. We shall examine briefly the results of studies into the Lorentz-Doppler profile and the shape of the far wings of the line.

The theoretical study of Rautian and Sobel'man /72/ is concerned with different line profiles produced by the combined Doppler and Lorentz effects, treated both as statistically independent and statistically dependent. It is shown /72/ that the real line profile should have less intense wings than the Lorentz profile. It follows from the dispersion profile that the intensity in the wings drops in proportion to the fourth power of the frequency, and not the second. Asymmetric line profiles, shifted relative to the center, were obtained for statistically dependent Doppler and Lorentz line broadening /72/.

Drayson /62/ calculated the integral absorption of lines for Lorentz and [mixed] Lorentz-Doppler profiles. The calculations were carried out for several values of the product $S\omega$, where S is the line strength and ω is the precipitable layer of the absorbing gas. It was found that for average values of $S\omega$ the integral absorption of a line with a mixed profile at low pressures (high atmospheric altitudes) can exceed the integral absorption of a Lorentz-broadened line by more than a factor of 2. Thus, as the pressure changes from 100 to 0.5 mb (altitudes from 16 to 55 km) the error in calculating the integral absorption of a line due to neglect of the Doppler effect ranges from 0.05 to 60%. At 10 mb (30 km) this error is about 5%. It was shown in /73/ that for limiting cases of small and large $S\omega$ the Doppler effect can be neglected up to an altitude of 50 km when calculating the integral absorption of a line.

It thus follows from /62/ that in calculating the integral absorption of individual lines, and consequently of a series of lines up to altitudes of 30 km to within 5%, one may neglect the Doppler effect. It should be stressed that this conclusion pertains to determination of the integral absorption of the lines and cannot be extended to computation of the absorption of monochromatic radiation in atmosphere. Methods of calculating the absorption coefficient for an incorrect line profile are examined elsewhere /74—77/.

Winters et al. /78/ list results of experimental studies of absorption in the CO_2 band at 4.3μ. More than 5 cm^{-1} from the line center the absorption factor $k(\nu)$ decreases more rapidly than follows from the dispersion profile. The following expression for $k(\nu)$, which satisfactorily describes their experimental results, is proposed by Winters et al. /78/:

$$\left.\begin{aligned} k(\nu) &= \left(\frac{\gamma}{\pi}\right)[(\nu-\nu_i)^2+\gamma^2]\,e^{-a(|\nu-\nu_i|-\Delta\nu_0)^b};\\ &|\nu-\nu_i|>\Delta\nu_0, \end{aligned}\right\} \qquad (3.12)$$

where γ is the half-width of the line, while a and b are empirical constants which depend on the kind of expanding gas, the value of $\Delta\nu_0$ according to estimates of Benedict does not exceed 3 cm^{-1}.

Kyle et al. /79/ measured the spectral absorption of solar radiation in the 4.3μ band at altitudes from 10 to 30 km with a resolution of 0.5 cm^{-1}. This resolution is high enough to require a line-by-line calculation to interpret the spectra. Calculations and experimental data were found to be in satisfactory agreement, provided one uses the Lorentz profile at high altitudes and profile (3.12), proposed by Benedict, for lower altitudes.

Burch et al. /80, 81/ conducted experimental studies of spectral absorption by water vapor and carbon dioxide for wavelengths from 0.6 to 5.5μ. Their results suggested that not one of the existing expressions for the shapes of the extreme wings of the line is suitable for an exact description of experimental data. The experiments showed that for the same absorbing gas it depends on the absorption band, on the broadening gas, and on temperature.

Thus, while the central part of the line is more or less satisfactorily described by existing formulas for the line profile, this does not apply to far wings. New theoretical studies are required. Available experimental data indicate that we require such formulas for the far wings which make allowance for the absorbing and broadening gases, the properties of the absorption band, and the temperature-dependence.

2. Integral absorption and the absorption function of a single line. An expression for the integral absorption of a single nonoverlapping line with a dispersion profile was obtained in /82/:

$$\left.\begin{aligned} B &= \int_{-\infty}^{+\infty} A(\nu)\,d\nu = 2\pi\gamma f(x);\\ f(x) &= xe^{-x}[J(x)+J_1(x)], \end{aligned}\right\} \qquad (3.13)$$

where

$$x=\frac{S\omega}{2\pi\gamma},$$

$J_0(x)$ and $J_1(x)$ are zero-and first-order Bessel functions of pure imaginary argument; ω is the mass of the absorbing gas, sometimes termed the precipitable gas layer. The values of function $f(x)$ for different x were tabulated by Elsasser /83/, and for x between 0.01 and 10 they are listed in /84/. In limiting cases equation (3.13) yields

$$B = S\omega \quad \text{for} \quad x \ll 1, \tag{3.14}$$

$$B = 2\sqrt{S\gamma\omega} \quad \text{for} \quad x \gg 1. \tag{3.15}$$

The absorption function for a single dispersion-profile line (provided the intensity of incident radiation $I(\nu) = \text{const}$ within the spectral interval $\nu_2 - \nu_1$ occupied by the line) has the form

$$A = \frac{\int\limits_{\nu_1}^{\nu_2} (1 - e^{-k(\nu)\omega})\, d\nu}{\nu_2 - \nu_1}, \tag{3.7a}$$

where $k(\nu)$ is given by expression (2.57).

A general expression for function A was obtained elsewhere /85/:

$$A = 1 - e^{-x(1+\cos 2\delta)} - \frac{e^{-x}}{2b} \sum_{n=0}^{\infty} \delta_n (-1)^n J_n(x) B_n(\delta), \tag{3.16}$$

$$B_n(\delta) = -4n\left[\sum_{k=0}^{n} \delta_k (-1)^{n-k} \frac{\sin 2k\delta}{2k} - \frac{\sin 2(n-1)\delta}{2(n-1)}\right], \tag{3.17}$$

where $x = \frac{S\omega}{2\pi\gamma}$; S and γ are the strength and half-width of the line; ω is the mass of absorbing gas; $\delta = \arctan b$; $2b\gamma$ is the width of the spectral interval in which function A is determined; $J_n(x)$ is the nth order Bessel function of imaginary argument.

The following asymptotic formulas for A can be derived from (3.16):

1) $x \ll 1, \; A = 2x\delta/b;$ (3.18)

2) $x \gg 1, \; A = 1 - e^{-z^2} + \sqrt{\pi z}\,[1 - \Phi(z)];$ (3.19)

3) $b \gg 1, \; A = A_1 + A_2,$

$$A_1 = \frac{\pi}{b} x e^{-x} [J_0(x) + J_1(x)], \tag{3.20}$$

$$A_2 = 1 - e^{-z^2} - \sqrt{\pi z}\, \Phi(z); \tag{3.21}$$

4) $b \gg 1, \; x \ll 1, \; A = \frac{S\omega}{2\gamma b}.$ (3.22)

In (3.19) and (3.21), $\Phi(z)$ is the error function with argument $z = \frac{1}{b}\sqrt{2x}$.

We now compare the above asymptotic formulas for A with corresponding formulas for the integral absorption B of a single line (formulas (3.14) and (3.15)). Comparison of (3.22) and (3.14) shows that the expressions for A and B coincide within a constant factor. This means that the physical corollaries following from analysis of one of the expressions are applicable to the other. Both A and B are therefore independent of pressure in this case.

Comparison of (3.19) and (3.15) when $x \gg 1$ shows that here the physical corollaries following from analysis of the expression for the integral absorption of a line cannot be used when treating the absorption function in a specified spectral interval. In particular, B is given by the square root of the absorbing mass (the square root law for a single line), while A satisfies an entirely different expression and the square root law cannot be used for describing the absorption function of an individual line at high x (high ω or low pressures).

Note in conclusion that neither the absorption function A nor the integral absorption B of a single line when $x \ll 1$ for sufficiently large b, when arctan b can be replaced by $\frac{\pi}{2}$, depends on pressure (see expressions (3.18), (3.22) and (3.14)). When $x \gg 1$, mass ω and pressure P enter in a similar manner into the corresponding expressions for A and B (see (3.19) and (3.15), which contains the half-width $\gamma = \gamma^0 P$)).

3. ABSORPTION OF GROUPS OF OVERLAPPING LINES

The spectra of all atmospheric gases consist of groups of overlapping lines. Depending on the radiation source, the absorption of which in the atmosphere must be known, allowance for line overlapping must be made either for individual wavelengths (such as in the case of gas lasers) or in some spectral intervals. In the first case it is necessary to know the absorption coefficients which enable one to determine the spectral absorption, while absorption functions are determined in the second case.

One of three approaches is utilized in finding an absorption function: 1) calculation with allowance for absorption of radiation of given frequency by each spectral line; 2) calculation from spectrum-model formulas; 3) experimental determination of the required quantities under natural or laboratory conditions. We consider each of these approaches.

1. Direct calculation of the absorption function. The absorption function for a group of overlapping lines (band) occupying the spectral interval $[\nu_1, \nu_2]$, when the pressure on the ray path and the

intensity of incident radiation over this interval are constant, has the form

$$A = \frac{1}{\nu_2 - \nu_1} \int_{\nu_1}^{\nu_2} (1 - e^{-k(\nu)\omega})\, d\nu, \qquad (3.23)$$

where

$$k(\nu) = \sum_i S_i b_i(\nu), \qquad (3.24)$$

$b_i(\nu)$ is the profile, and i denotes the line normalized to unity. In the latter formula summation is over all lines contributing to the absorption of radiation with frequency ν. If it is assumed that all such lines have a dispersion profile, then expression (3.24) yields

$$k(\nu) = \sum_i S_i \frac{\gamma_i}{\pi} \frac{1}{(\nu - \nu_{0i})^2 + \gamma_i^2}. \qquad (3.25)$$

Direct calculation of the absorption function thus requires knowledge of the location of centers, strengths and half-widths of all the lines contributing to the absorption of radiation over the spectral interval under study. Calculation and experimental determination of the parameters of atmospheric-gas lines is extremely difficult. Only recently has progress been achieved in this direction. Several publications appeared in which the absorption function was calculated directly in certain atmospheric-gas bands, and a series of studies appeared on the experimental determination of line parameters. These investigations will be examined when presenting specific results of studies on individual gases in subsequent sections. Here we only emphasize that, in spite of the great mathematical and computational difficulties, direct calculation of absorption functions definitely holds great promise, since it yields most universal data, without any limitations as to the width of the spectral interval for which the absorption function must be known.

In conclusion we note the method of direct calculation of the absorption function presented in /86/. The expression for the absorption function obtained there is general in nature and can be used to compute the absorption of radiation in various narrow spectral sections where the absorption lines can be described by a dispersion profile with constant half-width. It is essential that this expression makes allowance for radiation absorption in the given interval by lines whose centers lie outside this interval.

2. Method of absorption-spectra models. Direct calculation of absorption functions is very labor-consuming and cannot always be used due to the unavailability of data on individual line parameters in the absorption spectra of atmospheric gases. In this connection spectrum models have been used extensively in calculating absorption functions. According to this method the spectra of atmospheric gases are simulated by a given ensemble of assumptions concerning the nature of line location and their strength. The line profiles in all models are taken as the dispersion profile and the half-widths of all lines are assumed to be identical. We shall examine briefly the principal models used in calculating absorption functions of atmospheric gases (this problem has been examined in more detail in /3, 88/).

1. The absorption function in the Elsasser model. This model assumes equality of line strengths S, their half-widths γ, and the distance d between them within the limits of the spectral interval under study. If this interval is so wide that the number of lines in it can be treated as large, then the absorption coefficient is derived in the form

$$k(\nu) = \frac{S \operatorname{sh} \beta}{d} \frac{1}{\operatorname{ch} \beta - \cos t}, \qquad t = \frac{2\pi\nu}{d}; \quad \beta = \frac{2\pi\gamma}{d}. \tag{3.26}$$

Substitution of (3.26) into (3.23) yields

$$A = \operatorname{sh} \beta \int_0^{x'} e^{-\xi \operatorname{ch} \beta} J_0(\xi)\, d\xi, \tag{3.27}$$

$$x' = \frac{S\omega}{d \operatorname{sh} \beta}, \tag{3.28}$$

where $J_0(\xi)$ is a zero-order Bessel function of pure imaginary argument.

Integration over the above limits shows that A is calculated over a spectral interval occupied by a single line. However, having reference to the fact that in this model $k(\nu)$ is a periodic function with period d, the value of A for any spectral interval containing n lines will also satisfy (3.27). This expression then gives the absorption function /88/:

$$A = \operatorname{sh} \beta x' e^{-x' \operatorname{ch} \beta} J_0(x') + \operatorname{th} \beta x' e^{-x' \operatorname{ch} \beta} J_1(x') + \frac{\operatorname{sh}^3 \beta}{\operatorname{ch} \beta} \int_0^{x'} \xi e^{-\xi \operatorname{ch} \beta} J_0(\xi)\, d\xi. \tag{3.29}$$

For the atmosphere $\beta \ll 1$ (the distance between the lines is much greater than their half-width). In this case (3.29) yields

$$A = \beta x e^{-x(1+\beta^2/2)} [J_0(x) + J_1(x)] + \\ + \beta^3 \int_0^x \xi e^{-\xi(1+\beta^2/2)} J_0(\xi)\, d\xi, \tag{3.30}$$

$$x = \frac{\omega S}{2\pi\gamma}. \tag{3.31}$$

We now consider asymptotic expressions for A, which follow from (3.30):

1) $x \ll 1$. Expansion of (3.30) yields

$$A = \beta x = \frac{S}{d}\omega. \tag{3.32}$$

The absorption function here is not a function of pressure. It should be noted that (3.32) is also obtained directly from (3.29) and consequently the conclusion reached here does not depend on the value of β.

2) $x \gg 1$. The use of asymptotic expressions of Bessel functions for large arguments yields

$$A = \Phi\left(\sqrt{\frac{1}{2}\beta^2 x}\right) = \Phi\left(\frac{1}{2}\sqrt{\pi S \gamma \omega}\right). \tag{3.33}$$

It follows that the absorption function in this case depends on the product ωP, rather than on ω and P individually. This result applies only when $\beta \ll 1$.

3) $x \gg 1$ and $\beta^2 x \ll 1$ (the lines virtually do not overlap). Expression (3.33) yields directly

$$A = \beta\sqrt{\frac{2x}{\pi}} = \frac{2}{d}\sqrt{\pi\gamma S\omega}. \tag{3.34}$$

4) $x \approx 1$. On this assumption $\beta^2 x \ll 1$, and series expansion of the integral in (3.30) in x yields

$$A = \beta x e^{-x} [J_0(x) + J_1(x)]. \tag{3.35}$$

In /89, 90/, the expression for the square root law (3.34) was derived from the Elsasser model in the limiting case $\omega \ll 1$ and $\beta \ll 1$. The error of this conclusion is shown elsewhere /88/.

2. Absorption function in the statistical model. The principal prerequisite of the statistical model, first proposed

by Goody /91/, consists in assuming a random distribution of line strengths in the spectrum at equal half-widths of the lines, obeying the laws of probability theory. The statistical model was developed further in /92—94/.

The analytical expressions obtained in /91—94/ actually give the magnitude of the spectral absorption for some frequency, rather than the value of the absorption function. A proper mathematical examination of the statistical model, based on the introduction of an "average measurable absorption coefficient," is given by Soshnikov /95/. Later, we shall present expressions for function A in which the shortcoming allowed in /91—94/ has been eliminated.

The absorption function in the statistical model for a sufficiently wide interval $\nu_2-\nu_1$ for an equally probable distribution of lines has the form

$$A=1-\frac{1}{\nu_2-\nu_1}\int\limits_{\nu_1}^{\nu_2} d\nu \exp\left\{-\frac{1}{d}\int\limits_{\nu_a}^{\nu_b} d\nu' \times \right. \\ \left. \times \int\limits_0^\infty Q(S)\left[1-e^{-\frac{\omega S\gamma}{\pi}\frac{1}{(\nu-\nu')^2+\gamma^2}}\right]dS\right\}, \tag{3.36}$$

where $Q(S)$ is the probability density function of the line intensity distribution normalized to unity; d is the average line-to-line spacing; $\nu_a-\nu_b=D$ is the interval within which the lines contribute to the absorption in the interval $\nu_2-\nu_1$ under study. Assuming further that $|\nu_b-\nu_a|\gg|\nu_2-\nu_1|$, then

$$A=1-\exp\left\{-\frac{1}{d}\int\limits_{-\infty}^{\infty} d\nu' \int\limits_0^\infty Q(S)\left[1-e^{-\frac{S\gamma\omega}{\pi}\frac{1}{\nu'^2+\gamma^2}}\right]dS\right\}. \tag{3.37}$$

Goody /91/ investigated the case

$$Q(S)=\frac{1}{\overline{S}}e^{-S/\overline{S}}, \tag{3.38}$$

where $\overline{S}$ is the average line strength, and obtained

$$A=1-\exp\left[-\frac{\beta\overline{x}}{(1+2\overline{x})^{1/2}}\right], \tag{3.39}$$

where

$$\overline{x}=\frac{\overline{S}\omega}{2\pi\gamma}; \quad \beta=\frac{2\pi\gamma}{d}. \tag{3.40}$$

Plass /92/ and Godson /93/ considered the case when the lines possess the same strength:

$$Q(S)=\delta(S-\bar{S}_0), \tag{3.41}$$

and obtained

$$A=1-\exp\{\beta\bar{x}_0 e^{-\bar{x}_0}[J_0(\bar{x}_0)+J_1(\bar{x}_0)]\}, \tag{3.42}$$

where

$$\bar{x}_0=\frac{\bar{S}_0\omega}{2\pi\gamma}. \tag{3.43}$$

Equation (3.39) has the following asymptotic expansions:

1) $\bar{x}\ll 1$,

$$A=\beta\bar{x}=\frac{\bar{S}}{d}\omega; \tag{3.44}$$

2) $\bar{x}\gg 1$,

$$A=1-\exp\left(-\sqrt{\frac{1}{2}\beta^2\bar{x}}\right). \tag{3.45}$$

Equation (3.42) has similar asymptotic expressions, except that $\bar{x}$ in (3.44) should be replaced by $\bar{x}_0$, while in (3.45) it should be replaced by $4\bar{x}_0/\pi$.

As in the Elsasser model, the absorption function at small $\bar{x}$ does not depend on pressure, while at high $\bar{x}$ it is a function of the product ωP.

Expressions (3.38) and (3.41), which define the strength distributions of lines, are quite arbitrary. However, calculations /95/ for a Boltzmann line strength distribution showed that the values of A in this case are virtually identical with those obtained from (3.39) and (3.42).

3. The absorption function in the combined model. According to this model /94, 92/, it is assumed that all the lines in the corresponding spectral region can be subdivided into several groups, within which they are situated at equal distances from one another and have the same strength and half-width, i.e., they are described by the Elsasser model. The groups proper are located randomly, in the same manner as lines in the statistical model. In this case the expression for A assumes the form

$$A=1-\prod_{j=1}^{N}\operatorname{sh}\beta_j\int_{x_j}^{\infty}e^{-\xi\operatorname{ch}\beta_j}J_0(\xi)\,d\xi, \tag{3.46}$$

where

$$\beta_j = \frac{2\pi\gamma}{d_j}; \quad x_j = \frac{S_j\omega}{d_j \operatorname{sh} \beta_j}; \tag{3.47}$$

d_j is the interline spacing in the jth group, described by the Elsasser model with line strength S_j; N is the number of groups.

The limiting expressions at $\beta_j \ll 1$ are obtained similarly to that of the Elsasser model:

1) all $x_j \ll 1$,

$$A = \sum_{j=1}^{N} x_j\beta_j = \omega \sum_{j=1}^{N} S_j/d_j; \tag{3.48}$$

2) all $x_j \gg 1$,

$$A = 1 - \prod_{j=1}^{N} \left(1 - \Phi \sqrt{\frac{1}{2}\beta_j^2 x_j}\right); \tag{3.49}$$

3) all x_j of order unity,

$$A = \prod_{j=1}^{N} \{1 - \beta_j x_j e^{-x_j} [J_0(x_j) + J_1(x_j)]\}. \tag{3.50}$$

As in the preceding models, at small x_j A does not depend on P, while at high x_j it depends on the product ωP.

4. The quasi-random model. This model was suggested by Wyatt et al. /96/ and is based on the main idea of the statistical model. The spectral interval $\Delta\nu$, within which δ must be determined, is divided into smaller intervals δ, within each of which the statistical model is utilized. This eliminates to a large extent the principal shortcoming of the statistical model: the assumption of a random distribution of the location and strength of lines in the given interval. In fact, the smaller the spectral interval, the more valid the assumption of the statistical model.

The lines within each interval δ are subdivided with respect to strength into five groups, for each of which the transmittance $f_i(\nu)$ is determined at the center of the interval. Simple multiplication of these values of $f_i(\nu)$ then yields the transmittance at the center of interval δ, which is treated as the average transmittance of the interval. Hence it is easy to find the transmission function over the wider interval Δ. This procedure makes it possible to quite correctly take into account the contribution to radiation absorption over the given spectral interval of lines whose centers lie outside this interval.

Note that the technique of subdividing the spectral interval under study when using the statistical model was also used by Soshnikov /95, 97/.

5. Some general conclusions. Comparison of the principal formulas for the absorption functions in the models under study shows that in all cases the argument is

$$x = \frac{S\omega}{2\pi\gamma^0 P}, \tag{3.51}$$

where S is the integral line strength in the Elsasser model; $S = \bar{S}$ or $S = \bar{S}_0$ in the statistical model, and $S = S_j$ in the combined model.

Depending on the value of x, a certain asymptotic expression for the absorption function will be valid for each of the models. It is of importance that x depends on the ratio ω/P, and not on each of these two variables separately.

When $x \ll 1$, the absorption function in all spectral models does not depend on pressure. It can be shown that inequality $x \ll 1$ is equivalent to the condition $\omega k(\nu) \ll 1$ for all ν contained in the given interval $\Delta\nu = \nu_2 - \nu_1$ /88/. The term equivalent here expresses the fact that both these conditions yield the same result.

When $x \gg 1$, the absorption function in all the models is a function of $\sqrt{\omega P}$. It can be shown that the corresponding asymptotic formulas are obtained from general expression (3.23) for the absorption function when $(\nu - \nu_i)^2 \gg \gamma^2$. The physical meaning of this condition consists in the fact that the integral in (3.23) is evaluated by integration over sections occupied by the line wings.

The applicability limits of the different expressions for the absorption function in the Elsasser, statistical, and combined models were examined by Plass /98/. Numerical calculations by the present author showed that the transition region between the asymptotic cases $x \ll 1$ and $x \gg 1$ is relatively small. It can be assumed approximately that the relationship of the first type is directly transformed into that of the second.

Expressions for the absorption function in all the models contain characteristics of the spectral lines contained in the interval under study, such as interline spacing, line strength and half-width. Insufficient theoretical or experimental data on these line characteristics exist at present. Hence the coefficients in formulas for the absorption functions containing the line parameters are usually treated in practice as adjustment parameters and are determined from experimental data. Techniques for determining these parameters for the statistical model have been developed in detail by Godson /99, 100/.

A second factor, limiting the possibilities of spectrum models, is the assumption of constant line width, i.e., neglect of the dependence of the half-width on the rotational quantum number.

In spite of the above shortcomings, calculations of absorption functions utilizing model concepts yield an accuracy sufficient for practical purposes. A substantial advantage of the models, in addition to the simplicity of calculations, is that they can be used for quantitative estimation of infrared-radiation absorption in the atmosphere for situations which either cannot be reproduced experimentally, or are so expensive and labor-consuming that their cost does not justify the results (for example, study of radiation propagation along horizontal and slant directions, etc.). The models also possess an important feature, that they yield formulas for absorption function A expressible in terms of elementary functions. This makes possible analysis of the dependence of this function on its arguments in general form.

It should also be noted that in experimental studies of absorption, due to the limited resolving power of spectral instruments, one does not measure the spectral absorption $A(\nu)$ or transmission $f(\nu)$, but the absorption (transmission) function in the given spectral interval, which is a function of the spectrometer parameters. Finally, formulas of the model theories can be quite useful in direct calculations of the absorption functions. They substantially reduce the volume of computational work after the direct method has been applied for calculating the location of centers, strength and half-width of lines in the corresponding spectral regions.

3. Experimental determination of absorption functions. Experimental determination of absorption functions of various atmospheric gases under laboratory and natural conditions have been dealt with in many studies, the principal results of which will be considered below. Here we consider only the advantages and shortcomings of experimental determination of absorption functions compared with the previously examined theoretical approaches.

Experimental determination of absorption functions is unencumbered by the necessity of allowance for line overlapping, since the latter occurs automatically. Techniques for measuring absorption functions are relatively simple, which, however, cannot be said of the equipment used in these measurements.

In addition to complicated equipment, the experimental method of finding absorption functions has some additional shortcomings. The most important of these is due to the limited resolving power of spectral apparatus, which distorts the actual spectrum pattern and results in errors when determining the absorption function in relatively narrow (of width smaller than the total absorption band) spectral intervals. In a number of cases experimental determination of absorption functions is extremely difficult (for example, measurement of absorption functions on horizontal paths at different altitude, or along slant directions when the terminal points of the measuring paths are situated in the atmosphere).

It should be noted in conclusion that each of these methods has its advantages and disadvantages, and one is as useful as the other. The development of all the above methods will yield a rapid solution to the problem of obtaining sufficiently accurate data on absorption functions of atmospheric gases in different regions of the visible and infrared spectral intervals.

4. LIMITS OF APPLICABILITY OF THE BOUGUER-BEER LAW

The definitions of the absorption coefficient $k(\nu)$ (see equation (3.1)) and absorption function A (see equation (3.7a)) assume satisfaction of the Bouguer-Beer law. The same assumption is utilized in obtaining analytic expressions for absorption functions in all the spectrum models. Finally, the assumption of satisfaction of this law also underlies the processing of experimental data. In this connection it is necessary to examine the limits of applicability of this law. Here a distinction should be drawn between actual and apparent divergences from this law, both of which will be treated.

The Bouguer-Beer law has been derived on the basis of two main assumptions, which were subsequently analyzed by Gershun /101/ and Strong /102/. To characterize these assumptions we express this law in the form

$$I(\nu) = I_0(\nu)\, e^{-k(\nu)cl}, \tag{3.52}$$

in the derivation of which it is assumed that the absorption coefficient $k(\nu)$ does not depend on the incident radiation intensity $I_0(\nu)$ or on the concentration c. The first assumption was checked experimentally by Vavilov /103/, who showed its validity when $I_0(\nu)$ varied by almost 20 orders of magnitude for electron transitions with short lifetimes τ of electrons in the excited state. In the study of electron transitions with large τ it was found /103/ that the absorption coefficient does depend on $I_0(\nu)$.

As far as is known, no check was made of the dependence of $k(\nu)$ on $I_0(\nu)$ for vibration-rotation transitions in absorption spectra of atmospheric gases. However, some conclusions in this respect can be drawn concerning the fact that the integral absorption of solar radiation in the terrestrial atmosphere, calculated for individual lines or groups of lines, with consideration of the dependence of $k(\nu)$ on $I_0(\nu)$, was found to be in satisfactory agreement with experimental data /104/. Similar results were also obtained in other studies for a wide spectral interval, containing entire absorption bands. This shows that for the lines and bands under study at values of $I_0(\nu)$ characteristic of the solar radiation spectrum the dependence of $k(\nu)$ on $I_0(\nu)$ is inconsequential. Without additional experiments it is impossible to draw any more conclusions. It is probably true that $k(\nu)$ will depend on $I_0(\nu)$ when powerful lasers are used as radiation sources. Here the dependence of $k(\nu)$ on $I_0(\nu)$ should manifest itself differently for different lines.

The independence of $k(\nu)$ on concentration c means that each molecule absorbs radiation independently of other molecules. It has been shown by numerous experiments that this assumption is valid at low absorbing-substance concentrations. Increasing the concentration of the absorbing gas and addition of a foreign gas results in amplification of the effects of intermolecular interaction, which alter the values of the absorption coefficient, and results in actual departure from the Bouguer-Beer law. This effect was first observed experimentally in the infrared region by Angström /105/ in the study of integral absorption by CO_2. Schaefer /166/ investigated spectral absorption in CO_2 bands of 2.7 and 4.3μ and also found that $k(\nu)$ is a function of partial pressure. Eva von Bahr /107/ investigated the absorption of infrared radiation by mixtures of various absorbing and foreign gases and concluded that the observed departures from the Bouguer-Beer law are due to intermolecular collisions. These collisions result in broadening the spectral absorption lines, which depends on the partial pressures of the absorbing and foreign gas and on the temperatures. Broadening of spectral lines results, in its turn, in changing $k(\nu)$.

Under conditions for absorption of infrared radiation prevailing in the atmosphere, where the total pressure varies little with time at given altitude while the partial pressures of absorbing gases are small compared with the total pressure, in the majority of cases when the problem at hand need not be solved with a high degree of accuracy the Bouguer-Beer law can be regarded as applicable. It should be remembered here that $k(\nu)$ for radiation propagating horizontally at different altitudes will be different for the same frequency ν, since the total and partial pressures of the absorbing components are a function of altitude. When it is necessary to determine accurately given absorption characteristics of infrared radiation propagating horizontally, it should be borne in mind that for a concentration of the principal absorbing atmospheric constituent (water vapor) equal to 1% by volume, the contribution to the broadening of absorption lines by mutual collisions of water molecules can amount to approximately 5% of the contribution during collision between water and air molecules.

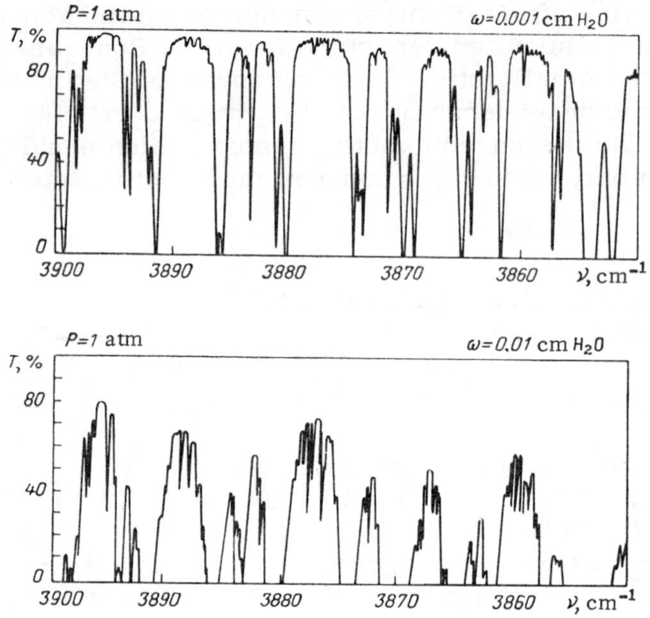

FIGURE 3.1. Fully resolved absorption spectrum of water vapor at 3850 — 3900 cm^{-1} (2.56 — 2.6 μ) when $P = 1.0$ atm and precipitable water vapor layer $w = 0.001$ cm (upper curve) and $w = 0.01$ cm (bottom curve) /189/.

We now consider the apparent deviation from the Bouguer-Beer law, due to the limited resolving power of instruments used in experimental studies of absorption. It is known that the light beam leaving any spectrometer contains a frequency interval determined by the resolving power of the instrument. The highest resolution obtained in spectrometers with diffraction gratings in the near infrared is 0.025 cm^{-1} /108—110/. On the other hand, the half-width of lines of rotation-vibration bands (see Sec. 5) of absorption in the lower atmosphere ranges from several hundredths to several tens cm^{-1}. This means that, even when using instruments with high resolving power, the spectral width of the instrument's slit is of the same order of magnitude as the width of absorption lines. Several lines or even several line groups pass through the output slit of standard spectrometers. The spectral transmission $T(\nu)$ recorded by the instrument is not equal to its actual value $f(\nu)$. Actually $T(\nu)$ characterizes the transmission function distorted by the operation of the slit function of the spectrometer for interval $\Delta\nu$, equal to the spectral width of the output slit of the instrument (Figures 3.1 and 3.2).

The distorting effect of the spectrometer on the actual spectral pattern and its associated departure from the Bouguer-Beer law is the greater, the smaller the resolving power of the instrument and the sharper the dependence of the absorption coefficient on the frequency. The measured spectral transmission equals the actual transmission only when $k(\nu) = \text{const}$ in the spectral interval under study.

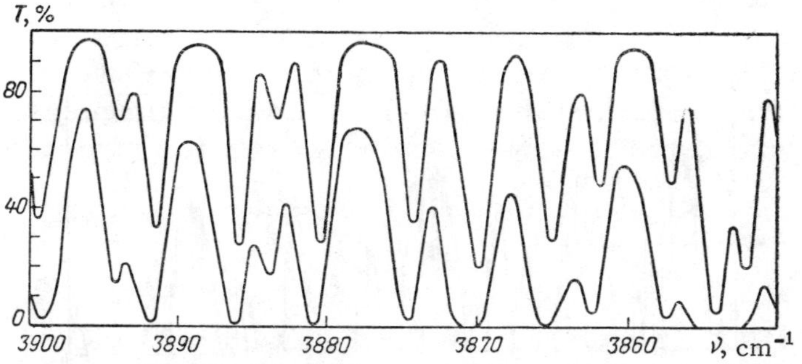

FIGURE 3.2. Absorption spectrum of water vapor obtained by multiplying $T(\nu)$ (Figure 3.1) by the slit function with width $2a = 2.0\ \text{cm}^{-1}$ /189/.

The measured spectral absorption $A'(\nu)$ is related to the real absorption $A(\nu)$ and the slit function S by the expression

$$A'(\nu) = \int_{-\infty}^{\infty} S(\nu - \nu') A(\nu')\, d\nu'.$$

For a Gauss-shaped slit function, $A'(\nu)$ takes the form

$$A'(\nu) = \frac{\sqrt{\ln 2}}{a^2 \sqrt{\pi}} \int_{-\infty}^{\infty} \exp\left\{-\frac{\ln 2}{a^2}(\nu - \nu')^2\right\} A(\nu')\, d\nu',$$

where a is the half-width of the slit function. Applying Fourier transforms we derive

$$A(\nu) = \frac{1}{2\pi} \int_{-\infty}^{\infty} \frac{A'(\omega)}{S(\omega)} \exp(-i\omega\nu)\, d\omega,$$

where $A'(\omega)$ and $S(\omega)$ are the transforms of $A(\nu)$ and $S(\nu)$.

Extensive calculations of $A(\nu)$ were carried out /784/ for cases when the absorption satisfies a dispersion shape, while the slit function was Gauss-shaped.

Detailed theoretical studies into the effect of the spectrometer on the measured spectral and integral transmission were carried out by Nielsen et al. /111/, who first showed that the integral absorption of the band as a whole does not depend on the instrument slit width. This was subsequently confirmed experimentally by Howard et al. /112/ and Zuev /113, 114/. A formula was derived /115/ enabling one to estimate the maximum error in determining integral transmission by a spectrometer over an interval $\Delta\nu$ from its known slit function.

The effect of the spectrometer on the measured spectral absorption, caused by the slit function, the spectral width of the slit and the scanning rate, and the selection of optimum measurement conditions were investigated extensively (see, for example, /116—123, 785/).

5. FINE STRUCTURE OF THE VIBRATION-ROTATION AND PURE ROTATIONAL SPECTRUM OF WATER VAPOR

We now consider the current status of theoretical and experimental studies into the location of line centers, their strengths and half-widths in the vibration-rotation and pure rotational spectrum of water vapor.

1. Locations of line centers. 1. Theoretical investigations. The locations of absorption line centers ν_{0i} can be calculated on the basis of the quantum theory of absorption. Determination of ν_{0i} involves knowing the energy levels of the molecules. The combination of transitions between these levels in line with known selection rules makes it possible to obtain the sought quantities. The vibration-rotation molecular levels can be derived approximately by simple summation of vibrational and rotational energies determined separately.

Detailed study of the vibrational molecular energy /124—126, etc./ showed that the solution of the quantum mechanical problem of the anharmonic oscillator requires knowledge of the anharmonic constants. These constants are found with the aid of corresponding experimental data. Proceeding in this way Khachkuruzov /127/ calculated the energy for the first 100 vibrational levels of water molecules.

The rotational energy of asymmetric-rotator molecules, such as the water molecule, was determined by Wang /128/ and Ray /129/. The latter introduced the concept of the asymmetry parameter and showed that the problem can be reduced to finding some quantity which is a function of this parameter alone. Ray's method was developed and computed numerically by Hainer et al. /130/, who give tables for calculating the rotational energy up to $j=12$. A technique has been found for calculating the rotational energy when $j>12$ /131/; this approach was used to calculate the rotational energy up to $j=40$ /52/.

The apparatus for the approximate calculation of the vibration-rotation energy of the H_2O molecule has thus been developed. The accuracy in determining ν_{0i} is here quite sufficient for solving many problems involving transmission functions for not too narrow spectral sections. When a problem requires data on the spectral transmission $T(\nu)$ (such as the use of a laser as a radiation source), this approximation is not suitable, particularly for lines produced by transitions with high j. Exact determination of the vibration-rotation energy levels of molecules, and hence values of ν_{0i}, requires consideration of the interaction between the vibrations and rotations. A first attempt at this, using a perturbation method, was made by Wilson and Howard /132/, who showed that allowance for this interaction results in the appearance in the rotational-energy operator of new terms, the coefficients of which are centrifugal distortion constants. Wilson /133/ found an analytical relationship between these constants and molecular parameters. This very problem of determining the rotation-vibration energy was considered by El'yashevich /134/, who obtained general expressions for the centrifugal distortion constants.

Shaffer and Nielsen /135/ found values of the vibration-rotation energy and centrifugal distortion constants of type XY_2 molecules and derived an equation for determining most anharmonic constants. The solution of the Schrödinger differential equation for an asymmetric rotator has also been examined /136, 137/: in particular, the vibration-rotation energy for $j=10$ was computed numerically. Comparison of these results with other data /138/ indicates that their difference does not exceed 3.49 cm^{-1} for energy values from 1114.6 cm^{-1} ($j_\tau = 10_{-10}$) to 2698.6 cm^{-1} ($j_\tau = 10_{10}$), which points to satisfactory agreement between theoretical and experimental results.

However, this accuracy in calculating the location of lines is quite unsatisfactory for solving problems involving a quantitative estimate of the absorption of laser radiation in the atmosphere, and problems of finding absorption functions in narrow spectral intervals. The theoretical results obtained until recently have thus been shown not to provide the necessary accuracy in determining the location of line centers. In this connection relevant investigations were carried out in the author's laboratory /139—147, 343, 779/.

To find a system of vibration-rotation levels one must solve the complete Schrödinger equation for the vibration-rotation motion of molecular nuclei:

$$H\psi = E\psi, \tag{3.53}$$

where H and ψ are the vibration-rotation Hamiltonian and wave function of the molecule, while E represents the eigenvalues of the energy operator.

The vibration-rotation Hamiltonian for type XY_2 molecules is

$$H = H_v + \frac{1}{2}\sum_{\alpha\beta}\mu_{\alpha\beta}P_\alpha P_\beta - \frac{1}{2}g_z P_z, \tag{3.54}$$

where H_v is the Hamiltonian of anharmonic vibrations of the system of nuclei, expressed either in exact form /132, 135, 148, 149/ or approximate form /145/; P_α is the operator of the component of the total angular momentum along the α axis of the coordinate system fixed to the molecule ($\alpha = x,\ y,\ z$); $\mu_{\alpha\beta}$ is some known /132/ function of normal coordinates q_k ($k = 1, 2, 3$); $g_z = \mu_{zz}P_z + P_z\mu_{zz}$; $P_z = \sum_{ik} G^z_{ik} q_i P_k$ is the component of the vibrational angular momentum along the z axis; G^z_{ik} are Coriolis constants.

Since it is impossible to solve the Schrödinger equation with a Hamiltonian of type (3.54) in closed form, various approximate methods are introduced. The usual procedure is as follows. By means of a unitary transformation /134/ or by a sequence of contact transformations /150—153/ the matrix of operator H is reduced to diagonal form relative to vibrational quantum numbers. Then the rotational-energy operator is singled out from the diagonal elements and the eigenvalue problem of a nonrigid asymmetric rotator is solved.

We too shall follow this general procedure, except that the matrix of operator H will be diagonalized by means of perturbation theory. In expression (3.54), we select as zero approximation $H^0 = H_v$, in which case the perturbation operator is

$$w = \frac{1}{2}\sum_{\alpha\beta}\mu_{\alpha\beta}P_\alpha P_\beta - \frac{1}{2}g_z P_z, \tag{3.55}$$

with matrix elements $\langle v|w|v'\rangle$ being noncommuting operators relative to the rotational variables. If one uses tabulated formulas of ordinary perturbation theory, then difficulties arise as early as the third approximation due to indefiniteness of the order of the matrix elements. However, in the course of writing the formulas for the perturbation energy it is possible to determine uniquely the sequence of matrix elements.

Consider now the limits of applicability of perturbation theory in this case. The condition of applicability of perturbation theory assumes the form /154/

$$\left| \frac{\langle v \,|\, w \,|\, v' \rangle}{E_v - E_{v'}} \right| \ll 1. \tag{3.56}$$

It has been shown /134/ that in the expansion of the vibrational-energy operator the ratio of the following to the preceding term is approximately equal to λ. For light molecules, such as the water molecule, $\lambda \approx 10^{-1}$.

Simple estimates show that one may set

$$\left| \frac{\langle v \,|\, w \,|\, v' \rangle}{E_v - E_{v'}} \right| \approx \lambda^3 j^2, \tag{3.57}$$

where j is the rotational quantum number, E_v are the eigenvalues, while $\langle v|$ are the eigenfunctions of operator $H^0 = H_v$.

The condition of applicability of perturbation theory for rotational levels with quantum number $j < 31$ is thus satisfied in this case. It should be noted that for those vibrational levels whose energies are approximately equal (random degeneracy), the above estimates may be incorrect, and hence care must be taken in using this method.

Relationship (3.57) can be used for estimating the order of magnitude of contributions of the different approximations in the rotational energy operator $H_R^{[v]}$ relative to the zero approximation. As zero approximation we take the Hamiltonian of the effective rigid asymmetric rotator:

$$(H_R^{[v]})^{(0)} = \frac{1}{2} \sum_\alpha a_{\alpha\alpha} P_\alpha^2, \tag{3.58}$$

where $a_{\alpha\alpha} = \langle v_1 v_2 v_3 \,|\, \mu_{\alpha\alpha}^0 \,|\, v_1 v_2 v_3 \rangle_0$ are the effective rotational constants.

Due to specific features of the perturbation operator, the corrections of each approximation are not uniform. The nth approximation has $n+1$ terms for n even and n terms for n odd. The orders of magnitude of these terms for each j are generally different. For example, terms involving the eighth power of the angular momentum component from the fourth approximation become, with increasing j (for $j \gg 10$), larger than terms involving sixth powers of the angular momentum component from the third approximation, i. e., the contribution of individual terms is redistributed.

Such a redistribution is also observed within the limits of each approximation. For example, in the second approximation for $j < 10$ the greatest contribution to the Hamiltonian of a nonrigid rotator should be expected from terms involving P_α^2, while for $j \approx 30$ the greatest contribution is given by terms involving P_α^4.

A detailed analysis of this problem yielded an estimate of the absolute error made in calculating the matrix elements of operator $H_R^{[v]}$ for $1 \leqslant j \leqslant 30$, and the maximum error in determining the energy of rotational levels. Thus, in order that the error for $j = 10$ should not exceed 0.1 cm^{-1} it is necessary to account for all terms containing P^2, P^3, P^4, P^5, P^6, P^8 up to the fourth approximation. At the same time it is unnecessary to incorporate terms involving P^7. After transformation by means of commutation relationships, Hamiltonian $H_R^{[v]}$ of the nonrigid rotator in the fourth approximation can be expressed in the form

$$H_R^{[v]} = \sum_\alpha b_{\alpha\alpha} P_\alpha^2 + \sum_{\alpha\beta\gamma\delta} \tau_{\alpha\beta\gamma\delta} P_\alpha P_\beta P_\gamma P_\delta + \\ + \sum_{\alpha\beta\gamma\delta\varepsilon\eta} [\alpha\beta\gamma\delta\varepsilon\eta] P_\alpha P_\beta P_\gamma P_\delta P_\varepsilon P_\eta + \\ + \sum_{\alpha\beta\gamma\delta\varepsilon\eta\nu\xi} [\alpha\beta\gamma\delta\varepsilon\eta\nu\xi] P_\alpha P_\beta P_\gamma P_\delta P_\varepsilon P_\eta P_\nu P_\xi. \tag{3.59}$$

Simple, but cumbersome calculations enable one to find analytic expressions for effective rotational constants $b_{\alpha\alpha}$ and centrifugal distortion constants $\tau_{\alpha\beta\gamma\delta}$, $[\alpha\beta\gamma\delta\varepsilon\eta]$ and $[\alpha\beta\gamma\delta\varepsilon\eta\nu\xi]$. For example,

$$[\alpha\beta\gamma\delta\varepsilon\eta\nu\xi] = \sum_k \mu_{\alpha\beta}^k \mu_{\nu\xi}^k \frac{\tau_{\gamma\delta\varepsilon\eta}^k}{2\hbar\omega_k^2} + \\ + \sum_{ik} \mu_{\alpha\beta}^k \mu_{\gamma\delta}^i \mu_{\varepsilon\eta}^k \mu_{\nu\xi}^i \frac{1}{\omega_i^2 - \omega_k^2} \left(v_k + \frac{1}{2} \right), \tag{3.60}$$

where ω_i is the principal frequency, $\tau_{\gamma\delta\varepsilon\eta}^k$ is the linear coefficient of expansion of second-order centrifugal distortion constants with respect to vibrational quantum numbers, while $\mu_{\alpha\beta}^k = \left(\frac{\partial \mu_{\alpha\beta}}{\partial q_k} \right)_{q=0}$. We shall now write the total vibration-rotation Schrödinger equation as follows:

$$[H_v + H_R^{[v]} + H_{Rv} - H_R^{[v]}]\,\psi = E\psi, \tag{3.61}$$

where H_{Rv} represents the second and third terms in (3.54).

Equation (3.61) is solved by the method of perturbations. Operator $H_v + H_R^{[v]}$ is taken as zero approximation. It then follows from the method of obtaining operator $H_R^{[v]}$ that the total vibration-rotation energy is

$$E=E_v+E_R^{[v]} \tag{3.62}$$

to within the omitted terms in $H_R^{[v]}$.

The normalized total vibration-rotation functions in the second approximation of the perturbation theory are represented in the form

$$\psi_{vj\tau}=(1+C^{(2)}_{vj\tau,\, vj\tau})\,\psi_v\psi^{[v]}_{j\tau}+\sum_{v'j'\tau'}{}'\,(C^{(1)}_{vj\tau,\, v'j'\tau'}+ \\ +C^{(2)}_{vj\tau,\, v'j'\tau'})\,\psi_v\psi^{[v']}_{j'\tau'}, \tag{3.63}$$

where

$$C^{(1)}_{vj\tau,\, v'j'\tau'}=\langle v'j'\tau'\,|\,RH_{Rv}\,|\,vj\tau\rangle; \tag{3.64}$$

$R=\dfrac{1}{E_v-H_v}$ is the operator of the resolvent of the problem of vibrations;

$$C^{(2)}_{vj\tau,\, v'j'\tau'}=[\langle v'j'\tau'\,|\,R_{Rv}\,|\,v'j'\tau'\rangle- \\ -\langle vj\tau\,|\,H_{Rv}\,|\,vj\tau\rangle]\,\langle v'j'\tau'\,|\,R^2H_{Rv}\,|\,vj\tau\rangle+ \\ +\frac{1}{4}\,b_{zz}\,(vv')\,\langle j\tau'\,|\,P_z^2\,|\,j\tau\rangle+ \\ +\frac{1}{4}\sum_{\alpha\beta}[b_{\alpha\beta z}\,(vv')\,\langle j\tau'\,|\,P_\alpha P_\beta P_z\,|\,j\tau\rangle+ \\ +b_{z\alpha\beta}\,(vv')\,\langle j\tau'\,|\,P_zP_\alpha P_\beta\,|\,j\tau\rangle+ \\ +\frac{1}{4}\sum_{\alpha\beta\gamma\delta}t_{\alpha\beta\gamma\delta}\,(vv')\,\langle j\tau'\,|\,P_\alpha P_\beta P_\gamma P_\delta\,|\,j\tau\rangle; \tag{3.65}$$

$$C^{(2)}_{vj\tau,\, vj\tau}=\frac{1}{2}\sum_{\alpha}b_{\alpha\alpha}\,(vv)\,\langle j\tau\,|\,P_\alpha^2\,|j\tau\rangle+ \\ +\frac{1}{4}\sum_{u\beta\gamma\delta}t_{\alpha\beta\gamma\delta}\,(vv)\,\langle j\tau\,|\,P_\alpha P_\beta P_\gamma P_\delta\,|\,j\tau\rangle. \tag{3.66}$$

Constants $b_{\alpha\alpha}\,(vv)$ are analogous to corrections to the effective rotational constants, while constants $t_{\alpha\beta\gamma\delta}$ are analogous to corrections to the centrifugal distortion constants:

$$b_{zz}\,(vv)=\frac{1}{2}\,\langle v\,|\,g_zR^2g_z\,|\,v\rangle; \\ b_{\alpha\alpha}\,(vv)=\pm\,i\hbar\,\langle v\,|\,g_zR^2\mu_{xy}\,|\,v\rangle,\quad \alpha=x,\ y; \\ t_{\alpha\beta\gamma\delta}\,(vv)=\langle v\,|\,\mu_{\alpha\beta}R^2\mu_{\gamma\delta}\,|\,v\rangle. \tag{3.67}$$

Constants $b_{\alpha\beta z}\,(vv')$, $b_{z\alpha\beta}$, b_{zz} and $t_{\alpha\beta\gamma\delta}$ have the form

$$A\,(vv')=(-1)^n\,\langle v'\,|\,R_aR_b\,|\,v\rangle, \tag{3.68}$$

a and b being different combinations of operators g_z and $\mu_{\alpha\beta}$, while n is the number of operators g_z among a and b.

The above expressions were used to estimate the contributions of the different terms of the 2nd, 3rd and 4th approximations to the Hamiltonian of the rigid rotator for $j = 1(1)30$. These data can be used to determine the maximum order of approximation for $H_R^{[v]}$, which must be used (for known j) when computing, to some degree of accuracy, the location of the center of the corresponding line. The subsequent solution procedure reduces to the following. All the coefficients contained in the expression for $H_R^{[v]}$ are found, the operator matrix is constructed and is diagonalized numerically. Then the eigenvalues of operator $H_R^{[v]}$ are found and (3.67) is used to derive the values of the total vibration-rotation energy from which (employing selection rules) it is easy to find the centers of the corresponding lines.

2. Experimental investigations. The fine structure of the pure rotational and vibration-rotation spectra of water vapor has been studied extensively /206/. The first studies of vibration-rotation bands were carried out by Sleator /155/ and Sleator and Phelps /156/; however, their resolution did not suffice for analyzing the rotational structure. Plyler and Sleator /157/ and then Plyler /158/ investigated the principal bands with improved resolution. Utilizing other data /157, 158/, as well as the results of their own measurements in the photographic spectral region, Mecke and co-workers /159—161/ were the first to analyze the rotational structure of individual bands and determined the moments of inertia and geometric parameters of water molecules.

Freudenberg and Mecke /162/ studied the water vapor spectrum with a high resolution in the region of principal vibration frequencies and of the combination bands $(\nu_1 + \nu_3)$ and $(\nu_2 + \nu_3)$. The spectral region from 1.34 to 1.97 μ was studied /163/ with a slightly better resolution. Benedict and Plyler /164/ investigated the 2.7 μ band with a high resolution; they gave a new identification to the lines of the ν_1 band, expanded the analysis of the spectrum of the ν_3 band and presented values of the (100) and (001) rotational terms of states up to $j = 13$. Additional data on analysis of the fine structure of the water vapor spectrum are listed by Benedict /165/. The vibration-rotation spectrum of water vapor was studied /60/ with a high resolution, with the result that several wavelengths needed for spectrometer calibration have been determined with a high degree of accuracy.

The water vapor spectrum in the 4.5—5.7 and 6.6—13 μ region was investigated in detail by Benedict et al. /166/. As a result of this study a more detailed analysis was carried out of the rotational structure of the ν_2 band, the rotational terms of the ground state were determined up to $j = 18$, terms of the (010) and (020) states were calculated, and the values were listed of a number of rotational constants of the H_2O molecule. The spectral region of 5.26—7.14μ was treated in detail with a high resolution /167/. The water vapor spectrum in the 1.9μ region in flames was studied with a high resolution /168/, yielding data on the high levels of the vibrational state and on absorption bands between the high vibrational levels: (021)→(010) and (031)→(020).

The pure rotational absorption spectrum of water vapor was investigated with a high resolution /169/, thus refining Mecke's data. The most detailed study in the 18—135μ region was carried out by Randall et al. /170/, who recorded more than 200 absorption bands. The rotational spectrum range explored with a high resolution was substantially widened /172/, and measurements were made in the 20—2500μ region. New data on the pure rotational water vapor spectrum at wavelengths of 55—1000μ were obtained by Furashov /173/.

We shall briefly examine investigations into the fine structure of the water vapor absorption spectrum under natural conditions. In the majority of cases the sun was used as radiation source. In measurements under natural conditions, one records spectra of other atmospheric gases alongside that of water vapor. It is hence not always possible to completely identify the lines due to spectral overlap. Nevertheless, most of the lines have been identified.

The first atlas of the solar radiation absorption spectrum in the terrestrial atmosphere was published in 1940 /174/. This atlas pertained to the 3612—8771 Å spectral region, which is easily accessible for photographic recording of the spectrum. Data on the solar spectrum in the 6600—13,495 Å spectral range were published in 1947 /175/. Mohler and co-workers measured and compiled an atlas of the solar spectrum in the region from 0.8 to 2.6μ /176/, and the lines of this atlas were identified /177, 178/. Another group published atlases of the solar spectrum for the 3.0—5.2 /179/ and 7—13μ regions /180/. A total of more than 1600 lines were detected in the 1—13μ region, of which more than 70% pertained to water vapor.

The solar spectrum over the wavelength range 2.8—23.7μ was recorded by Migeotte, Neven and Swenson at the Jungfrau high-mountain station (altitude 3580 m) /181/. More than 3600 individual lines were identified /182/. Using more refined equipment, Delbouille and Roland /183/ obtained at the same station a new spectral atlas of the sun. About 10,500 lines were detected in the relatively narrow 0.75—1.2μ range.

The solar radiation spectrum was recorded by Farmer and Key /184/ in the wavelength range 7—400μ in the Bolivian Andes (altitude 5200 m). In the 7—30μ region the spectrum was obtained with a resolution of 0.3 cm^{-1}; new tellurium lines of H_2O absorption were recorded. In the 40—400μ region, solar radiation in the atmospheric column above 5.2 km is absorbed almost completely, with the exception of the relatively narrow atmospheric "window" detected near the 345μ wavelength /184/.

The absorption spectrum of the surface layer of the atmosphere for the 16.3 km altitude, recorded with a high resolution in the region 5400 to 8520 Å, was published in 1963 /185/. Due to the sufficiently high resolution and the large atmospheric thickness, a large number of weak absorption lines in the visible spectral region were detected /185/. Tables of the absorption spectrum of the terrestrial atmosphere, containing about 24,000 lines in the 2935—8770 Å range, were published in 1966 /186/. Oppenheim and Goldman /187/ composed an atlas of the water vapor absorption spectrum in regions of the 1.9 and 2.7μ bands at 1200°K. The atlas contains lines due to transitions between high-energy vibration-rotation levels of the H_2O molecule.

We shall now summarize the main experimental studies of the absorption spectrum of water vapor under laboratory and natural conditions. At present the location of many lines of the vibration-rotation as well as the pure rotational spectrum of H_2O have been determined. It should, however, be noted that in spectra recorded even with the highest resolution, many close lines often remain unresolved. Sometimes groups of lines are taken as single lines. This is especially true of spectral regions in which strong and weak lines lie close to one another. The accuracy in determining the line centers in the ultraviolet, visible and near infrared lies between several hundredths and 0.1—0.2 Å.

2. Line strengths. 1. Theoretical studies. Calculation of the line strengths of rigid asymmetric-rotator molecules does not involve any difficulties. At present there exist sufficiently detailed data /188, 878/ which give an approximate pattern of line strengths of the pure rotational spectrum of water vapor. The true spectrum pattern can be obtained by solving the problem of the rotation of a nonrigid rotator.

In exact calculations of line strengths in the vibration-rotation spectrum of water vapor one must take into account the nonrigidity of the molecule and the interaction of various kinds of its motions. Usually the line strengths in the vibration-rotation H_2O bands are found in the following manner /879, 880/. It is assumed that the line strength is

$$S = S_{\text{rot}} S_{\text{vib}}, \tag{3.69}$$

where S_{rot} is the strength of the corresponding line from the pure rotational spectrum, the value of which is assumed known; S_{vib} is a factor making allowance for the interaction between vibrations and rotations. The value of S_{vib} is determined experimentally. Here it is assumed that S_{vib} is the same for all lines of the given vibration-rotation band. On this assumption integration of (3.69) over the band from the left yields the experimentally determined integral strength of the band, and from the right it yields the product of S_{vib} and the total line strengths due to pure rotational transitions. This scheme was used for calculating line strengths of the ν_1, $2\nu_2$ and ν_3 vibration-rotation absorption bands in the 2.7μ region (2857—4444 cm^{-1}) /879, 880/.

Benedict and Calfee /881/ computed data on the line strengths of absorption bands in the 1.9 and 6.3μ region. These calculations make partial allowance for the interaction between the vibrational and rotational motions of the H_2O molecule. The tables in /881/ contain information on the locations of centers, on the strength and half-width of lines of the $\nu_1\nu_2$, $\nu_2\nu_3$ and $3\nu_2$ bands in the 1.9μ region, and of the ν_2 band in the 6.3μ region. The tabulation was carried out for lines with $j \leqslant 20$. Data were also obtained /881/ for all three parameters of the fine structure of 4 "hot" bands (021—010, 100—010, 020—010, 001—010), of bands of the H_2O^{17} and H_2O^{18} isotopes, and of the ν_2 band of HDO. Reference /881/ contains 4110 band absorption lines in the 1.9μ region (4212—6254.3 cm^{-1}) and 5054 lines in the 6.3μ region (640—2885 cm^{-1}).

The tables pertain to lines whose strengths exceed 0.0001 $cm^{-1}/g\cdot cm^{-2}$. The exception is the central region of the ν_2 band (1400—1850 cm^{-1}), which contained lines with strengths in excess of 0.001 $cm^{-1}/g\cdot cm^{-2}$. Parameters of lines of the pure rotational spectrum of H_2O were calculated in the 640—1000 cm^{-1} range, so that the data would be as far as possible more complete over the entire wavelength range under study. Note that the maximum line strength in the 1.9 and 6.3μ band region was found to be respectively 1042.7 (to 5413.90 cm^{-1} line) and 10,976.2 $cm^{-1}/g\cdot cm^{-2}$ (the 1684.84 cm^{-1} line).

Strictly speaking, the assumption of constancy of S_{vib} for all the lines in the band is incorrect. It disregards the specific properties of interaction between the vibrational and rotational motion of the molecule for individual lines. The latter circumstance can result in substantial errors when calculating line strengths due to transitions between high-energy molecular levels.

In our laboratory an attempt was made to develop a new scheme for calculating the line strengths of the vibration-rotation spectrum of water vapor with subsequent allowance for interaction between the vibrational and rotational molecular motions /139—141, 147, 190, 220, 779, 907/. We shall now describe the results.

The line strength is expressed in the form /191/

$$S_{vR}^{v'R'} = \frac{8\pi^3 N}{hcz} \Delta\omega \left[2 - (-1)^{|\tau|}\right] e^{-\frac{hc}{kT}\omega_{\tau R}} \times$$
$$\times \left[1 - e^{-\frac{hc}{kT}\Delta\omega}\right] |\langle v'R' | M_z | vR \rangle|^2. \quad (3.70)$$

Here $\Delta\omega = \omega_{v'R'} - \omega_{vR}$, $\omega_{v'R'}$ and ω_{vR} are the levels of the higher and lower states, M_z is the component of the vector of the electron dipole moment in the stationary coordinate system, N is the number of particles in the absorbing gas per unit volume, z is the statistical sum, τ is the rotational quantum number, h is Planck's constant, c is the speed of light, k is Boltzmann's constant, T is temperature, and $|vR\rangle$ is the total vibration-rotation wave function. Component M_z can be written in the form

$$M_z = \Phi(zx)\mu_x + \Phi(zy)\mu_y, \quad (3.71)$$

where μ_x and μ_y are components of the dipole-moment vector along the coordinate axes moving with the molecule, while $\Phi(zx)$ and $\Phi(zy)$ are directional cosines.

It follows from the symmetry of the H_2O molecule /134/ that when no allowance is made for the interaction between vibrations and rotations a contribution to parallel bands in M_z is made only by the second term. Quantities μ_x and μ_y are expanded as a series in normal coordinates.

Function (3.63) yields quantity $\langle vR | M_z | v'R' \rangle$ in (3.70):

$$\langle vR | M_z | v'R' \rangle = \langle vR | M_z | v'R' \rangle_0 + C^{(1)}_{v'R', vR} [\langle vR | M_z | vR \rangle_0 -$$
$$- \langle v'R' | M_z | v'R' \rangle_0] + \sum_{v''R''} [C^{(1)}_{vR, v''R''} \langle v''R'' | M_z | v'R' \rangle_0 +$$
$$+ \langle vR | M_z | v''R'' \rangle_0 C^{(1)}_{v'R', v''R''}]. \quad (3.72)$$

Terms containing $C^{(2)}$ have been omitted, since their contribution to the matrix element are highest-order infinitesimals. The second term in (3.72) is identically equal to zero, since $\langle R | \Phi | R \rangle \equiv 0$ due to symmetry of the molecule. The matrix elements in the right-hand side of (3.72) are calculated by means of zero-approximation functions $\psi^0_{vR} = \psi_v \psi_R^{[v]}$. Here the rotational wave function $\psi_R^{[v]}$ is represented by a wave-function expansion of the rigid symmetric rotator:

$$\psi_R^{[v]} = \sum_{k=-j}^{j} g_{k\tau}^{[v]} | jkm \rangle. \quad (3.73)$$

In the third term of equation (3.72) it becomes necessary to sum over the vibrational structure of intermediate vibrational states. We shall consider this term in more detail. Utilizing (3.64) and (3.61) it can be represented in the form

$$\frac{1}{2}\sum_{R''v''}\sum_{g}\Big\{\sum_{\alpha\beta}\Big[\frac{\langle v'\,|\,\mu_g\,|\,v''\rangle\langle v''\,|\,\mu_{\alpha\beta}\,|\,v'\rangle}{E_v-E_{v''}}\times$$
$$\times\langle R'\,|\,\Phi(z_g)\,|\,R''\rangle\langle R''\,|\,P_\alpha P_\beta\,|\,R\rangle+\frac{\langle v'\,|\,\mu_{\alpha\beta}\,|\,v''\rangle\langle v''\,|\,\mu_g\,|\,v\rangle}{E_{v'}-E_{v''}}\times$$
$$\times\langle R'\,|\,P_\alpha P_\beta\,|\,R''\rangle\langle R''\,|\,\Phi\,|\,(zg)\,|\,R\rangle\Big]-\frac{\langle v'\,|\,\mu_g\,|\,v'\rangle\langle v''\,|\,g_z\,|\,v\rangle}{E_v-E_{v''}}\times$$
$$\times\langle R'\,|\,\Phi(zg)\,|\,R''\rangle\langle R''\,|\,P_z\,|\,R\rangle-$$
$$-\frac{\langle v'\,|\,g_z\,|\,v''\rangle\langle v''\,|\,\mu_g\,|\,v\rangle}{E_{v'}-E_{v''}}\langle R'\,|\,P_z\,|\,R''\rangle\langle R''\,|\,\Phi(zg)\,|\,R\rangle\Big\}. \tag{3.74}$$

Summation over the rotational states with consideration of (3.73) gives

$$\sum_{\substack{k''\tau''\\ kk'}} g_{k\tau}^{[v]}\,(g_{k\tau''}^{[v'']})^2\,g_{k'\tau'}^{[v']}\,\langle jkm\,|\,P_\alpha P_\beta\,|\,jk''m\rangle\times$$
$$\times\langle jk''m\,|\,\Phi\,|\,j'k'm'\rangle, \tag{3.75}$$
$$\sum_{\substack{\tau''k\\ k'}} g_{k\tau}^{[v]}\,(g_{k\tau''}^{[v'']})^2\,g_{k'\tau'}^{[v']}\,\langle jkm\,|\,P_z\,|\,jkm\rangle\times$$
$$\times\langle jkm\,|\,\Phi\,|\,j'k'm'\rangle. \tag{3.76}$$

Summation over k'' in (3.75) is elementary, since $\langle jkm\,|\,P_\alpha P_\beta\,|\,jk''m\rangle$ are nonzero only when $k''=k,\ k\pm 2$. Subsequently, since matrix $g_{k\tau}^{[v]}$ is normalized /130/, $\sum_{\tau''}(g_{k\tau''}^{[v'']})^2=1$.

The matrix elements of directional cosines are expressed in terms of products of Clebsch-Gordan coefficients. The approximation under study does not contradict the selection rules for rotational quantum numbers j and τ.

We now pass on to calculation of summations over intermediate quantum states. The matrix elements contained in these summations are calculated from the function of the three-dimensional anharmonic oscillator. Functions of normal coordinates μ, $\mu_{\alpha\beta}$, g_z and v are represented in terms of series in normal conditions, such as

$$v=\sum_i\omega_i^2 q_i^2+\sum_{ik}a_{iik}q_i^2 q_k+$$
$$+\sum_{ikl}a_{iikl}q_i^2 q_k q_l+\ldots \tag{3.77}$$

In this series the ratio of the subsequent to the preceding term is λ. According to estimates /134/, $\lambda\sim 10^{-1}$. Further, the orders of magnitudes in expansions of v, g_z, μ and $\mu_{\alpha\beta}$ increase in the same manner and this makes it possible to postulate the validity of this estimate for all these quantities. This is confirmed by direct calculations using other data /190, 193, 194/. In order to estimate the order of magnitude of the contribution to the matrix element of the dipole moment, given by (3.74), it is necessary to calculate the order of magnitude of

$$\mu^0_{\alpha\beta}P_\alpha P_\beta \frac{1}{E_v - E_{v''}} = \varepsilon j^2.$$

Direct calculations in accordance with estimates /134/ yield $\varepsilon \sim \lambda^2$ for small j. Note that quantity $\frac{g_z P_z}{E_v - E_{v'}}$ is also of the order of magnitude of λ^2.

In calculating the contribution of (3.74) we restrict ourselves to terms of order not higher than λ^3. The contribution of (3.74) can be estimated from formulas for matrix elements μ and operator g_z by functions of the three-dimensional anharmonic oscillator /142, 144/. Due to the cumbersomeness of the final results, we present as an illustration only the contribution of (3.74) for the 000—010 vibrational transition:

$$\begin{aligned}
&\langle 000j\tau \,|\, M_z \,|\, 010j'\tau'\rangle = \langle 000j\tau \,|\, M_z \,|\, 010j'\tau'\rangle_0 \times \\
&\times \left[1 + \alpha + \beta\frac{S'}{S} + \gamma\frac{S(k)}{S}\right]; \quad \beta = \frac{\mu_0\mu^{(2)}_{xx}\hbar}{\sqrt{2}\,\mu_2\omega_2}; \\
&\alpha = \frac{\hbar\mu_0}{2\mu_2\omega_2}\left(\mu^2_{xx} + \mu^2_{zz}\right) + \frac{\mu_3\mu^{(0)}_{zz}\xi^z_{23}\hbar}{2\omega_3\mu_2}\,\frac{\omega_2 + \omega_3}{\omega_3 - \omega_2}; \\
&\gamma = -\frac{\mu_0\mu^{(2)}_{zz}\hbar}{\mu_2\omega_2} - \frac{\mu_3\mu^{(0)}_{zz}\xi^z_{23}\hbar}{\omega_3\mu_2}\,\frac{\omega_3^2 + \omega_2^2}{\omega_3^2 - \omega_2^2}; \\
&S = \sum_{kk'} g^{[v]}_{k\tau}(j)\, g^{[v']}_{k'\tau'}(j')\left[(1j-1k\,|\,j'k') + (1j1k\,|\,j'k')\right]; \\
&S(k) = \sum_{kk'} k g^{[v]}_{k\tau}(j)\, g^{[v']}_{k'\tau'}(j')\left[(1j-1k\,|\,j'k') - (1j1k\,|\,j'k')\right]; \\
&S' = \sum_{kk'} g^{[v]}_{k\tau}(j)\, g^{[v']}_{k'\tau'}(j')\Big[\sqrt{(j-k)(j+1+k)}\,(1j0k+1\,|\,j'k') + \\
&\qquad + \sqrt{(j+k)(j-k+1)}\,(1j0k-1\,|\,j'k')\Big].
\end{aligned} \tag{3.78}$$

The approximation at hand makes allowance for anharmonicity of vibrations, nonrigidity of the molecule, and Coriolis interaction.

Note that, unlike the zero approximation, the vibration-rotation interaction results in the appearance in the matrix elements of the dipole moment of its components along the x and y axes. For parallel bands the component is directed along the x axis, while for perpendicular bands it is directed along the y axis. Here the selection rules in the rotational spectrum which are brought about by molecular symmetry are not violated.

2. Experimental investigations. Experimental determination of line strengths is a very difficult task. The main difficulty is due to the need to make allowance for line overlapping. At present only a limited number of lines of the water vapor absorption spectrum have been determined experimentally.

Benedict et al. /166/ found under laboratory conditions the strengths of several rotational lines and lines from the vibration-rotation ν_2 band at relatively high temperature and H_2O concentration. The strength of a series of lines of the vibration-rotation H_2O spectrum in the region of bands centered about 2.7μ was found by Vasilevskii /195—197/. The strength of three lines of the pure rotational water vapor spectrum located near 92, 170 and 188 cm^{-1} was measured /198/. The strength of three rotational and two vibration-rotation H_2O lines was determined by Saidy /199/. Bolle /200/ reports on the determination of the strength of a series of weak H_2O lines in the 8—20μ region from atmospheric emission spectra by comparing the results of measurements with calculations of monochromatic absorption. This comparison showed that in the case of the weakest lines there exists a discrepancy between experimental and theoretical results of approximately 30%, the theoretical results being less than experimental data. This discrepancy is apparently due to neglect of interaction between the vibrational and rotational molecular motions in calculating line strengths.

The strength of a line in the 6.3μ band was measured by Oppenheim and Goldman /187, 201/, who then utilized their results to determine the integral intensity of the band in the same manner as in /202, 203/ for the 2.7μ band. Burch and Gryvnak /81/ communicate the results of measurements of water vapor and CO_2 absorption in the region from 0.6 to 5.5μ, from which it is possible, in particular, to extract information on the strengths and half-widths of several hundreds of the most important H_2O lines. However, no data on line parameters are given in this study. The strengths and half-widths of a number of lines in the wings and in the central part of the ν_2 band were determined elsewhere /885—887/. The strengths of several lines in the visible part of the spectrum in the 0.69μ region were measured by Kirienko et al. /888/.

3. **Half-widths of lines.** 1. Theoretical investigations. Benedict and Kaplan /57/ used the Anderson theory /56/ to compute the half-widths γ of lines of the rotational spectrum of water vapor up to $j = 13$ and several lines from the ν_2 band of the vibration-rotation spectrum during collisions between water and nitrogen molecules. They also investigated the dependence of half-widths of lines in the 220—2400°K range and the effect of interaction between vibrations and rotations.

The principal results of these calculations reduce to the following. At 300°K the half-width of lines ranges from 0.11115 $cm^{-1}\cdot atm^{-1}$ for the 1_1-1_{-1} line to 0.032 $cm^{-1}\cdot atm^{-1}$ for the $14_{-13}-13_{-13}$ line. The results showed that γ decreases with increasing j, and at given j the value of γ decreases with increasing orbital rotational quantum number τ. As regards the temperatures characteristic of the stratosphere, it was found that the temperature-dependence of γ or T satisfies $\gamma(T)/\gamma(300)=(T/300)^n$. Here the value of n changes from one line to another, being on the average equal to 0.62 at $T=260$°K. For temperatures in the 300—2400°K range the temperature-dependence of γ varies greatly from line to line.

Benedict and Kaplan /58/ present the results of computing the half-width of pure rotational lines of water vapor during collisions between H_2O-H_2O and H_2O-O_2 molecules. It was found that at 300°K the half-widths of lines lie between 0.5663 and 0.6607 $cm^{-1}\cdot atm^{-1}$.

The calculated ratios of line half-widths during H_2O-H_2O and H_2O-N_2 collisions vary from line to line within the limits 1.90 to 6.85. For the widest and strongest lines, corresponding to small j, this ratio lies between 4.5 and 6.0. The average value of this ratio for the entire rotational band was found to equal 5.49.

The half-widths of the vibration-rotation spectrum of water vapor were calculated without subsequent incorporation of interaction between the vibrational and rotational motions /879, 880, 189, 881/.

Zuev et al. /139—141/ attempted to allow for this interaction. The substance of their attempts consists in the following. If the total vibration-rotation eigenfunction is given by

$$\psi_{vj\tau}=\psi_v\psi_{j\tau}^{[v]}, \tag{3.79}$$

where ψ_v is the vibrational wave function, obtained with allowance for anharmonicity of vibrations, while $\psi_{j\tau}^{[v]}$ is the natural wave function of a nonrigid rotator, then it is possible to use the contribution scheme of the known Anderson theory /56/ to derive the half-width of lines.

Using the Dirac rotation of wave functions, the half-width of the spectral line is

$$\gamma(vj\tau m\longrightarrow v'j'\tau'm')=\frac{Nv}{2c}\sum_{j_2}B_{j_2}\sigma_{j_2}; \tag{3.80}$$

$$\sigma_{j_2}=\int_0^\infty 2n\rho S_{j_2}(\rho)\,d\rho; \tag{3.81}$$

$$S_{j_2}(\rho)=\frac{1}{2}\sum_{M_2}\frac{\langle vj\tau m;\ j_2M_2\,|\,P^2\,|\,vj\tau m;\ j_2M_2\rangle}{(2j+1)(2j_2+1)}+$$

$$+\frac{1}{2}\sum_{M_2}\frac{\langle v'j'\tau'm';\ j_2M_2\,|\,P^2\,|\,v'j'\tau'm';\ j_2M_2\rangle}{(2j'+1)(2j_2+1)}; \tag{3.82}$$

$$P=\frac{1}{\hbar}\int_{-\infty}^{+\infty} e^{\frac{-H_0 t}{i\hbar}} C(t) e^{\frac{H_0 t}{i\hbar}} dt, \tag{3.83}$$

where N is the particulate concentration per unit volume; v is the average particulate velocity; ρ is the impact parameter; vjm are quantum numbers designating the level from which the transition occurs; $v'j'm'$ are the quantum numbers of the final level; j_2M_2 are the quantum numbers of the excited molecular states; Bj_2 is the population of state j_2 obeying the Boltzmann distribution; $C(t)$ is the potential energy of interaction between the absorbing and perturbing molecules; H_0 is the Hamiltonian of an isolated absorbing molecule.

The following expression can be derived for the dipole-quadrupole interaction of two charge systems /204/:

$$C(t)=\sum_{\alpha\beta}\left[\frac{5}{2}\frac{D_{\alpha\beta}R_\alpha R_\beta(\vec{M}\vec{R})}{R^7}-\frac{D_{\alpha\beta}R_\alpha M_\beta}{R^5}\right], \tag{3.84}$$

where α and β assume values x, y, z; D are the components of the tensor of the quadrupole moment of the perturbing molecule; M_β and R_α are the components of the dipole moment and distance R between interacting molecules on the axis of the stationary coordinate system with origin at H_2O molecule.

Using matrices of transition from components of the dipole vector and of the tensor of the quadrupole momenta in a stationary system to a system moving with the molecule, it is easy to derive an expression for $C(t)$ on the basis of (3.84):

$$C(t)=\sum_{\substack{m_1k_1\\ m_2k_2}} C^{j_1j_2}_{m_1k_1m_2k_2} D^{j_1}_{m_1k_1} D^{j_2}_{m_2k_2}, \tag{3.85}$$

where $m_1 = 1, 0, -1$; $m_2 = 2, 1, 0, -1, -2$; $k_1 = 1, -1$; $k_2 = 0$; $j_1 = 1$; $j_2 = 2$; D^j_{mk} are generalized spherical harmonics.

Substitution of (3.83) into (3.84), with allowance for (3.85) yields

$$S_{j_2}(\rho)=\frac{1}{2}\sum|a^{12}_{m_1k_1m_2k_2}|^2\langle j_2M_2|D^2_{m_2k_2}|j_2'M_2'\rangle\times$$
$$\times\langle j_2'M_2'|D^2_{m_2'k_2'}|j_2M_2\rangle\langle j\tau m|D^1_{m_1k_1}|j''\tau''m''\rangle\times$$
$$\times\langle j''\tau''m''|D^1_{m_1'k_1'}|j\tau m\rangle\frac{1}{(2j+1)(2j_2+1)}+$$
$$+\text{ a similar term for the final state,} \tag{3.86}$$

where

$$a^{12}_{m_1k_1m_2k_2}=$$
$$=\left\langle v\left|\frac{1}{\hbar}\int_{-\infty}^{\infty} e^{-i\omega_{\alpha\beta}t} C^{12}_{m_1k_1m_2k_2}(t)\right|v\right\rangle, \tag{3.87}$$

$$\omega_{\alpha\beta}=\frac{E_{v''j''\tau''}-E_{vj\tau}+E_{j_2}-E_{j'_2}}{\hbar}. \tag{3.88}$$

In expression (3.86) summation is over all the subscripts, with exception of v, j, m and j_2 for the initial stage, and v', j', m' and j_2 for the final state.

Using expression (3.73), the relationship between wave functions of the rotor and D functions, the expression for the integral of the product of three D functions in terms of Clebsch-Gordan coefficients /205/, and summing in (3.86) over m, m'', M_2 and M_2' we obtain

$$S_{j_2}(\rho)=\sum|a^{12}_{m_1k_1m_2k_2}|^2\frac{(g_k^{[vj\tau]})^2\,(g_{k''}^{[v''j''\tau'']})^2}{15}\times$$
$$\times(jj_1k-k_1\,|\,j''k'')^2\,(2j_200\,|\,j_2'k_2')+$$
$$+\text{ a similar term for the final state,} \tag{3.89}$$

where $(abcd\,|\,ef)$ are Clebsch-Gordan coefficients.

Summation in (3.89) over m_1, m_2, and k_1 can be carried out utilizing appendix C of /71/. Finally

$$\sum_{m_1k_1m_2}|a^{12}_{m_1k_1m_2k_2}|^2=\frac{D^2\mu^2_{vv''}}{\hbar^2v^2\rho^6}f(x), \tag{3.90}$$

where

$$f(x)=\frac{1}{288}\{x^6[18k_3^2+12k_2^2+78k_1^2+84k_0^2]+$$
$$+x^4[-32k_2^2-96k_1^2]\};$$
$$x=\frac{\rho}{v}\omega_{\alpha\beta};$$
$$\mu^2_{vv''}=|\langle v\,|\,\mu_{x'}+i\mu_{y'}\,|\,v''\rangle|^2,$$

$k_n(x)$ are Hankel functions for which the representation (8, 432, 5) /207/ was utilized; μ_x and μ_y are components of the dipole moment of the H_2O molecule in the moving coordinate system. After equation (3.90) is used when substituting (3.89) into (3.81), the latter is found to diverge at small ρ, because the distance between charge systems in obtaining (3.84) is assumed to be greater than the dimensions of the systems themselves.

Following /71/ we assume that

$$S(\rho)=1 \text{ for } 0\leqslant\rho\leqslant\rho_0.$$

Hence (3.82) yields

$$\sigma_{j_2} = \pi\rho_0^2 + 2\pi \int_{\rho_0}^{\infty} S_{j_2}(\rho)\, d\rho. \tag{3.91}$$

After various manipulations, (3.80) and (3.89)—(3.91) give the following expressions for the half-width of lines:

$$\begin{aligned}
&\gamma(vj\tau m \to v'j'\tau'm') = \\
&= \frac{Nv}{2c} \sum_{j_2} B_{j_2}\rho_{j_2}^2 (vj\tau m \to v'j'\tau'm') \times \\
&\times \rho_{j_2}^2 (vj\tau m \to v'j'\tau'm') = \\
&= \rho_0 \left\{ 1 + \frac{1}{45\hbar^2 v^2 \rho_0^6} \sum D^2 \mu_{vv''}^2 (g_k^{[vj\tau]})^2 \times \right. \\
&\times (g_{k''}^{[v''j''\tau'']})^2 (j \mid k - k_1 \mid j''k'')^2 (j_2 j_2 00 \mid j_2' 0)^2 f(x_0) + \\
&\left. + \text{a similar term for the final state} \right\},
\end{aligned}$$

$$\begin{aligned}
f(x_0) &= \frac{1}{32} x_0^6 [3k_2(x_0)\, k_4(x_0) + 2k_1(x_0)\, k_3(x_0) + \\
&+ 13k_0(x_0)\, k_2(x_0) - 3k_3^2(x_0) - 2k_2^2(x_0) + k_1^2(x_0) - 14k_0^2(x_0) - \\
&- \frac{x_0^4}{3} \int_{x_0}^{\infty} \frac{1}{x} [k_2^2(x) + 3k_1^2(x)]\, dx,
\end{aligned}$$

where

$$x_0 = \left| \frac{\rho_0}{v} \omega_{\alpha\beta} \right| > 0,$$

since its sign is seen to have no effect whatever on the results.

In the last expressions D is the quadrupole moment of the perturbing molecule; $\mu_{vv''}^2$ is the square of the matrix elements of the dipole moment of the $v \to v''$ transition, for which analytic expressions can be derived / 142/.

2. Experimental investigations. Experimental investigations of line half-widths are usually carried out simultaneously with the determination of line strengths and constitute quite a complex problem. This explains why until now a very limited volume of data was obtained on the half-widths of absorption lines of H_2O.

Becker and Autler / 208/ determined the half-widths of H_2O lines in the microwave region of the rotational spectrum of H_2O ($\nu_0 = 0.744\ \text{cm}^{-1}$). They found that collisions between water molecules broadened the line by a factor of 4.7 more than collisions between water and air molecules.

The half-widths of a number of lines of the 2.7μ band were measured by Vasilevskii /195—197/. A detailed study of the broadening effect of various gases was carried out for lines with center at $\nu_0 = 4025.38$ cm^{-1}. The experimental data are satisfactorily described by the expression

$$\gamma = \gamma^0 (P_a + \sigma_{a,b} P_b), \tag{3.92}$$

where γ^0 is the half-width of lines at effective pressure $P = P_a + \sigma_{a,b} P_b = 1$ atm and temperature $T = 293°$K; P_a and P_b are the partial pressures of the absorbing and foreign gases; $\sigma_{a,b}$ is the relative effectiveness of optical collisions between molecules a and b.

The values of $\sigma_{a,b}$ for collisions between molecules of H_2O and CO_2, air, N_2, and Ar were found to be, respectively, 0.239, 0.170, 0.165 and 0.123, which points to the great dependence of γ on the nature of the broadening gas. Line broadening by collision of water molecules with one another was found to be sixfold greater than during collisions between water and air molecules.

The half-width of a line with center at $\nu_0 = 914.03$ cm^{-1} was measured by Saidy /199/. The half-widths of three lines were measured /198/, and it was found that these are approximately 50% greater than would follow from the Anderson theory. The half-width of the $1_1 - 1_{-1}$ line ($\nu_0 = 18.6$ cm^{-1}) for collisions between water molecules with one another and with molecules of air, oxygen and nitrogen was determined by Bastin et al. /786/. Its value was found to be equal, respectively, to 0.5 ± 0.2, 0.05 ± 0.01, 0.03 ± 0.02 and 0.06 ± 0.02 cm$^{-1}\cdot$atm^{-1}. The half-widths of lines $4_0 - 5_{-4}$ ($\nu_0 = 10.86$ cm^{-1}), $2_2 - 3_{-2}$ ($\nu_0 = 6.1$ cm^{-1}) and $3_1 - 4_{-3}$ ($\nu_0 = 12.67$ cm^{-1}) for collisions between molecules of water and of air were measured /787—789/; the half-widths of these lines were found to be 0.101, 0.107 ± 0.005 and 0.105 ± 0.004 cm^{-1} in the surface layer of the atmosphere.

6. ABSORPTION FUNCTION IN NARROW REGIONS OF THE VIBRATION-ROTATION AND PURE ROTATIONAL WATER VAPOR SPECTRUM

1. **Direct calculation of the absorption function.** Absorption functions for regions 25 cm^{-1} wide in the pure rotational water vapor spectrum from 100 to 350 cm^{-1} were calculated /209/ utilizing data /170/ on line locations and strengths. The line half-width was assumed to be constant and to range from 0.05 to 0.4 cm^{-1}. Theoretical results on line parameters of the pure rotational

spectrum of H_2O were used /210/ for calculating absorption functions for different frequency intervals. It was found that for sufficiently wide intervals the values of these functions differ little from those calculated using spectral models.

Absorption functions in sections of the vibration-rotation spectrum of water vapor in the infrared were calculated by Calfee /189/ with the aid of experimental data on the locations of line centers. The line strength and half-width were determined by the technique described in the preceding section. Consistent allowance was not made for the vibrational and rotational motions of the H_2O molecule.

The dependence of spectral absorption on wavelength was multiplied by the slit function of the spectrometer in order both to compare theoretical and experimental results, and to study the effect of spectrometer parameters on the recorded spectrum.

Pokrovskii /882/ derived algorithms and calculated the spectral absorption and absorption functions in a narrow spectral section (3795—3836 cm^{-1}) and for an individual line ($\nu_0 = 1677.95$ cm^{-1}). The main advantage of such algorithms is that Doppler as well as dispersion line profiles are taken into account. Data on absorption at different altitudes for individual lines clearly point to the need for taking into account Doppler broadening in the central part of the line starting with altitudes as low as 10 km.

Absorption functions for 8 narrow spectral sections, from 5.3 to 8.5 cm^{-1} wide, in the 20—50μ region for pressures of 1000, 500, 100 and 10 mb and temperatures of 300, 260 and 220°K have been calculated /211/. Analysis of the temperature-dependence of the absorption functions showed that it is primarily determined by the effect of temperature on the line strengths and differs for each specific spectral section. Thus, the value of m in the temperature correction of the transmission function

$$f(T) = (T/T_0)^m$$

ranges from 0.86 to 4.0 for the sections under study.

Direct calculation of absorption function A was used by Timofeev /883/ to estimate the errors when A is calculated by approximate methods. He showed that the greatest error of these methods occurs when the mixing ratio of the absorbing gas increases with altitude.

The feasibility of the direct calculation of transmission functions in a narrow region of the absorption spectra of atmospheric gases at different atmospheric stratification was analyzed /884/, and it was shown that this method is promising for overlapping bands and for solving inverse problems.

Absorption functions for individual segments of the vibration-rotation spectrum of water vapor in the visible and infrared were calculated /780/ using theoretically-obtained data on the location of line centers and on line strengths and half-widths.

2. Calculation of absorption functions by spectrum-model methods. Plass and Stull /212/ computed absorption functions for infrared radiation by H_2O and CO_2 vapor for relatively narrow spectral regions (10—20 cm^{-1} wide) utilizing the statistical, combined and Elsasser models. The parameters were determined from experimental data provided by Howard et al. /213/.

New calculations of absorption functions for narrow regions of the spectrum of the fundamental vibration-rotation bands of water vapor /214/ and CO_2 /215/ were carried out utilizing the quasi-random model postulated by Wyatt et al. /96/ (see Sec. 3). Absorption functions of water vapor were calculated in the 1 to 10μ range for spectral intervals 2.5 cm^{-1} wide in the 997.5—3400 cm^{-1} region, and 20, 50 and 100 cm^{-1} wide in the 1000—10,000 cm^{-1} region. In the last three cases data were obtained in the respective intervals of 10, 25 and 50 cm^{-1}. The absorption functions for the 2.5 cm^{-1} intervals were calculated using results supplied by Benedict et al. /166/. The remaining data were obtained by computing relative line strengths of the vibration-rotation spectrum of water vapor /214/ and reduced to absolute scale by means of experimental results /216/.

The absorption functions for the above spectral intervals were calculated for 15 values of the precipitable water vapor layer ω from 0.001 to 50 cm for 7 values of pressure P: 1.0, 0.5, 0.2, 0.1, 0.05, 0.02 and 0.01 atm, for 3 values of temperature T: 200, 250 and 300°K.

Some of the data obtained were published, with the calculation method, in /217/, which includes a table of transmission functions for regions 100 cm^{-1} wide at T = 300°K and P = 1 atm for 15 values of ω.

Figure 3.3 compares theoretical /214/ and experimental /216/ results for the most intensive and widest vibration-rotation water vapor band: the 6.3μ band. Satisfactory agreement between these two kinds of data is demonstrated by the figure and points to the feasibility of sufficiently accurate calculation of absorption functions in narrow spectral regions using band models, provided their parameters are correctly determined with the aid of experimental data /216, 218/.

We attempted to obtain absorption functions for spectral intervals 0.1μ wide which would be convenient for practical application. The appropriate calculations were conducted /219/ using parameters obtained by Plass and Stull /212/ with the exception of the 3.2, 1.87, 1.38 and 1.1μ absorption bands, for which we used the formula of the statistical model with parameters obtained by Howard et al. /213/:

$$T = \exp\left[\frac{-\frac{\omega}{\omega_0} k\gamma}{\bar{d}\left(\gamma^2 + \frac{\omega}{\omega_0}\frac{k\gamma}{\pi}\right)^{1/2}}\right] = \exp\left[\frac{-\frac{\omega}{\omega_0}\alpha}{\left(1 + \frac{\omega}{\omega_0}\delta\right)^{1/2}}\right], \qquad (3.93)$$

where $\alpha = k/d = 1.97$ and $\delta = k/\pi\gamma = 6.67$ are parameters determined from experimental data, and ω_0 is the precipitable water-vapor layer corresponding to $T = 0.5$. The calculations were conducted for total pressures of 1.0, 0.9, 0.7, 0.5, 0.3, 0.1, 0.05 and 0.01 atm corresponding to altitudes of 1, 3, 5, 10, 16, 21 and 31 km. The values of ω were selected for each altitude in such a manner that a distance from 0.1 to 100 km would be covered at humidities prevailing in the atmosphere.

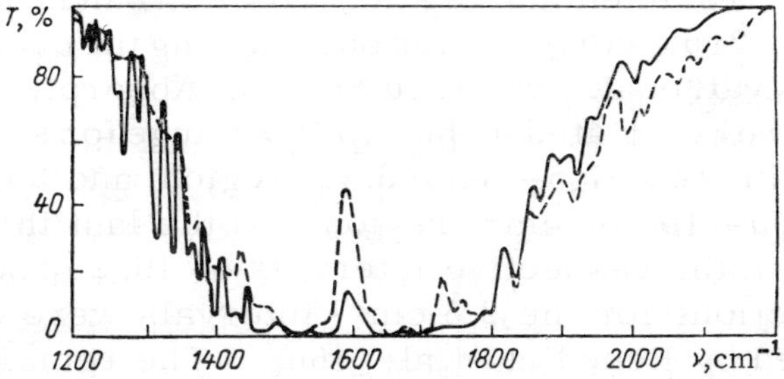

FIGURE 3.3. Transmission functions in narrow sections of the spectrum of the 6.3 μ band of H_2O from experimental /216/ and theoretical /214/ data:

—— - experimental, $P = 805$ mm Hg, $w = 0.0707$ cm;
----- - theoretical, $P = 760$ mm Hg, $w = 0.1$ cm.

All the calculated data were plotted on graphs, from which one can directly find the value of the transmission function for a given interval 0.1μ wide for any P and ω encountered in horizontal layers of the atmosphere at distances from 0.1 to 100 km and altitudes from 0 to 30 km /790/.

3. **Experimental determination of absorption functions.** This has been treated by many investigators, especially by the Howard /112, 213/, Burch /216, 221/ and Golubitskii /791–793/ teams.

The Howard team investigated the spectral absorption of the fundamental vibration-rotation H_2O bands in the infrared spectral region (6.3, 3.2, 2.7, 1.87, 1.38, 1.1 and 0.94μ) and the 3.7μ band of HDO. Processing of experimental data yielded the following empirical relations for the integral absorption of bands:

$$\int_{\nu_1}^{\nu_2} A(\nu)\,d\nu = c\omega^{1/2}(P+P_a)^k, \tag{3.94}$$

$$\int_{\nu_1}^{\nu_2} A(\nu)\,d\nu = C + D\log\omega + K\log(P+P_a), \tag{3.95}$$

where ω and P_a are the precipitable layer (cm) and partial pressure of water vapor (mm Hg); $P = P_{N_2} + P_a$ is the total pressure; c, k, C, D

and K are empirical constants. Expressions (3.94) and (3.95) describe the integral absorption of the band respectively at low and high ω, given by Howard et al. /112/ for each band. For intermediate values of ω the integral absorption of the band is determined by extrapolating data obtained from (3.94) and (3.95).

The results obtained by the Howard team were very rapidly incorporated in various calculations. They were used /222/ to construct a series of tables for radiation transmission by the atmosphere at sea level and at different altitudes over 0.1μ intervals. Transmission in the principal atmospheric "windows" was calculated by Langer /223/. Roach /224/ computed solar radiation absorption in cloudless atmosphere. These data were used extensively in calculating the integral absorption function and the absorption function in narrow intervals occupied by individual bands /225/.

The Burch team used spectral apparatus of higher resolution than that used by Howard et al. /213/. This yielded more reliable data on absorption of radiation in narrow spectral regions and enabled them to study the effects of band broadening on the variation in the total and partial pressures of absorbing and nonabsorbing gases. The absorbing gases under study were H_2O, CO_2, N_2O, CO and CH_4. The problem consisted in determining the quantity $B = C_{a,a}/C_{a,b}$, characterizing the ratio of the ability of gases as regards line broadening by collision of its molecules with one another (self-broadening) to their corresponding ability when colliding with molecules of a foreign gas, which in this case was N_2.

Experiment shows that the value of B varies over the spectrum (Table 3.1). This becomes evident when it is remembered that the diameters of optical collisions vary from one line to another. For example, for the 2224 cm^{-1} N_2O band within the limits from 2250 to 2175 cm^{-1} B ranges from 0.9 to 1.3, i.e., increases by 50%.

Processing of measurements /216/ yielded expressions which approximate satisfactorily experimental data on the integral absorption of bands:

$$\int_{\nu_1}^{\nu_2} A(\nu)\,d\nu = c\,[\omega P_e^a]^b, \tag{3.96}$$

$$\int_{\nu_1}^{\nu_2} A(\nu)\,d\nu = C + D\log[\omega P_e^a], \tag{3.97}$$

where a, b, c, C and D are empirical constants, $P_e = P + (B-1)\,P_a$ is the effective pressure, involving total pressure P, partial pressure P_a and self-broadening coefficient B.

An important result is that these studies apply not only to water vapor and CO_2, but also to minor atmospheric constituents.

Therefore, their role in the absorption of infrared radiation under natural atmospheric conditions can be estimated quantitatively.

TABLE 3.1. Coefficients of self-broadening of absorption lines and bands (B) for H_2O, N_2O, CO_2, CO and CH_4

Gas	Band, cm^{-1}	Source	B
Water vapor H_2O	5332, 3756, 1595	Burch et al. /226/	5 ± 1.5
	500 region	Palmer /227/	$11-6T$ *
	4025.4 line	Vasilevskii and Neporent /195/	6
	Four lines in the 500–600 region	Izatt /228/	3.6–5.5
N_2O	2224, 1285, 1167	Burch et al. /226/	1.12 ± 0.07
	1285	Goody and Wormell /229/	1.27 ± 0.04
	1167	"	1.35 ± 0.07
	2224	Cross and Daniels /230/	1.29
CO_2	3716, 3609, 2350, 961, 875–495	/226/	1.30 ± 0.08
	3716, 3609, 2350, 961, 875–495	Edwards /231/	$2+0.5P_a$ **
CO	2143	Burch et al. /226/	1.02 ± 0.06
	4260	"	1.08 ± 0.06
CH_4	3020	"	1.30 ± 0.08
	1306	"	1.38 ± 0.08

* T is the spectral transmission $T(\nu)$.

** P_a is the partial pressure of CO_2 in the atmosphere.

Table 3.2 lists ranges of ω (cm) attained for different absorption bands in experiments conducted by the Howard and Burch teams.

A new series of measurements of water vapor absorption spectra in the 0.6–5.5μ region /81/ makes it possible to determine the absorption function in very narrow spectral regions in the visible and near infrared. These measurements were carried out at different temperatures up to 1800°K.

Experimental studies were conducted of absorption functions of water vapor for spectral regions from 3 to 20 cm^{-1} wide in the 1–14μ wavelength range /791–793/. The measurements were carried out at different water vapor temperatures and values of ω, equivalent to the value of the latter in the surface layer of a real atmosphere which can be as large as 12 cm. An empirical expression was derived which satisfactorily describes the dependence of the transmission function on ω and P:

$$T_\lambda = \exp\{-\beta_\lambda \omega^{m_\lambda} P^{n_\lambda}\}, \tag{3.98}$$

where β_λ, m_λ and n_λ are empirical parameters, determined for spectral intervals of different width.

TABLE 3.2. Ranges of precipitable water vapor layers for different integral absorption bands of H_2O /213, 216/

Band, μ	Precipitable layer ω, cm	
	Howard /213/	Burch /216/
0.94	0.01—3.85	
1.1	0.01—3.85	
1.38	0.01—3.85	
1.87	0.01—3.85	0.003—0.11
2.7	0.002—2.4	0.003—0.11
3.2	0.002—2.4	0.003—0.11
6.3	0.02—1.2	0.002—0.07

Some authors /785—787/ claim that equation (3.98) describes the measured absorption functions with sufficient accuracy (the relative error does not exceed 6—8% when $\omega \leqslant 12$ cm, $P \leqslant 1$ atm). A comparison /785—787/ shows that available experimental data are in satisfactory agreement with results obtained from equation (3.98).

Kiseleva /794/ proposed empirical formulas which describe approximately the absorption function in the given spectral interval for $0.05 \leqslant A \leqslant 0.9$.

The transmission of a water vapor — nitrogen mixture was measured in the 14—20μ region /232/. A series of curves was obtained for various total and partial pressures. The value of ω ranged from 0.0037 to 1.163 cm, the pressure of nitrogen lay in the 0.00286—0.805 atm range, while the water vapor pressure ranged from $0.264 \cdot 10^{-3}$ to $11.5 \cdot 10^{-3}$ atm. Measurements were found to be in satisfactory agreement with Benedict's calculations /210/.

The transmission functions in the region of the pure rotational spectrum of H_2O were measured /233/ for different values of partial pressures of H_2O and nitrogen, which was used as the foreign gas.

The transmission function for intervals 10 cm^{-1} wide were found in the 500—320 cm^{-1} region, for intervals 5 cm^{-1} wide in the 340—200 cm^{-1} region, and for seven overlapping intervals 50 cm^{-1} wide in the 200—500 cm^{-1} region. For the wide intervals these data are in satisfactory agreement with Benedict's results /210/. As the spectral interval is reduced, the data differ the more from the theoretical values the narrower the interval. The absorption of water vapor and of its mixture with nitrogen in the region of the long-wave transparency window near the 340μ wavelength was measured by Burch and Reisman /234/.

4. Integral strengths of absorption bands. Knowledge of integral strengths of absorption bands is of greater importance in the study of the fine structure of spectra and in the determination of the absorption functions of entire bands, as well as of their component parts. This circumstance explains the quite large number of studies dealing with measurements of the integral strengths of bands of different atmospheric gases, including water vapor. In recent years new measurements were carried out of the strengths of a number of bands when recording spectra at high temperatures.

Goldstein /235/ measured integral strengths of the 1.38, 1.87, 2.7 and 6.3μ bands at temperatures of 200, 400, 600 and 727°C. Emission spectra of water vapor at different temperatures have also been measured /236—239/. The 6.3μ band was recorded at temperatures of 540, 1030, 1535, 1850 and 2200°K. At temperatures in excess of 1800°K the fine structure of the spectrum virtually fades out, facilitating measurements of the integral strength. The spectral emission in the 10—22μ region was recorded at 590, 850, 1040, 1640, 1830 and 2200°K. At 2750°K the transmission spectra of H_2O were measured over a wide range of wavelengths from 1 to 25μ. The integral strengths of various bands were found from the measured emission spectra /236—239/.

Goldman and Oppenheim /240/ determined the integral strengths of the 1.87 and 2.7μ bands from emission spectra of H_2O at 1200°K. The measured values of water vapor absorption at temperatures between 300 and 1100°K are listed by Edwards et al. /241/; these data were used to derive the integral strengths of the 1.38, 1.87 and 6.3μ vibration-rotation bands and the short-wave wing of the rotational band up to the 20μ wavelength. Integral strengths of H_2O bands obtained by various investigators are tabulated in Table 3.3.

5. Absorption by far wings of strong lines. The role of far wings of strong lines in the absorption of radiation in atmospheric "windows" is quite substantial. In determining the absorption function in these windows the continuous absorption due to far wings of strong lines can be the dominating factor. Hence it is very important to possess sufficiently exact data on the absorption coefficients and functions of the continuum. It is also noteworthy that radiation transfer in the atmosphere occurs primarily through spectral intervals of these windows.

TABLE 3.3. Integral strengths of absorption bands of H_2O

Band, μ	Integral strength, $cm^{-2} \cdot atm^{-1}$	Temperature during measurement, °K	Source
6.3	300 ± 20%	273	/236/
	350 ± 15%	2750	/239/
	220 ± 22 *		/201/
	247 ± 10%	273	/235/
2.7	230 ± 15%	2750	/239/
	206 ± 21**	1000	/235/
	177 ± 18†	1000	/235/
	182 ± 36**	530—2200	/242/
	165 ± 33†	530—2200	/242/
	198 ± 20†	1200	/240/
	180 ± 18**	1200	/240/
	192 ± 28	300	/202/
	172 ± 7	300	/203/
1.87	22.3 ± 2.2**	873	/235/
	23.7 ± 1.7**	1000—2200	/243/
	18.0 ± 2.7**	1200	/240/

* — reduced to normal conditions.
** — reduced to a temperature of 300°K.
† — reduced to a temperature of 300°K and corrected by Hirshfeld et al. /203/.

The idea that continuous absorption by far wings of strong lines may play a role in attenuating radiation in these windows was put forward by Elsasser as long ago as 1938. Moller /244/ set the coefficient of continuous absorption in the 8—13μ range arbitrarily at 0.1 per unit air mass, and estimated the value of this absorption for meteorological purposes. The existence of continuous absorption in the 8—13μ window was confirmed in the surface layer /245, 246/ and with solar radiation used as the source /247, 248/. Kastrov represented the absorption coefficient in the 8—13μ interval as the sum of the coefficients of continuous absorption in line with data cited by Adel /247/ and of absorption by weak lines in line with data cited by Feigel'son /249/.

A quantitative experimental study of continuous absorption in the 8—13μ window was carried out by Roach and Goody /250/ in 1958. The measurements were carried out in London and Ascot (a country location) of infrared solar radiation in individual narrow spectral sections with a resolution of about 7 cm^{-1}. It was thus possible to single out from the total attenuation the selective absorption by weak lines contained in this section. The measurements were conducted under cloudless skies. The values obtained for continuous extinction coefficients in Ascot exceed approximately 50-fold their values due to aerosol attenuation in this weather /251/. The attenuation in London exceeded severalfold that in Ascot, which is attributed to the heavy pollution in London. A substantial shortcoming of the study of Roach and Goody /250/ is that they did not measure the attenuation in the visible region, which would have made it possible to single out the aerosol attenuation.

A second quantitative study of continuous absorption in the 8—13μ window was carried out in 1963 by Bignell et al. /252/, who eliminated methodological shortcomings of the first study which included absence of measurements in the visible region and the low resolving power of the spectrometer. The studies were carried out by two methods for 19 frequencies in the 1202—479 cm^{-1} range (8.32—20.9μ): 1) using the sun as source (1202—790 cm^{-1} range); 2) horizontal paths of 200 and 400 m using a Nernst lamp (716—479 cm^{-1} range). Measurements showed that the coefficient of continuous absorption in the wavelength range under study varies from 0.06 to 4.1 $g^{-1} \cdot cm^2$. More detailed data are listed in Table 3.4.

A special study of CO_2 absorption under laboratory conditions showed that outside the limits of the spectral range 790—560 cm^{-1} (12.7—18μ) the effect of CO_2 absorption on the measurement results can be neglected.

Table 3.5 lists values of coefficients of continuous absorption obtained by measuring solar spectra at the High-Mountain Station of the Atmospheric Physics Department of Leningrad State University at El'brus /253/. Data on the coefficient of continuous absorption were also obtained in the 8—25μ range /874, 875/ and in the 8—14μ range /876/.

TABLE 3.4. Values of coefficient k_ν of continuous absorption by far wings of strong water vapor lines for different wavelengths /252/ (Pressure $P=1$ atm, absolute humidity 12 mb)

λ, μ	8.32	8.66	8.86	9.13	10.14	10.40	11.10
k_ν, $g^{-1} \cdot cm^2$	0.074	0.0601	0.0595	0.0625	0.0666	0.0685	0.0925

λ, μ	11.44	11.82	12.02	12.20	12.66	13.96
k_ν, $g^{-1} \cdot cm^2$	0.105	0.114	0.119	0.119	0.158	0.6±0.2

λ, μ	16.38	17.02	17.88	18.83	20.12	20.88
k_ν, $g^{-1} \cdot cm^2$	0.8±0.1	1.4±0.1	1.7±0.1	2.0±0.1	3.0±0.1	4.1±0.1

TABLE 3.5. Values of coefficient k_ν of continuous absorption by far wings of strong water vapor lines in different narrow spectral regions /253/

λ, μ	3.58	3.70	8.15	8.30	8.53	8.76	8.86
k_ν, cm^{-1}	0.07	0.062	0.15	0.12	0.10	0.10	0.095
λ, μ	9.10	10.18	10.40	10.62	10.92	11.11	
k_ν, cm^{-1}	0.090	0.085	0.085	0.090	0.11	0.12	
λ, μ	11.51	11.98	12.18	12.72	12.95	13.95	
k_ν, cm^{-1}	0.15	0.15	0.16	0.21	0.22	0.225	

In conclusion we emphasize the necessity of continuing quantitative experimental studies of the phenomenon at hand. Such studies are important primarily because the theoretical solution of this problem is made difficult by the absence of analytical expressions which describe, with the required degree of accuracy, the far wings of spectral absorption lines.

7. FINE STRUCTURE OF THE ABSORPTION SPECTRUM OF CARBON DIOXIDE

The fine structure of the vibration-rotation spectrum of CO_2 has been studied extensively both experimentally and theoretically. A very detailed experimental analysis was carried out of bands in the 15 and 4.3μ regions. The bands in the 15μ region play an important role in the absorption of long-wave radiation in the atmosphere. The 4.3μ bands are so strong that their absorption can be detected even in the thermosphere. This circumstance was utilized in attempts to reconstruct vertical temperature profiles to high altitudes using corresponding measurements of upward radiation /254/.

The feasibility of utilizing the CO_2 absorption bands in the 4.3, 15.0μ, etc., regions for solving inverse problems of satellite meteorology has resulted in the recent appearance of many publications in which the fine structure of all the fundamental bands has been investigated. There follows a brief review of various investigations of individual absorption bands.

1. **Bands in the 15.0μ region.** The detailed structure of the band spectrum in this region was first studied theoretically by Kaplan /255/ and Yamamoto and Sasamory /256/, who obtained data on the location

of lines and on their strengths. The calculations were conducted for the principal isotope without consistent consideration of interaction between the vibrational and rotational molecular motions and of Coriolis interaction. The relative strengths of band lines in the region under study were calculated elsewhere /212, 215/.

The most exact experimental data on the location of line centers in the region of the 15.0μ bands were obtained /257/ with very high spectral resolution, virtually corresponding to the limit for echelette interferometer equipment. This enabled the location of line centers to be determined to within several thousandths of cm^{-1}. Comparison of experimental and theoretical data showed, for the majority of lines, agreement to within 0.01 cm^{-1}, with the maximum discrepancy not exceeding 0.027 cm^{-1}.

The half-width γ of lines in these bands has been determined experimentally /258, 61/. Kaplan and Eggers studied only strong lines, for which it was found that $\gamma = 0.064$ cm^{-1}. However, Kaplan subsequently revised the results of this study and found that under standard conditions $\gamma = 0.0075$ cm^{-1}. Madden /61/ investigated the γ on the rotational quantum number, which varies for each of the bands in the region under study. The half-widths and strengths of a number of lines of the P and R branches of the 10^00-00^01 band when utilizing a CO_2-N_2 laser as radiation source were found in /795–797/. In this case the centers of the emission and absorption lines coincide, the expressions for the Doppler and Lorentz line profiles simplify and can be used for sufficiently accurate determination of the parameters of the corresponding lines.

2. Bands in the 4.3μ region. The positions of line centers and strengths in these bands have been studied extensively by many investigators /259, 260, 78, 212, 215, 261, 262, 79, 182/. Nielsen and Yao /259/ analyzed ν_3 bands for the $C^{12}O_2^{16}$ and $C^{13}O_2^{16}$ isotopes, while Benedict et al. /260/ studied the R branches of bands. Courtoy /261/ obtained the most detailed recording of the spectrum in the 4.3μ region. Line strengths in eight isotopes of CO_2 are listed in /212, 215/. The relative line strengths for transitions from the 00^00 and 01^10 molecular states were obtained /262/, while the positions and strengths of about 300 lines were calculated by Kyle et al. /79/. These calculations showed that 80 of the strongest lines make a 97.1% contribution to the integral band strength. Table 3.6 lists data on these bands /79/.

The half-widths γ of lines in the 4.3μ band region have not been sufficiently investigated. When computing absorption functions and coefficients, γ is usually assumed to be constant and equal to 0.066 cm^{-1} under standard conditions.

3. Bands in the 2.7 and 2.05μ regions. The positions of line centers and line strengths in the regions of the 2.7 and 2.05μ bands have been studied in most detail (compared to other bands) by Calfee and Benedict /263/, who calculated the locations of centers and the strengths of 2823 lines in the 2.05μ band region and of 4790 lines in the region of the 2.7μ band. All the molecular parameters used in calculations of each band were tabulated in detail /263/. The line positions were calculated with allowance for interaction between vibrational and rotational molecular motions, induced by centrifugal distortion and Coriolis forces. Line strengths were calculated without consistent allowance for interaction between vibrations and rotations. The calculations were based on the scheme for calculating line strengths described in Sec. 5, utilizing band strength data obtained by the Burch team.

TABLE 3.6. Data on CO_2 bands in the 4.3μ region /79/

Isotope	State	Allowed values of j	Band center, cm^{-1}	Maximum value of j
$C^{12}O_2^{16}$	$0\ 0^0\ 0$	Even	2349.5	70
$C^{13}O_2^{16}$	$0{,}1^{\pm1}0$	Even and odd	2339.2	54
$C^{13}O_2^{16}$	$0\ 0^0\ 0$	Even	2283.7	70
$C^{13}O_2^{16}$	$0\ 1^{\pm1}\ 0$	Even and odd	2273.4	54
$C^{12}O^{18}O^{16}$	$0\ 0^0\ 0$	Even and odd	2331.8	48
$C^{13}O^{18}O^{16}$	$0\ 0^0\ 0$	Even and odd	2266.0	48

The table of line centers and intensities /263/ includes all the rotational molecular levels for the rotational quantum number $j \leqslant 100$. This means that the table contains virtually all lines which appear in measurements in the terrestrial atmosphere. It was stated by Williams /264/ in reviewing /263/ that he is ready to offer a reward to anyone who would discover lines not tabulated in /263/. Individual line strengths in the above bands were calculated elsewhere /212, 215/.

The fine structure of the spectrum and of its individual elements in the region of the 2.7 and 2.05μ bands was studied in depth under experimental conditions. A bibliography of these studies can be found in /798, 3, 265/. The most detailed experimental study of the position of lines in the 2.7μ region was carried out in /265/, another paper /265/, which gives measurements of the positions of centers in 10 bands of the $C^{12}O_2^{16}$ isotope, in one band of the $C^{13}O_2^{16}$ isotope, and in two bands of the $C^{16}O^{16}O^{18}$ isotope. All the measurements were made to within 0.005 cm^{-1}.

The half-widths of several tens of absorption lines of the $4\nu_2 + \nu_3$ band (2.05μ) were measured by Vasilevskii et al. /266/. The measurements were carried out with pure CO_2 as well as with CO_2+N_2. The strengths and half-widths were determined of almost all the lines of the *P* and *R* branches of the band. In pure CO_2 the line half-width γ ranges from 0.126 to 0.06 $cm^{-1}\cdot atm^{-1}$ for rotational quantum numbers from 4 to 50—56. In the mixture of gases the value of γ ranges from 0.122 to 0.09 $cm^{-1}\cdot atm^{-1}$.

4. Weak absorption bands. In addition to the previously described most intensive CO_2 bands there exist a number of quite weak bands, the fine structure of which can be determined either theoretically or in the laboratory using large masses of CO_2. The bands lie in the spectral regions near 10.4, 9.4, 5.2, 4.8, 1.6 and 1.4μ, as well as intermediate between them and strong absorption bands, or in atmospheric windows, including the visible part of the spectrum.

The fine structure of weak CO_2 bands has not been sufficiently studied because their role in atmospheric processes is insignificant, and their study is extremely difficult. A bibliography of investigations is given in /1, 3, 267/.

To summarize, we can draw the following conclusions from theoretical and experimental studies of the fine structure of the absorption spectrum of CO_2. Most complete data on the position of line centers were obtained experimentally. Here the positions of centers for the majority of lines in all the fundamental bands were found between 0.005 and 0.1 cm^{-1}. The positions of line centers were also calculated for the overwhelming majority of lines. In all cases no consistent allowance was made for interaction between molecular vibrations and rotations for individual lines, for which the corresponding error has yet to be estimated.

The greatest quantity of data on line strengths was obtained theoretically, and only for a very limited number of lines were these data obtained experimentally. The line strengths as well as their locations were calculated without consistent allowance for interaction between molecular vibrations and rotations.

There is available a limited quantity of theoretical as well as experimental data on the half-widths of lines in all the vibration-rotation CO_2 bands. In all calculations the half-width of lines is assumed to be constant within the limits of a given band which, strictly speaking, is incorrect. The inaccuracies in determining all the three line parameters are most evident in the value of the absorption coefficient in the central part of the lines.

8. ABSORPTION FUNCTIONS IN DIFFERENT REGIONS OF THE CO_2 SPECTRUM

The absorption functions in different regions of the CO_2 spectrum, as in the case of water vapor, are calculated by three methods examined below.

1. **Direct calculation of absorption functions.** A detailed calculation of absorption functions in the 12 to 18μ range was carried out by Drayson /62/. A total of about 2000 lines were used in calculations, where it was assumed that their half-width at 1 atm and 298°K is constant and equal to 0.064 cm^{-1}. The dependence of the line half-width on the rotational quantum number was neglected. On the basis of data on the strength center position and half-width of lines the absorption coefficient was calculated and was subsequently used to compute absorption functions.

In calculating the absorption coefficient the term γ^2 in the expression for the line profile

$$k(\nu)=\frac{S}{\pi}\,\frac{\gamma}{(\nu-\nu_0)^2+\gamma^2}$$

was omitted for frequencies satisfying $(\nu-\nu_0)>7.0$ cm^{-1}. The resulting error did not exceed 10^{-4} of the value of $k(\nu)$ at all pressures. The absorption band was subdivided into small intervals with minimum width 0.001 cm^{-1} in central regions of lines and 0.1 cm^{-1} in their wings. The absorption functions were obtained for regions 0.1 cm^{-1} wide and then averaged for 5 cm^{-1} wide intervals.

The computation results were compared with experimental data of the Burch team and Yamomoto's results. Satisfactory agreement between these data was obtained for pressures of 0.2 and 1.0 atm. For low pressures marked discrepancies were found for the central parts of bands which, according to Drayson /62/, can be attributed to inaccurate determination of the integral strength of the band (assumed by Drayson /62/ to equal194 cm^{-1}) or neglect of the dependence of the line half-width on the rotational quantum number. It should be noted however that at very low pressures the data of Drayson /62/ and of Yamomoto differ so markedly that it is difficult to attribute this discrepancy to the above factors. For example, in the spectral region with center near 15.1μ for $P=1$ mb this discrepancy exceeds two orders of magnitude of the absorption function /268/. As of now it is difficult to assess the actual causes of such high discrepancies.

2. **Calculation of absorption functions from spectrum model formulas.** Calculations of absorption functions in narrow regions of vibration-rotation CO_2 bands have been carried out /212, 215, 270/. Plass and Stull obtained data on absorption functions in regions 10—20 cm^{-1} wide near the 15, 4.3, 2.7 and 2.0μ bands. Stull et al. (215/ employed the quasi-random model to compute absorption functions in regions 2.5 cm^{-1} wide in the 500—1000 cm^{-1} (1—20μ) range for three temperatures (200, 250, and 300°K) for the same seven pressure values used in water vapor calculations: 1.0, 0.5, 0.2, 0.1, 0.05, 0.02 and 0.01 atm, and for fifteen values of the CO_2 mass from

0.02 to 10,000 atm·cm. Note that the bottom kilometer of the surface layer of the atmosphere has a CO_2 mass of approximately 30 atm·cm. Stull et al. /215/ also derived absorption functions for the above values of parameters in intervals of 20, 50 and 100 cm^{-1} with centers, respectively, in 10, 25 and 50 cm^{-1} intervals. Some of the data in /215/ are published in /269/. Comparison of theoretical /215/ and experimental /216/ results points to their satisfactory agreement.

To obtain results convenient for practical applications, we used the numerical values of parameters obtained by Plass and Stull /212/ for calculating absorption functions of CO_2 for spectral intervals 0.1μ wide. The absorption functions in spectral regions of this width within the limits of the 15, 4.3, 2.7 and 2.0μ bands were carried out for pressures of 1.0, 0.9, 0.7, 0.5, 0.3, 0.1, 0.05, and 0.01 atm and a CO_2 mass corresponding to horizontal radiation propagation paths from 0.1 to 100 km at every altitude. These results have been plotted /1, 790/.

Gray /271/ reports on calculations of absorption functions in the region of 9.4 and 10.4μ cm^{-2} bands. The line half-width was taken as $\gamma = 0.07$ cm^{-1}, while the integral strength of bands was assumed to be 0.016 and 0.030 cm^{-1}·atm^{-1} at 300°K. Comparison of theoretical and experimental results for temperatures from 296 to 555°K showed satisfactory agreement.

The absorption functions of CO_2, N_2O and CO mixtures in the 4.2—4.8μ spectral region were calculated by Gray and McClatchey /272/, whose results are listed for different altitudes of the homogeneous atmosphere. Initially the calculations were made for 99 vibrational levels. However, it was subsequently found that no significant differences result if only 27 levels are included.

3. Experimental determination of absorption functions. The most detailed data on absorption functions in regions of the fundamental vibration-rotation CO_2 bands were obtained by the Howard /213, 220/, Burch /216, 273/ and Golubitskii /792, 868/ teams.

The first team obtained a series of recordings of absorption spectra of CO_2—N_2 mixtures in regions of absorption bands near 1.4, 1.6, 2.0, 2.7, 4.3, 4.8, 5.2 and 15.0μ at different partial pressures of the constituents. The measurements were carried out over a wide range of absorbing mass of CO_2. The absorbing masses of the second team in recording spectra of the 2.7, 4.3, and 15.0μ bands were much smaller than in /213, 220/, but the weak 9.4 and 10.4μ bands were studied with large masses, from 48 to 11,200 atm·cm. We recall that the CO_2 mass in the kilometer horizontal thickness of the surface air is 30 atm·cm.

As in the study of water vapor, processing of measured absorption spectra in both groups of studies made it possible to find empirical formulas for the integral absorption of entire bands. These relationships were found to be the same as for water vapor (see (3.94)—(3.97)). The empirical constants were found seperately for each band.

Experimental studies of absorption functions in CO_2 spectral ranges from 5 to 15 cm^{-1} wide carried out by the Golubitskii team /792, 867, 868/, yielded an empirical formula for the absorption functions at different absorbing masses of CO_2 and air pressures.

A detailed study of CO_2 absorption in the 0.6—5.5μ spectral region over a wide pressure range (from several thousandths mm Hg to 15 atm) was carried out /81/ and 30 new absorption bands were identified. Integral strengths were determined for the most important bands. Experimental data on the integral absorption of different CO_2 bands at elevated temperatures were obtained by Edwards et al. /231, 274, 275/.

The integral strength of the 15μ CO_2 band was determined most exactly in /276/, where it was found to be equal to 200± 10 $cm^{-2}\cdot atm^{-1}$. Comparison with other results was also carried out there. In five other studies values between 170± 18 and 218± 5 $cm^{-2}\cdot atm^{-1}$ were obtained for this quantity. However, in two other studies the integral strengths of the band were found to be 330± 90 and 146± 18 $cm^{-2}\cdot atm^{-1}$.

9. FINE STRUCTURE AND ABSORPTION FUNCTIONS IN THE OZONE SPECTRUM

The fine structure of the vibration-rotation and pure rotational spectrum of ozone was studied theoretically and experimentally, a bibliography of which is cited in /1, 3, 277/. The most detailed analysis of the fine structure of the fundamental overlapping O_3 bands (9.1 and 9.6μ) was carried out by Clough and Kneizys /277, 278/. In spectral regions 995—1070 cm^{-1} and 1065—1155 cm^{-1} these investigators calculated the positions of the centers and strengths of more than 1200 and 1800 absorption lines, respectively. The calculations made allowance for Coriolis interaction between the lines of the ν_1 and ν_3 bands. The results of these calculations were used for finding the spectral absorption spectrum of O_3 and for its subsequent comparison with corresponding experimental curves of White et al. /279/, recorded with a resolution of 0.08—0.06 cm^{-1}. In calculating line strengths they used the value of the integral strength of the ν_3 band, found by Walshaw /280/ to be 455 $cm^{-2}\cdot atm^{-1}$ under standard conditions.

The structure of the ν_2 ozone band near 14μ was measured by Nexsen /281/, who first attempted to analyze this band. An experimental study of the structure of the $\nu_1 + \nu_3$ combination bands with center near 4.7μ was carried out under natural and laboratory conditions /282/. These results were used /283, 284/ when analyzing the fine structure of this band.

The fine structure of the pure rotational spectrum of O_3 was studied by Gora /285/. New data on the high-resolution spectrum of O_3 in the 80—92 cm^{-1} region were obtained by Gebbie et al. /286/. The strengths and locations of line centers in the 4.7μ band, as well as absorption functions in 10 cm^{-1} wide regions of the 9.6 and 4.7μ bands, were calculated elsewhere /889/.

The most complete experimental data on the absorption function of the 9.6μ absorption band of O_3 were obtained by Walshaw /287/. The measurements were carried out under laboratory conditions with the precipitable layer of O_3 ranging from 0 to 1.5 cm and at pressures from 0.5 to 760 mm Hg. Gal'tsev /890/ used precipitable layers of O_3 up to 4.7 cm, and eight pressure values from 10 to 400 mb and more. Data processing yielded quite a complex empirical formula /287/ for the absorption function of the entire band; Gal'tsev /890/ selected a simpler formula for absorption functions in three spectral intervals into which the band was subdivided: 990—1100 (band center), 960—990 and 1080—1110 cm^{-1} (band wings).

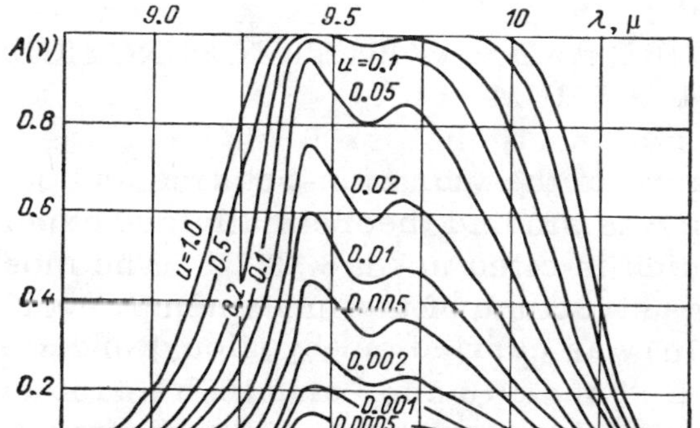

FIGURE 3.4. Spectral absorption curves in the 9.65μ O_3 band for different values of the precipitable layer u /289/.

Walshaw /287/ also obtained quantitative data for absorption in the 4.7μ ozone band. Altshuller /288/ used data cited in /280/ on absorption of infrared radiation by the 9.6 and 4.7μ bands, and constructed graphs which can be used to estimate the absorption of radiation by these bands under atmospheric conditions. Using the same data cited in /280/ very useful curves were constructed /289/ for characterizing the spectral transmittance of this band with low resolution for various values of the precipitable layer of ozone (Figure 3.4).

10. FINE STRUCTURE AND ABSORPTION FUNCTIONS OF THE OXYGEN SPECTRUM

The position of lines of the fine structure of the red system of the absorption spectrum bands of atmospheric oxygen was studied by Babcock and Herzberg /799/. The line strength distribution in the 0.76μ band (0—0 transition) and the half-widths of lines were measured in /800/, where the oscillator strength of the band was also determined. The average values of half-widths of lines during $O_2 \leftrightarrow O_2$ collisions and during collisions between oxygen and air molecules were found to be, respectively, 0.085 and 0.077 $cm^{-1} \cdot atm^{-1}$. No substantial dependence of the line half-width on rotational quantum number was found /800/.

Using data on the oscillator strength of the band and the average line strength, as well as data on line positions /799/ and their strength distribution in the band /801/, Adiks and Dianov-Klokov /800/ calculated absorption functions of the band in the surface layer of the atmosphere.

The transmission function of the 0.76μ band was also calculated /802/ using data on the integral band absorption. It has been noted /800/ that in order to obtain agreement between theoretical and experimental data Wark and Mercer /802/ had to modify the value of the integral band strength. A great deal of attention has recently been paid to the feasibility of utilizing the 0.76μ oxygen absorption band for determining the cloud top height from meteorological satellites /802—804/.

The dependence of the equivalent widths S of tellurium O_2 lines in the 0.69μ band on the temperature of the atmosphere was studied by Kozhevnikov et al. /891/. These studies showed that as the temperature rises the values of S change differently, depending on the rotational levels from which they start.

11. FINE STRUCTURE AND ABSORPTION FUNCTIONS OF SPECTRA OF MINOR ATMOSPHERIC CONSTITUENTS

1. Methane. The vibration-rotation absorption spectra of methane have been investigated in more than twenty studies, the bibliography of

which up to 1964 is given by Gray and Penner /237/. The ν_3 band was investigated by Allen and Plyler /290/. A high-resolution spectrum of the ν_2 band was obtained by Herranz et al. /292/. The appearance in the spectrum of the ν_2 forbidden band is due to vibrational-rotation interaction. In addition to analyzing the line positions, Herranz et al. /292/ found their relative strengths with the aid of known expressions /291/.

TABLE 3.7. Values of relative broadening coefficients F^b and relative optical collision diameters $\frac{d_{a,a}}{d_{a,b}}$ for different gases /216/

Absorbing gas	Band, cm^{-1}	P_a, mm Hg	P_b, mm Hg	Line-broadening gas	F^b	$\frac{d_{a,a}}{d_{a,b}}$
N_2O	2224	10	15—990	O_2	0.83	0.93
				CH_4	1.08	0.94
	1285	100	80—900	O_2	0.72	0.87
				CO_2	1.17	1.14
				CO	0.97	0.99
CO	2143	100	100—2900	CH_4	1.12	0.96
CO_2	2350	50	30—650	O_2	0.81	0.92
CH_4	3020	50	50—2450	CO_2	1.25	1.15

Data on the absorption functions of the 3.33, 6.45, 7.65μ CH_4 bands, of the 3.9, 4.05, 4.5, 7.78, 8.55, 14.5, 17.0μ N_2O bands, and the 2.32 and 4.67μ CO bands were obtained by the Burch team /216, 293, 294/. Measurements were carried out with pure gases as well as in mixtures with nitrogen and other foreign gases. The N_2O, CH_4 and CO concentrations ranged over limits much greater than those under actual atmospheric conditions.

As a result of these studies data were obtained not only on the integral absorption, but on coefficients B, characterizing the relative effectiveness of collisions between molecules of the absorbing gas among themselves and with those of the foreign gas (N_2) (see Table 3.1). In addition Burch et al. /216/ determined values of the relative diameters $d_{a,a}/d_{a,b}$ of optical collisions (subscripts a and b pertain, respectively, to the absorbing and foreign gases) and of the relative broadening coefficients $F^b = P_{N_2}/P_b$, where P_b is the gas pressure needed to produce the same line-broadening effect as that of N_2 (Table 3.7).

Data on the integral absorption of bands of test gases, obtained by the Burch team, make it possible to determine the absorption functions of these bands and to assess quantitatively the contribution of each band to the attenuation of infrared radiation from continuous-spectrum sources at different temperatures.

Experimental studies of absorption functions in narrow regions of the methane spectrum were carried out by the Golubitskii team /793, 869/. Processing of these results yielded parameters contained in the empirical formula for absorption functions. Relationships for the integral absorption in the 7.6 and 3.3μ CH_4 bands were obtained by Edwards and Menard /274/ from models. The parameters of these relationships were obtained for 300, 555 and 833°K.

2. Nitrous oxide. The positions of centers of the lines of the fine structure of the vibration-rotation and pure rotational N_2O spectra have now been explored in great detail. The corresponding bibliographies are listed in /3, 296/.

The strengths and half-widths of the absorption lines of N_2O have not been studied in sufficient detail. The harmonic oscillator approximation was used to compute the relative strengths of lines of this band (4.6μ) for the 200—300°K temperature range /297/. The absolute values of line strengths were determined using 1850 $cm^{-2} \cdot atm^{-1}$ as the integral band strength at 300°K. Using data of line strengths and assuming that the half-widths of the lines are the same and equal to 0.05 cm^{-1} under standard conditions, Gray /297/ calculated the absorption function for narrow regions of the band spectrum, using the statistical model.

The strengths and half-widths of more than 70 lines in the 5.3μ N_2O band were measured /298/ and it was found that the half-widths are a function of the rotational quantum number. An approximate method, based on formulas for the absorption function in the Elsasser model and experimental data on the absorption function with low resolution, was employed /299/ to determine the half-widths of lines. It was found that for a part of the 4.5μ band the half-width is $\gamma^0 = 0.08$ $cm^{-1} \cdot atm^{-1}$ at 300°K when nitrogen is used as the broadening gas.

The strengths of individual lines in the region of the 3.9μ N_2O band were found by Rank et al. /300/. Various molecular and force constants of this molecule have also been found /301, 302/.

The integral strengths of N_2O bands were investigated by the Burch team /216/ and elsewhere /303, 304/. Absorption functions in narrow regions of the fundamental N_2O bands were measured by the Golubitskii team /869, 870/. Absorption functions of the overlapping N_2O and CO absorption bands in the 2200 cm^{-1} region were measured by Hathaway et al. /305/.

3. Carbon monoxide. Due to their relative simplicity, the vibration-rotation and pure rotational spectra of CO have been studied in great detail. A bibliography of these studies can be found in /3, 66/, where detailed data are presented on the molecular and force constants of the molecule.

The strength and half-widths of lines of vibration-rotation CO bands were studied experimentally /307/. The half-widths and strengths of individual lines of the fundamental 4.6μ were found by Shaw and France /308/. The half-widths of lines of the pure rotational CO spectrum in the 100—3 cm^{-1} range were also measured /309/. It was found /307, 308/ that the half-width γ of a line depends substantially on the rotational quantum number j (for variation of the latter from 1 to 25, γ decreases from 0.1 to 0.05 cm^{-1} under standard conditions). The fine structure of the spectrum for "hot" CO molecules, consisting of lines with high vibrational quantum numbers, was studied by Weinberg et al. /310/.

The integral absorption of CO bands at different partial pressures of CO and of foreign gases were measured by the Burch team /216/, and by Gryvnak and Shaw /311/. An experimental study of the integral absorption of the fundamental vibration-rotation CO band (4.6μ) at 300, 273 and 198°K over a wide range of pressure and absorbing masses was carried out by Hathaway et al. /312/. The integral absorption of the fundamental band (1900—2450 cm^{-1} interval) at 300, 600, 900, 1200 and 1500°K for pressures from 0.25 to 3 atm for a path length of 1, 5, 10 and 20 cm in a gas-filled container was measured in /313/. In this study the integral absorption of the fundamental band and its first overtone (4050—4450 cm^{-1} range) was computed for different temperatures and pressures.

TABLE 3.8. Absorption bands of atmospheric gases for which empirical coefficients were obtained in (3.98) for calculating the absorption functions in /791—793, 867—870/

Gas	H_2O										
Band, μ	0.94	1.1	1.38	1.87	2.7	6.3	8—14	14—40*	Continuum 1—14		
Gas	CO_2										
Band, μ	1.4*	1.6*	2.0	2.7	4.3	4.8	5.2	9.6	10.4	15.0	
Gas	N_2O										
Band, μ	2.12	2.3	2.94	3.04	3.57	3.86	4.0	4.5	7.8	8.6	15*
Gas	CH_4		CO		O_3		O_2	N_2			
Band, μ	2.3	3.3	7.6	4.7	9.6*	4.75*	1.26*	4.3*			

* The empirical coefficients were obtained by processing other data.

Shaw and Houghton /314/ calculated the integral absorption of the 4.6 μ band on the basis of data on the strength, half-width and shape of lines /307/. Edwards /306/ calculated the integral absorption of this band. Absorption functions in narrow regions of the CO spectrum were measured by Moskalenko /793/.

To complete our examination of absorption functions of atmospheric gases, we present the main data (Table 3.8) obtained by the Golubitskii team /791—793, 867—870/. A distinguishing feature of this series of studies consists in the fact that it yielded results convenient for practical utilization, ensuring rapid determination of absorption functions in relatively narrow spectral regions at different pressures and absorbing masses of the gases. A particularly valuable aspect of these studies is that the measurements covered wide ranges of absorbing gas masses. Note that the absorption of N_2 in the 4.3μ region is due to collisions /871, 872/. On the basis of theoretical data on absorption functions in narrow spectral regions presented in Table 3.8, nomograms were constructed /873/ for calculating the spectral transparency of the atmosphere in the 2.8—5.6μ region.

12. ABSORPTION OF OVERLAPPING LINES AND BANDS

In a real atmosphere one frequently deals with absorption of overlapping lines and bands of the same gas and of different gases. For this reason, in determining absorption coefficients and functions, it is necessary to take into account the overlapping of lines. As to the absorption coefficient, this problem is solved quite simply theoretically for known parameters of lines contributing to absorption of radiation with given frequency ν. In this case the resulting absorption coefficient is calculated as the sum of the absorption coefficients of all the lines.

At present there is insufficient data on the strength and half-width of absorption lines of atmospheric gases. It is therefore impossible in the majority of cases to determine theoretically the absorption coefficients of monochromatic radiation. These quantities can be determined experimentally only when using lasers as radiation sources.

The transmission function of the $\Delta\nu = \nu_2 - \nu_1$ section of a combination band consisting of overlapping bands of n gases has the form

$$T = \frac{1}{\Delta\nu}\int_{\nu_1}^{\nu_2} T(\nu)\,d\nu = \frac{1}{\Delta\nu}\int_{\nu_1}^{\nu_2}\prod_{i=n} T_i(\nu)\,d\nu, \qquad (3.99)$$

where $T(\nu)$ is the spectral transmission of a gas mixture for radiation with frequency ν; $T_i(\nu)$ is the same for the ith gas.

For the most prevalent case of overlapping of the bands of two gases in the atmosphere, expression (3.99) yields the following expression for the absorption function in section $\Delta\nu$:

$$A=\frac{1}{\Delta\nu}\int_{\nu_1}^{\nu_2} A(\nu)\,d\nu=\frac{1}{\Delta\nu}\int_{\nu_1}^{\nu_2} A_1(\nu)\,d\nu+$$
$$+\frac{1}{\Delta\nu}\int_{\nu_1}^{\nu_2} A_2(\nu)\,d\nu-\frac{1}{\Delta\nu}\int_{\nu_1}^{\nu_2} A_1(\nu)\,A_2(\nu)\,d\nu, \qquad (3.100)$$

where $A_1(\nu)$, $A_2(\nu)$ and $A(\nu)$ is the spectral absorption of radiation with frequency ν for the first and second gases and their mixture, respectively.

In the case at hand the absorption function of a combination band is thus smaller than the sum of absorption functions of the individual gases by the amount $A' = \frac{1}{\Delta\nu}\int_{\nu_1}^{\nu_2} A_1(\nu)\,A_2(\nu)\,d\nu$, due to overlapping of the bands of the two gases. Data for determining the spectral absorption of atmospheric gases, required when calculating A', has only recently begun to appear. The first studies devoted to determination of the absorption function for overlapping bands were experimental.

Before presenting the results of individual studies, we shall write down an approximate expression for the transmission of a two-gas mixture, which follows from (3.99):

$$T=T_1\cdot T_2, \qquad (3.101)$$

where T_1 and T_2 are transmission functions for the first and second gases in interval $\Delta\nu$, for which the value of T is found.

Already in the first series of his studies Howard attempted to find integral absorption functions of a water vapor — CO_2 mixture in the region of 2.7μ bands from known data on these functions for H_2O and CO_2 separately. Processing of the corresponding measurements showed that the transmission function of section $\Delta\nu$ of a band consisting of H_2O and CO_2 lines can (with an accuracy sufficient for practical purposes) be expressed as the product of the transmission functions of the H_2O and CO_2 bands taken separately.

The spectral transmission of the CO, N_2O, CH_4 bands and of their mixture was studied experimentally /305, 315/. For the CO and N_2O bands in the 2200 cm^{-1} region and for CH_4 and N_2O bands in the 1300 cm^{-1} region limits of applicability of (3.101) were found for describing the absorption function of the respective gas mixtures. These experimental data are compared with calculations by Gray and McClatchey /316/, who also computed absorption functions in narrow regions of the spectrum of the CO_2—N_2O — CO mixture from 4.2 to 4.8μ using formulas of the Elsasser model /272/.

Absorption in the overlapping bands of water vapor and CO_2 near 2.7 and 15μ at temperatures above 800°K was calculated by Penner and Varanasi /317/. Expressions for the equivalent width of two overlapping lines were obtained in /318—320/. The method proposed in /318/ was extended by Sakai /321/ to the case of overlapping bands.

13. ABSORPTION OF RADIATION FROM SELECTIVE SOURCES

Exact calculation of radiative heat transfer in the atmosphere requires knowledge of absorption functions for sources with continuous as well as selective emission spectra. In accordance with the Kirchhoff law all gases absorbing radiation in the visible and infrared have corresponding electronic, vibration-rotation and pure rotational emission spectra. In calculating radiative heat transfer allowance must be made for the solar and terrestrial radiation fluxes passing through the respective regions of the atmosphere, as well as the radiation fluxes emitted by the atmosphere itself. The last problem requires knowing the laws governing the absorption of selective emission in a selectively absorbing medium.

The second group of problems, for which the absorption of non-selective sources in the atmosphere must be known, is associated with the development of missiles and jet aircraft. The high flare temperature induces the highest energy transitions in molecules and radiation of "hot" bands of atmospheric gases. The study of emission spectra of flares at different distances from them involves knowledge of the absorption of the emission of these sources.

Depending on the type of emitting and absorbing gas and on the temperature of the former or latter gas, we deal with one of the following cases or their combination: 1) the centers of emitting and absorbing lines coincide; 2) the centers of emitting and absorbing lines are shifted relative to one another; 3) the absorbing and emitting lines do not coincide at all.

The first case is attained when both emission and absorption are due to the same gas and at the same temperature. In the second case the emitting and absorbing gas is the same, but the emission conditions differ from those of absorption. The third case can be encountered when one gas emits while another absorbs, or when the given gas emits bands which, under atmospheric conditions, are virtually not encountered in absorption. The latter refers primarily to the "hottest" absorption bands. The absorption function for emission from a selective source in the spectral interval $\Delta\nu = \nu_2 - \nu_1$ can be expressed in the form /322/

$$A = \frac{\int\limits_{\nu_1}^{\nu_2} [1 - \exp(-\sum_i k_{e_i}\omega_e)][1 - \exp(-\sum_i k_{a_i}\omega_a)]\, d\nu}{\int\limits_{\nu_1}^{\nu_2} [1 - \exp(-\sum_i k_{e_i}\omega_e)]\, d\nu}, \qquad (3.102)$$

where k_{e_i} and k_{a_i} are the absorption coefficients of the ith spectral line in the emitting and absorbing gases, while ω_e and ω_a are the precipitable layers of the emitting and absorbing gases.

For direct calculation of exact quantitative data on absorption coefficients and functions for selective emission sources one requires all the elements of the fine structure of both the emission and absorption spectra. At present such information is scarce. Hence the absorption functions are found either from spectrum models or experimentally.

Plass /322, 323/ derived a series of expressions for selective emission source functions pertaining to different cases of line overlapping in the emitter as well as in the absorber (nonoverlapping lines, overlapping lines described by the Elsasser model, overlapping lines described by the statistical model). Cases were also considered when the pressure, temperature and concentration of the absorbing gas vary along the beam path /322/.

Numerical calculations /322/ showed that when the same lines participate in emission and absorption, the absorption function of the flame is much higher than for a source with a continuous emission spectrum.

The absorption of selective emission by carbon dioxide was also calculated /324, 325/. The absorption of emissions of hot H_2O and CO_2 by these gases at standard temperature was examined experimentally in /81/. Curcio and Buttrey /326/ studied experimentally the absorption function of CO_2 for a flame at 2800°K, different pressures and emission path lengths in a cell at 24°C. Relatively high transmittance was detected in the 2.5—3.5 and 4.0—5.5μ spectral regions /326/. Thus, in a CO_2 layer equivalent by the magnitude of its absorbing mass to the 50-km surface layer of the atmosphere, the transmission function in the 4.0—5.5μ region was found to be 0.2.

14. ABSORPTION OF LASER RADIATION IN THE ATMOSPHERE

The high monochromaticity, sharpness of focusing, coherence, feasibility of obtaining ultrashort-period pulses and attaining gigantic emitting powers of lasers have led to a number of specific features in the propagation of laser radiation through the atmosphere, compared with the propagation of radiation from other sources.

The half-width of the radiation lines of gas lasers is much smaller than the half-width of lines in absorption spectra of atmospheric gases. For a quantitative estimate of the absorption of laser radiation in the atmosphere one therefore requires (with a very high degree of accuracy) the positions, strengths and shapes of lines in spectra of atmospheric gases. All quantitative data on the absorption spectra of atmospheric gases obtained before the advent of lasers were found to be unsuitable when estimating the absorption of laser radiation in the atmosphere /1/.

The interaction of a spatially coherent light beam with aerosol results in modifications of the light intensity in the direction of propagation, as a result of direct attenuation of radiation according to Bouguer's law and of single and multiple scattering. The contribution of direct attenuated and scattered light varies with the depth of radiation penetration into the scattering medium. The light field in the medium is found to depend not only on its optical properties, but also on the geometric parameters of the source.

The small diameter of laser beams and their small divergence, together with the coherence of radiation are responsible for the fact that a beam propagating in turbulent atmosphere has a spotty space and time structure.

The gigantic power of laser radiation gives rise to a number of nonlinear effects accompanying laser radiation in the atmosphere. These effects include: 1) spectroscopic saturation effect (dependence of absorption coefficient on the intensity of radiation incident on the absorbing layer); 2) multiphoton effects (excitation, ionization and dissociation of atoms and molecules of atmospheric gases, which do not absorb radiation at the given frequency when the radiation intensity is low); 3) effects due to incidence of radiation on atmospheric aerosol particles (light pressure, evaporation, explosion, etc.); 4) combined induced scattering; 5) self-focusing of the beam.

The great promise held by lasers in various devices intended for operation through the atmosphere (communications, transmission of information, radar devices, telemetry, etc.) requires intensive study of all features of laser radiation propagation in the atmosphere.

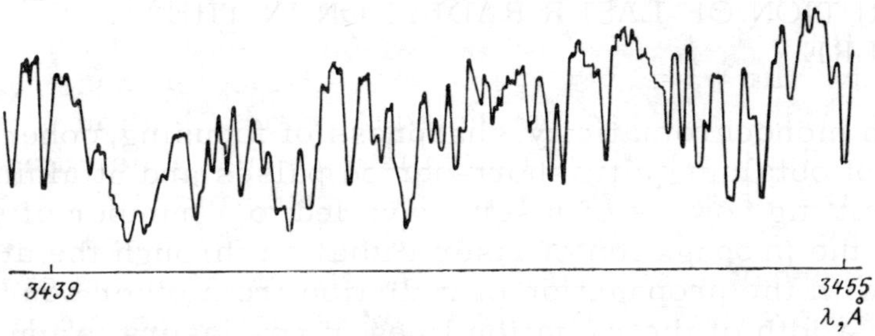

FIGURE 3.5. Recording of the fine structure of the absorption spectrum of the terrestrial atmosphere near λ= 0.345 μ /174/.

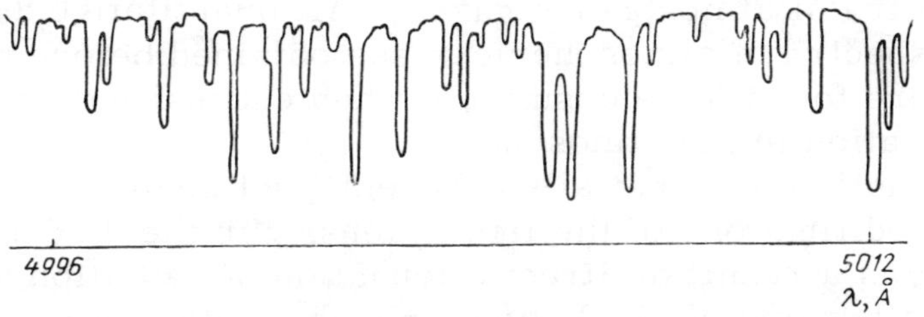

FIGURE 3.6. Recording of the fine structure of the absorption spectrum of the terrestrial atmosphere near 0.5μ /174/.

1. General pattern of the fine structure of the absorption spectrum of the terrestrial atmosphere in laser emission regions. A general idea about the fine structure of the radiation spectrum of the terrestrial atmosphere in spectral emission regions of the most prevalent lasers can be obtained from published data. Figures 3.5 through 3.15 present various absorption spectra of the terrestrial atmosphere with high resolution in regions with the most prevalent laser emissions: first harmonic of a ruby laser (λ = 0.345μ) and of an argon laser (λ = 0.5μ), first harmonic of a glass-neodymium laser (λ = 0.53μ), of a helium-neon laser (λ = 0.63μ), of a ruby laser (λ = 0.69μ), of a gallium-arsenide laser (λ = 0.84μ), of a glass-neodymium laser (λ = 1.06μ), of a helium-neon laser (λ = 1.15μ), of a helium-neon laser (λ = 3.39μ), of a xenon-helium laser (λ = 3.51μ), and of a carbon dioxide—nitrogen laser (λ = 10.6μ).

Figures 3.5 through 3.15 show clearly the very sharply expressed selectiveness of the absorption of the terrestrial atmosphere for any laser emission. All the figures represent the recordings of the absorption spectrum of the terrestrial atmosphere with high, but finite resolution, which smoothes out the true, even sharper selectivity of the absorption. In every figure, for example, many of the absorption lines have not been resolved as individual lines. It is evident that if the wavelength of laser emission changes for some reason, then this can be accompanied by a modification of the absorption of this radiation in the atmosphere. This circumstance can be illustrated by the example of the radiation of the ruby laser, which is used most extensively. In Figure 3.9 the ordinate axis displays the spectral absorption of the entire thickness of the atmosphere, recorded with high resolution using the sun as radiation source. The abscissa axis has a wavelength scale in angstroms and a temperature scale for corresponding wavelengths of ruby-laser emission. It is clear from the figure that the absorption of ruby-laser radiation in the atmosphere can vary markedly, depending on the temperature of the working substance of the generator. There are no strong absorption lines in the emission range of the ruby laser.

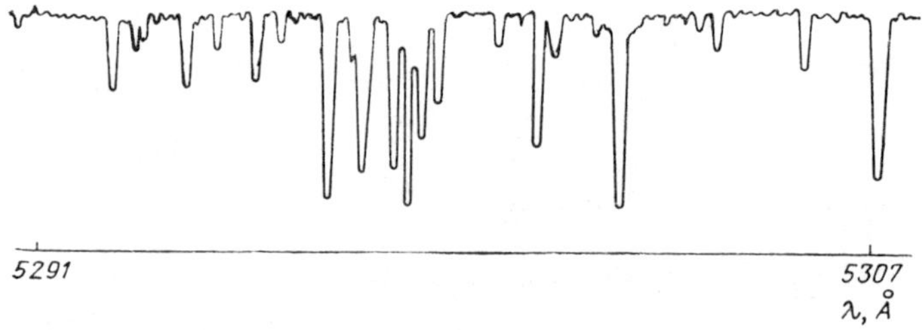

FIGURE 3.7. Recording of the fine structure of the absorption spectrum of the terrestrial atmosphere near $\lambda = 0.53\,\mu$ /174/.

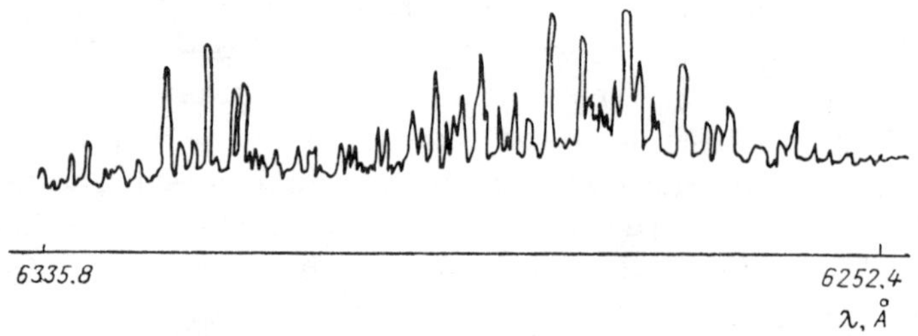

FIGURE 3.8. Recording of the fine structure of the absorption spectrum of the terrestrial atmosphere near $\lambda = 0.63\,\mu$ /185/.

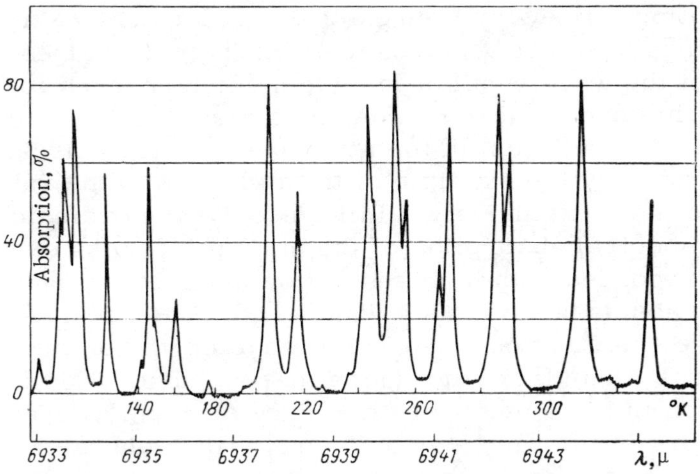

FIGURE 3.9. Recording of the fine structure of the absorption spectrum of the terrestrial atmosphere near $\lambda = 0.69\,\mu$ /295/.

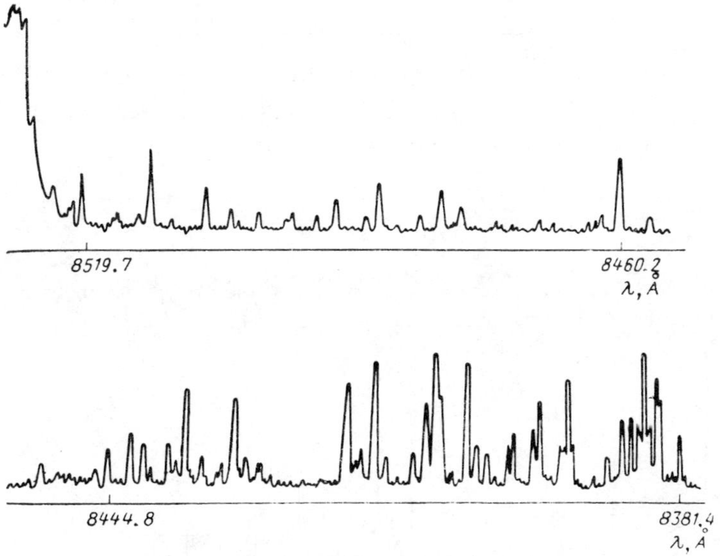

FIGURE 3.10. Recording of the fine structure of the absorption spectrum of the terrestrial atmosphere near $\lambda = 0.84\,\mu$/185/.

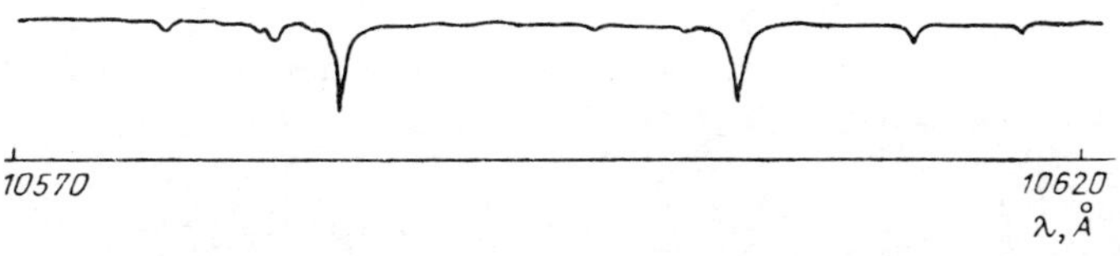

FIGURE 3.11. Recording of the fine structure of the absorption spectrum of the terrestrial atmosphere near $\lambda = 1.06\,\mu$ /183/.

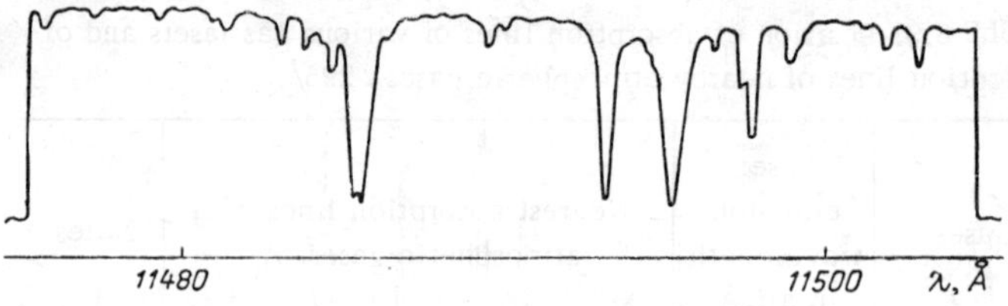

FIGURE 3.12. Recording of the fine structure of the absorption spectrum of the terrestrial atmosphere near $\lambda = 1.15\,\mu$ /183/.

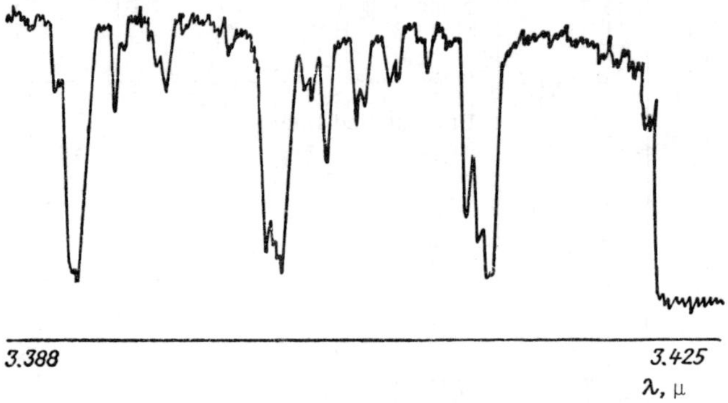

FIGURE 3.13. Recording of the fine structure of the absorption spectrum of the terrestrial atmosphere near $\lambda = 3.39\,\mu$ /181/.

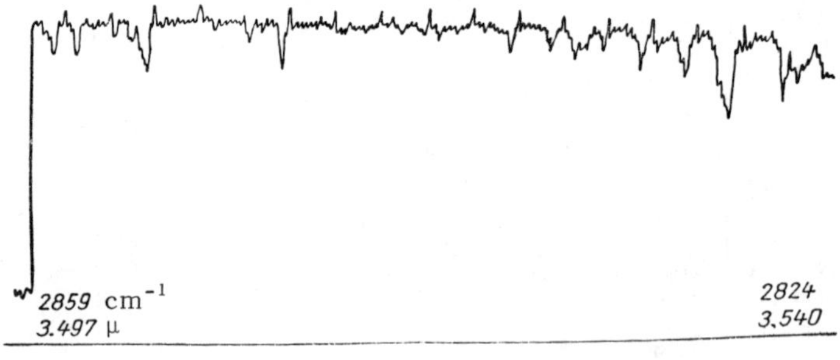

FIGURE 3.14. Recording of the fine structure of the absorption spectrum of the terrestrial atmosphere near $\lambda = 3.51\,\mu$ /181/.

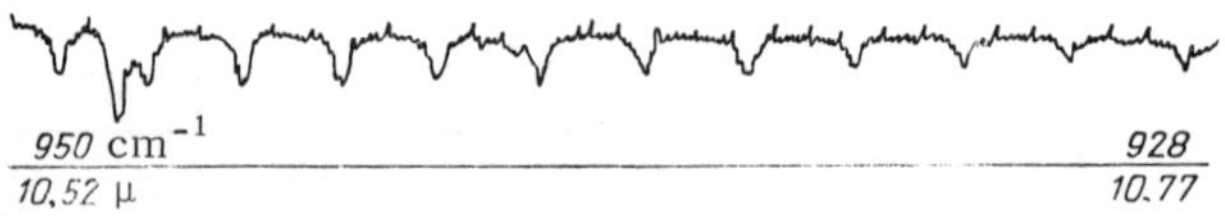

FIGURE 3.15. Recording of the fine structure of the absorption spectrum of the terrestrial atmosphere near $\lambda = 10.6\,\mu$ /181/.

TABLE 3.9. Position of absorption lines of various gas lasers and of absorption lines of nearby atmospheric gases / 295/

Gases	Laser emission wavelength in air, Å	Nearest absorption lines of atmospheric gases, Å	Notes
He—Ne	6328.172	None	
O_2	8446.2	None in terrestrial atmosphere	1
He—Ne	10798	None	
He—Ne	10845	None	
He—Ne	11143	11143.32 — H_2O	
He—Ne	11177	11175.59 — H_2O	
		11177.45 — ?	
		11177.49 — H_2O	
He—Ne	11390	Unidentified line	
He—Ne	11409	None	2
He—Ne	11522.76	11522.77 — H_2O	
		11523.19 — H_2O	
		11523.73 — H_2O	
		11524.20 — H_2O	
		11524.23 — H_2O	
He—Ne	11601	11600.21 — H_2O	
		11601.05 — H_2O	
He—Ne	11614	11613.84 — H_2O	
He—Ne	11767	11768.37 — H_2O	
He—Ne	11985	11984.8 — ?	
He—Ne	12066	12067.24 — H_2O	
Ne	15231	15230.59 — H_2O	
Ar	16180	None	
Kr	16900	16900.97 — H_2O	
		16901.41 — CH_4	
		16901.3 — H_2O	
Kr	16935.81	16935.1 — CH_4	
		16936.25 — CH_4	
		16937.22 — H_2O	
Ar	16940.59	16940.13 — H_2O	
Kr	17843	17843.37 — H_2O	3
Ar	17930	17934.71 — H_2O	4
Kr	18185	18180 — H_2O	5
Kr	19211	— — H_2O	3
Xe	20262.28	20260.75 — CO_2	
		20263.21 — CO_2	
He	20603	20601.57 CO_2	
		20615.75 — H_2O	
		20618.04 — CO_2	
Ne	21019	21018.45 — CO_2	6
Kr	21165	21165.76 — $OC^{13}O$	4
Kr	21902.53	21899.5 — H_2O	4
Xe	23193.28	23192.89 — H_2O	
		23194.44 — CH_4	
Kr	25234	25234.39 — H_2O	3
	25414	— — H_2O	3
Xe	26268.95	— H_2O	3
Xe	26510.96	— H_2O	3
Xe	26600.6	— H_2O	3
	31069.14	31053.8 — H_2O	2
		31081.5 — H_2O	
Ne	33333	33336.3 — CH_4	
Xe	33666.5	33658.6 — CH_4	
		33666.6 — CH_4, H_2O	
		33676.8 — CH_4	

TABLE 3.9 (contd.)

Gases	Laser emission wavelength in air, Å	Nearest absorption lines of atmospheric gases, Å	Notes
Ne	33913.17	33913.0 — CH_4	
Xe	34341	34344.1 — CH_4	
		34348.7 — CH_4	
		34352.1 — CH_4	
		34357.2 — CH_4	
Xe	35070.4	35069.5 — ?	4
Xe	36209	36205.5 — CH_4	
Xe	36503	36501.0 — HDO, CH_4	6
Xe	36788.36	36790.1 — CH_4	6
Xe	36848.56	—	6
Ne	37734	37732.9 — CH_4, HDO	4
Xe	38686	38686.7 — N_2O	
Xe	38939.74	38937.5 — N_2O	
Xe	41516		4
Xe	41516		4
Xe	46096	46096.8 — N_2O, CO	
Ne	54072	54077.1 — CO, H_2O	7
Xe	55738	— — H_2O	3
Cs	71821	71600.9 — H_2O	
		71801.4 — H_2O^{17}	
Xe	73147	73042.3 — H_2O, CH_4	
		73143.8 — CH_4	
		73181.3 — CH_4, H_2O	
Xe	90040	90057.6 — O_3	
Xe	97003.36	96979.1 — O_3	
Xe	122990		4
Xe	122630	122564 — CO_2	4
		122731 — H_2O	
Xe	129130	129134 — CO_2	6
Ne	182760	182505 — H_2O	5
Ne	182820	182989 — H_2O	
Ne	183040	182989 — H_2O	3
Ne	183970		4
Xe	185010	184771 — H_2O	5
Ne	185910	185805 — CO_2, H_2O	4
Ne	185970		
Ne	203510	203215 — H_2O	5

Notes. 1 — absorption by atomic oxygen; 2 — between two strong H_2O lines; 3 — total absorption; 4 — virtually no absorption; 5 — wing of H_2O lines; 6 — weak lines; 7 — solar CO.

These figures also explain why a substantial contribution to the absorption of laser radiation can be made by very weak lines when their centers coincide with the centers of radiation lines of the respective lasers. Hence neglect of absorption by weak lines, which is justified in calculating absorption functions over wide spectral regions, is in no way applicable when calculating the absorption of laser radiation. In calculating the absorption function in relatively wide spectral regions, until recently consideration was usually given to absorption of water vapor, carbon dioxide and ozone. The absorption of radiation in integral vibration-rotation bands of such minor constituents as methane, nitrous oxide, and carbon monoxide is neglected. This is more justified the wider the spectral regions for which the absorption functions are determined, since the absorption bands of minor atmospheric constituents occupy quite narrow regions.

In calculating the absorption of laser radiation one cannot neglect the absorption of bands or lines on the basis of their strengths alone. It is necessary to first know, with a high degree of accuracy, the position of centers of the absorption lines and of the laser emission lines.

Table 3.9 /295/ lists data on the position of the emission lines of 67 gas lasers and of the nearby absorption lines of atmospheric gases, detected in absorption spectra recorded with high resolution.

2. Requirements governing the accuracy in determining the centers of absorption lines. We assume that the centers of laser emission lines are known with a high degree of accuracy and that the half-widths of the emission lines are substantially smaller than the half-width of the absorption lines. The latter assumption is fully justified for gas lasers.

We shall estimate the accuracy in determining the centers of absorption lines of spectra of atmospheric gases needed for quantitative estimates of laser-radiation absorption in the atmosphere /327/. This accuracy depends on: 1) the total and partial pressures and temperature, or the altitude at which laser radiation propagates; 2) the distance from the center of the absorption line to the laser emission line.

It was shown in Chapter 2 that in the lower 20—30 km layer of the atmosphere the central part of the absorption line of atmospheric gases is described satisfactorily by the dispersion profile:

$$k(\nu) = \frac{S}{\pi} \frac{\gamma}{(\nu - \nu_0)^2 + \gamma^2},$$

where $k(\nu)$ is the monochromatic absorption coefficient at frequency ν; S, $\gamma = \gamma^0 (P_a + \sigma_{ab} P_b)$, and ν_0 are the strength, half-width and position of the center of the absorption line; γ^0 is the half-width of lines for standard conditions; P_a is the pressure of the nonabsorbing gas; P_b is the pressure of the absorbing gas; σ_{ab} is the effectiveness of collisions between molecules of the absorbing and nonabsorbing gas. In the lower 30 km layer of the atmosphere for all gases absorbing in the infrared, the product $\sigma_{ab} P_b$ can be neglected compared with P_a. One may therefore assume that $\gamma = \gamma^0 P$, where P is the total pressure.

Bearing in mind the slight temperature-dependence of γ^0 $\left(\gamma^0 \sim \frac{1}{\sqrt{T}}\right)$ and the not too extensive variation in the latter with altitude, one may further assume (with accuracy sufficient for practical purposes) that γ^0 is independent of altitude.

TABLE 3.10. Values of $k(\nu)$ at different distances from the center of a line with strength $S=1$ for various line half-widths

$\Delta\nu$, cm^{-1}	$\gamma^0 = 0.1\ cm^{-1}\cdot atm^{-1}$, $P=1$ atm	$\gamma^0 = 0.03\ cm^{-1}\cdot atm^{-1}$, $P=1$ atm	$\gamma^0 = 0.1\ cm^{-1}\cdot atm^{-1}$, $P=0.1$ atm	$\gamma^0 = 0.03\ cm^{-1}\cdot atm^{-1}$, $P=0.1$ atm
0	3.18	10.6	31.8	106.1
0.005	3.17	10.3	25.4	28.1
0.01	3.15	9.5	15.9	8.8
0.02	3.06	7.3	6.4	2.4
0.03	2.92	5.3	3.2	1.1
0.04	2.74	3.8	1.9	0.6
0.05	2.54	2.8	1.2	0.4
0.1	1.59	0.9	0.3	0.1
0.2	0.6	0.2	0.1	0.02
0.3	0.3	0.1	0.04	0.01
0.4	0.2	0.06	0.02	0.01
0.5	0.1	0.04	0.01	
1.0	0.03	0.01		
2.0	0.01			

On the basis of the above assumptions we calculated values of $k(\nu)$ for $S = 1$, $\gamma^0 = 0.03$ and $0.1\ cm^{-1}\cdot atm^{-1}$, $P = 1$ and 0.1 atm. The values of γ^0 assumed in the calculations correspond to the minimum and maximum half-widths of the absorption lines of water vapor, the principal absorbing component of the atmosphere. Pressures of 1.0 and 0.1 atm correspond to altitudes of 1 and 16 km above sea level. The results are tabulated in Table 3.10.

Table 3.11 presents values of coefficient $k(\nu)$ for the above four cases, but not for unit line strength. For convenience in interpreting the results, the values of $k(\nu)$ at the center of the line ($\Delta\nu = 0$) in all four cases were taken to be 100.

These tables indicate that in the surface layer of the atmosphere ($P = 1$ atm) for the widest lines (these are usually also the most intensive lines), the value of $k(\nu)$ decreases 10-fold over a distance $\Delta\nu = 0.3\ cm^{-1}$ from the line center; for narrow lines such a reduction occurs as soon as $\Delta\nu = 0.1\ cm^{-1}$. At a 16-km altitude $k(\nu)$ decreases by an order of magnitude over distances of $0.03\ cm^{-1}$ and

0.01 cm^{-1} for wide and narrow lines, respectively. When $\Delta\nu = 0.1$ and 0.03 cm^{-1} for wide and narrow lines at this altitude, $k(\nu)$ decreases 100-fold.

TABLE 3.11. Values of $k(\nu)$ at different distances from the center of lines for various half-widths, reduced to a single scale with $\Delta\nu = 0$

$\Delta\nu$, cm^{-1}	$\gamma = 0.1$ $cm^{-1}\cdot atm^{-1}$ $P = 1$ atm	$\gamma = 0.03$ $cm^{-1}\cdot atm^{-1}$ $P = 1$ atm	$\gamma = 0.1$ $cm^{-1}\cdot atm^{-1}$ $P = 0.1$ atm	$\gamma = 0.03$ $cm^{-1}\cdot atm^{-1}$ $P = 0.1$ atm
0	100	100	100	100
0.005	99.8	97.3	80.0	26.5
0.01	99.1	89.5	50.0	8.3
0.02	96.2	69.2	20.1	2.3
0.03	91.7	50.0	10.1	1.0
0.04	86.2	36.0	6.0	0.6
0.05	80.0	26.5	3.8	0.4
0.1	50.0	8.3	1.0	0.1
0.2	20.1	2.2	0.3	0.02
0.3	10.1	1.0	0.1	0.01
0.4	6.0	0.6	0.06	0.01
0.5	3.8	0.4	0.03	—
1.0	1.0	0.1	—	—
2.0	0.3	—	—	—

If the monochromatic laser emission impinges on the central part of the absorption line of water vapor, then to estimate the value of $k(\nu)$ one requires not only the precise radiation frequency of the laser but also the position of the absorption line center with a very high degree of accuracy. This accuracy in the surface layer of the atmosphere has a magnitude of several hundredths of cm^{-1} and increases markedly with altitude. All existing recordings of absorption spectra of the terrestrial atmosphere (atlases of spectra) do not provide such an accuracy in determining the position of line centers. In this connection the utilization of such atlases, even for a rough quantitative estimate of $k(\nu)$ in the region of the central parts of lines, should be approached with the utmost care.

As the laser emission frequency moves further away from the center of the absorption line of the atmospheric gas, the requirements as to the accuracy in determining the location of this center becomes less rigid. A quantitative measure of these requirements can be gleaned from the above tables.

These conclusions are valid not only for the absorption spectrum of water vapor. They can be applied also to absorption spectra of other atmospheric gases, since the half-width of lines of the latter are of the same order of magnitude.

3. Results of studies of laser-radiation absorption in the atmosphere. 1. Theoretical studies. Rigorous requirements are put to the determination of line positions, as well as their strengths and half-widths. Thus in theoretical solutions to the problem of computing absorption spectra of atmospheric gases it is necessary to take into account all factors affecting the characteristics of spectra. The greatest effect on all these characteristics is had by the interaction of vibrational and rotational molecular motions, which is more substantial, the higher the rotational quantum number involved in calculating absorption lines. It should be mentioned in passing that lines with high rotational quantum numbers are as a rule contained in spectral intervals between absorption bands which are of practical importance (in atmospheric windows).

The most complex vibration-rotation spectrum of all atmospheric gases is the spectrum of water vapor, the principal absorbing atmospheric constituent. The first attempt to correctly calculate the positions, strengths and half-widths of lines, and the absorption coefficients with consistent allowance for interaction between vibrational and rotational molecular motions was directed at water vapor.

The method for calculating the positions of line centers, their strengths and half-widths was treated in detail in Sec. 5. Data obtained on these three parameters of H_2O absorption lines were then used to find the monochromatic absorption coefficients in individual narrow regions of the vibration-rotation spectrum of water vapor in the emission ranges of the most extensively used lasers. The calculations were carried out for several pairs of temperatures and pressures, characteristic for specific altitudes. The lines were assumed to be dispersion [Lorentz] broadened. The monochromatic absorption coefficient was therefore taken in the form

$$k(\nu)=\sum_i \frac{S_i(T)}{\pi}\,\frac{\gamma_i(T,\ P)}{(\nu-\nu_{0i})^2+\gamma_i^2(T,\ P)}\,.$$

Analysis of the spectral variation in the absorption coefficient of H_2O in the spectral intervals under study points to the extremely high selectivity of $k(\nu)$, even in regions free of strong absorption lines. Thus, in laser emission regions with wavelengths 0.84μ (gallium-arsenide laser) and 1.06μ (glass-neodymium laser) the absorption coefficient changes by several orders when the wavelength changes by only several cm^{-1}. The unavailability of corresponding experimental data on monochromatic absorption coefficients unfortunately makes it impossible to estimate the accuracy of the calculations.

The absorption of the radiation of the majority of lasers is due to water vapor. In a number of cases, however, laser radiation is incident on spectral regions of other atmospheric gases. For example, the absorption of radiation of $CO_2 + N_2$ (spectral region of 10.6μ) and He + Ne lasers (3.39μ wavelength) is due in the first case to carbon dioxide and water vapor simultaneously, while in the second it is due only to methane.

Calculation of monochromatic absorption coefficients of $CO_2 + N_2$ lasers near the 10.6μ wavelength is not too difficult. We have obtained components of this coefficient which are due to absorption of carbon dioxide and water vapor. At the centers of lines of the P branch of the CO_2 band with rotational quantum number j from 14 to 28 the component of the absorption coefficient of carbon dioxide in the surface layer of the atmosphere ranges from 0.12 to 0.19 km^{-1}. The component of the absorption coefficient of water vapor in the narrow spectral region occupied by these lines is due to continuous absorption by wings of far strong lines and virtually do not depend on the wavelength. Its value is 0.08 km^{-1} at a surface-layer humidity of $10\,g/m^3$ /192/.

At present there is insufficient data on the parameters of individual lines for theoretical calculation of monochromatic absorption coefficients in the region of methane bands and in the majority of the spectral regions of other atmospheric gases.

It should be noted in conclusion that, in spite of known successes in the theoretical determination of absorption coefficients of monochromatic radiation in the atmosphere, we are still very far from being able to exactly predict the absorption of monochromatic or quasi-monochromatic radiation in the atmosphere, even if the spectral composition of this radiation is known a priori with high accuracy. Hence the absorption characteristics of various laser emissions in the atmosphere can be determined most precisely from corresponding experimental studies.

2. Experimental investigations. A series of experimental investigations into the absorption of laser radiation of different types was carried out in the author's laboratory /328—331, 782, 806, 893, 908—910/.

Direct experimental determination of the absorption coefficients of atmospheric gases in laser emission regions is virtually impossible, due to the insufficient resolving power of even the best spectrometers. However, if lasers proper are used as radiation sources in these measurements, then one can determine directly from experiments the absorption coefficients or functions for specific lasers. In the case of gas lasers absorption coefficients can actually be obtained from such measurements. For other types of generators one determines directly the absorption function or the average absorption coefficient.

Experimental studies on the absorption of the radiation of various lasers in the author's laboratory were carried out 1) by means of a multipass absorption cell, and 2) under natural conditions at experimental grounds. In the first case it is possible to study pure absorption, while in the second absorption must be separated from the overall attenuation. However, in absorption cell measurements it is impossible to attain large thicknesses. Thus both types of measurements supplement one another.

In the absorption-cell studies we obtained absorption coefficients for helium-neon lasers with wavelengths of 1.15 and 3.39μ. The absorption coefficient for radiation with wavelength 1.15μ in the surface layer was found to be 0.3 mm^{-1}, where mm^{-1} denotes the reciprocal of a millimeter of precipitable water. Since this radiation is absorbed by water vapor, the value of the absorption coefficient in reciprocal units of distance, or in dB/km, depends on humidity. At a humidity of 12 g/m^3 in the surface layer the attenuation coefficient is 16 dB/km. Table 3.12 lists the transmission of the surface layer of nonturbid atmosphere for laser radiation with 1.15μ wavelength for various humidities.

The absorption coefficient for 3.39μ laser emission for the surface layer was found to be 1.4 km^{-1}, or 6 dB/km. This radiation is absorbed by methane. It should be emphasized that in calculating the absorption of radiation from thermal sources the absorption of methane is usually neglected, due to the very low concentration (millionths of parts per volume) of this gas. The substantial absorption of 3.39μ laser emission detected in our experiments show that in this case neglect of methane absorption should not even be mentioned. The relatively strong absorption is here attributable to the fact that the centers of absorption lines of methane and of the laser radiation virtually coincide. This circumstance is confirmed by bright-

TABLE 3.12. Transmission *T* of clear atmosphere for 1.15μ laser emission at different absolute humidity *a* for path lengths of 0.1, 0.5, and 1.0 km in the surface layer of the atmosphere

D, km	*T*			
	a = 1 mm/km	*a* = 5 mm/km	*a* = 10 mm/km	*a* = 15 mm/km
0.1	0.97	0.86	0.74	0.64
0.5	0.86	0.47	0.22	0.11
1.0	0.74	0.22	0.05	0.01

ening of the atmosphere in the 3.39μ laser emission region when the total pressure of the gas mixture is increased while the methane content remains constant (Figure 3.16).

The absorption of laser radiation with different wavelengths at experimental grounds was measured for much greater thicknesses than in the absorption cell. The path lengths over which the measurements were carried out ranged from several hundred meters to 3.5 km (the maximum path length in the absorption cell was 96 m).

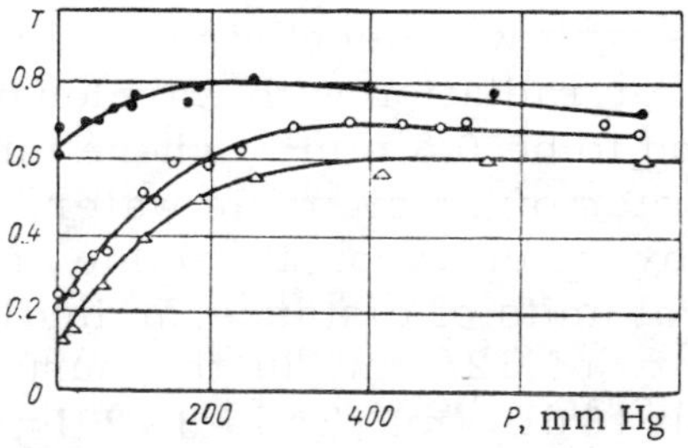

FIGURE 3.16. Transmission of 3.39 μ laser emission as a function of total pressure at different constant partial pressures of methane and for a geometric layer thickness $L = 64$ m.

The radiation sources for experimental-grounds measurements consisted of lasers whose parameters are tabulated in Table 3.13.

The absorption coefficients for 1.15μ laser emission in experimental-grounds measurements was found to be the same as with the absorption cell. For 3.39μ laser emission under natural conditions the absorption coefficients varied from one series of measurements to another by 1.4—5.0 km^{-1}. The minimum absorption coefficient thus coincides with that measured under

TABLE 3.13. Radiation parameters of lasers employed in experimental-grounds measurements

Laser type	Emitted wavelength, μ	Emitted power	Emission mode	Beam divergence
Helium-neon	0.63	10 mW	Continuous multimode	1'
Ruby	0.69	1 kW	Pulsed	1.5'
Gallium arsenide	0.84	1 mW	Pulsed	20'
Glass-neodymium	1.06	0.1 kW	Pulsed	3'
Helium-neon	1.15	10 mW	Continuous multimode	1'
Helium-neon	3.39	20 mW	Continuous multimode	1'
Helium-xenon	3.51	20 mW	Continuous multimode	2'
Carbon dioxide-nitrogen	10.6	1 W	Continuous multimode	2'

laboratory conditions. The cause for the variation in the absorption coefficients has yet to be found. It may be due to a corresponding change in methane concentration at the location of the measurements.

During measurements with the $CO_2 + N_2$ laser (10.6μ) the measured attenuation was not subdivided into components due to molecular absorption and aerosol scattering. The attenuation coefficients obtained ranged from 0.12 to 4 dB/km, due to different prevailing meteorological conditions (haze with visibility ⩾2 km, snowfall, light fog).

The absorption of ruby laser radiation varied due to variations in the ruby temperature. The absorption coefficient at a humidity of 10 g/m^3 varied from 1.1 to 4.7 dB/km. The absorption of 0.63, 0.84, 1.06 and 3.51μ laser emission in experimental-grounds measurements was fully masked by measurement errors, which were as high as 10—30% depending on the meteorological conditions and distance.

The absorption of a 3.39μ helium-neon laser emission was studied by Edwards and Burch /332/ using a small [absorption] cell. They determined the distance between centers of laser emission lines and methane absorption lines and the half-width of the CH_4 lines, which were found to be equal to 0.003 ± 0.002 cm^{-1} and 0.13 ± 0.04 cm^{-1} under standard conditions. This indicates that the centers of the emission and absorption lines coincide.

The absorption of a 3.39μ gas laser over a path length of 2.9 km under natural conditions was measured by Chu and Hogg /333/, who found that the attenuation coefficient is 5.5 ± 0.5 dB/km. This value is in satisfactory agreement with the above data /332/ and with our measurements in the absorption cell, but differs from our experimental-grounds measurements. It should be mentioned here that in a later series of measurements we again obtained absorption coefficients 50—200% higher than the figures given in /332, 333/ and than our measurements in the absorption cell. The reasons for these discrepancies are still unclear.

The absorption of a 1.15μ helium-neon laser was measured /334/, and the values obtained are in satisfactory agreement with this author's measurements.

McCubbin and Darone /335/ used a $CO_2 + N_2$ laser to measure the absorption in a gas mixture at different pressures of the absorbing and foreign gases in several lines of the 10.4μ CO_2 band. The results of these measurements were used when determining the strengths and half-widths of lines. Similar measurements were carried out elsewhere /795—797/.

The absorption of a $CO_2 + N_2$ laser in a long [absorption] cell was measured by Long and McCoy /336/. The radiation was generated on the line of the P branch with rotational quantum number $j = 20$. At a CO_2 pressure of 0.25 mm Hg and nitrogen pressure of 760 mm Hg, corresponding to conditions of the surface layer of the atmosphere, the absorption coefficient of the center of the emission line was found to be 0.125 dB/km /336/. It should be noted that emission from a $CO_2 + N_2$ mixture can be induced at different lines. For example, Howe and McFarlane /337/ measured the positions of 17 lines of the $(01^11) \rightarrow (11^10)$ band in the interval from 897.00 to 911.33 cm^{-1}, which give laser emission. In the 9—11μ interval more than 100 lines generated by CO_2 and N_2O molecules were detected /805/. The strengths of lines in the absorption spectrum of CO_2 change with variation in the rotational quantum number, and therefore one should expect different absorption coefficients for generators which emit at different lines.

Thus when dealing with the absorption coefficient for laser radiation one requires, with a high degree of accuracy, not only the position of absorption lines of atmospheric gases, but also the position of the laser emission lines. It should again be emphasized that it makes sense to refer to absorption coefficients for laser radiation only when the half-width of the emission line is much smaller than the half-width of absorption lines. This condition can be regarded as satisfied (with accuracy sufficient for practical purposes) in gas lasers. If this condition is not satisfied, then the Bouguer law cannot be applied to the description of the absorption of laser radiation. In this case one must determine the absorption function for selective radiation sources.

4. *ABSORPTION OF VISIBLE AND INFRARED RADIATION IN AN INHOMOGENEOUS ATMOSPHERE*

1. SPECTRAL TRANSMISSION (ABSORPTION). TRANSMISSION (ABSORPTION) FUNCTION

The expression describing the variation of emission intensity in an absorbing medium can be expressed in the differential form (see Chapter 3)

$$dI_\nu = -I_\nu k(\nu)\,\rho(l)\,dl, \tag{4.1}$$

where I_ν is the emission intensity of frequency ν, $k(\nu)$ is the absorption coefficient, ρ is the density of the absorbing gas, and dl is a path length element in the absorbing medium.

If the absorption coefficient is a function of points lying on the ray path, one speaks of propagation of radiation in an inhomogeneous medium. The reason why the atmosphere is inhomogeneous in this sense is that the half-width and strength of lines defining quantity $k(\nu)$ depend on the pressure and temperature, which vary markedly with altitude and depend on the geographic coordinates of the point under study (see Chapter 1). Hence the absorption coefficient is a function of points lying on the ray path for propagation along slant paths in the atmosphere or in the case of horizontal propagation over distances for which the earth's curvature cannot be neglected.

Integration of equation (4.1) yields the spectral transmission in the form

$$F(\nu) = \frac{I_\nu}{I_{0\nu}} = \exp\left(-\int_{(l)} \rho(l)\,k(\nu,\ l)\,dl\right), \tag{4.2}$$

where $I_{0\nu}$ is the transmission intensity for frequency ν at the start of the path. The exponent contains a curvilinear integral of the first kind, evaluated along the ray path points. Rigorously speaking, the integration should be taken over the actual ray path with allowance for refraction. However, refraction is usually ignored when determining spectral transmission.

Spectral absorption $A(\nu)$, as in the case of the homogeneous atmosphere, is given by the simple expression

$$A(\nu) = 1 - T(\nu). \tag{4.3}$$

Transmission function F and absorption function H for an inhomogeneous path over the spectral interval $\Delta\nu = \nu_2 - \nu_1$ are, by definition, given by

$$F = \frac{1}{\nu_2 - \nu_1} \int_{\nu_1}^{\nu_2} \exp\left[-\int_{(l)} \rho(l)\, k(\nu,\, l)\, dl\right] d\nu, \tag{4.4}$$

$$H = 1 - F. \tag{4.5}$$

We recall that expressions (4.4) and (4.5) are valid for constant spectral emission density of the source over interval $\Delta\nu$. Function F represents the ratio of the radiant flux passing through the given layer of inhomogeneous atmosphere, to the radiation incident on the layer over spectral interval $\Delta\nu$. Function H is defined similarly.

2. DIRECT COMPUTATION OF SPECTRAL TRANSMISSION AND THE TRANSMISSION FUNCTION

It is evident from equation (4.2) that direct computation of $F(\nu)$ requires a knowledge of functions $\rho(l)$ and $k(\nu,\, l)$. The concentration distribution of absorbing gases in the atmosphere has been examined more or less thoroughly (see Chapter 1). In any case function $\rho(l)$ can be specified as a statistical-average vertical concentration profile of some gas.

The absorption coefficient for radiation at frequency ν is the sum of coefficients of the individual spectral lines $k(\nu,\, l) = \sum_i S_i(l) b_i(\nu,\, l)$, where S_i and b_i are the total strength and profile of the line. Summation in the above expression can only be carried out numerically. The exponent in equation (4.2) is now $\sum_i \int_{(l)} \rho(l)\, S_i(l)\, b_i(\nu,\, l)\, dl$, i.e., the terms are different functions of frequency than in the case of a homogeneous atmosphere, and depend on the shape of the distributions of the meteorological variables. This means that in each specific case this sum must be predicted. In addition, the integrals in the latter expression are not as a rule expressible in terms of elementary functions.

If it is assumed that the spectral lines have a dispersion profile, then

$$k(\nu,\ l) = \sum_i \frac{S_i\,[T\,(l)]}{\pi}\ \frac{\gamma_i\,[T\,(l),\ P\,(l)]}{(\nu-\nu_{0i})^2+\gamma_i^2\,[T\,(l),\ P\,(l)]}\,, \tag{4.6}$$

where $T(l)$ and $P(l)$ are the temperature and effective pressure at point l.

Direct calculation of the spectral transmission (absorption) of the inhomogeneous atmosphere from (4.2) and (4.6) has thus been shown to involve the following information: 1) the distribution of the concentration of absorbing gases and temperature, and the distribution of the effective pressure along the ray path; 2) the positions of lines ν_{0i} contributing to the absorption of radiation with frequency ν; 3) the strengths and half-widths of these lines at the corresponding temperature and effective pressure. It should be emphasized that for exact calculation of $F(\nu)$ one requires data on the line parameters determined with an extremely high degree of accuracy (see Chapter 3, Sec. 13). In particular this pertains to computation of the spectral transmission in the region of the central parts of lines.

The hope has only recently arisen of obtaining data on the spectral transmission by direct calculation. Such a calculation for a limited number of wavelengths was carried out in the author's laboratory using data on the parameters of water vapor lines, obtained by the method described in Chapter 3, Sec. 5. The problem of direct computation of absorption functions for an inhomogeneous atmosphere is even more complicated.

In addition to the difficulties encountered in calculating the absorption function discussed in Chapter 3, associated with the evaluation of an integral over some frequency interval, it is now also necessary to evaluate the integral contained in the exponent. This clearly complicates the problem, since integration with respect to ν and l cannot be interchanged. With the exception of particular distributions of meteorological elements and absorption spectra, which are not of special interest, expression (4.4) can be evaluated only numerically. However, the large volume of computations, the need for calculating each particular case and the large number of variables make this approach extremely complicated. At present several studies exist in which absorption functions were obtained for slant directions in the atmosphere by direct calculations, which make allowance for the absorption of each line.

Kyle et al. /79/ calculated absorption functions in the 4.3μ band (2235–2415 cm^{-1} interval). First the transmission spectrum was calculated with a very small step. Calculations over the above spectral region were carried out for 1800 lines. Then the resulting dependence of the spectral transmission on wavelength was multiplied by the slit function of the spectrometer used in the measurements. The positions and strengths of CO_2 lines were calculated without subsequent allowance for interaction between the vibrational and rotational molecular motions. The half-width of lines was assumed to be identical and equal to 0.066 cm^{-1} under standard conditions for all the lines, while the CO_2 concentration was taken to be 0.033% by volume at all latitudes. In spite of the approximate nature of the calculations, the results were in satisfactory agreement with experimental data. This shows that the requirements as to accuracy in determining line parameters when calculating absorption functions are not as rigorous as when calculating the spectral absorption.

Drayson /62/ conducted similar computations for the 15.0μ CO_2 band and took into account about 2000 lines in the 12—18μ range. He used the mixed Lorentz-Doppler profile for pressures below 100 mb. The dispersion profile was used for pressures above 100 mb. The atmosphere was subdivided into 34 layers in such a manner that the temperature within each layer would not vary by more than several degrees. The calculations were carried out for six zenith angles. Averaged data for spectral intervals 5 cm^{-1} wide are presented for different atmospheric thicknesses.

The fully resolved absorption spectrum of water vapor in the 3848—3860 cm^{-1} region was calculated by Calfee and Gates /338/. This interval does not contain absorption lines of other gases. The computational procedure, as in /62/, involves subdivision of the atmosphere into layers. Ten layers were selected for altitudes above 13.7 km (tropopause). The data, after transformation using the slit function of the spectrometer, were compared with corresponding measured results, so enabling the vertical humidity profiles to be determined for a large number of cases. The starting data on the half-width and strength of absorption lines were obtained without subsequent allowance for interaction between the vibrational and rotational motion of the H_2O molecule. The positions of line centers were determined experimentally.

In all the above studies /79, 62, 338/ the absorption function was in general determined from data on spectral transmission obtained for a large number of wavelengths. The information on line parameters underlying the calculations could not yield sufficiently accurate values of the true spectral transmission for the corresponding specific wavelengths. Nevertheless, the values of the resulting absorption functions describe quite satisfactorily the true absorption pattern within finite spectral intervals, because the exact position of line centers is not of great importance in calculating absorption functions. It is important that the line at hand be contained in the interval $\Delta\nu$ under study. In addition, neglect of interaction between vibrations and rotations of molecules is reflected mostly in the parameters of weak lines, which cannot make a large contribution to the value of the absorption function.

3. APPROXIMATE COMPUTATION OF ABSORPTION FUNCTION H

The underlying idea behind approximate methods of computing function H reduces to finding the absorption function when the absorbing-gas concentration, pressure and temperature are constant along the ray path. This is usually carried out either by introducing an effective absorbing mass, or by the Curtiss-Godson approximation.

In the first method an effective absorbing mass is selected so as to satisfy the equation

$$H = A(P_0, \omega^*), \tag{4.7}$$

where P_0 is the pressure at the earth's surface, while ω^* is the effective absorbing mass.

Usually ω^* is given by the expression

$$\omega^* = \int_{l_1}^{l_2} \rho(l) \left[\frac{P(l)}{P_0}\right]^{(n)} dl. \tag{4.8}$$

The value of n is suitably selected. Sometimes ω^* assumes the form

$$\omega^* = \int_{l_1}^{l_2} \rho(l) \frac{P(l)}{P_0} \left(\frac{T_0}{T(l)}\right)^{1/2} dl. \tag{4.9}$$

In (4.8) and (4.9), $\rho(l)$ is the density of the absorbing gas, $P(l)$ and $T(l)$ are the pressure and temperature at point l, P_0 and T_0 are the pressure and temperature at the earth's surface.

According to the Curtis-Godson approximation, or by introducing the weighted-mean pressure, absorption function H becomes

$$H = A(\overline{P}, \omega), \tag{4.10}$$

$$\overline{P} = \frac{\int_{l_1}^{l_2} \rho(l) P(l)\, dl}{\int_{l_1}^{l_2} \rho(l)\, dl}, \tag{4.11}$$

$$\omega = \int_{l_1}^{l_2} \rho(l)\, dl. \tag{4.12}$$

The method involving the weighted-mean pressure was proposed independently by Curtiss in 1952 and Godson in 1953 /3/. Since then quite a large number of calculations were carried out for the absorption functions of atmospheric gases for slant directions in the atmosphere. For the absorption-band intervals of water vapor and carbon dioxide the method yields satisfactory results not only in the weak and strong lines approximation, for which (4.10)—(4.12) were derived, but also for intermediate values of x (see equation (3.52)). The applicability of the weighted-mean pressure method for all values of x was demonstrated elsewhere /339/. In the case of the 9.6 μ ozone absorption band the Curtiss-Godson approximation is not sufficiently accurate for average values of x.

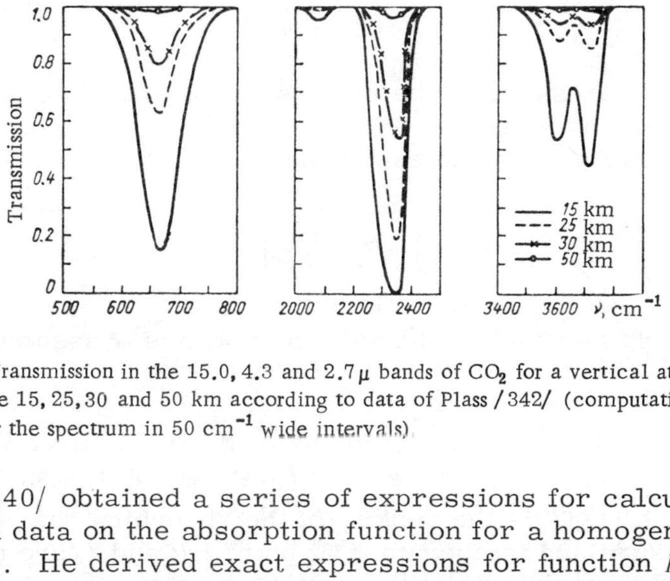

FIGURE 4.1. Transmission in the 15.0, 4.3 and 2.7 μ bands of CO_2 for a vertical atmospheric thickness above 15, 25, 30 and 50 km according to data of Plass /342/ (computation involved averaging over the spectrum in 50 cm^{-1} wide intervals).

Plass /340/ obtained a series of expressions for calculating H from known data on the absorption function for a homogeneous atmosphere. He derived exact expressions for function H for the weak- and strong-line approximation, when the absorbing gas concentration does not vary along the beam path. Similar expressions were obtained for the case of varying pressure, concentration and temperature along the beam path. In addition, a procedure was developed /340/ for calculating function H approximately. The starting data on absorption functions of the homogeneous atmosphere, needed when determining function H, can be expressed in the form of theoretical or empirical relationships, or can be tabulated.

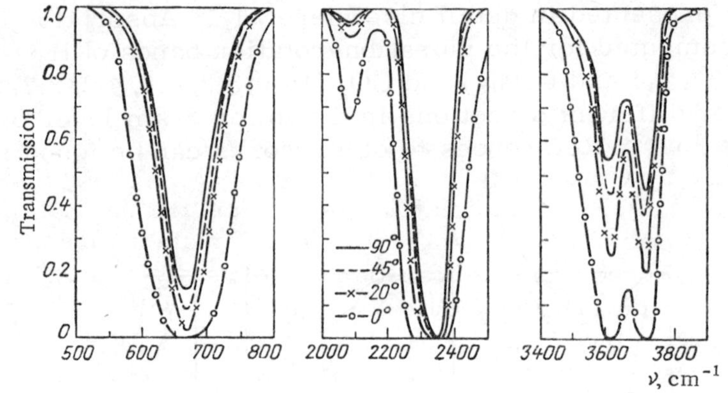

FIGURE 4.2. Transmission in the 15.0, 4.3 and 2.7 μ bands of CO_2 for atmospheric thicknesses penetrated by a beam from an altitude of 15 km at angles $\theta = 0, 20, 45$ and 90° to the local horizon according to data of Plass /342/ (computation involved averaging over the spectrum in 50 cm^{-1} wide intervals).

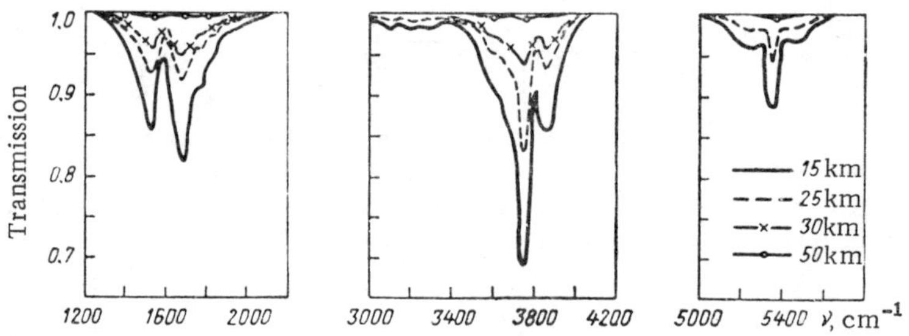

FIGURE 4.3. Transmission in the 6.3, 2.7 and 1.87 μ bands of H_2O for a vertical atmospheric thickness above 15, 25, 30 and 50 km according to data of Plass /342/ (computation involved averaging over 50 cm^{-1} wide intervals for a wet-atmosphere model).

Formulas and tabulated data on absorption functions for a homogeneous atmosphere were used by Plass /341, 342/ to conduct extensive calculations of H in narrow spectral regions of the fundamental vibration-rotation water vapor and carbon dioxide bands with allowance for vertical variation in pressure, absorbing-gas concentration, and temperature. Some of his results /342/ are plotted in Figures 4.1 through 4.5.

An absorption function H was calculated in our laboratory over a 0.1μ wide spectral region of water vapor and carbon dioxide gas for some model of the atmosphere incorporating variation in pressure, temperature and the H_2O and CO_2 concentrations with altitude. The calculations pertained to all the fundamental vibration-rotation bands of H_2O and CO_2. The values of H were obtained with a step of 1 km for the lower 30-km atmospheric layer. The corresponding results are presented in detail elsewhere /1/. Absorption functions have been computed for the vibration-rotation bands of H_2O (6.3, 3.2, 2.7, 1.87, 1.38, 1.1 and 0.94μ) and CO_2 (15.0, 5.2, 4.8, 4.3, 2.7, 2.0, 1.6 and 1.4μ) for different directions in the lower 8-km layer of the atmosphere /59/. References to other works can be found in /1/ and /342/.

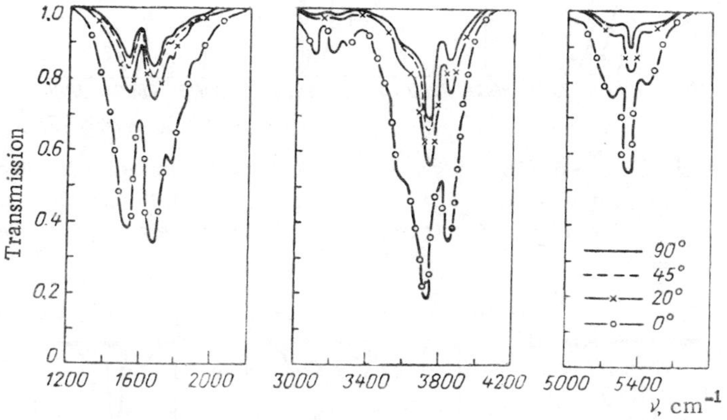

FIGURE 4.4. Transmission in the 6.3, 2.7 and 1.87μ bands of H_2O for atmospheric thicknesses penetrated by a beam from an altitude of 15 km at angles $\theta = 0, 20, 45$ and 90° to the local horizon according to data of Plass /342/ for a wet-stratosphere model (computation involved averaging over the spectrum in 50 cm^{-1} wide intervals).

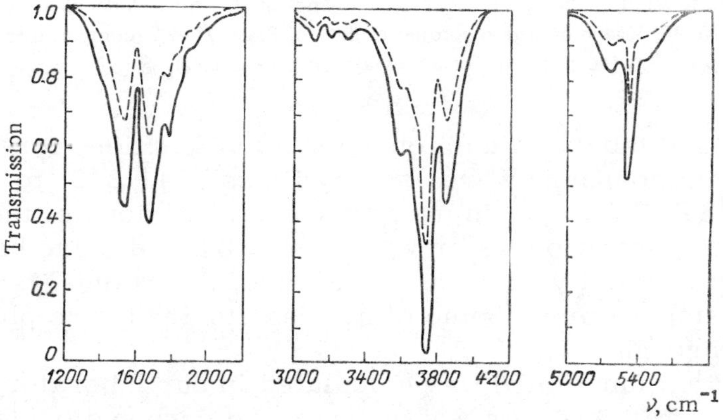

FIGURE 4.5. Transmission in the 6.3, 2.7 and 1.87μ bands of H_2O calculated by Plass /342/ for a dry- and wet-stratosphere model for atmospheric thicknesses penetrated by a beam, propagating horizontally from a point at an altitude of 30 km:

——— – wet stratosphere; — — — — dry stratosphere.

4. APPROXIMATE ALLOWANCE FOR TEMPERATURE IN COMPUTATION OF FUNCTION H

In the computation of function H one frequently neglects its temperature-dependence in view of the fact that the temperature does not vary vertically over a wide range. Nevertheless, it is often necessary to allow for the temperature-dependence of H. We now present an approximate method which allows for this dependence and is suitable for application to the vibration-rotation spectra of water vapor and carbon dioxide.

The temperature-dependence of H is defined by the temperature-dependence of the half-widths and strengths of absorption lines. The latter dependence is given by

$$\gamma_i (T) = \gamma_{0i} \frac{P(l)}{P_0} \left(\frac{T_0}{T(l)} \right)^{1/2} = \gamma_{0i} \frac{T_0^{1/2}}{P_0} \beta(l), \tag{4.13}$$

$$S_i (T) = S_i (T_0) \left[\frac{T_0}{T(l)} \right]^{3/2} \times$$
$$\times \exp\left[-E_i (T(l) - T_0)(kT(l) T_0)^{-1} \right], \tag{4.14}$$

where γ_{0i} is the half-width of line i at standard conditions $P = P_0$ and $T = T_0$; $S_i (T_0)$ is the strength of line i at standard temperature; E_i is the energy of the lower level from which the transition producing the line starts; $\beta(l) = \frac{P(l)}{[T(l)]^{1/2}}$. Quantity γ_{0i} will subsequently be assumed independent of temperature.

Quantity E_i for lines of vibration-rotation spectra can be approximately expressed in the form

$$E_i = E_{iv} + E_{iR}, \tag{4.15}$$

where E_{iv} and E_{iR} are the vibrational and rotational energies of the lower state of line i. Since $E_{iv} \gg E_{iR}$ (see Chapter 2), we can set

$$E_i = E_{iv}. \tag{4.16}$$

The value of E_{iv} does not depend on the rotational quantum number within the limits of a given vibration-rotation band. Further, for the fundamental vibration-rotation H_2O and CO_2 bands one can identify E_{iv} with the energy of the ground vibrational state. Analysis of the fundamental vibration-rotation bands of these gases shows that each of them consists of lines pertaining to different vibrational transitions. However, the majority of such overlapping transitions has the ground state as its fundamental state and, most important, their strength considerably exceeds that of transitions from other than the ground state. Hence in calculating the absorption function for rotational lines these transitions can be neglected. Quantity E_v could have been replaced by its average $\bar{E}_v = \Sigma E_{vk}\tau_k$, where E_{vk} is the vibrational energy state of the lower state of the kth transition (forming the given band), while τ_k is its relative intensity ($\Sigma\tau_k=1$). However, estimates show that $\bar{E}_v$ is close to the ground-state energy, in which case

$$\left.\begin{aligned} S_i(T) &= S_i(T_0)\varphi, \\ \varphi &= (T_0/T)^{3/2}\exp(-C(T-T_0)(TT_0)^{-1}), \end{aligned}\right\} \quad (4.17)$$

where φ is not a function of i, and C is some quantity corresponding to lines of the H_2O and CO_2 spectrum. It should be emphasized that the approximation at hand (that φ is not a function of i) is not permissible for lines of the rotational spectrum, since it can only be made on the assumption that $E_{iv} \gg E_{iR}$. In the above approximation

$$k_\nu = \sum S_i(T_0)\, b_i(\nu, \beta), \quad (4.18)$$

$$A = A(\omega', \beta), \quad \omega' = \omega\varphi, \quad (4.19)$$

where

$$\beta = P\sqrt{\frac{T_1}{T_2}}.$$

Hence the technique for calculating H /1, 339/ is applied without any changes after P has been replaced by β and ρ' by $\rho\varphi$. The final result now assumes the form

$$\left.\begin{aligned} H &= A(m', \bar{\beta}), \\ m' &= \int \varphi(l)\rho(l)\,dl, \\ \bar{\beta} &= \frac{\int \beta(l)\varphi(l)\rho(l)\,dl}{m'}. \end{aligned}\right\} \quad (4.20)$$

5. APPROXIMATE DETERMINATION OF THE SPECTRAL TRANSMISSION OF AN INHOMOGENEOUS ATMOSPHERE

Equation (4.2) enables the spectral transmittance of the inhomogeneous path to be computed in the general case. However, the above difficulties force one to seek some approximate methods.

The method involving the weighted-mean pressure can be extended to the case at hand; the final expression for $F(\nu)$ then becomes

$$F(\nu) = I_\nu/I_{0\nu} = \exp[-\omega k(\nu, \bar{P})], \quad (4.21)$$

where ω and $\bar{P}$ are given by (4.11) and (4.12).

If allowance is made for the temperature variation along the beam path, then

$$F(\nu) = \exp(-m'k(\nu, \bar{\beta})) \tag{4.22}$$

for the vibration-rotation spectrum where m' and $\bar{\beta}$ can be calculated from (4.20). It should be noted that equations (4.21) and (4.22) are now apparently rougher approximations than similar results for the absorption function.

It is easy to show that these relationships are asymptotically correct for spectral regions in which absorption is due to line wings. Expressions (4.10)—(4.12) and (4.20) follow automatically from equations (4.21) and (4.22), but this clearly does not mean that the converse is also true.

However, it is easy to find cases which cannot be treated within the framework of the above technique, such as: 1) absorption in centers of lines; 2) absorption at frequency ν due to lines with different lower vibrational energy levels. At present no numerical data exist which could be used to determine the accuracy of the approximate expressions derived in Secs. 4 and 5.

6. COMPUTATION OF THE ABSORPTION FUNCTION WITH ALLOWANCE FOR THE EARTH'S CURVATURE

In many cases it is necessary to have available absorption functions for the inhomogeneous atmosphere with allowance for the earth's curvature. Corresponding exact formulas will be treated, as well as an approximate method for solving this problem. We start by considering some geometric relationships to be used below.

Suppose the geographic coordinates of the radiation source and receiver and their altitude are known, and that all quantities pertaining to the source are designated by subscript 1 (point 1) and all quantities pertaining to the receiver by subscript (point 2). We introduce a Cartesian coordinate system with origin at the center of the earth; the z axis is directed through the North Pole and the x axis passes through Greenwich. The x, y and z axes form a right-handed orthogonal system. In the same system we designate in the standard manner polar coordinates r, θ, φ, both sets of coordinates being related in the usual manner:

$$\left.\begin{array}{c} x=r\sin\theta\cos\varphi;\ \ y=r\sin\theta\sin\varphi;\ \ z=r\cos\theta; \\ r=\sqrt{x^2+y^2+z^2};\ \ \cos\theta=z/r;\ \ 0\leqslant\theta\leqslant\pi; \\ \sin\varphi=(y/r)\sin\theta;\ \ \cos\varphi=(x/r)\sin\theta. \end{array}\right\} \tag{4.23}$$

The altitude of a point is designated by h. It is easy to show that:

$$\left.\begin{array}{l} \text{if } \xi \text{ is the eastern longitude of the point, then } \varphi=\xi; \\ \text{if } \xi \text{ is the western longitude of the point, then } \varphi=2\pi-\xi; \\ \text{if } \eta \text{ is the northern latitude of the point, then } \theta=\pi/2-\eta; \\ \text{if } \eta \text{ is the southern latitude of the point, then } \theta=\pi/2+\eta; \end{array}\right\} \tag{4.24}$$

$$r=R+h,$$

R is the radius of the earth.

It follows from (4.10)—(4.12) that in order to compute function H it is necessary to evaluate integrals of the type

$$B=\int F(l)\,dl, \tag{4.25}$$

where $F(l)$ is some function of the points on the beam path, with integration carried out along this path. In calculating ω, function F coincides with the distribution of the absorbing-gas concentration when calculating $\bar{P}$:

$$F=\rho(l)\,P(l).$$

In general, F is a function of the geographic location of the point and of its altitude. Function $F(r,\theta,\varphi)$ can now be constructed easily with the aid of (4.24). It will be assumed that this procedure has been carried out, where r, θ, and φ are functions of the points on path l. Applying the standard rules for operating with curvilinear integrals of the first kind, one can derive the following exact formulas which make possible calculation of function H:

$$B=\int_0^s F[r(l),\ \theta(l),\ \varphi(l)]\,dl, \tag{4.26}$$

$$s=\sqrt{r_1^2+r_2^2-2r_1r_2\cos\delta}, \tag{4.27}$$

$$\cos\delta=\cos\theta_1\cos\theta_2+\sin\theta_1\sin\theta_2\cos(\varphi_2-\varphi_1), \tag{4.28}$$

$$r(l)=\left(r_1^2+l^2+2l\,\frac{r_1r_2\cos\delta-r_1^2}{s}\right)^{1/2}, \tag{4.29}$$

$$\cos\theta(l)=\frac{r_1\cos\theta_1+l\cos\gamma}{r(l)},\quad 0\leqslant\theta\leqslant\pi, \tag{4.30}$$

$$\cos\varphi(l)=\frac{r_1\sin\theta_1\sin\varphi+l\cos\beta}{r(l)\sin\theta(l)}, \tag{4.31}$$

$$\sin\varphi(l) = \frac{r_1 \sin\theta_1 \sin\varphi_1 + l\cos\alpha}{r(l)\sin\theta(l)}, \tag{4.32}$$

$$\cos\alpha = \frac{r_2 \sin\theta_2 \cos\varphi_2 - r_1 \sin\theta_1 \cos\varphi_1}{s}, \tag{4.33}$$

$$\cos\beta = \frac{r_2 \sin\theta_2 \sin\varphi_2 - r_1 \sin\theta_1 \sin\varphi_1}{s}, \tag{4.34}$$

$$\cos\gamma = \frac{r_2 \cos\theta_2 - r_1 \cos\theta_1}{s}. \tag{4.35}$$

Integrals B can be evaluated from the exact formulas (4.26)–(4.35), provided sufficiently extensive information is available on the meteorological situation in different geographic regions.

We now examine an approximate method for evaluating integrals along beam paths and which is sufficient in the majority of practical applications. It is assumed for definiteness that the radiation receiver is located above the source. This is clearly not detrimental to the generality of the result, since the integrals under study are invariant relative to changes in the limits of integration.

We designate by x_3, y_3 and z_3 the coordinates of a point (point 3) lying on a straight line passing through the locations of the receiver and situated at some minimum distance from the coordinate origin. Then

$$x_3 = x_2 + t^*(x_1 - x_2),$$

$$y_3 = y_2 + t^*(y_1 - y_2),$$

$$z_3 = z_2 + t^*(z_1 - z_2),$$

$$t^* = \frac{x_2(x_2 - x_1) + y_2(y_2 - y_1) + z_2(z_2 - z_1)}{(x_2 - x_1)^2 + (y_2 - y_1)^2 + (z_2 - z_1)^2}. \tag{4.36}$$

The expression for t^* can also be expressed in the form

$$t^* = \frac{r_2(r_2 - r_1\cos\delta)}{r_1^2 + r_2^2 - 2r_1 r_2 \cos\delta}. \tag{4.37}$$

It follows that $t^* > 0$.

Two locations of points 1, 2 and 3 are possible, depending on the value of t^*:

$$\left.\begin{array}{l} 1)\ 0 \leqslant t^* \leqslant 1 \text{ — point 3 is situated between points 1 and 2;} \\ 2)\ t^* > 1 \text{ — point 1 is situated between points 2 and 3.} \end{array}\right\} \tag{4.38}$$

We now utilize the following two facts: 1) the vertical profile of meteorological variables is much more variable than the horizontal profile; 2) infrared radiation is absorbed primarily in that part of the atmosphere where the concentration of absorbing components is maximum. With reference to the above, it is possible to replace $\theta(l)$ and $\varphi(l)$ in exact equations (4.26)—(4.35) by the coordinates of the point lying on the path and which is closest to the ground. These will be the coordinates of point 3 in the first version of (4.38), and those of point 1 in the second version. It then follows from (4.10)—(4.12) and relationships describing the pressure distribution and concentration of H_2O and CO_2 in the troposphere that function F has the form exp $(-\alpha h)$, where α is some constant. Expression (4.26) is integrated starting from a point located distance r' from the origin of the above coordinate system to a point located distance r'' from this origin, with angle σ between these directions. The resulting integral is designated by $B(\alpha, \sigma, h', h'')$, where h' and h'' are the altitudes of points r' and r''. If r' and r'' are sufficiently small compared with R, the expression for $B(\alpha, \sigma, h', h'')$ is given by

$$B(\alpha, \sigma, h', h'') =$$
$$= e^{-\alpha h'} \sqrt{\frac{\pi R}{2\alpha}} \frac{1}{\sin \psi} e^{\xi_0^2} [\Phi(\xi^*) - \Phi(\xi_0)]. \quad (4.39)$$

Here Φ is the error function defined by

$$\Phi(z) = \frac{2}{\sqrt{\pi}} \int_0^z e^{-z'^2} dz', \quad (4.40)$$

$$\xi^* = \xi + \xi_0, \quad \xi = l \sqrt{\frac{\alpha}{2R}} \sin \psi,$$
$$\xi_0 = \left(\frac{\alpha R}{2}\right)^{1/2} \cot \psi,$$
$$\cos \psi = \frac{r'' \cos \sigma - r'}{l}, \quad \sin \psi = \sqrt{1 - \cos^2 \psi}. \quad (4.41)$$

Angle σ is easily determined from the relations

$$\sigma = \sigma'' - \sigma', \quad \cos \sigma' = \frac{r_3}{R + h'}, \quad \cos \sigma'' = \frac{r_3}{R + h''}.$$

The geometric meaning of ψ is the angle between the direction of r', and of the straight line connecting points r' and r''. Particular cases of (4.39) can be found elsewhere /1/.

It follows from data in Chapter 1 that quantity α introduced there generally varies for different altitudes, if one considers the vertical distribution of any absorbing gas in the troposphere and stratosphere. Under these conditions the integration path should be subdivided into segments along which α has a constant value.

7. EXPERIMENTAL STUDY OF ABSORPTION FUNCTION *H*

Under laboratory conditions it is virtually impossible to obtain an inhomogeneous atmosphere with continuous variation in pressure, temperature and concentration of absorbing gases. Therefore absorption functions H are measured under natural conditions using the sun as radiation source.

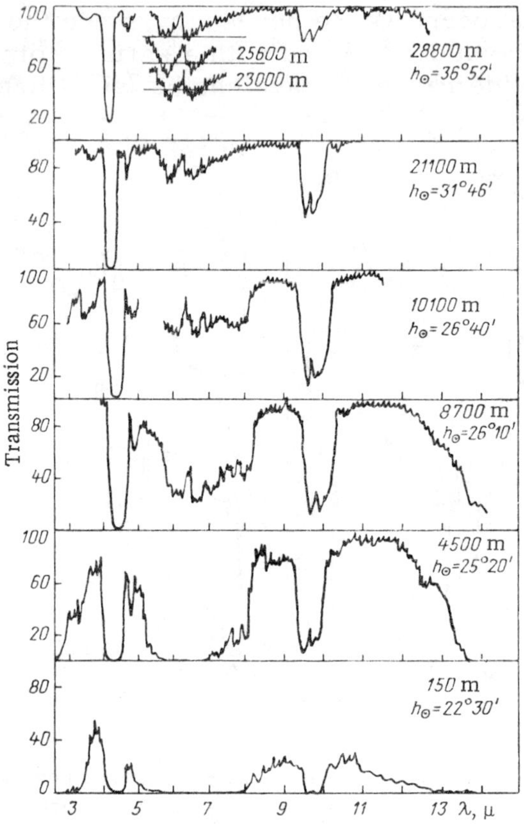

FIGURE 4.6. Transmission of different atmospheric thicknesses from various levels according to data of /344/ in the 3—13μ region (absorption in the 6.3μ band above 20 km is ascribed in /344/ to the moisture contained in the spectrometer).

At present, a large number of measurements of H have been carried out from the ground and from various altitudes with differen spectral resolution in all the ground vibration-rotation absorption bands of atmospheric gases (Figures 4.6—4.12). The most thoroughly explored absorption bands are the 6.3, 3.2, 2.7, 1.87, 1.38 and 1.1μ bands of H_2O, the 4.3 and 2.7μ bands of CO_2, and the 9.6μ band of O_3. Measurements are conducted from different altitudes by apparatus carried aloft by aircraft or balloons. In the former case altitudes up to 20 km are attained, while in the latter they extend to 30 km.

In investigations under natural conditions one does not measure the pure absorption of solar radiation by the atmosphere, but rather its attenuation due to absorption, aerosol and molecular scattering. To reduce the effect of the aerosol, the measurements are conducted only on clear, cloudless days. The effect of molecular scattering can be taken into account where necessary, since the corresponding coefficients have been determined with a satisfactory degree of accuracy over a wide range of wavelengths (see Chapter 6).

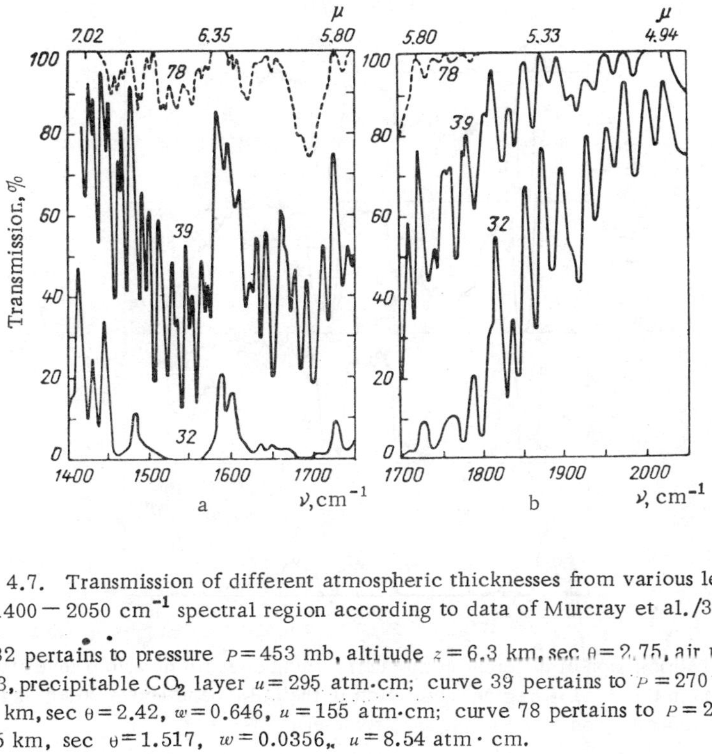

FIGURE 4.7. Transmission of different atmospheric thicknesses from various levels in the 1400 — 2050 cm^{-1} spectral region according to data of Murcray et al. /345/.

Curve 32 pertains to pressure $p = 453$ mb, altitude $z = 6.3$ km, sec $\theta = 2.75$, air mass $w = 1.23$, precipitable CO_2 layer $u = 295$ atm·cm; curve 39 pertains to $p = 270$ mb, $z = 9.9$ km, sec $\theta = 2.42$, $w = 0.646$, $u = 155$ atm·cm; curve 78 pertains to $p = 23.8$ mb, $z = 25.5$ km, sec $\theta = 1.517$, $w = 0.0356$, $u = 8.54$ atm · cm.

The most complete data on functions H were obtained by the Kondrat'ev /253, 344, 807/, Neporent and Fedorova /348, 808, 809, 877/, Murcray /79, 345—47/, Yates /349, 350/, Houghton /351—353/ and Cumming /354/ teams. Some results were also obtained in /355—357/.

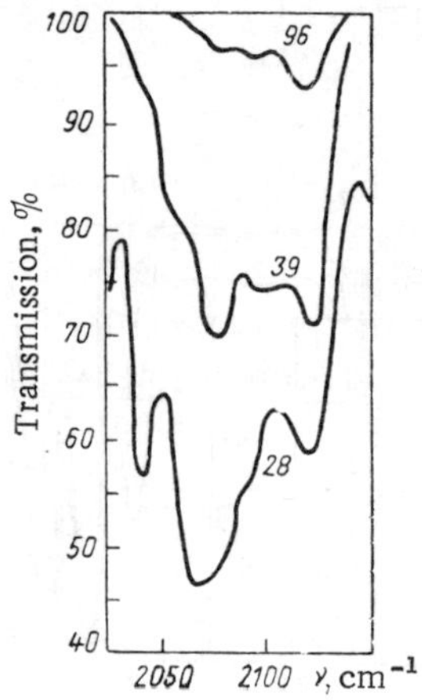

FIGURE 4.8. Transmission of different atmospheric thicknesses from various levels in the 2000 — 2150 cm^{-1} spectral region according to data of Murcray et al. /345/.

Curve 28 pertains to pressure $P = 557$ mb, altitude $z = 4.8$ km, sec $\theta = 2.97$, air mass $w = 1.63$, precipitable CO_2 layer $u = 395$ atm·cm; curve 39 pertains to $P = 27.0$ mb, $z = 9.9$ km, sec $\theta = 2.42$, $w = 0.646$, $u = 155$ atm·cm; curve 96 pertains to $P = 10.3$ mb, $z = 30.9$ km, sec $\theta = 1.321$, $w = 0.0135$, $u = 3.24$ atm·cm.

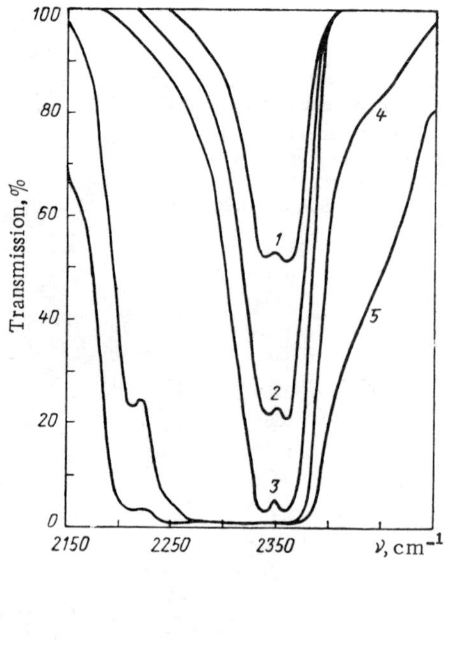

FIGURE 4.9. Transmission of different atmospheric thicknesses from various levels in the 2150 — 2500 cm^{-1} spectral region according to data of Murcray et al. /345/.

Curve 1 pertains to pressure $P = 10.3$ mb, altitude $z = 30.9$ km, sec $\theta = 1.273$, air mass $w = 0.0130$, precipitable CO_2 layer $u = 3.12$ atm.cm; curve 2 pertains to $P = 18.6$ mb, $z = 27.1$ km, sec $\theta = 1.476$, $w = 0.0272$, $u = 6.53$ atm·cm; curve 3 pertains to $P = 41.5$ mb, $z = 21.9$ km, sec $\theta = 1.63$, $w = 0.0668$, $u = 16.0$ atm·cm; curve 4 pertains to $P = 453$ mb, $z = 6.3$ km, sec $\theta = 2.75$, $w = 1.23$, $u = 295$ atm·cm; curve 5 pertains to $P = 764$ mb, $z = 2.3$ km, sec $\theta = 3.45$, $w = 2.60$, $u = 624$ atm·cm.

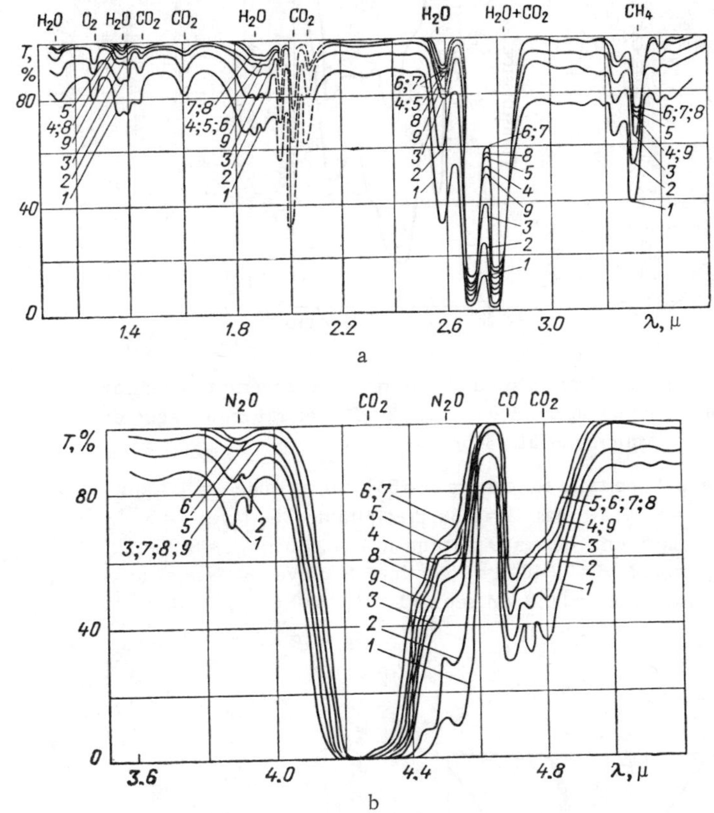

FIGURE 4.10. Transmission of the atmosphere at different altitudes in the 1 — 3.5 μ (a) and 3.5 — 5.5 μ (b) regions with the summer sun in a low position according to data of Saiedy et al. /004/.

Curve No.	Altitude, km	Sun altitude
1	12	1°57'
2	13	2°39'
3	14	3°24'
4	15	4°24'
5	15.6	4°54'
6	15	5°54'
7	14.6	7°06'
8	13.8	8°15'
9	12.8	91°8'

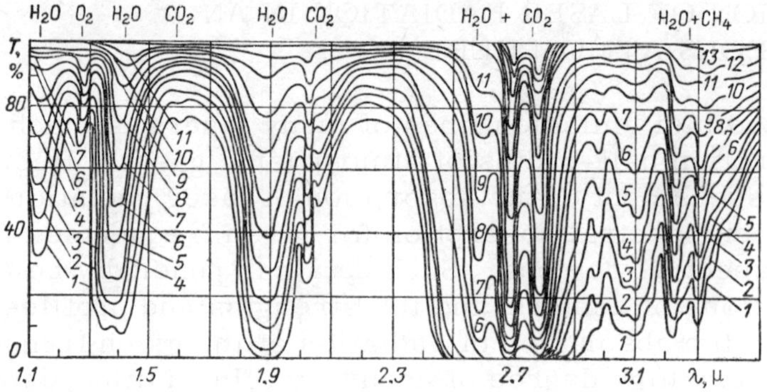

FIGURE 4.11. Transmission of the atmosphere at different altitudes in the $1-3.5\,\mu$ region according to data of Saiedy et al /804/:

Curve No.	Altitude, km	Sun altitude
1	0	20°10'
2	1.0	20°50'
3	2.3	21°20'
4	3.4	22°00'
5	4.4	22°40'
6	5.6	23°20'
7	6.8	24°00'
8	8.5	24°30'
9	9.8	25°10'
10	11.5	25°50'
11	12.5	26°30'
12	15.4	27°40'
13	17.6	28°20'
14	19.6	29°30'
15	22.1	30°40'
16	24.0	32°40'

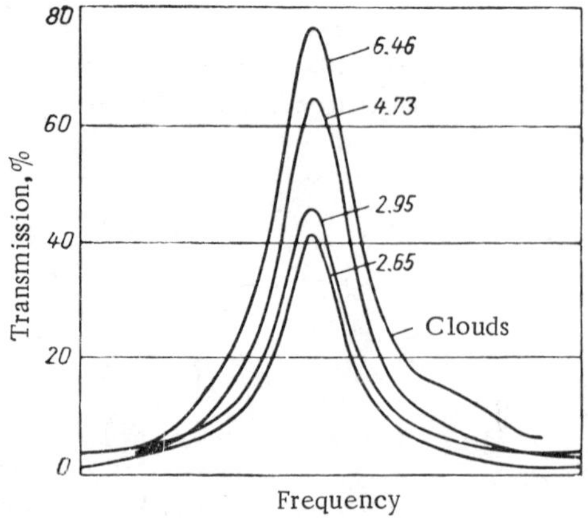

FIGURE 4.12. Transmission of the entire atmosphere in the vicinity of the H_2O absorption line with center near 6943.8 Å for different air masses according to measurements of Long /295/ on 26 December 1965.

8. ABSORPTION OF LASER RADIATION IN AN INHOMOGENEOUS ATMOSPHERE

The half-width of emission lines of many lasers is much smaller than the absorption half-widths of atmospheric gases. Hence for a quantitative assessment of the absorption of laser radiation one requires data on spectral absorption for given frequencies in certain, very narrow spectral regions. Such data can generally be calculated when the positions of centers, and the strengths and profiles of lines contributing to the absorption of radiation of the given frequency are known with a very high degree of accuracy. In addition, data must be available on the vertical distributions of pressure, temperature and concentration of absorbing gases.

A quantitative estimate of the absorption of laser radiation can be obtained in principle by measuring the absorption spectrum of solar radiation in the earth's atmosphere. The spectral resolution of the measuring instruments should be extremely high so that distortions in the spectrum due to the slit function of the spectrometers can be neglected. In addition, the locations of absorption lines and of laser emission lines must be known to within $10^{-2}-10^{-4}$ cm^{-1}, depending on the distance between the centers of the absorption and emission lines (see Chapter 3, Sec. 14). The absorption of laser radiation in an inhomogeneous atmosphere can also be measured by using the laser as an emission source.

In Sec. 5 we considered an approximate method of computing spectral absorption in an inhomogeneous atmosphere. This method consists in reducing the problem to that of computing spectral absorption in a homogeneous atmosphere, the solution of which is assumed known.

A number of useful expressions for assessing the absorption of laser radiation propagating in slant directions in the atmosphere was obtained by Plass /409/. When the volumetric concentration of the absorbing gas can be taken as constant at all altitudes, while the temperature-dependence of the line strengths and half-widths can be neglected, Plass proposes the following expression for the spectral transmission of monochromatic radiation at frequency ν in a slant direction:

$$F(\nu)=\prod_i [(1+\xi_{0i}^2)/(1+\xi_{1i}^2)]^{\vartheta}, \qquad (4.42)$$

where

$$\xi_{0i}=\gamma_{0i}(\nu-\nu_{0i})^{-1}, \qquad (4.43)$$

$$\xi_{1i}=\gamma_{1i}(\nu-\nu_{0i})^{-1}, \qquad (4.44)$$

$$\vartheta_i=\frac{S_i c P_{s_i}\sec\theta}{2\pi\gamma_{s_i} g}. \qquad (4.45)$$

Here γ_{0i} and γ_{1i} are half-widths of the ith line at the start and end of the slant path, c is the volumetric concentration of the absorbing gas, P_{s_i} and γ_{ji} are the effective pressure and half-width of lines under standard conditions, θ is the angle between the direction of radiation propagation and the vertical, g is the acceleration of gravity, ν_{0i} and S_i are the center and strength of the ith line.

Plass /409/ also examined the case when the temperature and concentration of the absorbing gas vary along the beam path. The product of the line strength and gas concentration c has the form

$$Sc = S_s c_s \gamma_s^{-m} \gamma^m. \tag{4.46}$$

Hence $F(\nu)$ is given by the expression

$$F(\nu) = \exp\left\{-2\vartheta_s \gamma_s^{-m} \int\limits_{\gamma_0}^{\gamma_1} \gamma^{m+1} [(\nu-\nu_0)^2 + \gamma^2]^{-1} d\gamma\right\}, \tag{4.47}$$

where

$$\vartheta_s = S_s c_s P_s \sec\theta / 2\pi\gamma_s g, \tag{4.48}$$

S_s, c_s, γ_s and P_s being the values of S, c, γ and P under standard conditions, and m a number which is generally different for each line and is a function of altitude. For the water vapor line in the troposphere $m \simeq 4$, in which case expression (4.48) yields

$$F(\nu) = \exp\Big(\vartheta_s \xi_s^{-4} \Big\{\frac{1}{2}(\xi_1^4 - \xi_0^4) - (\xi_1^2 - \xi_0^2) + \\ + \ln[(1+\xi_1^2)(1+\xi_0^2)^{-1}]\Big\}\Big), \tag{4.49}$$

where

$$\xi_0 = \gamma_s (\nu - \nu_0)^{-1}. \tag{4.50}$$

Finally, Plass /409/ considers the case when the center of an absorption line with a dispersion profile can be shifted by pressure. In this case the absorption coefficient is

$$k(\nu) = \frac{S}{\pi} \frac{\gamma}{(\nu - \nu_0 + A\gamma)^2 + \gamma^2}, \tag{4.51}$$

where quantity $A\gamma$ defines the shift of the line center and is a function of the pressure, since γ is a function of pressure.

The latter expression can be used to derive the following expression for $F(\nu)$:

$$F(\nu)=\{[(1+A\xi_0^2)+\xi^2]/[(1+A\xi_1^2)+\xi_1^2]\}^{\vartheta(1+A^2)^{-1}}\times$$
$$\times\exp[(1+A^2)^{-1}\cdot 2\vartheta A\{\tan^{-1}[(1+A^2)\xi_1+A]-$$
$$-\tan^{-1}[(1+A^2)\xi_0+A]\}], \qquad (4.52)$$

where ν_0 is the position of the line center at zero pressure.

Formulas (4.51) and (4.52) are valid for atomic lines, the shift of which in the visible region of the spectrum was determined experimentally. It was found that $A = 0.725$. No pressure shift of line centers has as yet been observed in vibration-rotation absorption spectra of atmospheric gases. The use of lasers for this purpose will definitely make it possible to resolve this problem. If the shift is substantial, then it must be taken into account in finding $F(\nu)$.

Experimental studies of atmospheric transmission at high resolution in the emission region of ruby lasers with the sun as radiation source were carried out by Long /295/. The spectral absorption curve obtained for solar heights from 17.4 to 19.3° is shown in Figure 3.9. The spectrum was recorded at 6933—6946 Å, which corresponds to the emission of a ruby laser operating at different ruby temperatures. The figure shows that the spectral absorption can span the range from 0 to 80% when the emission wavelength changes by less than 1 Å.

Figure 4.12 presents results of more detailed investigations of atmospheric transmission in the H_2O absorption region at 6943.8 Å, obtained by Long /295/ and corresponding to the emission of a ruby laser at ruby temperatures of 305—310°K. The figure shows transmission curves for air masses of 6.46, 4.73, 2.95 and 2.65 on 26 December 1965. The absorption in the central part of the line for an air-mass range of 2.65—6.46 increases approximately from 10 to 80%; this result pertains to atmospheric absorption in winter. Absorption in all the H_2O lines increases markedly in summer.

Part Two

SCATTERING OF VISIBLE AND INFRARED RADIATION IN THE EARTH'S ATMOSPHERE

5. THE EARTH'S ATMOSPHERE AS A SCATTERING MEDIUM

Electromagnetic waves of the visible and infrared ranges are attenuated when propagating through the earth's atmosphere due to absorption by gas constituents and by scattering 1) on molecular density fluctuations (molecular scattering), 2) on aerosol particles (aerosol scattering), and 3) on inhomogeneities, induced by turbulent motion in the atmosphere. Attenuation due to scattering on micro- and macroturbulent inhomogeneities is negligible in comparison with molecular and aerosol scattering.

Attenuation of visible and infrared radiation by molecular scattering in the atmosphere has been studied quite extensively. At present a large number of tables of coefficients of molecular or Rayleigh scattering are available and cover a wide range of wavelengths in the ultraviolet, visible and infrared.

Rayleigh scattering coefficients are virtually independent of time and space. Aerosol scattering coefficients, however, depend markedly on the dimensions, chemical composition and concentration of aerosol particles, which are highly variable in time and space. Quantitative estimates of the attenuation of visible and infrared waves in the atmosphere by aerosol scattering thus requires reliable data on all the principal characteristics of atmospheric aerosols.

In spite of the fact that the dimensions, concentration and chemical composition of atmospheric-aerosol particles vary over a wide range, it is possible to single out certain characteristic types of aerosol (such as cloud, fog, haze, precipitation, dust). The present chapter is concerned with the spectra of the dimensions, chemical composition and concentration of particles of the most characteristic atmospheric aerosols.

1. CLOUD AND FOG

1. Particulate dimension spectra. The processes of cloud and fog formation are associated with a wide variety of factors which determine the droplet size (dependence of growth rate on concentration, size distribution and nature of condensation nuclei, on temperature and cooling rate of air, on turbulence scales and intensity). It has therefore not been possible to solve theoretically the problem of particulate size distribution functions.

The existing analytic expressions were derived as approximations while processing the corresponding experimental histograms. The overwhelming majority of experimental data obtained by different investigators in the USSR /358—360, etc./ and in the West /361, 362, etc./ show that the distribution function of particulate sizes of clouds and fogs is a single-peak asymmetric curve similar to that depicted in Figure 5.1.

We shall discuss the most extensively used approximate expressions for particulate size distribution functions in droplet clouds and fogs.

In 1951, Best /363/ processed experimental data and used the formula

$$1-F=\exp\left[\left(\frac{d}{b}\right)^{c}\right], \tag{5.1}$$

where F is the fraction of liquid water in the cloud comprising water droplets with diameter below d, while b and c are parameters.

It was found that c ranges from 12 to 29, while b varies from 1.92 to 4.90. Rittberger /362/ investigated the size distribution of 300,000 droplets and found that b actually varies much more widely, from 0.4 to 12.5. He concluded that Best's formula should not be used in cloud physics.

In 1952, Khrgian and Mazin /364/ used the following expression to process the large quantity of systematized data on cloud particle sizes collected by Borovikov /365/:

$$f(a)=Aa^{2}e^{-ba}, \tag{5.2}$$

where A is a normalizing factor and b a parameter.

In 1954, Levin /366/ proposed applying a lognormal law governing the size spectrum of repeatedly crushed gold sand particles in placers, to describe the size distribution of cloud particles:

$$f(a)=\frac{1}{\sqrt{\pi t}}\frac{1}{r}\,e^{-\frac{1}{4}-\frac{1}{t}(\ln\tau)^{2}}, \tag{5.3}$$

where $\tau=a/r$, r is the most probable radius, and t is a parameter.

However, the latter lognormal law is complex and rather inconvenient to use in theoretical calculations. Therefore in 1958

Levin /368/ suggested describing the spectrum of cloud droplets using a gamma distribution:

$$f(a)=\frac{1}{\Gamma(\mu+1)}\mu^{\mu+1}\frac{a^{\mu}}{r^{\mu+1}}e^{-\mu\frac{a}{r}}, \tag{5.4}$$

where $\Gamma(\mu+1)$ is a gamma function, equal to $\mu!$ for integral μ; r and μ are parameters.

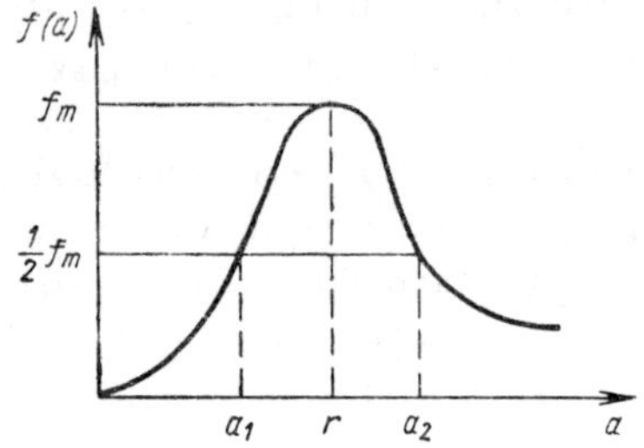

FIGURE 5.1. Typical particulate size distribution of droplet clouds and fogs.

The Khrgian-Mazin formula is a particular case of the gamma distribution when $\mu = 2$, while the gamma distribution itself is a particular case of the four-parameter Shifrin distribution /369/

$$f(a)=Aa^{\mu}e^{-\beta a^{\gamma}}. \tag{5.5}$$

The above expressions for $f(a)$ are the most widely used; others can be found elsewhere /358, 370, 371/.

The Khrgian-Mazin formula was used to describe the droplet spectrum when processing a large quantity of statistical data (660,000 droplets were measured in stratiformis), and its simplicity is a great advantage. Khrgian and Mazin /372/ showed that in the case of stratiformis which do not produce precipitation, their formula, Best's formula and the lognormal law approximate experimental curves of the particulate spectrum with errors of the same order.

Levin /358/ studied the applicability of various expressions for $f(a)$ for approximating experimental data on the dimensions of cloud and fog droplets on the basis of a large volume of statistical data obtained by the El'brus expedition. He found that the Khrgian-Mazin formula yields satisfactory results for droplet sizes $a > 5\mu$, while the lognormal law and gamma distribution describe satisfactorily the droplet spectrum pattern for virtually the entire size range that he investigated in clouds and fogs. An investigation of the microstructure of thick cumulus showed that the gamma distribution is suitable for describing the size spectrum in these clouds too /360/.

2. Quantitative data on microstructural parameters. Methods of determining microstructural parameters have been considered in detail /359, 361, 370/. Depending on the problem at hand, the investigator may be interested in one or other combination of the following characteristics of the particulate size distribution curve /358, 361/: 1) arithmetic mean radius a_1; 2) root-mean-square radius a_2; 3) root-mean-cube radius a_3; 4) most probable or modal radius r; 5) radius of droplets making the greatest contribution to the geometric cross-section, a_Q; 6) radius of droplets making the greatest contribution to the water content of the volume, a_q; 7) median radius a_m; 8) minimum and maximum radii r_{min} and r_{max}; 9) particulate concentration N per 1 cm^3.

The first seven characteristics are in some way proportional to one another, and these relationships were expressed analytically for all the distribution functions under study /358, 359, 364/. To use the Khrgian-Mazin formula one requires only one of the characteristics, whereas the gamma distribution or the lognormal law requires any two. The quantities most frequently determined experimentally are a_1 and r.

The most significant studies into the microstructure of clouds, carried out primarily in the West up to 1957, were summarized by Mason /361/. The results of Soviet and Western studies conducted before 1960 were analyzed by Borovikov et al. /359/. Analysis of data on numerical values of cloud and fog microstructural parameters /358—360, 371, 373—384/ yields a number of specific conclusions on the applicability of a given approximate formula for describing particulate size spectra and on the variation of average values of individual parameters:

1. The simple and convenient single-parameter Khrgian-Mazin formula can be used for stratiformis. To determine the particulate spectrum from this expression one requires only the arithmetic mean radius of cloud droplets. Experimental statistical data /359/ show that it is possible to recommend the value $a_1 = 5\mu$ for *St, Sc, Ns* and *Ac* clouds. Other studies for the same cloud types yield values of a_1 between 5 and 10μ. On the one hand such a discrepancy can be attributed to a disregard in earlier experiments of the capture coefficients of small droplets, which can alter values of a_1 by 1—2.5μ /359/; on the other hand, the discrepancy may be a result of regional features and values of μ different from 2 /385/.

2. The Khrgian-Mazin formula cannot be used to approximate narrow particulate spectra. In this case, it is in general possible to use the single-parameter function, setting parameter μ for the gamma distribution equal to the value most frequently encountered in such clouds. However, if the value of μ is not fixed, then the gamma distribution can be used for describing (with accuracy sufficient for

practical purposes) the droplet spectra in any clouds, at any stage of their development and in different parts. Here the narrow size spectrum will have correspondingly higher values of μ. Thus, it was found that the value of μ for high-mountain clouds and fogs lies between 4 and 8, and in more than 70% of the cases $\mu = 8$ /358/. In warm cumulus, it was found that in 72% of cases $\mu = 8-10$ /378/. For thick cumulus μ was found to vary from 0.5 to 5.5, depending on the level above the cloud base /360/. Other studies /386/ also point to variations in the values of μ and r of the gamma distribution depending on the height above the cloud base.

If the relationship between the average (a_1) and most probable (r) particulate radii is used to determine μ for data presented by Mason /361/ on the microstructure of various clouds, then it is found to be 2 for fair-weather stratus, nimbostratus and cumulus, 7 for cumulonimbus, 9 for high cumulus, and 0.33 for thick cumulus and cumulonimbus. Hence all the above information implies that μ ranges from 0.33 to 10. The data on the most probable radius indicates that its value ranges approximately from 3 to 10 microns for all clouds. Note that reference is had to variations in average characteristics. For individual clouds, and especially for individual local sections and during different development phases of a given cloud, these variations are much greater due to inescapable fluctuations in the microstructural characteristics of the clouds /358, 360/.

3. As in the case of clouds, the overwhelming majority of studies on the microstructure of fogs show that their size spectra are approximated by a gamma distribution. Here the mean, most probable radius can lie approximately between 2 and 10 microns. Sufficient data are not available for estimating reliably the value of μ.

All the information examined above on the microstructure of clouds pertains to single-peak distribution curves. It is noteworthy that twin-peak distributions are sometimes observed /382, 387, 388/.

3. **Water content of clouds and fogs.** The complete description of the microstructure of a cloud or fog involves not only the particulate size distribution, but also the particulate concentration N per unit volume. The latter can be determined if function $f(a)$ and water content q are known, the latter being defined as the quantity of liquid water contained in the droplets per unit volume and usually expressed in grams per cubic meter. Due to the great importance of this characteristic and the relatively simple technique for its determination, it is measured in the USSR not only by individual investigators, but by a network of aerological sounding stations. Water-content measurement techniques are examined elsewhere /359, 361, 370/.

Table 5.1 lists data on the water content of various clouds and fogs. It also presents average water-content characteristics of different fogs and clouds, which can differ substantially from the water contents at different levels and at different development stages of a given cloud. In addition, the mean water content of clouds is a function of temperature. It is precisely this dependence which is responsible for the scatter of the average water-content values presented in Table 5.1 from data cited by Borovikov et al. /359/. On the other hand the water content of clouds of the same type can differ substantially at the same temperature under different conditions of their genesis and development /376/.

It is seen from Table 5.1 that the average water-content values of different cloud forms, with the exception of heavy cumulus, vary over the relatively narrow range 0.1—0.3 g/m^3. The water content may possess maxima of 3—4 g/m^3 and even 43 g/m^3. However, such water contents are only very rarely attained in *St*, *Sc*, *Ac*, *Ns* and *As* clouds. In this connection we consider now the frequency of the water content of various cloud forms at different temperatures, which is of great practical importance.

Minervin /391/ constructed frequency curves of the average water contents of air-mass and frontal clouds at different temperatures after processing about 5000 aircraft measurements. The maximum frequency of the average water content of the clouds under study at temperatures from −25 to +15°C lie between 0.1 and 0.3 g/m^3. The frequency of clouds with average water content 1.0 g/cm^3 is virtually zero. Similar data were obtained elsewhere /359/.

In the majority of experimental studies on particulate size spectra in clouds and fogs little attention is paid to obtaining data on the concentration of large particles, in spite of the fact that their contribution to the water content of clouds can be very large.

Some data on the concentration of large particles in clouds have been obtained /392—394/. The concentration of large particles (in these measurements particles with radii in excess of 75, 85, 130, 200, 300, 400, 500, 650 and 800μ were recorded) was found to depend to a large extent on the cloud form, and varies highly within a given cloud. The size distribution of large particles is sometimes exponential and sometimes satisfies an inverse power law. Frequently, not a single large particle is encountered over many kilometers.

4. Average values of microstructural parameters and water content of clouds and fogs. The average spectra of particle sizes in different cloud forms and fogs are satisfactorily approximated by a gamma distribution with values of μ from 1 to 10 and of r from 3 to 10 microns. The water content q ranges approximately from 10^{-2} to 43 g/m^3. In calculating extinction coefficients for real clouds and fogs of various forms and in different development stages, these

TABLE 5.1. Water content of various clouds and fogs

Cloud form	Number of measurements	Water content, g/m^3			Source	Region of measurements
		minimum	average	maximum		
Sc, St, Ac	752	—	0.17 * (0.08 — 0.23)	1.15	/359/	Riga
Sc, St, Ac	571	—	0.27 * (0.08 — 0.38)	1.79	"	Minsk
Sc, St, Ac	493	—	0.23 * (0.11 — 0.28)	1.89	"	Vnukovo
St, Sc, Ac	4897		0.21 * (0.09 — 0.33)	3.14	"	All points of the Central Aerological Observatory
Ns, As	184		0.22 * (0.07 — 0.23)	1.56	"	Riga
Ns, As	123		0.31 * (0.14 — 0.35)	1.11	"	Minsk
Ns, As	156		0.23 * (0.09 — 0.28)	0.87	"	Vnukovo
Ns, As	1654		0.22 * (0.145 — 0.356)	1.56	"	All points of the Central Aerological Observatory
Different forms	493		0.11 * (0.005 — 0.161)		"	Leningrad
Cu, Cb, Ac	} 2157			1.5	"	Canada
Sc, St, Ac				1.4	"	"
Ac — As						
St, Sc	327 **			0.9	"	U.S.A.
Ac, Ac — As	246 **			0.5	"	"

Cu, Cb	324**			1.8	"	"
Cu cong	102		1.7 †	4.1†	"	USSR
Sc	74 flights	0.06	0.30	1.30	/361/	U.S.A.
Cu	> 1000		1.2 (0.47—1.91)		/360/	USSR
Sc	Network data	0.08	0.17 — 0.32	0.6	/389/	"
Ns	Network data	0.03	0.18—0.35	0.6	"	"
As—Ns	} 2137		0.154	0.77	/376/	Arctic
St			0.10	0.58	"	"
Sc			0.12	0.67	/375/	"
Cu	21	0.05	0.14	0.20	/378/	USSR
	22 flights		7—12	43	/390/	Western Europe
Fog	21 flights	0.1		0.22	/361/	
Fog				0.4	"	Japan
Winter fog		10^{-4}		10^{-2}	/383/	"
Ocean fog			$4.2 \cdot 10^{-3}$		/380/	"
Fog			0.10—0.30	0.60	/313/	Arctic

* Data calculated on the basis of measurements of average cloud water contents at different temperatures /359/. The variation of the average quantities is shown in parentheses.

** Results of measurements in supercooled clouds.

† Data pertain to the most dense parts of thick cumulus.

ranges of μ, r and q should naturally overlap. However, in practical utilization of such computational data it is very important to know the average or the most probable values of these parameters (Table 5.2).

TABLE 5.2. Most probable values of the gamma distribution parameters μ and r and of the water content of different droplet clouds and fogs

Cloud form	r, microns	μ	q, g/m^3
Heavy cumulus *Cu cong*	6	3	1.2
Cumulus *Cu*	6	3	0.2
Cumulonimbus *Cb*	6	1	
Stratocumulus *Sc*	5	2	0.1
Stratiformis *St*	5	2	0.1
Nimbostratus *Ns*	5	2	0.2
Altostratus *As*	5	2	0.2
Altocumulus *Ac*	5	2	0.1
Radiation fogs	5	6	0.1
Advection fogs	5	3	0.1

5. Chemical composition of cloud and fog particles. It is usually assumed in calculating the coefficients of aerosol attenuation of clouds and fogs that the particles consist only of pure water. Such an approximation is valid in the majority of cases, since the impurity concentration in cloud and fog particles is very low. This can be illustrated quantitatively by the the results of Petrenchuk and Drozdova /395/, who tabulated measured concentrations of impurities in cloud droplets and precipitation of some parts of European USSR. They found that the total content of SO_4^{2-}, Cl^-, HCO_3^-, NO_3^-, NH_4^-, Na^-, K^-, Mg^{2-} and Ca^{2-} ions in the cloud water in the vicinity of Leningrad, in Northern European USSR and Southwestern European USSR (Kiev region) is on the average 11.0, 13.0 and 15.1 mg/liter, respectively; the corresponding content in precipitation is 13.7, 14.6 and 37.1 mg/liter. It was also found /395/ that cities have a marked effect on the impurity content in cloud water and precipitation. For example, in one flight over Dnepropetrovsk the total impurity content of cloud water was found to be 300 mg/liter, over Odessa about 120 mg/liter, and between these cities approximately 10 mg/liter. The impurities in precipitation consist of copper, iron, beryllium, vanadium, cobalt, chromium, molybdenum, manganese, aluminum, zinc, nickel, and silicon in concentrations in excess of $2 \cdot 10^{-3}$ mg/liter.

However, the impurity content does not exceed 1% even in the most contaminated clouds and precipitation. Bearing in mind that the overall refractive indices of aqueous solutions of the above impurities at this concentration hardly differ from the corresponding value for pure water, it can be assumed that the impurities contained in cloud water and in precipitation have no perceptible effect on their attenuation coefficients.

2. HAZE

The term haze is used to denote frequently encountered atmospheric turbidity produced by relatively small particles at $S_m > 1$ or 2 km. The particles may be solid, liquid, wetted or nonwetted. The solid particles can have arbitrary shape, while the liquid particles are spherical. Particles with a solid core and water shell are the most frequently encountered. The shape of such two-layered particles depends on the relationship between the core and shell dimensions.

Existing experimental methods are as yet incapable of detecting exactly the size of haze particles. Available data on particulate size are also not too exact. The dimensions of aerosol particles depend markedly on the condensation, sublimation and evaporation of atmospheric moisture. We present below available experimental data on the microstructural characteristics of haze subject to the previous remarks.

1. Size spectra of **haze particles.** Junge /396, 894/ subdivided aerosols into three categories depending on particle size: "Aitken nuclei" (particle radii $r = 10^{-7}-10^{-5}$ cm), large particles ($r = 10^{-5}-10^{-4}$ cm), and gigantic particles ($r > 10^{-4}$ cm). From the standpoint of attenuation of visible and infrared waves in the atmosphere, atmospheric aerosol particles can be subdivided into optically active and inactive. The former include those which have a marked effect on the attenuation of radiation. The term optically inactive particles is applied to those which effect negligible radiation attenuation. The maximum size of particles which can be regarded as optically inactive depends on the emitted wavelength λ: the larger λ, the larger the corresponding critical particle size. Analysis of extinction coefficients shows that for any visible and infrared wavelengths, particles with linear dimensions smaller than 0.1μ can be regarded as optically inactive.

One can single out from the majority of available data two characteristic cases, distinguished by the nature of the particulate size spectra: 1) multi-peak distribution for particle sizes smaller than approximately 0.1μ /397, 398/; 2) single-peak distribution for particle sizes $a > 10^{-2}\mu$. As in the case of clouds, different investigators suggest different analytic expressions for the particulate size distribution function. For example, experimental measurements of city aerosols are approximated satisfactorily by a lognormal distribution (see equation (5.3)) /399–401/. Makhon'ko /400/ noted that this distribution was detected experimentally for stone /402/ and uranium /403/ mine dust. The distributions of gigantic chloride and sulfate condensation nuclei /404, 405/ are described satisfactorily by the Khrgian-Mazin formula (see equation (5.2)).

Junge /406/ showed that the part of the distribution curve characterizing large particles is approximated satisfactorily by what is known as the Junge distribution:

$$N = Ca^{-\beta}. \quad (5.6)$$

Here a is the particle radius, C an experimentally determined constant which is a function of particle concentration, and β a constant which for continental aerosol at the earth's surface is equal to 3.

The Junge distribution is used quite extensively. The value of β characterizes the steepness of the large-droplet branch of the distribution curve. Different investigators obtained different values of β. In measurements over various parts of the USSR at 100 m, 1 km and 2–3 km Laktionov /407, 408/ found that β is respectively 3, 4, and 5 for large particles ($a > 4\mu$). For nonhygroscopic gigantic nuclei, values $\beta = 1.7-3.2$ were found by Okita /410/. The value $\beta = 2$ was derived from measurements in the stratosphere at altitudes from 12 to 21 km at different latitudes (40°S–70°N) for particle sizes from 0.1 to 2μ /411, 412/.

TABLE 5.3. Distribution functions of aerosol particles /413/

Particle diameter, μ	Tropospheric aerosols	Stratospheric aerosols
0.5–0.1	$n \approx a^{-5/2}$	$n \approx a^{-2}$
> 0.5	$n \approx a^{-19/4}$	$n \approx a^{-4}$

Assuming dynamic equilibrium between the formation, coalescence and precipitation of aerosols and making other sensible assumptions, Friedlander /413/ obtained expressions for the distribution function which are given in Table 5.3. His theoretical values of β for the troposphere and stratosphere thus lie between 2 and 4.75. An expression similar to equation (5.6) was obtained by Idlis /414/ for the distribution function of cosmic dust particles when $\beta = 4$.

Measurements conducted in the surface layer of the atmosphere during summer 1967 in the vicinity of Ryl'sk (Kursk region). indicated that there exist two principal types of particulate size distribution /810/. Approximately every third spectra of the 100 processed spectra in the 0.1—1.5μ range is a Junge distribution with $\beta = 4$ (first type of distribution). In the second type of distribution there is a minimum concentration in the 0.25—0.6μ range and a moderate maximum for particle sizes between 2 and 3μ.

Deviations from the Junge particle size distributions were found by Fenn /811/: maxima were detected in the particle distribution in the region of 0.4μ and in the region of large particles, and minima in the 0.2—0.25μ and 0.4—0.5μ ranges. Measurements of sky brightness, carried out by Bullrich et al. /812/, point to the fact that the particulate size spectrum for sizes greater than 1μ obeys the Junge distribution for $\beta = 4$ with upper limit $a_2 = 10\mu$.

On the basis of results of the majority of investigators, it appears that the optically active part of the distribution curve of atmospheric aerosol particle sizes can be approximated by equation (5.6), where β can vary from 2 to 5. It should, however, be emphasized that individual particulate size spectra may differ from the smoothed distribution of (5.6) which, as is pointed out by Rozenberg /895/, reflects the properties of real distributions only in the most general outline. Real distributions frequently consist of curves with one or several maxima /811, 812, 896/. There is at present very little data on variations of the maxima.

It was found while measuring city dust that the dimensions of the most numerous particles range from 0.06 to 0.34μ /400/. In investigating different aerosols over the U. S. A. no particles smaller than 0.1μ were found, although the spectrometer used could measure particle sizes up to 0.08μ; the most probable particle size ranged from 0.2 to 0.5μ /415/. According to Junge /411/ the most probable size is 0.03 μ. On the basis of these data, and bearing in mind that small particles make virtually no contribution to the attenuation of radiation, one can recommend 0.1μ as the minimum particle size when using equation (5.6).

We note in conclusion that the Junge formula is only an approximation describing more or less correctly the real distributions of the larger haze particles, and does not embrace all possible distributions. However, available experimental data do not as yet suffice for carrying out the required statistical analysis and for establishing the frequency of the different particle size distributions.

2. Particulate concentration. Depending on the distribution function used, we are interested either in the total particulate concentration including the small droplets, or the particulate concentration over a given size range. In both cases it is necessary to know how the quantity under study varies as a function of location, time and height. It is well known that these variations are tremendous: in the atmosphere of large industrial cities the aerosol concentration may be as high as tens of thousands of particles per cm^3, while the purest air samples in rural areas may contain tens and in some cases single particles per cm^3 /416/.

Khrgian /416/ examined the vertical distribution of particulate concentrations in the atmosphere and showed theoretically that the concentration should decrease exponentially. A similar result was obtained elsewhere /417/, and confirmed experimentally for condensation nuclei at altitudes up to 4 km /418/. Penndorf /419/ summarized the data of various investigators and concluded that particulate concentrations over the $0.06-1\mu$ range decrease exponentially up to an altitude of 5 km. Theoretical and experimental verification of the exponential reduction in the concentration of condensation nuclei with dimensions of $10^{-2}-1\mu$ up to an altitude of 4—6 km,

$$N(z) = N(0)\, e^{-z/n}, \tag{5.7}$$

is given by Selezneva /420, 421, 900/. In equation (5.7), $N(z)$ and $N(0)$ are the concentrations at altitudes z and 0, and h is an empirical constant, which differs for different points. Numerical values of $N(0)$ and h, obtained in studies of condensation nuclei over various regions of the USSR, are given elsewhere /422, 376/.

Junge et al. /411, 412, 423—426/ conducted exhaustive studies on the concentrations of atmospheric aerosol particles at altitudes from 6 to 27 km. A major advantage of these studies is the individual study of particles of different size ($0.01-0.1\mu$, $0.1-1.0\mu$, and larger). The main results reduce to the following. The concentration of condensation nuclei in the upper troposphere is virtually independent of altitude, while in the lower troposphere and in the stratosphere it drops sharply with altitude. For $0.08-1\mu$ particles, an aerosol layer with maximum concentration was detected at altitudes from 15 to 23 km. The particulate concentration in this layer changes with time by not more than a factor of three. Measurements at different altitudes suggest that this layer is of a planetary nature. In the upper layers of the troposphere, as in the case of condensation nuclei, the large-particle concentration depends little on altitude; at altitudes of 10—15 and 23—27 km, however, a sharp rise and drop in concentration is observed.

Very few gigantic particles ($a > 1\mu$) are observed in the atmosphere. This is in agreement with data of Durbin et al. /427—429/, who show that these particles make a substantial contribution to the concentration only near the earth's surface. Its concentration decreases very rapidly with altitude. The greatest measured value, at 0.8 km, was found to be 2.5 cm^{-3}. Wright and Hodge /430/ found that the concentration of 3μ particles at 13—16, 20, and 17—29 km is, respectively, 4000, 2800 and 1000 m^{-3}.

A large quantity of data on concentrations of particles with $r > 4\mu$ over different parts of the USSR at altitudes from 0.1 to 3.0 km was obtained by Laktionov /431/. The concentration of such particles was found to depend essentially on the nature of the underlying surface. Thus, over various seas (Aral, Okhotsk, Barents) at an altitude of 0.2 km the particulate concentration was approximately identical and equal to 1.5—5 particles per liter. At the same altitude in adjoining shore regions the particulate concentration fluctuated greatly: 1.5 per liter over tundra, 30 per liter over forest-steppe, and 65 per liter over sands. In steppe regions the concentration changes sharply from winter to summer. For example, at 0.1 km in winter the concentration of particles with $r > 4\mu$ was 3 per liter, while in summer it was 60 per liter.

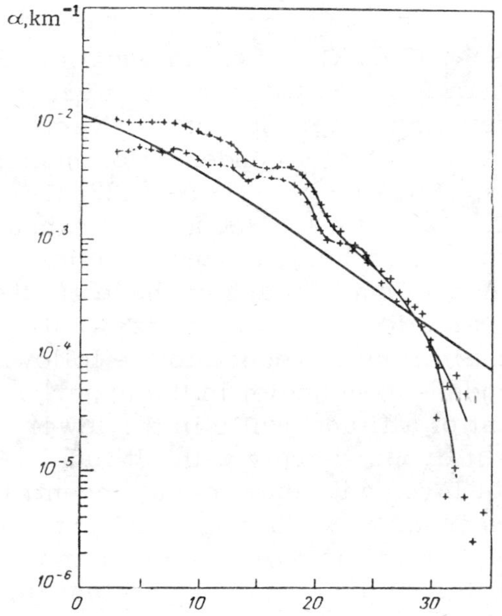

FIGURE 5.2. Examples of moderate-structured vertical profiles of the aerosol attenuation coefficient obtained on 9 April 1964 and 10 December 1964 /444/; the solid curve is the vertical variation of the Rayleigh scattering coefficient.

Direct measurements of aerosol particles above 30 km have not yet been carried out. However, analysis of optical studies /432/

indicated that aerosols are present up to altitudes of 100 km and more. Rossler and Vassy /433/ found that up to 50 km aerosol scattering exceeds Rayleigh scattering. An aerosol layer maximum was detected at 25 km, while a minimum was found to exist at the tropopause level. Data on the aerosol attenuation component from 48 to 102 km were obtained by Mikirov /434/. Aerosol attenuation is greater than Rayleigh scattering at all altitudes, and a new aerosol layer was detected at altitudes of 80 km and above. Rabinovich /436/ showed that aerosol attenuation of visible light by the atmospheric layer above 6 km amounts to 20—30% of attenuation by the lower-lying layer. These results point to the presence of a substantial concentration of aerosol particles in the upper layers of the atmosphere. At the same time it was found /446/ that the optical thickness of the atmosphere above 25 km is $2 \cdot 10^{-4}$, which indicates that the role of aerosol attenuation at high altitudes is negligible.

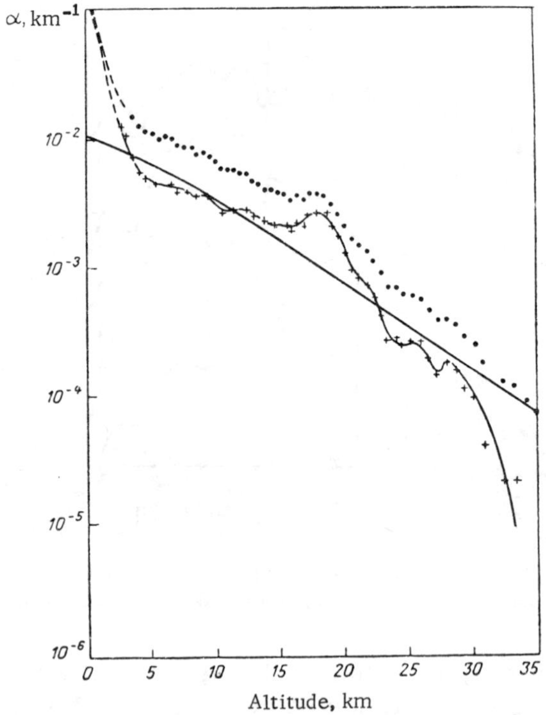

FIGURE 5.3. Example of a medium-structured vertical profile of the aerosol attenuation coefficient obtained on 8 May 1964 (curve with crosses) /444/; the solid curve is the vertical variation of the Rayleigh scattering coefficient, and the points denote values of the attenuation coefficient.

In recent years a large quantity of data has been obtained on the variation in the concentration of atmospheric aerosol particles on cloudless days by means of searchlight and laser probing of the atmosphere /437—444/. Very complete data were obtained by Elterman /441—444/. Probing was carried out in New Mexico up to altitudes of 70 km from December 1963 to December 1964, the distance between the source and collector being 30.2 km. Measurements were carried out at 0.55μ wavelength. Aerosol was detected at all altitudes, but above 35 km its density is very low. Altogether 105 vertical profiles of aerosol attenuation coefficients are listed in /444/. The aerosol profiles were categorized into moderate-structured (Figure 5.2), medium-structured (Figure 5.3), and full-structured (Figure 5.4). The vertical profile of the aerosol attenuation coefficient obtained by averaging the 105 individual profiles is shown in Figure 5.5.

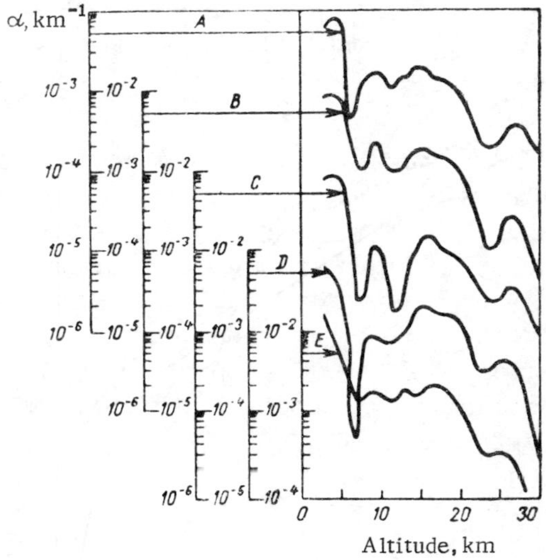

FIGURE 5.4. Examples of full-structured vertical profiles of the aerosol attenuation coefficient obtained between 2008 hrs on 12 April 1964 and 0219 hrs on 13 April 1964 / 444/.

Analysis of individual profiles of the aerosol attenuation coefficient points to the existence of aerosol stratification, which is not at all constant. The aerosol layer detected by Junge from 15 to 23 km does not appear sufficiently clear in all probings. In many cases the aerosol attenuation coefficient in the 5 to 18 km layer varies little with altitude. This is reflected also in the shape of the average profile.

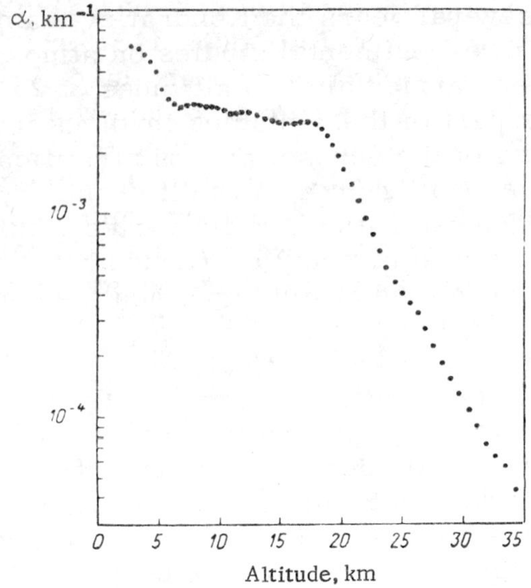

FIGURE 5.5. Average vertical profile of the aerosol attenuation coefficient obtained by averaging 105 individual profiles /444/.

Comparison of the average values of the aerosol attenuation coefficient with the values of the molecular scattering coefficient shows the following pattern. At altitudes from approximately 0 to 10 km and above 25 km, the molecular scattering coefficient is greater than the aerosol attenuation coefficient. The greatest differences occur from approximately 18 to 22 km and at 35 km. In the former case the aerosol attenuation coefficient is approximately 100—150% greater than the molecular scattering coefficient, the converse being true in the latter case.

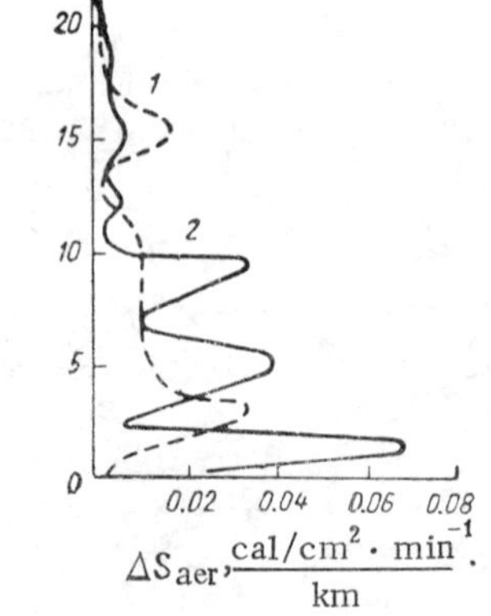

FIGURE 5.6. Vertical profiles of aerosol attenuation of direct solar radiation /445/:

1 – autumn; 2 – summer.

During the past several years the Kondrat'ev team have carried out a large number of experimental studies on atmospheric optics using instrumentation carried aloft to altitudes of 25—30 km by aerostats. A major part of these studies involves investigation of the optical properties of the aerosol and its stratification. Measurements showed that aerosol is present at different altitudes. The vertical particulate concentration profile varies within wide limits (Figure 5.6). These measurements confirmed both the altitude of aerosol layers, as well as the absolute particulate concentration at different altitudes /344, 445, 813/.

The above information shows that data on vertical profiles of atmospheric aerosol concentrations vary greatly. In spite of the fact that a large volume of data is presently available on concentrations of particles at different altitudes and in different geographic regions, this information still does not suffice for resolving the problem of average statistical profiles of particulate concentration. However, many practical problems require knowing certain average values of this quantity, which characterize the aerosol model of the atmosphere.

The experimental data of Junge on concentrations of 0.08—1.0μ particles at different altitudes led to the following aerosol model of the atmosphere /1/:

$$\left.\begin{aligned} &N(z)=N(0)\,e^{-bz}, \quad z\leqslant 5 \text{ km},\\ &N(z)=0.03, \quad 5 \text{ km}\leqslant z\leqslant 15 \text{ km},\\ &N(z)=0.03\,e^{0.06z}, \quad 15 \text{ km}\leqslant z\leqslant 20 \text{ km},\\ &N(z)=0.1\,e^{-0.09z}, \quad z\geqslant 20 \text{ km}, \end{aligned}\right\} \tag{5.8}$$

where $N(z)$ and $N(0)$ are the particulate concentrations at altitudes z and 0.

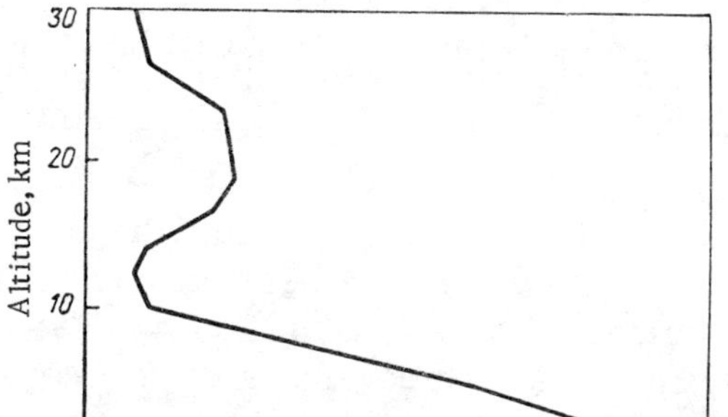

FIGURE 5.7. Vertical profile of aerosol particle concentration in the Elterman atmospheric model /447/.

Elterman /447, 448/ constructed an aerosol model of the atmosphere by summarizing available data on the aerosol attenuation coefficients. According to this model the particulate size distribution does not vary with altitude, and the particulate concentration at meteorological surface visibility $S_m = 25$ km is assumed to be 200 cm^{-3}. Concentration values at other altitudes are listed in Table 5.4.

Figure 5.7 depicts the vertical profile of aerosol particle concentration constructed from data of Table 5.3. Note that the Elterman model makes no allowance for possible variations in the parameters of the particulate size distribution function. However, the data presented here show that these variations can be quite substantial. It may hence be inferred that the particulate concentration can vary widely for the same meteorological visibility.

TABLE 5.4. Aerosol particle concentration at different altitudes in the Elterman model

Altitude, km	0	1	2	3	4	5	6	7
Concentration, cm^{-3}	200	87	38	16	7.2	3.1	1.1	0.4
Altitude, km	8	9	10	11	12	13	14	15
Concentration, cm^{-3}	0.14	0.05	0.026	0.023	0.021	0.023	0.025	0.041
Altitude, km	16	17	18	19	20	21	22	23
Concentration, cm^{-3}	0.067	0.073	0.080	0.090	0.086	0.082	0.080	0.076
Altitude, km	24	25	26	27	28	29	30	31–50
Concentration, cm^{-3}	0.052	0.036	0.025	0.024	0.022	0.020	0.019	0

It must be emphasized in conclusion that the atmospheric models considered are quite approximate, statistically averaged patterns of the aerosol stratification of a real atmosphere. In each specific case more or less significant departures of the real profiles of aerosol particle concentration from those assumed in the model are possible.

In addition to variations in the spectra of size and concentration of atmospheric aerosol particles, the question of the chemical composition of particles, which determines the components of their overall refractive index, is of great importance. A change in these components can have a substantial effect on the corresponding optical characteristics of the aerosol (scattering coefficients, scattering curves, etc.). A difference in these characteristics may be expected not only for sea and continental aerosols, but also for different types of both of these, depending on their origin.

At present, dozens of studies have been published dealing with particles of continental and sea aerosols (for example, /814—819/). Nevertheless, existing data are quite insufficient for more or less definite conclusions on the chemical compositions of particles of both sea and continental aerosols. In calculating scattering coefficients of atmospheric aerosols it is most frequently assumed that their particles consist of water, in which case the refractive index is 1.33. In a number of cases calculations are conducted for transparent particles, for which the refractive index assumes various discrete values between 1.2 and 1.5 /820/.

Future models of atmospheric aerosols should doubtlessly make allowance for the size spectra, concentration and chemical composition of particles, and also for variations in these characteristics under real conditions.

3. PRECIPITATION

All precipitation contains large particles, for which the values of parameter $\rho = \frac{2\pi a}{\lambda}$ (λ is wavelength and a particle radius) are such that the attenuation coefficients can be treated as proportional to the particle cross-sectional area (see Chapters 6 and 7). Hence in order to find the coefficients of aerosol attenuation in precipitation it is more important to know their intensity or water content, rather than their microstructural parameters. The latter have been studied in more or less satisfactory detail in rains.

Studies of rain microstructure show that, with rare exceptions, droplet size distributions are characterized by single-peak asymmetric curves, as in the case of clouds and fogs /449—453/.

The microstructural parameters vary as a function of rain intensity, and distance from the cloud which generated it. Kelkar /449/ showed that as one moved away from the cloud base to the ground, the relative number of small droplets increased regularly by evaporation during their fall. Substantial variations in microstructural parameters for different types of rain have been detected /450/. Very complete data on the microstructure of rains were obtained by Polyakova et al. /451—453/ (Table 5.5).

TABLE 5.5. Maximum and minimum values of rain characteristics /451–453/

Amount of rain	I, mm/hr	ω, g/m^3	n, m^{-3}	$a_{greatest}$, mm	a_{max}, mm	α, km^{-1}
Minimum	0.3	0.02	80	0.7	0.025	0.10
Maximum	57	1.98	19,750	3.2	0.3	3.34

The data of Table 5.5 characterize the scatter of microstructural parameters for the most frequently encountered rains. In very infrequent cases, intensity I and water content ω of a rain can be substantially greater than that given in the table. For example, the absolute maximum intensity in the USSR was recorded at Sochi in 1913 and was equal to 3.74 mm/min (224 mm/hr); the intensive torrential rain in Panama in 1911 yielded 12.6 mm/min, which corresponded to a water content of 38 g/m^3; and an intensity of up to 21.5 mm/min (1290 mm/hr) was recorded on the Hawaii Islands /416/.

The droplet size distribution in rains has been approximated by a gamma distribution with parameter $\mu = 2$ /451/. The same distribution with $\mu = 0$ was used by Marshall and Palmer /454/. Levin /368/ assumed that μ should vary from 0 to 2.

The microstructure of snowfalls was studied by Zel'manovich /455/. The particle sizes were identified with radii of spots left on filter paper after melting of the snowflakes. The particle size distribution function was given by $dn/dr = Ar^2 e^{-\beta r}$, where r is the particle radius. Zel'manovich /455/ made use of experimental data on the attenuation coefficient and the number of snowflakes to determine values of parameters A and β. The single-peak distribution curves constructed by using these values were found to be in satisfactory agreement with experimental data.

6. *SCATTERING OF VISIBLE AND INFRARED RADIATION BY A SINGLE PARTICLE*

1. REPRESENTATION OF POLARIZED LIGHT

Consider a plane monochromatic wave, the electric field intensity $\vec{E}$ of which is given in complex form by the expression

$$\vec{E} = \vec{E}_0 e^{-i\omega t + i\vec{k}\vec{r}}, \tag{6.1}$$

where ω is the frequency of the field, direction $\vec{k}$ coincides with the direction of wave propagation, $k = m_a\omega/c$, m_a is the refractive index of the medium, c is the speed of light, and $\vec{E}_0$ is a constant vector perpendicular to $\vec{k}$. The magnetic field intensity $\vec{H}$ is related to $\vec{E}$ in the usual manner:

$$\vec{H} = m_a [\vec{n}, \vec{E}], \tag{6.2}$$

where $\vec{n} = \vec{k}/k$.

The energy flow in the electromagnetic field across unit area is defined by the Poynting vector

$$\vec{P} = \frac{c}{4\pi} [\mathrm{Re}\,\vec{E}, \mathrm{Re}\,\vec{H}]. \tag{6.3}$$

Since radiation receivers react to the energy characteristics of the fields, the time-averaged components of (6.3) are optically measureable quantities. Time-averaging is conducted because the time of experimental analysis of field properties exceeds manyfold the period of light wave vibrations.

Consider now the entire ensemble of quantities being measured. The first quantity is the total wave intensity, defined as the time-averaged quantity P. Then it is also possible to measure the time-averaged vibrational intensity along a given straight line. The above vibrations can be singled out by placing along the wave path a suitably oriented analyzer. The above characteristics represent the entire experimentally determinable information pertaining to a light beam. This information can be described quantitatively as follows. Expressions (6.1) and (6.2) must be substituted into (6.3), and consideration must be given to the fact that on averaging with respect to time all terms with factors $\exp(\pm i2\omega t)$ vanish. The total intensity I is then given by

$$I = \frac{c}{8\pi} m_a (\vec{E}_0, \vec{E}_0^*). \tag{6.4}$$

We now introduce a Cartesian coordinate system x, y, z relative to which the wave under study propagates in the negative z direction. Then vector $\vec{E}_0$ is situated in the plane formed by unit vectors along the x and y axes. The direction of the straight line along which the intensity of vibrations is to be measured can be specified by angle η relative to the x axis. If $I(\eta)$ is the intensity of the vibrations in question, then relationship (6.4) yields

$$I(\eta) = \frac{c}{16\pi} m_a (S_1 + S_2 \cos 2\eta + S_3 \sin 2\eta). \tag{6.5}$$

If some device (such as a quarter-wave plate), which introduces a constant phase difference δ into the mutually perpendicular field components, is also placed in the path of the light beam, then

$$I(\eta) = \frac{c}{16\pi} m_a [S_1 + S_2 \cos 2\eta + (S_3 \cos\delta - S_4 \sin\delta) \sin\eta]. \tag{6.6}$$

In (6.5) and (6.6) S_j are Stokes parameters defined as

$$\left.\begin{aligned} S_1 &= E_{01}E_{01}^* + E_{02}^*E_{02}, \\ S_2 &= E_{02}E_{02}^* - E_{01}E_{01}^*, \\ S_3 &= E_{02}E_{01}^* + E_{02}^*E_{01}, \\ S_4 &= i(E_{02}E_{01}^* - E_{02}^*E_{01}), \end{aligned}\right\} \tag{6.7}$$

where E_{0j} are the first and second components of vector $\vec{E}_0$.

Using (6.4) and (6.7), the total intensity can be expressed in the form

$$I = \frac{c}{8\pi} m_a S_1. \tag{6.8}$$

It is easy to show that S_1 and S_4 are invariant relative to rotation about the z axis, while S_2 and S_3 are transformed according to the representation

$$S_2' = S_2 \cos 2\psi + S_3 \sin 2\psi, \quad S_3' = -S_2 \sin 2\psi + S_3 \cos 2\psi,$$

where S_2' and S_3' are the values of the Stokes parameter in a system rotated relative to the old system through angle ψ about the z axis.

It can further be shown that the end of vector $\vec{E}$ in (6.1) describes an ellipse, so waves of this type are termed elliptically polarized. The polarization ellipse can be described by angle φ between the major semiaxis of the ellipse and the first direction vector of the selected coordinate system, and degree of ellipticity β, where $\tan\beta = a_2/a_1$, a_1 and a_2 being respectively the minor and major semiaxes of the ellipse; β carries a positive sign for right-hand rotation (relative to $\vec{k}_0$) and a negative sign for the opposite sense. One can derive the relationships

$$\left.\begin{aligned} S_2 &= S_1 \cos 2\beta \cos 2\varphi, \\ S_3 &= S_1 \cos 2\beta \sin 2\varphi, \\ S_4 &= S_1 \sin 2\beta. \end{aligned}\right\} \qquad (6.9)$$

Formulas (6.6), (6.8) and (6.9) enable one to calculate the characteristics of the polarization ellipse with the aid of corresponding measurements.

2. SCATTERING OF A PLANE MONOCHROMATIC WAVE

We shall now identify the field examined in Sec. 1 with a wave incident on some object. Suppose the origin of this coordinate system is located at the center of the object and its first axis coincides with the major semiaxis of the polarization ellipse of the wave. The unit vectors of this system are designated by $\vec{l}_1$, $\vec{l}_2$, $\vec{l}_3$.

The material from which the object in question is made has a complex refractive index $m_b \neq m_a$ and consequently represents some optical heterogeneity. Hence a scattered field should exist outside the object. Subsequently all quantities characterizing the incident wave are assigned superscript 0, while those pertaining to the scattered wave are assigned superscript a.

Consider some point $\vec{r}$ at which the scattered field is analyzed. Denote by Q the plane containing the selected point and the direction of propagation of the incident wave, i.e., the third axis of the selected coordinate system.

Introduce system of coordinates g possessing a right-handed system of unit vectors $\vec{g}_1$, $\vec{g}_2$, $\vec{g}_3$ where $\vec{g}_3 = -\vec{l}_3$, $\vec{g}_1$ is perpendicular to plane Q, and $\vec{g}_2$ is parallel to this plane. In the g coordinate system the Stokes parameters $S_j^{(0)}$ of the scattered wave have the form of (6.9) and φ is the azimuthal angle at point $\vec{r}$.

The scattered wave is more convenient to analyze in the j coordinate system, the right-handed unit vectors of which are defined as follows: $\vec{j}_3 = \vec{r}/r$; $r = |\vec{r}|$, $\vec{j}_1$ is perpendicular to Q, $\vec{j}_2$ is parallel to this plane. It follows from the linearity of Maxwell's equations and from the emission principle that the scattered wave in the wave zone can be expressed in the form

$$\left.\begin{aligned} E_1^{(a)} &= \frac{e^{-i\omega t+ikr}}{ikr}\left(a_{11}(\theta,\varphi)E_1^{(0)} + a_{12}(\theta,\varphi)E_2^{(0)}\right),\\ E_2^{(a)} &= \frac{e^{-i\omega t+ikr}}{ikr}\left(a_{21}(\theta,\varphi)E_1^{(0)} + a_{22}(\theta,\varphi)E_2^{(0)}\right). \end{aligned}\right\} \tag{6.10}$$

Here r, θ, and φ are the polar coordinates of point $\vec{r}$, $E_i^{(0)}$ are the components of $\vec{E}_0^{(0)}$ in coordinate system g, $E_j^{(a)}$ are the scattered field components in the j coordinate system. The magnetic field intensity $\vec{H}$ is again calculated from (6.2). Functions $a_{ij}(\theta, \varphi)$ are determined by the particle shape and dimensions and its material properties.

Waves defined by (6.10) have the form examined in Sec. 1, since their characteristics are constant at all points of a sphere of given radius; $E_3^{(a)} = 0$ in the wave zone and factors $\exp(-i\omega t + i\vec{k}\vec{r})$ and $\exp(-i\omega t + ikr)$ are essentially the same. Hence in the j coordinate system we can introduce the Stokes parameters $S_j^{(a)}$ of scattered light and again determine expressions (6.7). If the Stokes parameters are written in the form of a single-column matrix $\{S\}$, then

$$\{S^{(a)}\} = \frac{1}{k^2r^2}\{M\}\{S^{(0)}\}. \tag{6.11}$$

The latter relationship defines scattering matrix $\{M\}$ which (as follows from the meaning of the Stokes parameter) contains in essence all the information on the scattered light. The components of the scattering matrix in explicit form are given by

$$M_{ij} = \frac{1}{2}\sum_{\alpha,\beta=1}^{2} |a_{\alpha\beta}|^2 \varepsilon_{\alpha\beta ij}, \quad i, j = 1, 2,$$

$$\varepsilon_{\alpha\beta ij} = \begin{pmatrix} 1 & 1 & 1 & 1\\ -1 & 1 & -1 & 1\\ -1 & -1 & 1 & 1\\ 1 & -1 & -1 & 1 \end{pmatrix};$$

$$\begin{aligned} M_{13} &= \mathrm{Re}\,(a_{12}a_{11}^* + a_{21}a_{22}^*), & M_{41} &= -\mathrm{Im}\,(a_{12}a_{22}^* + a_{21}a_{11}^*),\\ M_{23} &= \mathrm{Re}\,(a_{21}a_{22}^* - a_{12}a_{11}^*), & M_{43} &= -\mathrm{Im}\,(a_{11}a_{22}^* + a_{12}a_{21}^*),\\ M_{31} &= \mathrm{Re}\,(a_{12}a_{22}^* + a_{21}a_{11}^*), & M_{44} &= \mathrm{Re}\,(a_{11}a_{22}^* - a_{12}a_{21}^*),\\ M_{32} &= \mathrm{Re}\,(a_{12}a_{22}^* - a_{21}a_{11}^*), & M_{24} &= \mathrm{Im}\,(a_{21}a_{22}^* - a_{12}a_{11}^*),\\ M_{33} &= \mathrm{Re}\,(a_{11}a_{22}^* + a_{12}a_{21}^*), & M_{42} &= -\mathrm{Im}\,(a_{12}a_{22}^* - a_{21}a_{11}^*),\\ M_{14} &= \mathrm{Im}\,(a_{12}a_{11}^* + a_{21}a_{22}^*), & M_{34} &= -\mathrm{Im}\,(a_{11}a_{22}^* - a_{12}a_{21}^*). \end{aligned}$$

The explicit form of coefficients a_{ij} is determined after solving the corresponding boundary-value problem of electrodynamics, which reduces to solving Maxwell's equations for a continuous tangential field component at the interface between media with different electromagnetic properties. This problem is quite complicated, for which reason it was solved only for some particular cases. For spherical particles $a_{12} = a_{21} = 0$, a_{11} and a_{22} are functions of θ alone, and when $\theta = \pi$ (when the direction of scattered wave propagation is the same as that of the incident wave) $a_{11} = a_{22}$. In addition, M_{11} has the same meaning as the scattering indicatrix.

3. COEFFICIENTS OF SCATTERING, ABSORPTION AND EXTINCTION

It follows from (6.8) and (6.11) that intensity $I^{(a)}$ of scattered light is given by

$$I^{(a)} = \frac{m_a c}{16\pi k^2 r^2} (S_1^{(0)} M_{11} + S_2^{(0)} M_{12} + S_3^{(0)} 2M_{13} + S_4^{(0)} M_{14}). \tag{6.12}$$

Further, it follows from (6.2), (6.3) and (6.10) that the direction of the Poynting vector coincides with that of $\vec{j}_3$. Hence the energy flow $P^{(a)}$ of the scattered field through a sphere of sufficiently large radius is $\int I^{(a)} r^2 \sin\theta d\theta d\varphi$. The ratio of the latter quantity to the intensity of the incident light defines the scattering coefficient σ_{sc}.

Substitution of (6.12) into the definition of the flow, employing expression (6.8) for calculating the intensity of incident light, and also expressions (6.9), yields

$$\sigma_{sc} = \frac{1}{2k^2} \int (M_{11} + M_{12} \cos 2\beta \cos 2\varphi + \\ + 2M_{13} \cos 2\beta \sin 2\varphi + 2M_{14} \sin 2\beta) \sin\theta \, d\theta \, d\varphi. \tag{6.13}$$

Hence one can derive σ_{sc} for scattering of linearly polarized light ($\beta = 0$) and for circularly polarized light ($\beta = \pm\pi/4$ depending on the direction of polarization). If the particle is spherical, then, using the properties of a_{ij},

$$\sigma_{sc} = \frac{\pi}{k^2} \int_0^{\pi} M_{11} \sin\theta \, d\theta \tag{6.14}$$

irrespective of the polarization state of the incident beam.

Absorption coefficient σ_{abs} is defined as the ratio of the Poynting flow vector of the total field through a sphere with radius r to the intensity of incident light, with the sign reversed. The flow defined above is equal in absolute magnitude to the energy of the field absorbed in the volume under study. The minus sign indicates that we deal with an energy sink.

The Poynting vector $\vec{P}$ of the total field, in line with the principle of superposition of fields and expression (6.3), can be expressed in the form $\vec{P}=\vec{P}^{(0)}+\vec{P}^{(a)}+\vec{P}'$, where

$$\vec{P}' = \frac{c}{8\pi}\,\mathrm{Re}\{[\vec{E}^{(a)}, \vec{H}^{(0)*}]+[\vec{E}^{(0)}, \vec{H}^{(a)*}]\}. \tag{6.15}$$

Vector $\vec{P}'$ is here averaged over time.

From the definition of the absorption coefficient we have

$$\sigma_{\mathrm{abs}} = -\frac{1}{I^{(0)}}\int (\vec{P}, \vec{j}_3)\, ds, \tag{6.16}$$

where $ds = r^2 \sin\theta\, d\theta\, d\varphi$.

The sum of the scattering and absorption coefficients defines the extinction coefficient σ:

$$\sigma = \sigma_{\mathrm{abs}} + \sigma_{\mathrm{sc}}. \tag{6.17}$$

If the expression for $\vec{P}$ is substituted into (6.16), and if it is remembered that the integral involving the first term vanishes and that involving the second term yields the value of the scattering coefficient, then (6.17) yields

$$\sigma = -\frac{1}{I_0}\int (\vec{P}', \vec{j}_3)\, ds. \tag{6.18}$$

Upon substituting (6.15) into (6.18) one can write the integrand explicitly, using (6.1) and (6.10) for the electric field components and (6.2) for the magnetic field components. This gives rise to integral of the type $\int F(\theta, \varphi) \exp[ikr(1+\cos\theta)] \sin\theta d\theta$. Since $kr \gg 1$ in the wave zone, the above integrals can be evaluated by the steady-phase method. Computation yields /456/

$$\sigma = -\frac{1}{k^2}\,\mathrm{Re}\int\limits_0^{2\pi} [(a_{11}+a_{22}) + \cos^2\beta\cos 2\varphi\,(a_{22}-a_{11}) + (a_{12}+a_{21})\cos 2\beta\sin 2\varphi + i\,(a_{21}-a_{12})\sin 2\beta]\, d\varphi. \tag{6.19}$$

All functions a_{ij} in (6.19) have argument $\theta = \pi$.

Expression (6.19) represents in essence the known optical theorem generalized to the case of elliptically polarized light and scattering particles of arbitrary shape.

For spherical particles, the corresponding properties of a_{ij} together with (6.19) yield

$$\sigma = -\frac{4\pi}{k^2} \operatorname{Re} a(\pi) \tag{6.20}$$

irrespective of the polarization state of the incident wave. Here $a(\pi) = a_{11}(\pi) = a_{22}(\pi)$ represents the amplitude of a forward scattered wave.

Computation of the extinction coefficient involves knowing the amplitude of the wave scattered at a small angle to the direction of incident wave propagation. The reasons for this are quite simple, if it is borne in mind that any experimental scheme for determining σ is essentially associated with the study of interference between the incident wave and a wave scattered at small angles. Integrals of the above form are independent of the upper limit when evaluated by the steady-phase method. It can therefore be shown that the value of σ obtained from (6.19) coincides with the screening constant η. The latter is determined experimentally as $\eta = I_0^{-1}(\Pi_0 - \Pi)$ where Π and Π_0 are the quantities of energy recorded by the radiation receiver when the particle is located in the emission flux and when it is far from it. From a purely computational point of view formula (6.19) is useful in that it allows any approximate information on the amplitudes of light scattered at small angles to be used when determining σ.

4. SCALAR APPROXIMATION IN THE LIGHT-SCATTERING PROBLEM

The term scalar approximation is applied to the replacement of the solution of Maxwell's vectorial equations by the solution of Helmholtz's scalar equation for some function ψ. The latter has the form

$$\Delta\psi + k^2 V(\vec{r})\,\psi = 0. \tag{6.21}$$

The advantage of the scalar approximation is that it is sometimes much simpler to find an approximate solution of (6.21) than some asymptotic solution of Maxwell's equations. It is subsequently found that such solutions embrace the region of small scattering angles and hence allow some approximate representations to be sought for the extinction coefficient. Replacement of Maxwell's equations by the Helmholtz equation was examined elsewhere /457—459/ in order to gain insight into the meaning of the "scalar disturbance" which is a part of the diffraction theory of optics.

Consider first the scalar approximation for spherical particles. The solution of equation (6.21) is usually sought for the following boundary conditions at the surface of a sphere with radius a:

$$\psi_1+\psi_2=\psi_3, \quad P_0\,(\partial\,(a\psi_1+a\psi_2)/\partial a)=P_1\,\frac{\partial}{\partial a}(a\psi_3),$$

where ψ_1, ψ_2 and ψ_3 are functions which describe the incident, external and internal waves (here ψ represents a plane wave); P_1 and P_0 are constants.

It is possible to prove the following hypothesis /460/: in the region of small scattering angles the solution of Maxwell's equations for the potential field of the sphere can be represented as half the sum of the solutions of equation (6.21) for $P=P_0=1$ and for $P_0=1/P=m$, where m is the relative complex refractive index of the sphere material. The first solution corresponds to electrical vibrations, and the second to magnetic vibrations. This hypothesis can be established by simply comparing the respective solutions for $\theta\simeq\pi$.

Using the above hypothesis and some methods of the quantum theory of collisions, one can derive, for example, all the results cited in the second chapter of van de Hulst's book /461/. As an illustration we give the expression for the coefficient of light extinction by a two-layered transparent sphere, the dimensions of which markedly exceed the wavelength:

$$\sigma=\pi r_2^2\,(2-x_1-x_2), \quad x_1=[2-k\,(\rho_2,\,\beta)]\,[1-a_1/a_2],$$

$$x_2=4\int\limits_0^{a_1/a_2}\cos\{\rho_2\sqrt{1-\xi_2}+$$

$$+\rho_2\,[(n_1-1)/(n_2-1)-1]\sqrt{(r_1/r_2)^2-\xi^2}\}\,\xi\,d\xi.$$

Subscript "1" is assigned to quantities pertaining to the "core," and subscript "2" to the "shell"; $k=\sigma'/\pi a_2^2$, where σ' is the extinction coefficient for a sphere with radius a_2; $m=n_2$; $\rho_2=2\pi a_2\,(n_1-1)/\lambda$, where λ is the wavelength of the scattered light.

The refractive index α of a medium containing N particles with radii $a\gg\lambda$ is

$$\alpha=1+Na^2\,\frac{2\pi}{k}\left\{e^{-\rho\tan\beta}\,\frac{\cos\beta}{\rho}\cos(\rho-\beta)+\right.$$

$$\left.+\,e^{-\rho\tan\beta}\,\frac{\cos^2\beta}{\rho^2}\sin(\rho-2\beta)+\frac{\sin 2\beta\cos^2\beta}{\rho^2}\right\},$$

$$k=2\pi/\lambda, \quad \rho=2\,(n-1)\,ka, \quad \tan\beta=\varkappa/(n-1), \quad m=n+i\varkappa.$$

We now examine light scattering at small angles by particles of arbitrary shape with dimensions considerably exceeding the wavelength of the scattered light. In this case the field can be expressed in terms of some scalar function. This quantity (for known assumptions) coincides with the result of Kirchhoff's diffraction theory as applied to the description of Fraunhofer diffraction.

We introduce the vectorial and scalar potentials $\vec{A}$ and φ of the electromagnetic field satisfying $\vec{H} = m_a \operatorname{curl} \vec{A}$, $\vec{E} = -(m_a/c)\,(i\omega\vec{A})$ and express the Lorentz condition in the form $\operatorname{div} \vec{A} + (m_a/c)\,(i\omega\varphi) = 0$. The expedience of this form of $\vec{A}$ and φ in scattering problems will be noted later.

With the above expressions one can derive the following equation for $\vec{E}$ from Maxwell's equations:

$$\vec{E}(\vec{r}) = \vec{E}_0 e^{-i\vec{k}\vec{r}} + \\ + B(\vec{r}) \int U(\vec{r}\,')\,\vec{E}(\vec{r}\,')\,G(\vec{r}/\vec{r}\,')\,d\vec{r}\,'. \qquad (6.22)$$

Here $B(\vec{r}) = (4\pi)^{-1}(k^2 + \operatorname{grad}\ \operatorname{div})$, $G(\vec{r}/\vec{r}\,') = e^{-ikR}/R$, $R = |\vec{r} - \vec{r}\,'|$, $U = m^2 - 1$ inside the particle and $U = 0$ outside of it, and $m = m_j/m_a$.

The procedure of the subsequent solution is as follows. First, equation (6.22) is employed for finding field $\vec{E}_j$ inside the particle. Then, expressions

$$\left.\begin{aligned} \vec{E}_a &= \frac{e^{-ikr}}{r}\,(C_\theta i_\theta + C_\varphi i_\varphi), \\ C &= \frac{k^2}{4\pi} \int U(\vec{r}\,')\,\vec{E}_j(\vec{r}\,')\,e^{-i\vec{k}_a\vec{r}\,'}\,d\vec{r}\,' \end{aligned}\right\} \qquad (6.23)$$

following from equation (6.22), are used to determine the scattered field $\vec{E}_a$ in the wave zone. Here i_θ and i_φ are unit vectors of a spherical coordinate system with origin at the particle center.

The derivation of (6.22) and (6.23) shows that the free term in equation (6.22) can be represented in the form of an incident wave, and the region of integration can be reduced to the particle volume only when the field potentials and the Lorentz condition are expressed in the above form.

Solution of equation (6.22) by successive approximations yields $\vec{E}_j = \sum_{n=0}^{\infty} \vec{D}_n$, where $\vec{D}_0 = \vec{E}_0 e^{-i\vec{k}\vec{r}}$ and for $n \neq 0$

$$\vec{D}_n(\vec{r}) = B(\vec{r}) \int d\vec{r}_1 U(\vec{r}_1) G(\vec{r}/\vec{r}_1) \times$$
$$\times B(\vec{r}_1) \int d\vec{r}_2 U(\vec{r}_2) G(\vec{r}_1/\vec{r}_2) B(\vec{r}_2) \ldots B(\vec{r}_{n-2}) \times$$
$$\times \int d\vec{r}_{n-1} U(\vec{r}_{n-1}) G(\vec{r}_{n-2}/\vec{r}_{n-1}) \times$$
$$\times B(\vec{r}_{n-1}) \int d\vec{r}_n U(\vec{r}_n) G(\vec{r}_{n-1}/\vec{r}_n) \vec{E}_0 e^{-i\vec{k}\vec{r}_n}. \qquad (6.24)$$

We assume additionally that the complex refractive index of the particle material varies little over a distance of the order of a wavelength. Then, by evaluating the last integral in (6.24) by the steady-phase method and using operator B, it can be shown that (by virtue of the assumption made) the second term, involving the operator grad div, can be neglected. Since in this approximation $B(\vec{r}_{n-1}) \int d\vec{r}_n U(\vec{r}_n) G(\vec{r}_{n-1}/\vec{r}_n) \vec{E}_0 e^{-i\vec{k}\vec{r}_n}$ differs from vector $\vec{E}_0 e^{-i\vec{k}\vec{r}_{n-1}}$ only by a scalar factor, the above procedure can be continued. This will be equivalent to setting operator B equal to $k^2(4\pi)^{-1}$. Then transformation of (6.24) on our assumption becomes analogous to the calculations in the relevant part of Schiff's paper /462/. Therefore we present here only the final result for vector $\vec{C}$:

$$\vec{C} = -\vec{E}_0 \frac{ik}{2\pi} \int e^{-i(q_1 x_1 + q_2 x_2)} dx_1 dx_2 \times$$
$$\times \left(1 - e^{-\frac{ik}{2} \int U(\vec{r}) dx_3}\right), \qquad (6.25)$$

where $\vec{q} = \vec{k} - \vec{k}_a$, integration is taken over the particle volume, and the x_3 axis is directed along $\vec{k}$.

If $|m| \gg 1$, the second term in (6.25) can be neglected. Then the expression for the components of the diffracted field, following from expressions (6.23) and (6.25), will represent the results of the Kirchhoff method for Fraunhofer diffraction.

It is clear that (6.23) and (6.25) substantially widen the limits of the scalar approximation to include large particles of arbitrary shape.

If the exponential power in the parentheses of (6.25) is small, then we have immediately $\vec{C} = (k^2/4\pi) \times \int U(\vec{r}) e^{-i(q_1 x_1 + q_2 x_2)} d\vec{r}$. For small scattering angles this expression together with (6.25) is the starting equation in analyzing wave propagation in turbulent atmosphere (see, for example, /463/) and in examining Rayleigh-Gans scattering (according to the terminology of /461/).

Relationships (6.23), (6.21) and (6.19) yield the following expression for the light extinction coefficient σ:

$$\sigma = 2\mathrm{Re} \int (1 - e^{-\frac{ik}{2}\int U(\vec{r})\, dx_3})\, dx_1\, dx_2.$$

As an illustration we present an expression for the extinction coefficient of a nonabsorbing ellipsoid with near-unity relative refractive index:

$$\frac{\sigma}{S} = 2 - \frac{4}{\pi}\left(\frac{\cos b}{b^2} + \frac{\sin b}{b}\right) + \frac{4}{b^2}\sum_{n=0}^{\infty}(-1)^n (1-P^2) \times$$

$$\times \frac{\partial^{2n}}{\partial z^{2n}} J_0(z) + \frac{4\sqrt{1-P^2}}{b}\sum_{n=0}^{\infty}(-1)^n (1-P^2)\frac{\partial^{2n}}{\partial z^{2n}} J_1(z),$$

where

$$z = b\sqrt{1-P^2};$$
$$P = 4S/(SV\sqrt{a});$$
$$b = 2k(m-1)/\sqrt{a};$$
$$a = a_1^{-2}\cos^2\alpha_1 + a_2^{-2}\cos^2\alpha_2 + a_3^{-2}\cos^2\alpha_3,$$

S is the cross-sectional area of the projection of the ellipsoid on the perpendicular plane, v is the volume of the ellipsoid, α_j is the angle between k and the semimajor axis j of the ellipsoid with value a_j; J_0 and J_1 are Bessel functions.

If $\vec{k}$ is parallel to one of the major axes of the ellipsoid, then σ/S is equal to the corresponding expression for a sphere with radius equal to the length of the semiaxis along which propagates the wave incident on the ellipsoid.

The last expression plays the same role as the formula for the coefficient of extinction due to "soft" spheres, i.e., it provides a qualitatively true description of the behavior of the coefficient of extinction by ellipsoids of arbitrary size.

5. SCATTERING OF AN INHOMOGENEOUS FIELD

We apply the term inhomogeneous to a wave whose form differs from that described in Sec. 1. This can be either a polychromatic wave, or a nonplanar wave, or one having a nonconstant vector $\vec{E}_0$. Any combination of these factors is also possible.

In the case under study all the measured quantities can be constructed from the components of matrix $D(\vec{r}_1, \vec{r}_2, t_1, t_2)$, where $\vec{r}_1$ and $\vec{r}_2$ are arbitrary points of the field, and t_1 and t_2 are arbitrary times. Formally, matrix D can be written as follows. Vectors $\vec{E}$ and $\vec{H}$ are represented by single-column matrices, the elements of which are the components of the vector in some coordinate system. These matrices are denoted by E and H, in which case

$$D(\vec{r}_1, \vec{r}_2, t_1, t_2) = \langle E(\vec{r}_1, t_1)\,\overline{H}(\vec{r}_2, t_2)\rangle, \qquad (6.26)$$

where the bar above H denotes a transposed matrix while symbol $\langle\ \rangle$ means that each component of tensor $E\overline{H}$ averaged with respect to time according to the rule

$$\langle f_1 f_2\rangle = \lim_{T\to\infty} \frac{1}{2T} \int_{-T}^{T} f_1(\vec{r}_1, t_1+\xi)\, f_2(\vec{r}_2, t_2+\xi)\, d\xi,$$

f_1 and f_2 being some components of $\vec{E}$ and $\vec{H}$.

Subsequently we consider steady fields, for which D is a function only of the difference $t_2 - t_1$ and for which it is possible to introduce the expansion

$$D(\vec{r}_1, \vec{r}_2, t_1, t_2) = \int \Gamma(\vec{r}_1, \vec{r}_2, \omega)\, e^{-i\omega(t_1-t_2)}\, d\omega, \qquad (6.27)$$

where the components of matrix Γ are the Fourier components of the corresponding elements of matrix D.

The Stokes parameters at some point $\vec{r}$ are constructed from components of D at $\vec{r} = \vec{r}_1 = \vec{r}_2$ and $\tau = t_1 - t_2 = 0$. The total intensity, for example, according to (6.26) and (6.4), is $SpD(\vec{r}, \vec{r}, t_1, t_1)$. Experimental procedures for determining $D(\vec{r}_1, \vec{r}_2, t_1, t_2)$ were examined by Born and Wolf /464/. For example, components of $D(\vec{r}, \vec{r}, t_1 - t_2)$ arise in examining interference in Junge's experimental scheme, while $D(\vec{r}_1, \vec{r}_2, 0)$ is needed for diffraction analysis for sufficiently large openings. In the scalar version D is usually termed the coherence function.

Consider now an arbitrary electromagnetic field scattered by some object /465/. We assume that the medium, external relative to the object, is nonmagnetic and spatially homogeneous.

The general idea of the calculations presented below is well known. We denote by $\vec{F}(\vec{r}, t)$ and $\vec{F}_b(\vec{r}, t)$ the electric $\vec{E}(\vec{r}, t)$ or magnetic $\vec{H}(\vec{r}, t)$ vectors of the incident and total fields outside the object (t is time and $\vec{r}$ coordinates of some point). Relationship $\vec{F}_b(\vec{r}, t) = \hat{T}\vec{F}(\vec{r}, t)$ defines the interaction operator $\hat{T}$, which is linear within the framework of linear electrodynamics. We have to find this operator on the assumption that the solution to the scattering problem of plane monochromatic waves at the object under study is known. Operator $\hat{T}$ can be found by expanding the incident field with respect to plane waves and using the linearity of $\hat{T}$.

It follows from Maxwell's equations and from the relationship between $\vec{E}$ and the electrical induction vector that

$$\Delta\vec{F} - \frac{1}{c^2}\frac{\partial^2\vec{F}}{\partial t^2} - \frac{1}{c}\int \vec{F}(\vec{r}, \tau)\frac{\partial^2 f}{\partial\tau^2}d\tau = 0, \tag{6.28}$$

where $f(\tau)$ is the relaxation function (see, for example, /466/).

It can be shown using δ-functions that for real $\vec{F}$, satisfying equation (6.28), there exist expansions

$$\left.\begin{aligned} \vec{F}(\vec{r}, t) &= 2\mathrm{Re}\int \frac{\omega\sqrt{\varepsilon(\omega)}}{2c}\vec{G}\left(\frac{\omega\sqrt{\varepsilon}}{c}\vec{k}_{0\omega}\right) e^{\frac{-i\omega\sqrt{\varepsilon}\vec{k}_0\vec{r}}{c} + i\omega t} d\vec{k}_0\, d\omega, \\ \frac{\omega\sqrt{\varepsilon}}{2c}\vec{G}\left(\frac{\omega\sqrt{\varepsilon}}{c}\vec{k}_0, \omega\right) &= \frac{1}{(2\pi)^4}\int \vec{F}(\vec{r}', t')e^{-i\omega t + i\vec{k}\vec{r}'}k^2\, dk\, d\vec{r}'\, dt' \end{aligned}\right\} \tag{6.29}$$

where $\omega \in [0, \infty]$; $d\vec{k}_0 = \sin\beta d\beta d\psi$; β and ψ are coordinates of unit vector $\vec{k}_0$ on a unit sphere; $\varepsilon(\omega) = 1 + \int f(\tau)\exp[-i\omega\tau]\,d\tau$ is the dielectric constant for frequency ω; c is the speed of light.

It follows from the condition $\operatorname{div}\vec{F} = 0$ that $\vec{G}\left(\frac{\omega\sqrt{\varepsilon}}{c}\vec{k}_0, \omega\right)\vec{k}_0 = 0$, and Maxwell's equations yield $\vec{G}_H = \left(\frac{c}{\omega}\right)^2[\vec{k}_0\vec{G}]$. The latter expressions imply that equation (6.29) is an expansion of the incident field with respect to plane waves.

Expansion (6.29) presumes quadratic integrability of $\vec{F}(\vec{r}, t)$ with respect to time. However, in optics one frequently deals with the opposite case, since the quantities being measured are constructed from the components of tensor D.

Following Born and Wolf /464/ we can introduce the quadratically integrable function

$$\vec{F}^{(T)}(\vec{r}, t) = \begin{cases} \vec{F}(\vec{r}, t), & |t| \leqslant T, \\ 0, & |t| > T, \end{cases} \tag{6.30}$$

the tensor F_{pq} of which coincides with the respective tensor of function $\vec{F}$. It is assumed that this procedure has been followed in every case.

After commutation of $\hat{T}$ with the sign of the integral in equation (6.29) it will actually operate on a plane monochromatic wave. Upon the scattering of such a wave the field outside the object is composed of the incident and scattered fields, i.e., $\hat{T} = \hat{P} + \hat{T}_a$, where $\hat{P}$ is an operator which does not modify the plane wave, while $\hat{T}_a$ is the scattering operator. The latter is usually determined as follows (see /461/ and Sec. 2). The incident field is expressed in the coordinate system with unit vectors $\vec{g}_1, \vec{g}_2, \vec{g}_3$, where $\vec{g}_1$ is perpendicular to the plane containing vector $\vec{k}_0$ and vector $\vec{r}$, which relates the coordinate origin and the point of observation of the scattered field, $\vec{g}_2$ is parallel to this plane, while $\vec{g}_3 = \vec{k}_0$. The scattered-field components are then obtained in a coordinate system with unit vectors $\vec{m}_1, \vec{m}_2, \vec{m}_3$; here $\vec{m}_3 = \vec{r}_0 = \vec{r}/|r|$, while vector $\vec{m}_1$ is perpendicular, and $\vec{m}_2$ parallel, to the aforementioned plane.

In relationships (6.29) and (6.30), all the vectors are defined in an arbitrary coordinate system with unit vectors $\vec{l}_1, \vec{l}_2, \vec{l}_3$ and with origin at the center of the scattering object. Hence in this coordinate system $\hat{T}_a = \hat{N}^{-1}\hat{T}_g\hat{M}$, where $\hat{T}_g$ is the scattering operator of a plane wave in the $\vec{g}_1, \vec{g}_2, \vec{g}_3$ coordinate system; M is a transformation which transforms a vector from system $(\vec{l}_1, \vec{l}_2, \vec{l}_3)$, defined in the $\vec{k}$ space, to system $(\vec{g}_1, \vec{g}_2, \vec{g}_3)$; N is a transformation which transforms this vector from system $(\vec{l}_1, \vec{l}_2, \vec{l}_3)$ to system $(\vec{m}_1, \vec{m}_2, \vec{m}_3)$. It can be shown that the energy of the field is invariant under these transformations, i.e., the field obtained after transformation is equivalent to the starting field.

In order to facilitate the introduction of operator matrices in the usual manner, we represent the vectors by single-column matrices with elements which are components of vectors in the $(\vec{l}_1, \vec{l}_2, \vec{l}_3)$ system. Then it follows from (6.29) and (6.30) and from the expression for the scattering operator in $(\vec{l}_1, \vec{l}_2, \vec{l}_3)$ that

$$F_b(\vec{r}, t) = F(\vec{r}, t) + \frac{2\mathrm{Re}}{(2\pi)^4}\int U^{(\vec{E}, \vec{H})}(\omega, \vec{k}_0, \vec{r}) F(\vec{r}\,', t') \times \\ \times e^{i\vec{k}\vec{r}\,' - i\omega(t-t')}\, d\vec{k}\, d\omega\, d\vec{r}\,'\, dt' \equiv F + F_a, \tag{6.31}$$

where F and F_b are matrices corresponding to $\vec{F}$ and $\vec{F}_b$; $U = N^{-1}T_g^{(\vec{E}, \vec{H})}$; N and M are invariant matrices of operators $\hat{N}$ and $\hat{M}$, expressed below in explicit form: $T_g^{(\vec{E}, \vec{H})}$ is the matrix of operator $\hat{T}_g$ in the $(\vec{g}_1, \vec{g}_2, \vec{g}_3)$ coordinate system. The last matrix is equal to $T_g^{(\vec{E})}$ if $\vec{F}$ is taken as $\vec{E}$, and is equal to $T_g^{(\vec{H})}$ if $\vec{F}$ is taken as $\vec{H}$. Matrices $T_g^{(\vec{E})}$ and $T_g^{(\vec{H})}$ are determined on solving Maxwell's equations, obtained after eliminating the time-dependent factor.

It should be remembered in deriving (6.31) that the operation of the scattering operator on the complex conjugate terms in equation (6.29) yields complex conjugate expressions for the scattered fields. This can be demonstrated using the integral equation for the scattering amplitude presented in Sec. 4.

We denote the matrices D for the total and incident fields by D_b and D_0. Substitution of (6.31) into the definition of D and neglect of terms containing factor $\exp[i(\omega' + \omega'')t]$ in the integrand yield

$$D_b = \frac{2}{(2\pi)^8} \operatorname{Re} \lim_{T\to\infty} \frac{1}{2T} \int U^{(\vec{E})}(\omega', \vec{k}_0', \vec{r}_1)\, E(\vec{r}', t') \times$$
$$\times \overline{H(\vec{r}'', t'')}\, U^{(\vec{H})+}(\omega'', \vec{k}_0'', \vec{r}_2)\, e^{i\vec{k}'\vec{r} - i\omega'(t'-t_1-\xi) - i\vec{k}''\vec{r}'' + i\omega''(t''-t_2-\xi)} \times$$
$$\times\, d\vec{k}'\, d\vec{k}''\, d\vec{r}'\, d\vec{r}''\, d\omega'\, d\omega''\, dt'\, dt''\, d\xi +$$
$$+ \frac{2}{(2\pi)^4} \operatorname{Re} \lim_{T\to\infty} \frac{1}{2T} \int_{-T}^{T} E(\vec{r}_1, t_1+\xi)\, \overline{H(\vec{r}'', t'')} \times$$
$$\times\, U^{(\vec{H})+}(\omega'', \vec{k}_0'', \vec{r}_2)\, e^{-i\vec{k}''\vec{r}'' + i\omega''(t''-t_2-\xi)}\, d\vec{k}''\, d\vec{r}''\, d\omega''\, d\xi +$$
$$+ \frac{2}{(2\pi)^4} \operatorname{Re} \lim_{T\to\infty} \frac{1}{2T} \int_{-T}^{T} U^{(\vec{E})}(\omega', \vec{k}_0', \vec{r}_1)\, E(\vec{r}', t')\, \overline{H(\vec{r}_2, t_2+\xi)} \times$$
$$\times\, e^{i\vec{k}'\vec{r}' - i\omega'(t'-t_1-\xi)}\, d\vec{k}'\, d\vec{r}'\, d\omega'\, dt'\, d\xi + D_0.$$

Here and below the plus sign of matrix $U^{(H)+}$ designates Hermitian conjugation.

Substituting $t' = \tau' + \xi$, $t'' = \tau'' + \xi$, $\xi = \eta$ using the spectral expansion of D_0 and integrating with respect to τ', τ'', ω' and ω'', we obtain

$$D_b = \frac{2}{(2\pi)^6} \operatorname{Re} \int U^{(\vec{E})}(\omega, \vec{k}_0', \vec{r}_1)\, \Gamma(\vec{r}', \vec{r}'', \omega) \times$$
$$\times\, U^{(\vec{H})+}(\omega, \vec{k}_0'', \vec{r}_2)\, e^{i\vec{k}'\vec{r}' - i\vec{k}''\vec{r}'' + i\omega(t_1-t_2)}\, d\vec{k}'\, d\vec{k}''\, d\vec{r}'\, d\vec{r}''\, d\omega +$$
$$+ \frac{2}{(2\pi)^3} \operatorname{Re} \int \Gamma(\vec{r}_1, \vec{r}'', \omega)\, U^{(\vec{H})+}(\omega, \vec{k}_0'', \vec{r}_2)\, e^{-i\vec{k}''\vec{r}'' + i\omega(t_1-t_2)} \times$$
$$\times\, d\vec{k}''\, d\vec{r}''\, d\omega + \frac{2}{(2\pi)^3} \operatorname{Re} \int U^{(\vec{E})}(\omega, \vec{k}_0', \vec{r}_1) \times$$
$$\times\, \Gamma(\vec{r}', \vec{r}_2, \omega)\, e^{i\vec{k}'\vec{r}' - i\omega(t_1-t_2)}\, d\vec{k}'\, d\vec{r}'\, d\omega + D_0. \qquad (6.32)$$

In the case of quadratically integrable functions

$$\Gamma_{pq}(\vec{r}', \vec{r}'', \omega) = E_p(\vec{r}', \omega)\, H_q^*(\vec{r}'', \omega).$$

In the opposite case

$$\Gamma_{pq}(\vec{r}', \vec{r}'', \omega) = \lim \frac{1}{2T} E_p^{(T)}(\vec{r}', \omega)\, H_q^{(T)*}(\vec{r}'', \omega).$$

Here E_p, $E_p^{(T)}$, H_q and $H_q^{(T)}$ are the Fourier components of E, $E^{(T)}$, and H, $H^{(T)}$; sign $*$ denotes complex conjugation.

The explicit form of the matrices of operators $\hat{M}$ and $\hat{N}$ can be obtained from the expressions

$$\vec{g}_1 = (\sin\alpha)^{-1}[\vec{k}_0 \times \vec{r}_0],$$
$$\vec{g}_2 = (\sin\alpha)^{-1}[\vec{k}_0 \times [\vec{k}_0 \times \vec{r}_0]], \quad \vec{g}_3 = \vec{k}_0,$$
$$\vec{m}_1 = (\sin\alpha)^{-1}[\vec{k}_0 \times \vec{r}_0],$$
$$\vec{m}_2 = (\sin\alpha)^{-1}[[\vec{k}_0 \times \vec{r}_0] \times \vec{r}_0], \quad \vec{m}_3 = \vec{r}_0.$$

More detailed expressions are given by Sokolov and Tvorogov /465/.

6. SCATTERING OF VISIBLE AND INFRARED RADIATION BY A SINGLE SPHERICAL PARTICLE

1. Efficiency factors for extinction, scattering and absorption. In examining the scattering of electromagnetic waves by a spherical particle one introduces the efficiency factors for extinction K, scattering K_{sc} and absorption K_{abs}, where

$$K = \frac{\sigma}{\pi a^2}, \quad K_{sc} = \frac{\sigma_{sc}}{\pi a^2}, \quad K_{abs} = \frac{\sigma_{abs}}{\pi a^2}. \tag{6.33}$$

Hence functions K, K_{sc} and K_{abs} are equal numerically to the ratio of the energy respectively attenuated, scattered and absorbed by the particle, to the energy incident on its geometric cross-section πa^2.

1. Analytic expressions for functions K, K_{sc} and K_{abs}. General formulas for K, K_{sc} and K_{abs} are obtainable from the Mie theory, which is treated systematically and in great detail elsewhere /461, 467/. The expressions for functions K, K_{sc} and K_{abs} consist of infinite series; here these factors are a function of two arguments which characterize the relative dimension ρ and relative refractive index m of the particle:

$$\rho = \frac{2\pi a}{\lambda}, \tag{6.34}$$

$$m = \frac{m_1}{m_2}, \tag{6.35}$$

where λ is the wavelength of the scattered radiation, m_1 and m_2 are the complex refractive indices of the particle and of the medium.

As applied to radiation scattering conditions in the atmosphere the relative complex refractive index can be regarded as equal to the refractive index of the particle, i.e., it can be assumed that $m =$

$=m_1=n-i\varkappa$, where n is the ordinary refractive index and $\varkappa$ is the refractive index of the particle material.

In a number of asymptotic cases it is possible to obtain expressions for K, K_{sc} and K_{abs} in finite form. Some asymptotic expressions subsequently needed are presented below.

1. Small particles. The term small is applied to particles for which $\rho \ll 1$. In this case asymptotic formulas for K and K_{sc} involve expansions of Bessel and Hankel functions in powers of ρ in expressions for K and K_{sc}. Different formulas are obtained for different $|m|$. We are interested in the case when $|m|$ is small, for which the following expressions were obtained /467, 468/:

$$K(\rho, m)=\sum_{n=1}^{5} P_n\rho^n, \quad P_1=4\,\mathrm{Im}\left(\frac{1-m^2}{m^2+2}\right), \quad P_2=0,$$

$$P_3=2\,\mathrm{Im}\left[\frac{-2(m^2-1)(m^2-2)}{(m^2+2)^2}+\frac{m^2-1}{15}-\frac{m^2-1}{3(2m^2+3)}\right],$$

$$P_4=\frac{8}{3}\,\mathrm{Re}\left(\frac{m^2-1}{m^2+2}\right)^2,$$

$$P_5=2\,\mathrm{Im}\left(\frac{3}{175}\frac{m^2-1}{m^2+2}\frac{m^6+20m^4-200m^2+200}{m^3-2}-\right.$$
$$\left.-\frac{1}{56}\frac{m^2-1}{3m^2+4}-\frac{(m^2-1)(2m^2-3)}{315}\right). \qquad (6.36)$$

For transparent particles, setting $\varkappa=0$ in (6.36) yields

$$K(\rho, m)=K_{sc}(\rho, m)=\frac{8}{3}\rho^4\left(\frac{m^2-1}{m^2+2}\right)^2. \qquad (6.37)$$

In this latter particular case the scattering coefficient for a single particle (if its real refractive index is regarded as constant) is inversely proportional to the fourth power of the wavelength, as in molecular (Rayleigh) scattering. In other cases the relationship is quite complicated and not similar to that following from the Rayleigh law. The expressions for $K(\rho, m)$ and $K_{sc}(\rho, m)$ for large m and $m=\infty$ are given elsewhere /461, 467/.

2. Large particles. The term large is applied to particles for which $\rho \gg 1$. Analysis of the problem with direct application of the Mie formulas is quite difficult in this case since, as was shown by Debye, it is necessary to retain in the sums all terms with n ranging from 1 to ρ. For example, in the case of rain drops it is necessary to retain several thousand terms. This case was analyzed for a transparent particle by Shifrin /469/, who examined the scattered field consisting of beams of geometric optics and beams of radiation diffracted at the drop profiles. For an absorbing particle

similar to those investigated by Shifrin /469/, we can derive the following expression for $K(\rho, m)$:

$$K(\rho, m) = 2 - \frac{8\sqrt{n^2 - \varkappa^2}}{(n+1)^2 + \varkappa^2} \frac{n}{n-1} \times$$
$$\times \frac{e^{-2\rho\varkappa}}{\rho} \sin\left[2\rho(n-1) - 2\arctan\frac{\sqrt{n^2+\varkappa^2}}{\sqrt{n^2+\varkappa^2}+1}\frac{\varkappa}{\sqrt{n^2+\varkappa^2}+n}\right]. \qquad (6.38)$$

For transparent particles (6.38) yields

$$K_{sc}(\rho, m) = 2 - \frac{8n^2}{\rho(n+1)^2(n-1)} \sin[2\rho(n-1)]. \qquad (6.39)$$

It follows from expressions (6.38) and (6.39) that in the case $\rho \to \infty$, irrespective of the value of m, functions K and K_{sc} tend to the value 2. This means that the largest possible particles scatter twice as much radiation than is incident on their geometric cross-section.

3. Particles with $|m-1| \to 0$ ("soft" particles). This case was considered in detail by van de Hulst /461/. Note that water and water aerosol particles can be treated as "soft" with a degree of accuracy sufficient for practical purposes.

A quite simple and convenient formula was derived for large "soft" particles ($\rho \gg 1$) /461/:

$$K(\rho, m) = 2 - 4e^{-\rho' \tan\beta} \frac{\cos\beta}{\rho'} \sin(\rho' - \beta) -$$
$$- 4e^{-\rho'\tan\beta} \left(\frac{\cos\beta}{\rho'}\right)^2 \cos(\rho' - 2\beta) + 4\left(\frac{\cos\beta}{\rho'}\right)^2 \cos^2\beta, \qquad (6.40)$$

where

$$\rho' = 2\rho(n-1), \quad \tan\beta = \frac{\varkappa}{n-1}; \qquad (6.41)$$

and for a transparent particle:

$$K_{sc}(\rho, m) = 2 - \frac{4}{\rho'} \sin\rho' + \frac{4}{\rho'^2}(1 - \cos\rho'). \qquad (6.42)$$

Relationships (6.40)—(6.42) reproduce with sufficient accuracy the shape of curves of K and K_{sc} (including small values of ρ) obtained from tabulated exact Mie formulas.

2. Tabulation of functions K, K_{sc} and K_{abs}. When not one

of the above conditions is satisfied, functions K, K_{sc} and K_{abs} can be obtained only by computation on the basis of exact Mie formulas. In order to obtain sufficient computational accuracy one must retain in the series a number of terms of order ρ. The calculations are quite cumbersome; before the advent of high-speed computers they were carried out only for individual values of m and ρ.

As a result of the great interest expressed in functions K, K_{sc} and K_{abs} by specialists in a variety of fields, primarily in the fields of chemistry, geophysics and meteorology, a large number of calculations was carried out for different m and ρ. The overwhelming majority of calculations were made for real refractive indices: $0.60 \leqslant m \leqslant 2.105$ and $m = \infty$.

Of greatest interest here are the results of computing functions K, K_{sc} and K_{abs} for atmospheric aerosol particles, primarily for water whose absorption coefficient in the visible region is so small that it can be neglected compared with the refractive index. It is usually assumed that water in the visible region has a real refractive index, while in the infrared this index is complex (see Chapter 7).

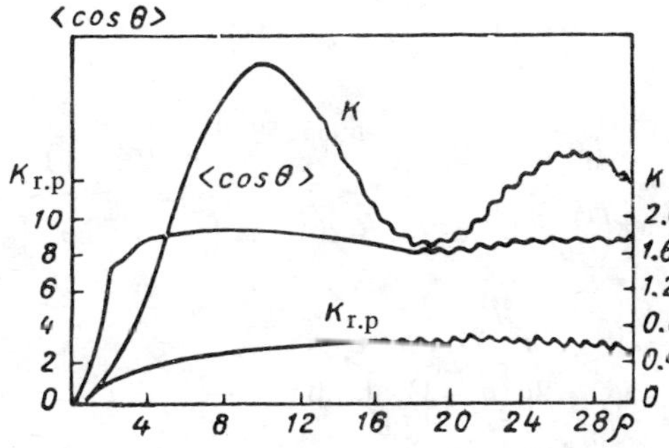

FIGURE 6.1. Asymmetry factor $\langle \cos\theta \rangle$ and cross-sections for extinction K and radiation pressure $K_{r.p}$ for dielectric spheres with $m = 1.20$ /477/.

Computations made for $\rho = 0.5(0.1)\ 10(0.05)\ 20(0.025)\ 30$.

Very comprehensive data on functions K, K_{sc} and K_{abs} for atmospheric aerosol particles are available /470 – 483/. Reports of other investigations can be found in /1, 470 – 483/.

Figures 6.1 through (6.20) present the most characteristic results of calculations of functions K, K_{sc} and K_{abs}.

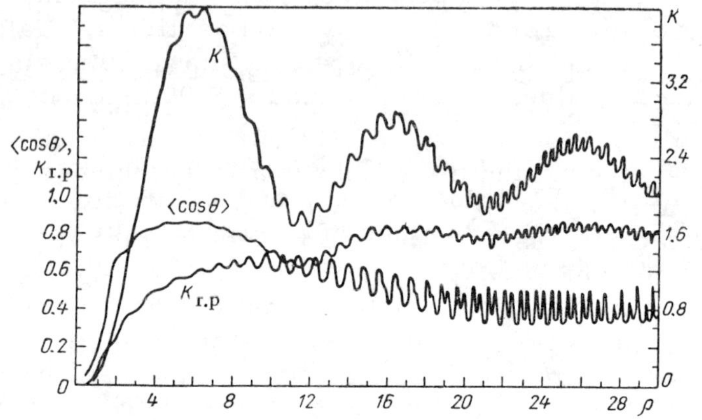

FIGURE 6.2. Asymmetry factor < cos θ> and cross-sections for extinction K and radiation pressure $K_{r.p}$ for dielectric spheres with $m = 1.33$ /477/.

Computations made for $\rho = 0.5(0.1)\ 10(0.05)\ 20(0.025)\ 30$.

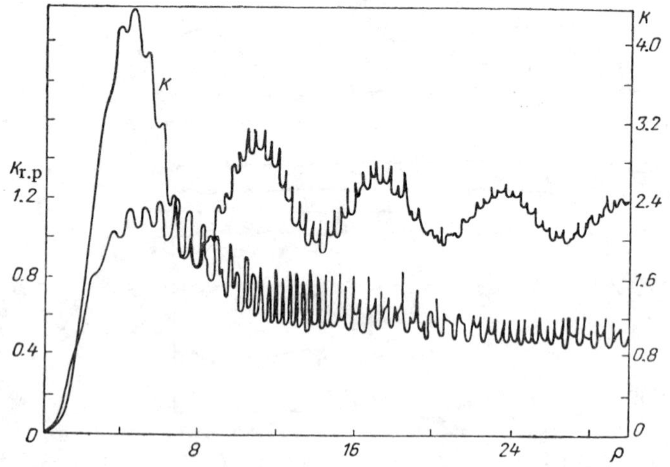

FIGURE 6.3. Cross-sections for extinction K and radiation pressure $K_{r.p}$ for dielectric spheres with $m=1.5$ /477/.

Computations made for $\rho = 0.5(0.1)\ 10(0.05)\ 20(0.025)\ 30$.

Figures 6.1— 6.3 show that the dependence of K on ρ for transparent dielectric spheres with values of m from 1.20 to 1.50 is characterized by the presence of a number of large-scale, and a large number of small-scale, peaks due to the interference nature of the scattering of electromagnetic waves on particles. Calculations by Irvine /477/, carried out with $\Delta\rho=0.01$, showed that small-scale peaks are found at values up to $\rho=50$ and $\rho=100$ (Figures 6.4 and 6.5).

Calculations of functions K, K_{sc} and K_{abs} with spheres possessing refractive index $n=1.01$, 1.33, 1.5 and 2 and absorption coefficient $\varkappa=10^{-4}$, 10^{-2}, 10^{-1} and 10 for values of ρ from 0 to 30 were carried out by Plass /479/ (see, for example, Figures 6.6— 6.11).

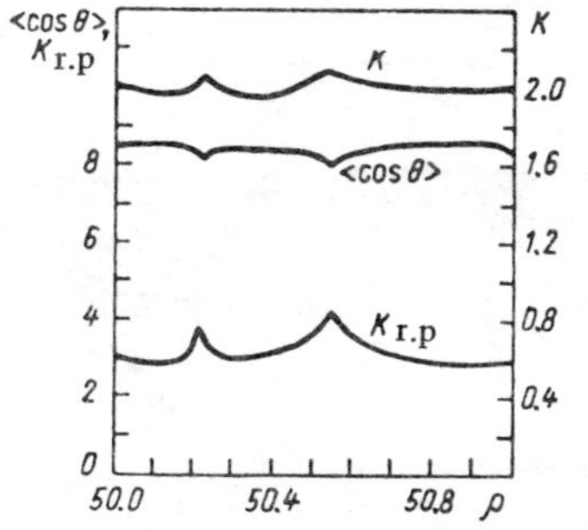

FIGURE 6.4. Asymmetry factor $<\cos\theta>$ and cross-sections for extinction K and radiation pressure $K_{r.p}$ for dielectric spheres with $m=1.33$ /477/.

Computations made for $\rho=50(0.01)\ 51.$/477/.

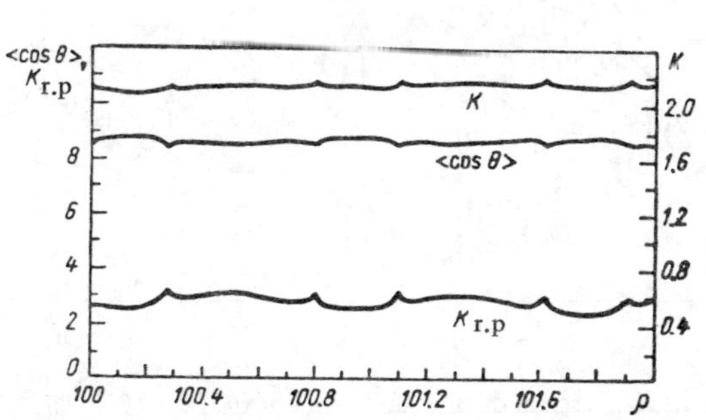

FIGURE 6.5. Asymmetry factor $<\cos\theta>$ and cross-sections for extinction K and radiation pressure $K_{r.p}$ for dielectric spheres with $m=1.33$ /477/.

Computations made for $\rho=100(0.01)\ 102.$

Figures 6.6 and 6.7 present, alongside with exact values of functions K_{abs}, approximate curves obtained from the van de Hulst formula /467/:

$$K_{abs} = 2Q(4\rho\varkappa), \tag{6.43}$$

where

$$Q(\omega) = \frac{1}{2} + \omega^{-1}e^{-\omega} + \omega^{-2}(e^{-\omega} - 1). \tag{6.44}$$

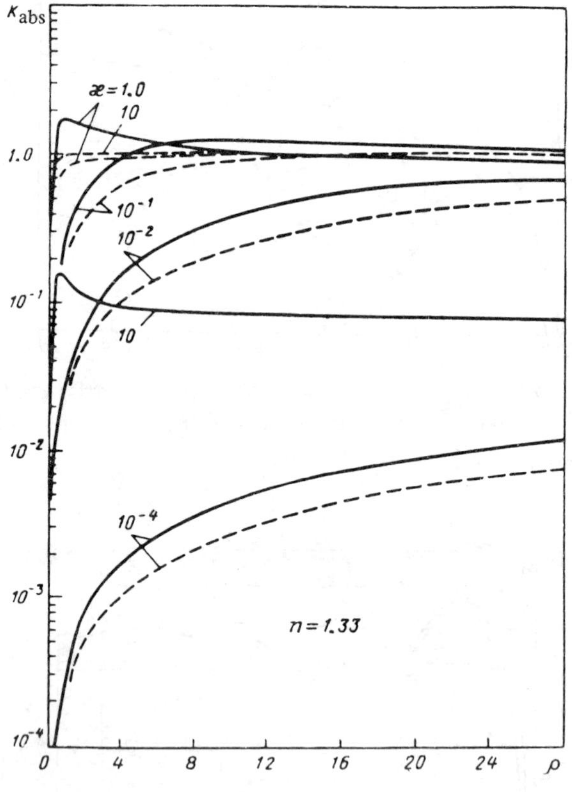

FIGURE 6.6. Functions K_{abs} for spheres with $n = 1.33$ and $\varkappa = 10^{-4}, 10^{-2}, 10^{-1}, 1, 10$ and different ρ, calculated by Plass /479/ from exact formulas of Mie theory (solid curves) and from approximate formula (6.43).

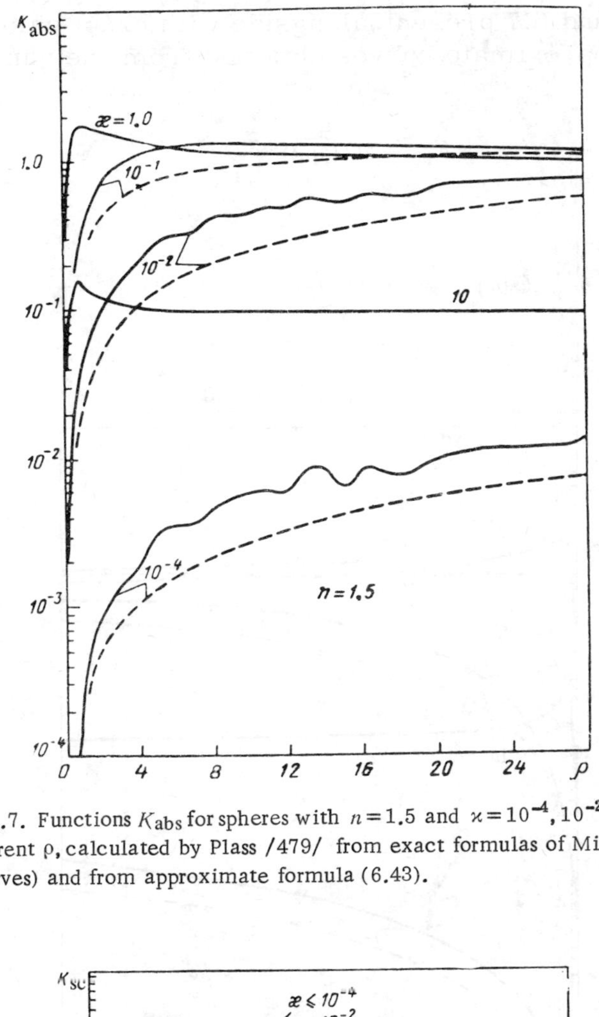

FIGURE 6.7. Functions K_{abs} for spheres with $n=1.5$ and $\varkappa=10^{-4}, 10^{-2}, 10^{-1}, 1, 10$ and different ρ, calculated by Plass /479/ from exact formulas of Mie theory (solid curves) and from approximate formula (6.43).

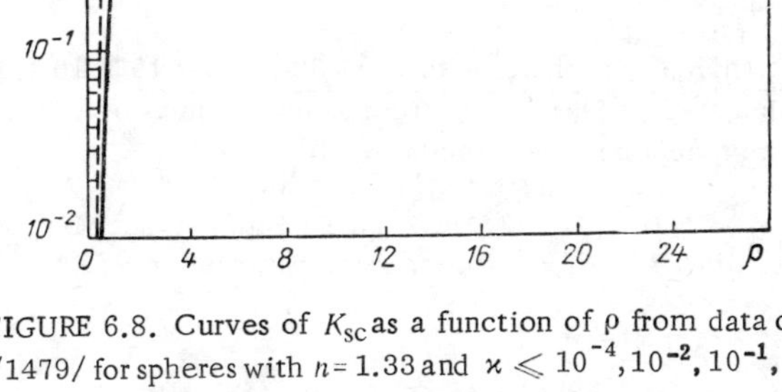

FIGURE 6.8. Curves of K_{sc} as a function of ρ from data of Plass /1479/ for spheres with $n=1.33$ and $\varkappa \leqslant 10^{-4}, 10^{-2}, 10^{-1}, 1, 10$.

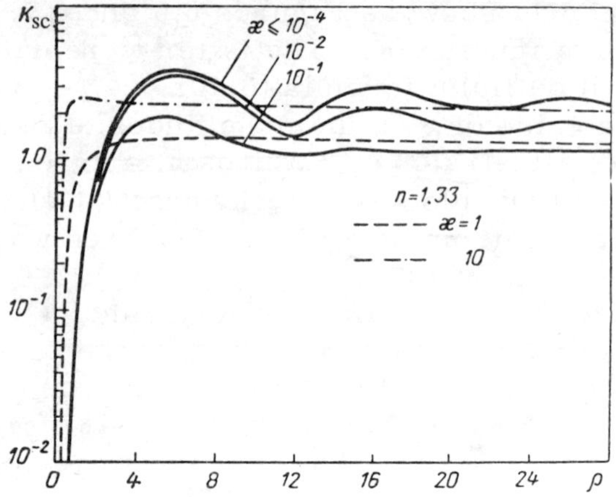

FIGURE 6.9. Curves of K_{sc} as a function of ρ from data of Plass /479/ for spheres with $n = 1.5$ and $\varkappa \leqslant 10^{-4}, 10^{-2}, 10^{-1}, 1, 10$.

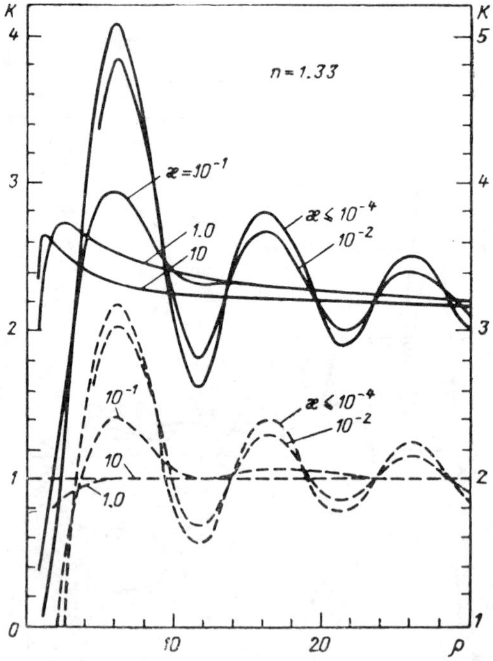

FIGURE 6.10. Curves of K as a function of ρ from data of Plass /479/ for spheres with $n = 1.33$, $\varkappa \leqslant 10^{-4}, 10^{-2}, 10^{-1}, 1, 10$, calculated from exact Mie formulas (solid curves, left ordinate axis) and from approximate formula (6.40).

Figures 6.10 and 6.11 show, as Figures 6.6 and 6.7, exact and approximate values of function K. The latter were calculated from the approximate van de Hulst formula (6.40).

The effect of the value of $\varkappa$ on the magnitudes of functions K_{abs} and K_{sc} for different n is illustrated by Figures 6.12 through 6.17.

Figures 6.18—6.20 are plots of functions K, K_{abs} and K_{sc} for water spheres with values of ρ from 0 to 30 and the three wavelengths $\lambda = 4$, 6 and 10μ from data of Herman /474/.

It is of interest to know the values of K, K_{sc} and K_{abs} when the atmospheric aerosol particles (for example, in the case of haze) can be treated as two-layer or multilayer spheres. The corresponding calculations were carried out elsewhere /484—487/. Fenn and Oser /486/ calculated K, K_{sc} and K_{abs} for concentric spherical water and soot particles. The computed values are plotted as graphs of K for values of ρ from 0.1 to 250 and various ratios of the particle and kernel radii. The results showed that, if the kernel radius is less than 0.1 of the radius of the entire sphere, the particle can be treated as a single layer. Kerker et al. /487/ considered a homogeneous sphere surrounded by a concentric spherical shell with different values of the refractive index. It is shown that these concentric spheres can be replaced by a homogeneous concentric shell with some average refractive index.

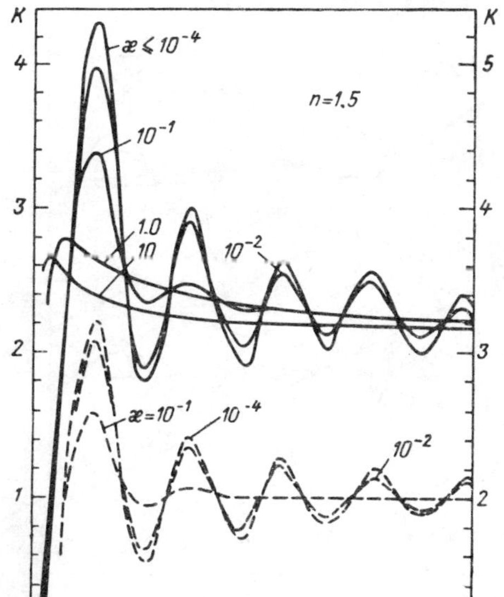

FIGURE 6.11. Plot of K as a function of ρ from data of Plass /479/ for spheres with $n = 1.5$, $\varkappa \leq 10^{-4}, 10^{-2}, 10^{-1}, 1, 10$, calculated from exact formulas (solid curves, left ordinate axis) and from approximate formula (6.40).

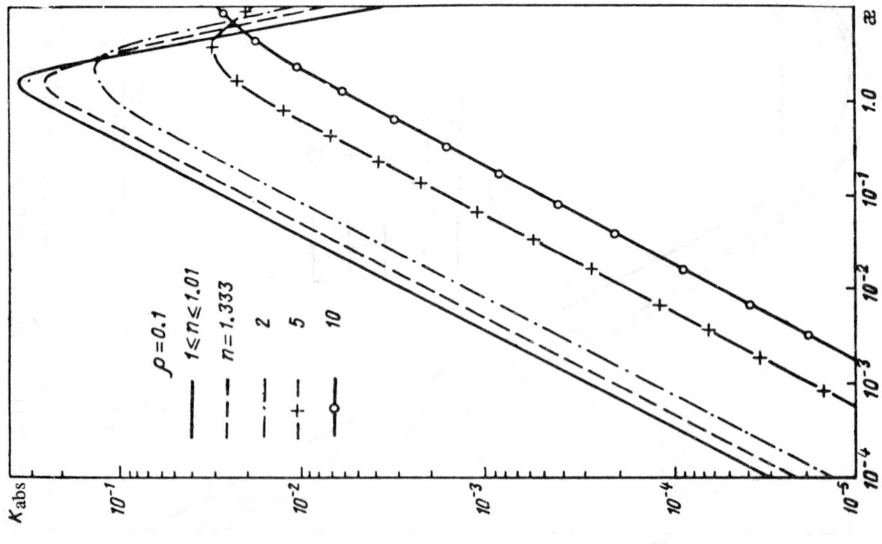

FIGURE 6.12. Function K_{abs} as dependent on $\varkappa$ for spheres with $\rho=0.1$ and different values of n /480/.

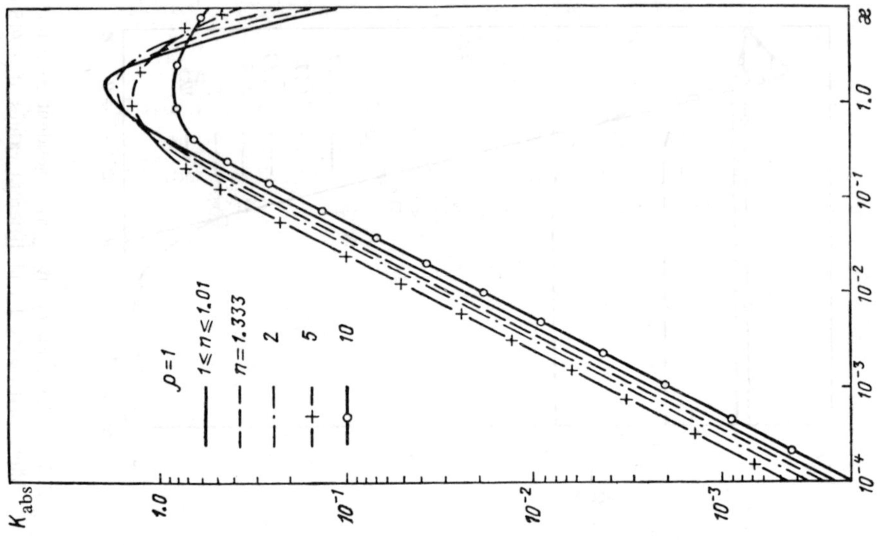

FIGURE 6.13. Function K_{abs} as dependent on $\varkappa$ for spheres with $\rho=1.0$ and different values of n /480/.

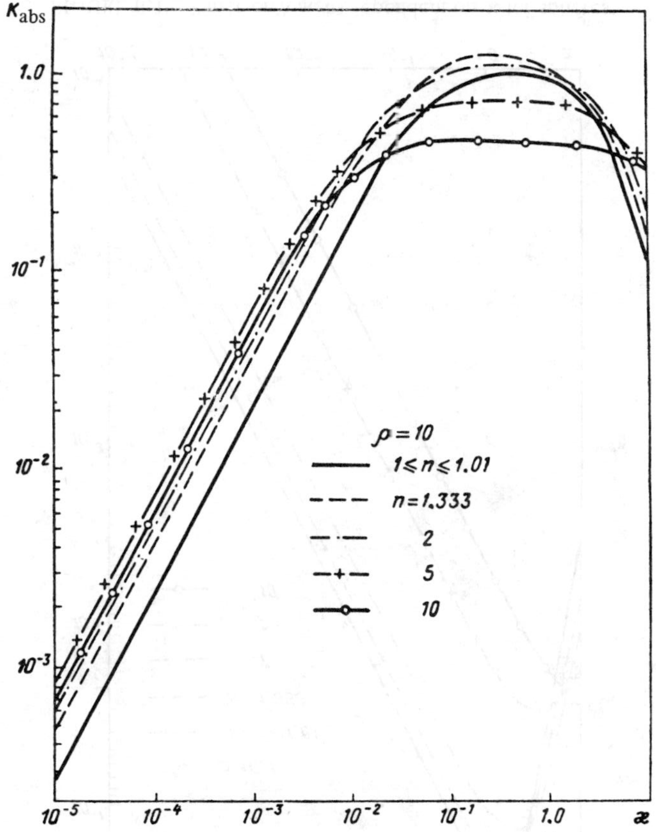

FIGURE 6.14. Function K_{abs} as dependent on $\varkappa$ for spheres with $\rho = 10$ and different values of n /480/.

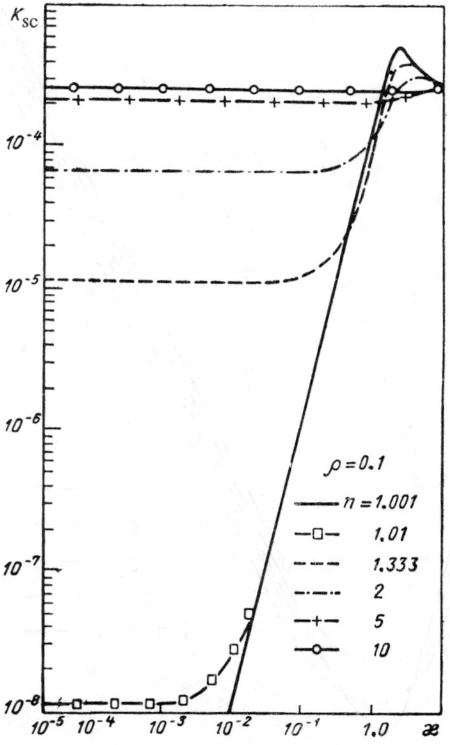

FIGURE 6.15. Function K_{sc} as dependent on $\varkappa$ for spheres with $\rho = 0.1$ and different values of n /480/.

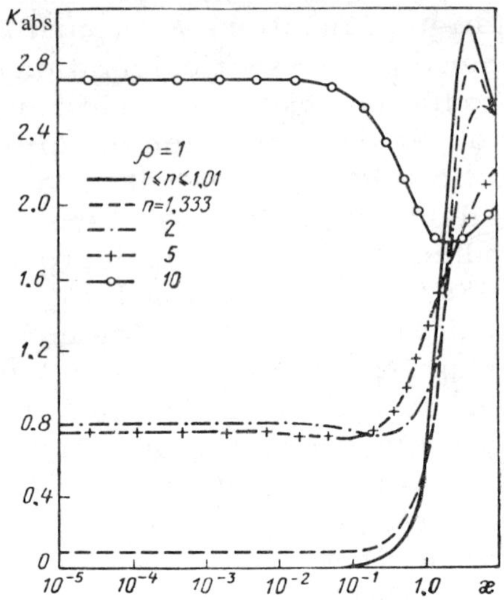

FIGURE 6.16. Function K_{abs} as dependent on $\varkappa$ for spheres with $\rho=1.0$ and different values of n /480/.

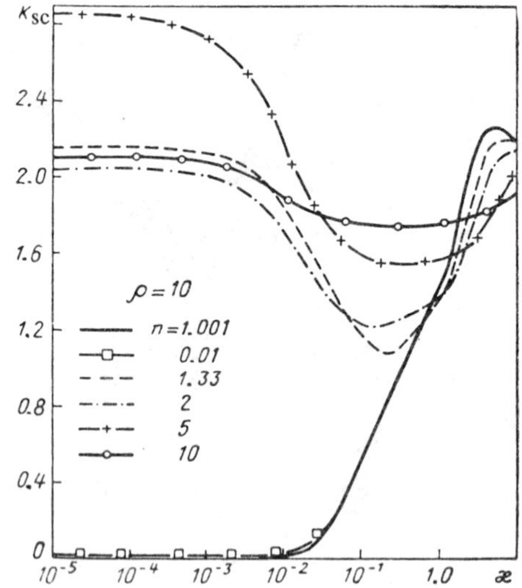

FIGURE 6.17. Function K_{sc} as dependent on $\varkappa$ for spheres with $\rho=10.0$ and different values of n /480/.

The above computation of functions K, K_{sc} and K_{abs} was carried out for standard temperature. Exact calculation of the extinction of visible and infrared radiation at atmospheric aerosol particles generally involves knowing these functions at different temperatures. However, such calculations have not as yet been carried out because no data have been available on the temperature-dependence of complex refractive indices of atmospheric aerosols. Only one study has been published /488/ in which functions K_{sc} and K_{abs} were calculated for spherical particles of aluminum oxide for the 1200— 2020°C temperature range. The particulate radii and wavelengths ranged from 0.1 to 10μ and from 0.5 to 6.0μ, respectively. It was found that K_{abs} for $\lambda = 2\mu$ increases by a factor of 40 as the temperature varies from 1200 to 2020°C. At wavelengths 0.5 and 6.0μ, K_{abs} over the same temperature range increases by factors of 7 and 2. Function K_{sc} under the same conditions changes only from 5 to 15% for different wavelengths.

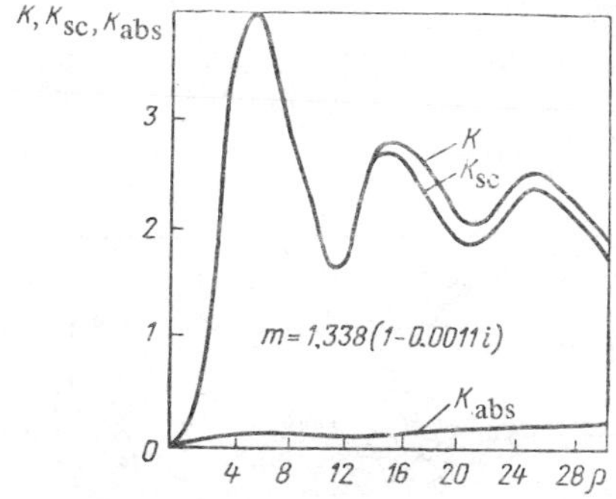

FIGURE 6.18. Functions K, K_{sc} and K_{abs} for water spheres with values of ρ from 0 to 30 and wavelength $\lambda = 4\mu$ /474/.

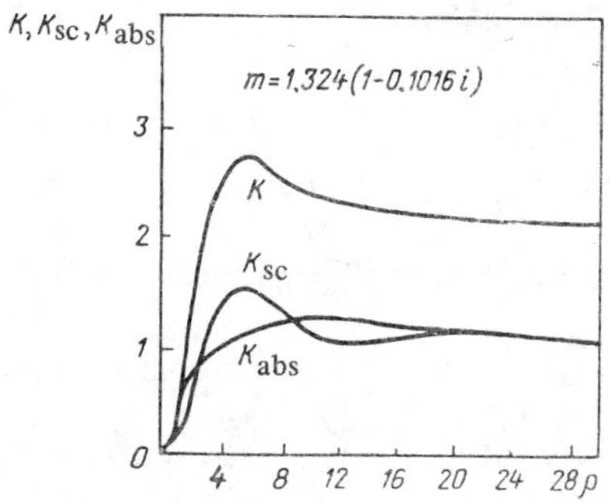

FIGURE 6.19. Functions K, K_{sc} and K_{abs} for water spheres with values of ρ from 0 to 30 and wavelength $\lambda = 6\mu$ /474/.

Approximating expressions for function K. A number of investigators found approximating expressions for the dependence of function K on m at different values of the latter. Levin /358/ obtained an expression for $K(\rho)$ for any ρ and $m=1.33$ (water spheres in the visible spectral region).

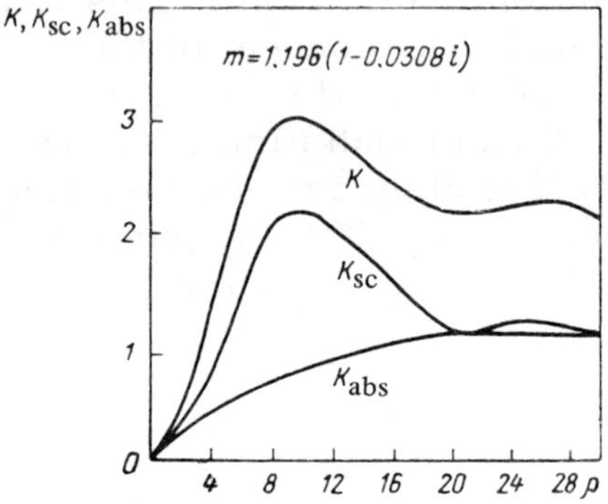

FIGURE 6.20. Functions K, K_{sc} and K_{abs} for water spheres with values of ρ from 0 to 30 and wavelength $\lambda=10\,\mu$ /474/.

An approximating expression for $K(\rho, m)$, selected by Deirmendjian /489/, is valid for any ρ and complex m, and satisfies the condition $|m|<2$. He expresses function K in the form

$$K=(1+D)K_1, \tag{6.45}$$

where K_1 is a function defined by formulas (6.40) and (6.42), while $(1+D)$ is an approximating factor.

Function $D(\rho, m)$ is given by:

$$\left.\begin{aligned} D_1 &= \frac{n-1}{2n}\frac{5(n-1)}{4.08}[d(\beta)+1]-\frac{5(n-1)-\rho'}{5(n-1)d(\beta)}, \\ &\rho'\leqslant 5(n-1)\leqslant\frac{4.08}{1+3\tan\beta}; \\ D_2 &= \frac{n-1}{2n}[d(\beta)+1]\frac{\rho'}{4.08}, \quad 5(n-1)\leqslant\rho'\leqslant\frac{4.08}{1+3\tan\beta}; \\ D_3 &= \frac{n-1}{2n}\frac{d(\beta)+1}{1+3\tan\beta}, \quad \frac{4.08}{1+3\tan\beta}\leqslant\rho'\leqslant\frac{4.08}{1+\tan\beta}; \\ D_4 &= \frac{n-1}{2n}\frac{d(\beta)+1}{d(\beta)}\frac{40.8}{\rho'}, \quad \rho'>\frac{4.08}{1+\tan\beta}; \\ &d(\beta)=(1+\tan\beta)(1+3\tan\beta). \end{aligned}\right\} \tag{6.46}$$

The values of ρ' and β are derived from formulas (6.42).

Deirmendjian /489/ shows that functions $K(\rho, m)$, obtained by means of (6.45) and (6.46), approximate computational data of

Johnson and Terrell /470/ on the basis of exact Mie formulas to within 4%. Calculations of $K(\rho, m)$ /470/ were carried out up to $\rho = 17.5$. Comparison of results obtained from formulas (6.45) with (6.46) and from asymptotic formula (6.38), carried out by this author, showed that the above approximation accuracy also exists when $\rho > 17.5$.

2. Scattering diagrams. In order to describe the properties of a particle scattering radiation at different angles to the direction of the incident wave, one introduces the concept of the scattering coefficient of a particle in a given direction. However, in practice one generally describes the spatial pattern of radiation scattered by a particle with the aid of the scattering diagram $f(\theta)$, specified as the ratio of the intensity of radiation scattered by the particle in a given direction to the energy flux scattered in all directions:

$$f(\theta) = \frac{I(\theta)}{\int\limits_{4\pi} I(\theta)\, d\omega}, \qquad (6.47)$$

where $I(\theta)$ is the intensity of light scattered by the particle in direction θ, while $d\omega$ is a solid-angle element.

The scattering diagrams of spherical particles of different sizes for different wavelengths of the visible and infrared spectra have been computed by many investigators; bibliographies on the subject can be found elsewhere /461, 467, 1/.

As of present scattering diagrams for particles have been computed for different real values of the refractive index. Calculations for particles with complex refractive indices over a wide range of refractive indices and absorption coefficients were carried out only recently. Very comprehensive and sufficiently exact data were obtained by Shifrin et al. /481, 482/.

Scattering diagrams for angles 0° (0.1°), 5° (1°) 90°, and in the region of the primary rainbow of water droplets 135— 140° with a step of 0.2° have been tabulated /481/. Values of all the components of the scattering matrix and components of the scattered wave field for water droplets in the visible and infrared spectral regions are presented in /482/. These values enable one to find all the characteristics of the scattered field, including the scattering diagram, degree of polarization, and degree of ellipticity of polarization. The data pertain to 12 spectral regions over the 0.4— 12μ wavelength range. The wavelengths were selected so as to embrace (with some approximation) the values of components of the complex refractive index of water over the entire range.

Calculations of all the characteristics of the scattered field were carried out for 51 values of parameter $\rho = 0.5(0.5)\ 6\ (1)\ 20\ (2)\ 50\ (5)$ 100, and 43 values of the scattering angle $\theta = 0, 0.05, 0.1, 0.2, 0.5, 1, 2,$

5, 10(10) 130, 135(0.5) 140(10) 170, 175, 178, 179, 179.5, 179.8, 179.9, 179.95, 180°. Thus very detailed data are available on small forward and back scattering angles.

Note that the values of $\Delta\lambda$ presented by Shifrin and Zel'manovich /482/ require substantial revision to bring them into conformance with new data on the components of complex refractive index n and $\varkappa$, obtained by the author (see Chapter 7) and presented in Appendix 2.

The behavior of function $i_1 + i_2$, contained in the expressions for the extinction coefficient and scattering diagram of a particle, was studied by Penndorf /490/ for angles $\theta = 0$, 90 and 180° and $m = 1.33$. He found that this function fluctuates markedly depending on the value of ρ. The frequency and amplitude of these fluctuations increase with increasing angle θ. Thus, if we assume that the amplitude of these fluctuations for $\theta = 0$ is 0.1, then for angles 90 and 180° it increases to 5 and 500, respectively. These fluctuations were detected while conducting calculations with a very small incremental step ($\Delta\rho = 0.1$).

Consider now analytic representations of the scattering diagram for spherical particles. Clark et al. /491/ introduced Legendre polynomials $P_n(\cos\theta)$ for this purpose:

$$f(\theta) = 1 + \sum_{n=1}^{\infty} a_n(\rho) P_n(\cos\theta), \tag{6.48}$$

where coefficients $a_n(\rho)$ were calculated for values of ρ from 1 to 30 and $m = 1.33$.

Shifrin /467/ obtained the following approximate expression for the scattering diagram for large particles:

$$f(\theta) = \frac{1}{2}\left[\frac{\tau(\theta)}{2} + \rho^2(1+\cos\theta)^2 \frac{J_1^2(\rho\sin\theta)}{(\rho\sin\theta)^2}\right], \tag{6.49}$$

where $\tau(\theta)$ is that part of the diagram due to beams of geometric optics, calculated with $\Delta\theta = 5°$ for spheres with $m = 1.33$; J_1 is a Bessel function.

We conclude by describing the qualitative variation of a diagram for a spherical particle as a function of its dimensions and the components of the complex refractive index. Small particles with $m \sim 1$ and $\rho \to 0$ have a symmetric Rayleigh diagram, while small particles with $m = \infty$ backscatter the radiation much more than they scatter it forward. For atmospheric aerosols in the visible spectral region the value of m assumed by different investigators ranges from 1.33 to 1.5 /423, 492, 820/.

As ρ increases from 0 to ∞ the scattering diagram for atmospheric aerosol particles continuously changes shape and becomes increasingly asymmetric and elongated forward (the Mie effect). Volz /492/ presents the following numerical illustration. For particles with $m = 1.5$ and radii $a = 0.5$, 1.5, 5.0 and 12.5μ, the ratio of radiation fluxes with wavelength 0.5μ which are scattered forward and backward is equal to 17, 74, 823 and 16,000 for $\rho = 6.28$, 18.84, 62.8 and 157, respectively. This ratio is the asymmetry factor of the diagram.

This pattern of variation in the scattering diagram with change in size of atmospheric aerosol particles is substantially affected by the complexity of the refractive index of the particles and by the fluctuation in the intensity of radiation scattered at different angles and depending on ρ, θ and m (discovered by Penndorf).

7. SCATTERING OF LIGHT BY NONSPHERICAL PARTICLES

Although many investigators note that atmospheric aerosols can usually be approximated by spherical particles, this question nevertheless requires particular attention, since dust particles, ice crystals and particles of ice clouds actually possess arbitrary shape. It is important to know how particle shape affects their optical properties. Many results appear in van de Hulst's book /461/ for ellipsoids, circular cylinders, round disks, and particles of other shapes with different ρ and m and orientation relative to the incident radiation. Optical properties of rod- and disk-like particles, which are approximated by ellipsoids, were investigated by Shifrin /467/.

Available results indicate that the optical characteristics of particles of different shapes depend substantially on the ratio of the maximum to minimum particle size, the orientation of particles relative to the incident radiation, the degree of polarization of the latter, and the values of the complex refractive index. For example, for a very long transparent rod whose linear dimensions are much smaller than the wavelength, the ratio of the scattering coefficients in directions opposite to and along the field is 6.25 when $m = 2$ and 1.64 when $m = 1.64$ /461/. It is natural that the ratios differ for the same absorbing rod. Changes in the rod dimensions involve corresponding changes in the ratio under study.

The optical characteristics of particles of different shapes can differ markedly from those of spherical particles possessing the same volume. It is hence necessary to exercise great care in applying the results of the Mie theory to nonspherical particles. At the same time if the particle shape does not differ markedly from spherical (such as ellipsoids with semiaxes ratio up to 1.5— 2, cubes, cylinders with height-to-diameter ratio close to unity), then such characteristics as the scattering diagram, extinction, scattering and absorption coefficients differ little from the corresponding characteristics of spherical particles possessing the same volume.

We point out one way of utilizing practically the results of the Mie theory for nonspherical particles. If any optical characteristic of a particle is averaged over all its possible spatial orientations, then it is subsequently possible to analytically express its volume in terms of the volume of a spherical particle with the same optical characteristic. Such an expression was obtained by Shifrin /467/ for the scattering diagram of ellipsoidal particles:

$$v^* = \sqrt{\varphi(t)}\, v,$$

where v^* and v are, respectively, the volumes of the sphere and ellipsoid with coincident scattering diagrams; $t = \frac{a}{b}$ is the ratio of the semiaxes of the ellipsoid: $\varphi(t)$ is a tabulated function /464/ and assumes values from 1 to 0.23 as t proceeds from 1 to 10. In particular, at $t=2$ and 3, $\varphi(t)=0.93$ and 0.71.

Relationships of this kind can be obtained for different scattering characteristics and for particles with other shapes.

7. SCATTERING OF VISIBLE AND INFRARED RADIATION IN THE ATMOSPHERE

Scattering of visible and infrared radiation in the earth's atmosphere is due primarily to fluctuations in air density and to aerosol particles. In the first case we deal with molecular or Rayleigh scattering, and in the second with aerosol scattering. In addition, at least two other kinds of scattering may occur in the atmosphere: scattering on small- and large-scale inhomogeneities caused by turbulent air motion, and Raman scattering. However, both these forms of scattering do not make a substantial contribution to the attenuation of visible and infrared radiation in the atmosphere compared with the previous two. Hence in this chapter, which is concerned primarily with a description of losses of energy transported by electromagnetic waves in the atmosphere, we shall not be concerned with scattering at turbulent inhomogeneities and Raman scattering.

Molecular scattering has been quite well investigated. At present extensive tables are available of scattering coefficients in the visible and infrared regions and enable one to determine with sufficient accuracy the energy losses of waves propagating in any direction in the atmosphere. Hence in describing molecular scattering we restrict ourselves primarily to some tabulated data on the scattering coefficients and optical thicknesses of the Rayleigh atmosphere.

In speaking about aerosol scattering one generally refers not only to scattering, but also to absorption of wave energy by the particles. It would therefore be more correct to refer to aerosol extinction, which is due to both scattering and absorption of radiation flux by aerosol particles.

In spite of the large variety of dimensional spectra and the wide range of variation in concentration and chemical composition of atmospherical aerosol particles, one can single out certain characteristic types of aerosols which differ widely in their ability to scatter electromagnetic waves of the optical region. These types include primarily clouds, fogs and haze. It is thus not accidental that a large part of this chapter is given to an examination of the scattering of visible and infrared radiation by these aerosols. Furthermore, although precipitation cannot be called aerosols, the attenuation of radiation on passing through them is also included in this chapter, since the nature of wave scattering at precipitation particles and particles of haze, fogs and clouds is the same.

1. MOLECULAR SCATTERING OF VISIBLE AND INFRARED RADIATION

The Cabannes-Rayleigh molecular scattering theory yields the following expression for the scattering coefficient in gases:

$$\sigma_{Ray}(\lambda)=\frac{8\pi^3(n^2-1)^2}{3N\lambda^4}\frac{6+3\delta}{6-7\delta}, \tag{7.1}$$

where N is the number density, n is the refractive index of the medium, λ is the emitted wavelength, δ is the depolarization factor of scattered radiation, according to recent measurements equal to 0.035 /447/.

TABLE 7.1. Molecular scattering coefficient σ_{Ray} at t = 15°C and P = 1013 mb and optical thickness τ_{Ray} of a vertical layer of the entire atmosphere /493, 494/

Wave-length λ, μ	σ_{Ray}, km^{-1}	τ_{Ray}	Wave-length λ, μ	σ_{Ray}, km^{-1}	τ_{Ray}
0.30	$1.446\cdot10^{-1}$	1.2237	0.65	5.893	0.0499
0.32	1.098	0.9290	0.70	4.364	0.0369
0.34	$8.494\cdot10^{-2}$	0.7188	0.80	2.545	0.0215
0.36	6.680	0.5653	0.90	1.583	0.0134
0.38	5.327	0.4508	1.06	$8.458\cdot10^{-4}$	0.0072
0.40	4.303	0.3641	1.26	$4.076\cdot10^{-4}$	0.0034
0.45	2.644	0.2238	1.67	$1.327\cdot10^{-4}$	0.0011
0.50	1.716	0.1452	2.17	$4.586\cdot10^{-5}$	0.0004
0.55	1.162	0.0984	3.50	$6.830\cdot10^{-6}$	0.0001
0.60	$8.157\cdot10^{-3}$	0.0690	4.00	4.002	0.0000

The molecular absorption coefficient was calculated from (7.1) by Penndorf /493/ for wavelengths in the 0.2–20μ range. Penndorf's data can be used to calculate the optical thickness τ_{Ray} of molecular scattering for different geometric atmospheric thicknesses.

TABLE 7.2. Coefficient σ_{Ray} for different altitudes z and optical thicknesses τ_{Ray} of atmospheric layers ($z-\infty$) for wavelengths $\lambda = 0.30$, 0.55 and 1.06 μ /447/

Altitude z, km	$\lambda = 0.30\ \mu$		$\lambda = 0.55\ \mu$		$\lambda = 1.06\ \mu$	
	σ_{Ray}, km^{-1}	$\tau_{Ray} \times (z-\infty)$	σ_{Ray}, km^{-1}	$\tau_{Ray} \times (z-\infty)$	σ_{Ray}, km^{-1}	$\tau_{Ray} \times (z-\infty)$
0	$1.446 \cdot 10^{-1}$	1.2237	$1.162 \cdot 10^{-2}$	0.0984	$8.458 \cdot 10^{-4}$	0.0072
5	$8.693 \cdot 10^{-2}$	0.6538	$6.988 \cdot 10^{-3}$	0.0526	5.085	0.0038
10	4.881	0.3212	3.924	0.0258	2.855	0.0019
15	2.999	0.1471	1.848	0.0118	1.345	0.0009
20	1.049	0.0672	$8.436 \cdot 10^{-4}$	0.0054	$6.138 \cdot 10^{-5}$	0.0004
30	$2.173 \cdot 10^{-3}$	0.0146	1.747	0.0012	1.271	0.0001
40	$4.716 \cdot 10^{-4}$	0.0035	$3.791 \cdot 10^{-5}$	0.0003	$2.758 \cdot 10^{-6}$	0.0000
50	1.212	0.0010	$9.743 \cdot 10^{-6}$	0.0001	$7.089 \cdot 10^{-7}$	0.0000

Tables 7.1 and 7.2 list data on σ_{Ray} and τ_{Ray} for different wavelengths and geometric thicknesses of the atmosphere. The dimension of the Rayleigh scattering coefficient calculated from (7.1) is the reciprocal of length.

2. SCATTERING OF WAVES BY A SYSTEM OF PARTICLES

The exact electrodynamic formulation of the problem of scattering of waves by a system of particles, methods of its solution and some results have been examined in the literature (see, for example, /494–503/). The main methodological principle of the solution of this problem consists in the following. It is assumed that the solution of the problem concerning scattering of light by a single particle is known, and that all the sought physical variables can be obtained after relevant statistical averaging over particle configuration.

In a particle-filled medium there exists some field composed of the field of the wave incident on the particle system and fields scattered by individual particles. The solution of the latter is known; here each particle is acted upon by its intrinsic, effective field. Hence

$$\left.\begin{aligned} \psi(\vec{r}) &= \psi_0(\vec{r}) + \sum_{j=1}^{N} T(\vec{r}, \vec{r}_j)\,\tilde{\psi}(\vec{r}_j), \\ \tilde{\psi}(\vec{r}_j) &= \psi_0(\vec{r}) + \sum_{j' \neq j} T(\vec{r}_j, \vec{r}_{j'})\,\tilde{\psi}(\vec{r}_{j'}), \end{aligned}\right\} \qquad (7.2)$$

where ψ is some field component, $\vec{r}_j$ are coordinates of the j th particle, ψ_0 is the corresponding component of the incident wave, $\vec{r}$ is the radius vector of an arbitrary point, T is the scattering operator, and $\tilde{\psi}$ is the effective field. The first of equations (7.2) yields the field in the medium, while the second in effect represents a system of equations defining the effective field acting upon each particle. The summation in equation (7.2) defines a fixed configuration of the scattering particles.

The next step consists in averaging equations (7.2) over all the configurations. Formally, this is carried out by multiplying both sides of the equations by the configuration distribution function and integrating with respect to the corresponding variables. Here the configuration distribution function in general depends essentially on the nature of interparticle interaction. Generally speaking, the respective calculations can be completed if the particles are treated as noninteracting (ideal-gas model), or by retaining the first terms of the expansion with respect to virial coefficients. It must also be assumed that the distances between the particles substantially exceed the light-wave length, in order to be able to employ the scattering operator in the wave zone in asymptotic form. The effective field is usually assumed equal to the average field at the point of particle location, since both fields differ by a quantity of the order of $1/N$, where N is the number of particles. This assumption makes it possible to consider only one equation for the average field.

A similar approach is used when considering the problem of the light intensity. In the scalar version this quantity can be defined as $|\psi|^2$. Upon multiplying both parts of equations (7.2) by the respective complex conjugate terms, the resulting equations are again averaged over configurations for a given distribution function.

If we consider noninteracting particles, then the ordinary transport equation is obtained for the radiation intensity in the medium. If interparticulate interaction is taken into account in the normal manner, this results in corrections to this equation.

When carrying out the analysis in this manner there arises automatically an expression for the extinction coefficient $\alpha(\lambda)$, calculated per unit beam path length:

$$\alpha(\lambda) = N \int_0^\infty \sigma(a, \lambda) f(a)\, da, \qquad (7.3)$$

where N is the number density, $\sigma(a, \lambda)$ is the coefficient of extinction of radiation with wavelength λ brought about by a particle of radius a

(here and subsequently the scattering particles are taken as spherical); $f(a)$ is a particle size distribution function defined by the expression $N_a da = Nf(a)\,da$, where N_a is the number of particles with radii between a and $a + da$.

The derivation of equation (7.3) can be easily followed by means of the theorem of optics (see Chapter 6, Sec. 2). The extinction coefficient of some volume can be determined from the scheme for finding the screening constant, described in Chapter 6, Sec. 3. This yields $\alpha = 4\pi/k^2 \operatorname{Im} A$, where A is the amplitude of the field scattered forward by the j th particle. By virtue of the superposition principle $A = \sum A_j$, where A_j is the amplitude of the wave scattered forward by the j th particle. Combination of the latter two expressions together with equation (6.20) shows that $\alpha = \sum \sigma_j$, where σ_j is the coefficient of extinction by the j th particle. After introducing distribution function $f(a)$ the last sum can be expressed in the form of equation (7.3).

Equation (7.3) and the equality $\sigma(\lambda, a) = \sigma_{sc}(\lambda, a) + \sigma_{abs}(\lambda, a)$, where σ_{sc} and σ_{abs} are coefficients of scattering and absorption of radiation at wavelength λ by a particle of radius a, automatically yield the following expressions for the polydisperse scattering $\alpha_{sc}(\lambda)$ and absorption $\alpha_{abs}(\lambda)$ coefficients:

$$\alpha_{sc}(\lambda) = N \int_0^\infty \sigma_{sc}(\lambda, a) f(a)\, da, \tag{7.4}$$

$$\alpha_{abs}(\lambda) = N \int_0^\infty \sigma_{abs}(\lambda, a) f(a)\, da. \tag{7.5}$$

Using coefficient $\alpha(\lambda)$ one can write in the usual manner the equation governing the variation in intensity I of radiation propagating along some path:

$$dI(\lambda) = -I(\lambda)\,\alpha(\lambda)\, dl. \tag{7.6}$$

Integration of (7.6) yields the known expression for the aerosol component of atmospheric transmittance

$$T(\lambda) = I/I_0,$$

where I_0 is the radiation intensity at the start of path:

$$T(\lambda) = \exp\left(-\int_{(l)} \alpha(\lambda, l)\, dl\right). \tag{7.7}$$

Integration in equation (7.7) is carried out over points of the path, while $\alpha(\lambda)$ in general is variable along the path. This variation is due to the corresponding change in particle size and concentration spectra.

Equation (7.7) describes the variation in intensity of direct radiation carrying given information. In experiments one cannot record direct radiation alone, since this would require receivers with zero angle aperture. The aperture angle of the receiving system can, in general, be made as small as desired, but not equal to zero. Hence receiving systems record not only direct radiation attenuated according to (7.7), but also light singly and multiply scattered by aerosol particles, which should be treated as one of the background interferences. This is responsible for the dependence of the measured quantities (scattering coefficient, $T(\lambda)$, etc.) on experimental conditions.

3. INVESTIGATION OF THE SCATTERING MATRIX

The scattering matrix M contains complete information on the light field scattered by particles (see Chapter 6, Sec. 2). Knowledge of the components of this matrix enables one to solve quantitatively any problem associated with wave scattering by particles.

In spite of the obvious importance of studying matrix M for atmospheric aerosols, insufficient data have been accumulated up to now on the components of this matrix. This is due, on the one hand, to the complexity of the corresponding theoretical and experimental studies, while on the other hand, many practical problems can be solved without knowing the components of the scattering matrix. For example, in many cases it suffices to know the scattering or extinction coefficients.

Matrix M has the simplest form in the case of molecular scattering:

$$M(\theta) = \frac{3}{4+3d}\begin{pmatrix} 1+\cos^2\theta+d & -\sin\theta & 0 & 0 \\ -\sin^2\theta & 1+\cos^2\theta & 0 & 0 \\ 0 & 0 & 2\cos\theta & 0 \\ 0 & 0 & 0 & 2\cos\theta \end{pmatrix}. \qquad (7.8)$$

For radiation scattering by atmospheric aerosol particles, all sixteen components of matrix M can in general differ. However, the symmetry of the various particles and their orientations in space results in a reduction in the number of independent components and in the vanishing of some of them. For example, matrix M for spherical particles has the form /504/

$$M(\theta) = \begin{pmatrix} M_{11} & M_{12} & 0 & 0 \\ M_{21} & M_{22} & 0 & 0 \\ 0 & 0 & M_{33} & M_{34} \\ 0 & 0 & M_{43} & M_{44} \end{pmatrix}, \qquad (7.9)$$

where $M_{11} = M_{22}$; $M_{12} = M_{21}$; $M_{33} = M_{44}$; $M_{34} = M_{43}$. For spherical particles we therefore have only four independent components, which can be calculated from the applicable formulas of Mie theory. An example of such a calculation is given by Malkova /505/, who calculated all four components of $M(\theta)$ for values of ρ from 0.1 to 1.0, $\theta = 0(5)180°$, and thirteen values of the complex refractive index of water, characteristic of wavelengths from 2 to 12μ.

The most comprehensive experimental data on the scattering matrix components were obtained by the Rozenberg team /16, 506, 507, 897, 898/, who in 1959 started their studies in the surface layer of the atmosphere using specially designed equipment. By processing many measurements over various spectral sections in different meteorological conditions, Rozenberg et al. derived quantitative data on all the components of matrix M. They were able to draw important conclusions on the presence of typical aerosol states in the ground layer of the atmosphere (fog, foggy haze, haze, haze with drizzle). Each of the aforementioned conditions has its corresponding ensemble of angular dependences of the components of matrix M. We do not refer here to a specific dependence of these components on θ, but to ensembles, since actually Rozenberg et al. obtained data for different specific conditions, each of which can be referred to a greater or lesser extent to conditions characteristic of one of the aerosol states of the ground layer of the atmosphere.

4. ENERGY CHARACTERISTICS OF SCATTERING OF VISIBLE AND INFRARED RADIATION BY CLOUDS AND FOG

In solving many problems concerning the propagation of visible and infrared radiation in clouds and fog, it is first necessary to know what part of the radiant flux incident on the layer is caused by the latter to change its initial direction due to aerosol extinction as a result of scattering and absorption by the particles of the layer. To resolve this question one must know the extinction coefficients $\alpha(\lambda)$. It is precisely for this reason that most attention in this section is paid to this quantity for water clouds and fog. Data on coefficient $\alpha(\lambda)$ for supercooled and ice clouds and fog have not yet been obtained.

1. Analytical expressions for the extinction coefficient and the averaged extinction factor. Let $f(a)$ be the particle size distribution function for a cloud or fog. Then, assuming that we are dealing with spherical particles, we can express the coefficient $\alpha(\lambda)$ and geometric cross-section Q of particles per unit volume in the form

$$\alpha(\lambda) = N \int_0^\infty \pi a^2 K(\rho, m) f(a)\, da, \tag{7.10}$$

$$Q = N \int_0^\infty \pi a^2 f(a)\, da. \tag{7.11}$$

Quantity

$$F = \frac{\alpha(\lambda)}{Q} = \frac{\int_0^\infty \pi a^2 K(\rho, m) f(a)\, da}{\int_0^\infty \pi a^2 f(a)\, da} \tag{7.12}$$

shall be termed the averaged efficiency factor for extinction. This designation is justified by equation (7.12), which defines F as the ordinary statistical mean of the efficiency factor for extinction for a single particle $K(\rho, m)$ for density function $\pi a^2 f(a)$. Function F is numerically equal to the ratio of the optical cross-section of particles per unit volume to the geometric cross-section.

It can be seen from equation (7.12) that function F is independent of N and, to within a constant factor, characterizes the bulk extinction coefficient. The values of F are determined by quantities m and λ $\left(F \text{ is a function of } \lambda \text{ through } \rho = \frac{2\pi a}{\lambda}\right)$. Equations (7.10) and (7.12) yield directly the expression

$$\frac{F(\lambda_1, m)}{F(\lambda_2, m)} = \frac{\alpha(\lambda_1)}{\alpha(\lambda_2)}, \tag{7.13}$$

which is valid for the given $f(a)$ and r.

The bulk extinction coefficient $\alpha(\lambda)$ and the averaged efficiency factor for extinction $F(\lambda, m)$ are the principal quantitative characteristics of radiation attenuation by polydisperse aerosols. It is natural that the expressions for $\alpha(\lambda)$ and $F(\lambda, m)$ differ markedly from the particle size distribution function.

To clarify the effect of the form of function $f(a)$ on the numerical values of $F(\lambda, m)$, this author used methods (presented later) to calculate values of F utilizing for $f(a)$ the lognormal and gamma distributions. Both these distributions approximate satisfactorily experimental curves of particle size spectra of clouds and fog. Calculations showed that these distributions can be used on an equivalent basis in calculating function F. Selection of equation (5.3) or (5.4) for $f(a)$ depends on the problem at hand.

An expression for F, valid for all ρ and $m = 1.33$, was obtained in /508/ by approximating the expression suggested there for function $K(\rho)$ and with a gamma distribution as function $f(a)$.

Using for "soft" transparent particles the approximate equation (6.42) for $K(\rho, m)$, cited by van de Hulst /461/, and a gamma distribution for function $f(a)$, Shifrin and Raskin /509/ derived an expression for the polydisperse scattering coefficient. Their tabulated values /509/ can be used to compute the spectral transmittance of water clouds and fog in the visible spectral region and for those sections of the infrared region where the complex nature of the refractive index of water can be neglected.

The complex nature of the refractive index of water in the infrared region was taken into account by Deirmendjian /489/ when calculating coefficients $\alpha(\lambda)$ from formulas (6.45) and (6.46) for $K(\rho, m)$. He thus derived values of $\alpha(\lambda)$ for one cloud model in the visible and infrared regions up to $\lambda = 13\mu$.

Tvorogov /510/ employed the lognormal law as the distribution function of aerosol particles and showed that function F in this case is the solution of a parabolic equation for given initial conditions. He presents three parabolic equations with initial conditions satisfied by function F.

It should be emphasized that in proving the above assertion no limitations were imposed on function $K(\rho, m)$. Hence the principal result of Tvorogov /510/ remains valid if $K(\rho, m)$ is replaced by the efficiency factor for scattering or absorption, the scattering diagram, or any other component of the scattering matrix. In this sense his result is universal.

Methods for solving the parabolic equations obtained in /510/ were developed later /511, 512/ by the same author. His results /510–512/ can be used to compute any averaged components of the scattering matrix.

Analytic expressions for function F when using a gamma distribution as function $f(a)$ were obtained later /513–515/. The expression for function F is

$$F = \frac{\int_0^\infty \tau^2 f(\tau) K(x\tau, m)\, d\tau}{\int_0^\infty \tau^2 f(\tau)\, d\tau} = F(x, \mu, m), \tag{7.14}$$

where

$$\tau = \frac{a}{r}, \quad x = \frac{2\pi r}{\lambda}, \tag{7.15}$$

r and μ being parameters of the gamma distribution. Equation (7.14) follows directly from (7.12) and (7.15).

Substitution of equation (5.4) into (7.14) yields

$$F(x, \mu, m) = \frac{\mu^{\mu+3}}{\Gamma(\mu+3)} \int_0^\infty \tau^{\mu+2} e^{-\mu\tau} K(x\tau, m)\, d\tau. \tag{7.16}$$

If we now substitute expressions (6.36)–(6.42) for function $K(\rho, m) = K(x\tau, m)$ into (7.16), we derive specific expressions for $F(x, \mu, m)$ in different cases. We present here without proof only the expression for $F(x, m, \mu)$, which will subsequently be used and which is obtained by substituting into equation (7.16) equations (6.40) and (6.41), describing function $K(x\tau, m)$ for large "soft" particles:

$$\begin{aligned} F(x, \mu, m) = 2 &- \frac{4}{\mu+2} \frac{\sin[(\mu+2)\varphi - \beta]\,[\cos(\varphi+\beta)]^{\mu+3}}{\sin\varphi\,[\cos\beta]^{\mu+2}} - \\ &- \frac{4}{(\mu+2)(\mu+1)} \frac{\cos[(\mu+1)\varphi - 2\beta]\,[\cos(\varphi+\beta)]^{\mu+3}}{\sin^2\varphi\,[\cos\beta]^{\mu+1}} + \\ &+ \frac{4\cos^2\beta\cos^2(\varphi+\beta)}{(\mu+2)(\mu+1)\sin^2\varphi}; \end{aligned} \tag{7.17}$$

$$\tan\varphi = \frac{u}{1 + u\tan\beta}; \ \tan\beta = \frac{\varkappa}{n-1}; \quad u = \frac{x'}{\mu}; \quad x' = 2x(n-1). \tag{7.18}$$

Equations (7.17) and (7.18) for real m assume the form

$$\begin{aligned} F(x, \mu, m) = 2 &- \frac{4}{(\mu+2)} \frac{\sin[(\mu+2)\varphi]\,[\cos\varphi]^{\mu+3}}{\sin\varphi} - \\ - \frac{4}{(\mu+2)(\mu+1)} &\frac{\cos[(\mu+1)\varphi]\,[\cos\varphi]^{\mu+3}}{\sin^2\varphi} + \frac{4\,\mathrm{cotan}^2\varphi}{(\mu+2)(\mu+1)}; \\ \tan\varphi &= u = \frac{2}{\mu} x(n-1). \end{aligned} \tag{7.19}$$

Equation (7.19) is also presented by Shifrin and Raskin /509/.

Note that

$$\lim_{x\to\infty} F(x, \mu, m) = 2, \quad \lim_{\mu\to 0} F(x, \mu, m) = 2,$$
$$\lim_{\mu\to\infty} F(x, \mu, m) = K(x, m). \tag{7.20}$$

We now examine an approximate method of calculating the averaged efficiency factor for extinction of radiation by water clouds and fog in the visible and infrared regions /513, 514/. The method is based on allowance for the complex nature of the refractive index of water and the real size spectrum of cloud and fog particles described by the gamma distribution. The method is based on utilization of expressions (6.45) and (6.46) for the efficiency factor for extinction of one particle $K(\rho, m)$, obtained elsewhere /489/ in an approximate manner.

Substituting expressions (6.45) and (6.46) into equation (7.16) we obtain

$$F(x, \mu, m) = F_1(x, \mu, m) + F_2(x, \mu, m). \tag{7.21}$$

Here function F_1 is defined by equations (7.17) and (7.18), while

$$F_2(x, \mu, m) = \frac{\mu^{\mu+3}}{\Gamma(\mu+3)} \sum_{j=1}^{4} \int_{h_j}^{h_{j+1}} D_j(x'\tau_m) \times$$
$$\times K(x'\tau, m)\, \tau^{\mu+2} e^{-\mu\tau}\, d\tau, \tag{7.22}$$

where

$$h_1 = 0; \quad h_2 = 5(n-1); \quad h_3 = \frac{4{,}08}{1+3\tan\beta}; \quad h_4 = \frac{4{,}08}{1+\tan\beta}; \; h_5 = \infty.$$

Function F was tabulated by this author for different values of φ, β and μ, selected to ensure obtaining bulk coefficients of extinction of visible and infrared radiation by clouds and fog with very different microstructural characteristics, actually encountered in nature.

Since $D(x'\tau)$ is a continuous function, the expression for F_2 can be written in terms of F_1 and D by applying the mean value theorem to equation (7.22):

$$F_2(x, \mu, m) = F_1(x, \mu, m)\, D(x', \bar{\tau}, m), \tag{7.23}$$

where $\bar{\tau}$ is a quantity which in general is a function of n, β, μ and x'.

Function F_2 was calculated from equation (7.22) by numerical integration for different x', μ, β and n, and then equation (7.23) was employed for finding τ. The results obtained from equations (7.22) and (7.23) agree to within 5% provided

$$\bar{\tau} = \frac{\mu+3}{\mu}. \tag{7.24}$$

Substitution of (7.23) into (7.21) finally yields for F

$$F(x, \mu, m) = F_1(x, \mu, m)\left[1 + D(x'\bar{\tau}, m)\right]. \tag{7.25}$$

Expressions (7.25) and (7.24) were utilized by this author for calculating functions $F(x, \mu, m)$ and bulk extinction coefficients $\alpha(\lambda)$ for clouds and fog. Estimates indicate that the above approximate method yields an error in values of $F(x, \mu, m)$ not exceeding 5–7% of its values obtained using tabulated values of function $K(\rho, m)$ /470/, derived from exact formulas of Mie theory. This accuracy is quite sufficient for practical purposes and for quantitative interpretation of numerical data.

2. **Calculation of extinction coefficients.** The most exhaustive theoretical data on the extinction coefficient $\alpha(\lambda)$ for water clouds and fog in the 0.5–14μ spectral region, with allowance for the real polydispersity of particles and the complex nature of the refractive index, were obtained in /516/, and described in detail in /1/. In the latter reference this author also examines the results of computations of $\alpha(\lambda)$ carried out elsewhere /489, 517/ for two specific cloud models in the 0.45–16.6μ spectral region, and computational data on the efficiency factor F obtained by Levin /508/ for different cloud models without allowance for the complex nature of the refractive index of water.

Calculations by this author and those by Deirmendjian /489, 517/ were based on data on the complex refractive index of water; these were published by Centeno already in 1941 /518/ and were until recently regarded as very comprehensive and reliable. However, detailed comparison of all known values of the components of the complex refractive index of water in the infrared region carried out by this author showed that the data of different investigators are not in satisfactory agreement. Bearing this in mind, and also that coefficients $\alpha(\lambda)$ depend essentially on the real part n as well as on the imaginary part $\varkappa$ of the complex refractive index of water, attempts were made in the author's laboratory to again determine these quantities. Data thus obtained on n and $\varkappa$ for different wavelengths in the 0.5–25μ region led to the recalculation of coefficients

$\alpha(\lambda)$. Below we present briefly the results of studies into the components of the complex refractive index of water and corresponding calculations of coefficient $\alpha(\lambda)$.

1. Components of the complex refractive index of water in the 0.5–25μ spectral region. The equipment and methods of measurement of the refractive index and of the absorption index of water in the 0.5–25μ spectral region have been described in detail elsewhere /519–521/.

Figure 7.1 depicts curves of the refractive index n and index of absorption $\varkappa$ as a function of wavelength in the 2.5–25μ spectral region. The values of n and $\varkappa$ in different spectral sections are compared in Figure 7.2.

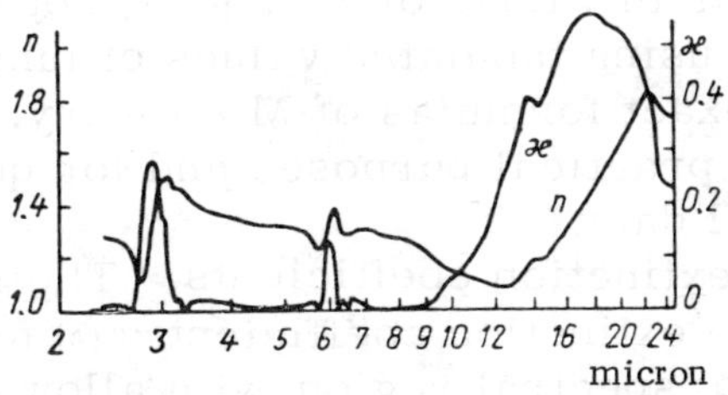

FIGURE 7.1. Refractive index n and index of absorption $\varkappa$ of water as a function of wavelength in the 2.5–25μ spectral region.

The technique for determining $\varkappa$ from reflection spectra /519, 520/ was not suitable for obtaining sufficiently accurate results in regions with small $\varkappa$. Therefore the value of $\varkappa$ in these regions was again determined /521/ using a technique based on utilization of transmission spectra. In addition, the spectral region in which n and $\varkappa$ were measured was expanded /521/. The results thus obtained are shown in Figures 7.3 and 7.4. The magnitude of the absorption index $\varkappa(\nu)$ can be found from known data on $\alpha(\lambda)$ with the aid of the expression

$$\varkappa(\nu) = \frac{\alpha(\nu)}{4\pi\nu}. \tag{7.26}$$

The data plotted in Figures 7.3 and 7.4 were determined with a relative error of not more than 1% for n and 4% for α.

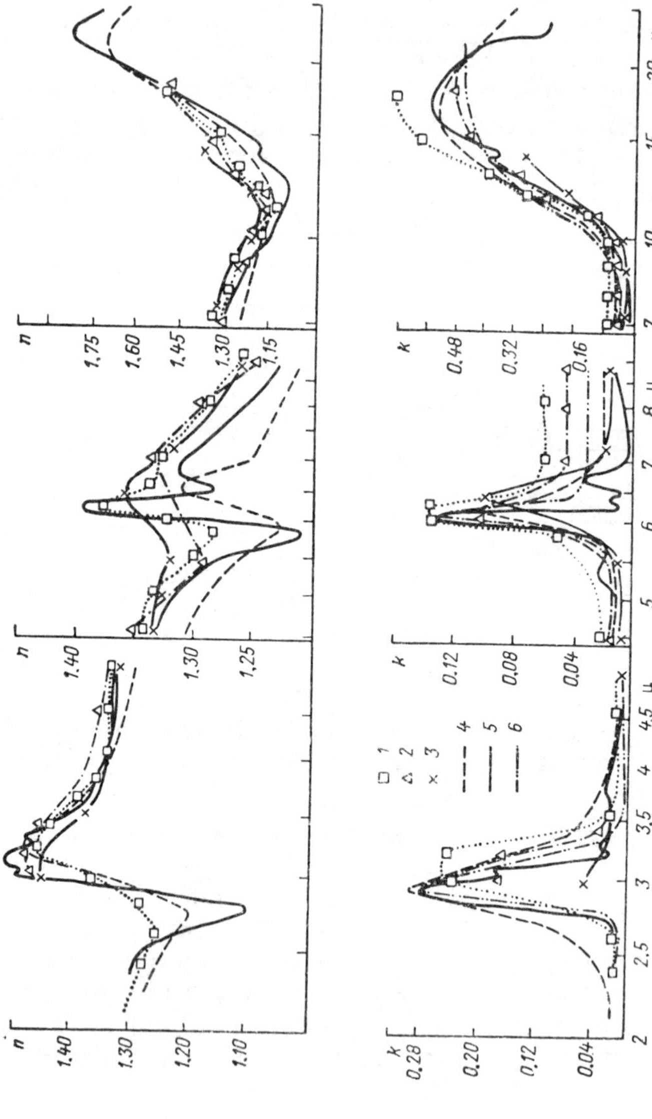

FIGURE 7.2. Index of refraction n and of absorption $\varkappa$ of water in the region from 2.5 to 25 μ:

1) data of /522/; 2) data of /467/; 3) data of /518/; 4) data of /523/; 5) data of /520/; 6) data of /524/.

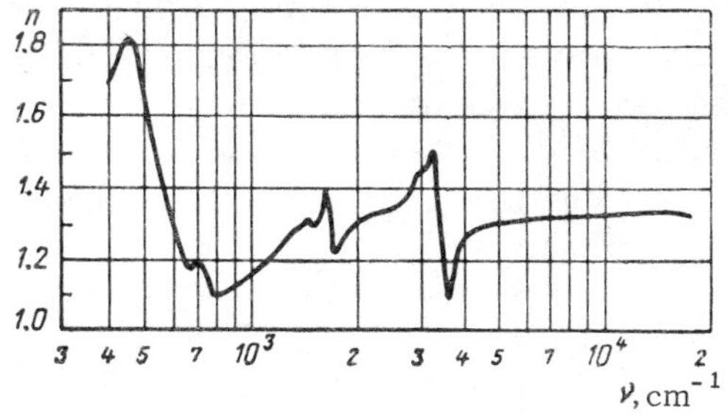

FIGURE 7.3. Index of refraction of water as a function of wavelength in the 0.5 − 25 μ spectral region.

Due to the inconvenience of using graphs in practical work, Appendix 2 lists detailed tabulated data on n and $\varkappa$ in the spectral region from 0.54 to 25μ, obtained by the author.

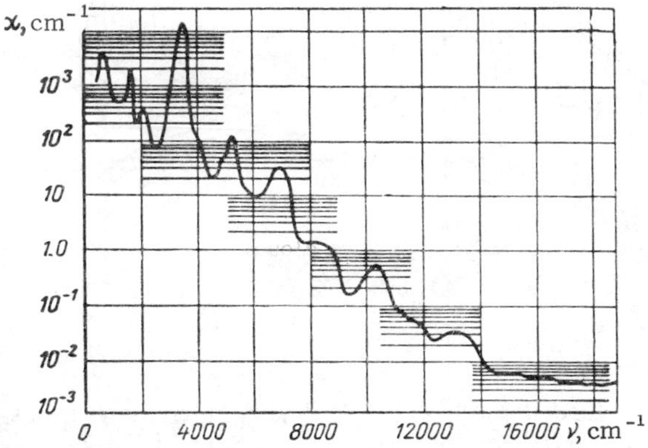

FIGURE 7.4. Index of absorption of water as a function of wavelength in the 0.5–25 μ spectral region.

2. Extinction coefficients. A new series of calculations of extinction coefficients α (λ) of water clouds and fog, with incorporation of real particle size spectra and data obtained by the author on the components of the complex refractive index of water, was carried out using the approximate method described in Sec. 1. Computation was conducted for values of the microstructural parameters r (most probable particle radius) and μ (characteristic of half-width of the distribution), equal respectively to 0.1, 0.5, 1.0, 2.0, 6.0, 10.0 microns and 1, 2, 3, 6 and 10. The ensemble of

wavelengths for which $\alpha(\lambda)$ was calculated was determined by the respective shape of the curve of the components of the complex refractive index of water. This author tried to incorporate all the features of this relationship. In addition, a series of wavelengths was selected corresponding to the emission frequencies of the most extensively used laser instruments.

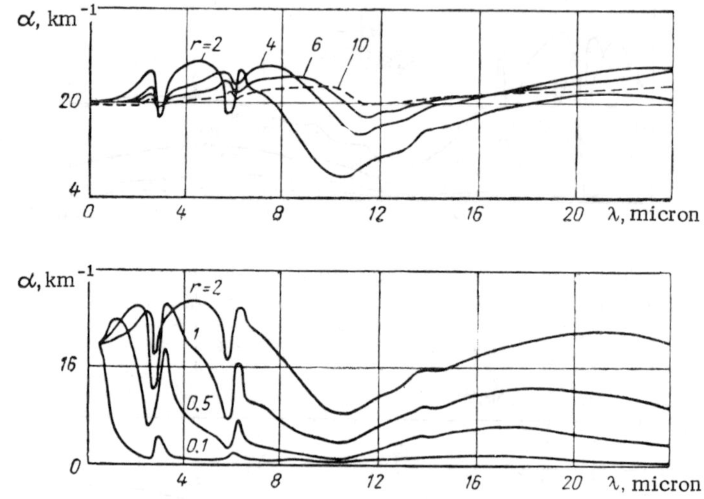

FIGURE 7.5. Extinction coefficient of water clouds and fog over the 0.5–25 micron wavelength range for $\mu = 2$, $r = 0.1, 0.5, 1, 2, 4, 6, 10$ microns, and $S_m = 0.2$ km.

Detailed computational data have been published elsewhere /525/. Figures 7.5—7.8 and Table 7.3 illustrate some of the results, showing the substantial effect of microstructural parameters and components of the complex refractive index of water on the extinction coefficient $\alpha(\lambda)$ and on its spectral variation.

Table 7.3 lists data on extinction coefficient $\alpha(\lambda)$ of clouds and fog for the following most frequent, naturally occurring characteristics: 1) most probable radius $r = 6$ microns; 2) parameter of half-width of the distribution $\mu = 2$; 3) meteorological visibility range $S_m = 0.2$ km. Coefficient $\alpha(\lambda)$ for other values of S_m can be obtained from data of Table 7.3, since the value of $\alpha(\lambda)$ is inversely proportional to S_m.

Figure 7.9 compares values of $\alpha(\lambda)$, calculated by this author on the basis of Centeno's data on the complex refractive index and on the basis of the author's data. In a number of spectral sections the later data on $\alpha(\lambda)$ obtained by the author differ markedly from the earlier data.

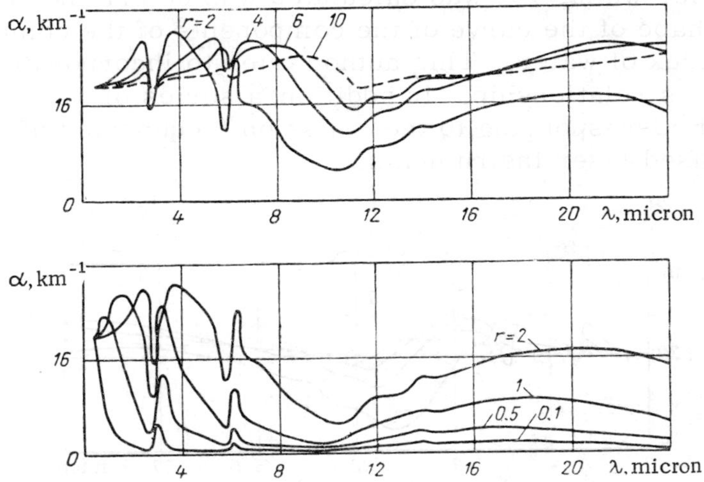

FIGURE 7.6. Extinction coefficient of water clouds and fog over the 0.5–25 micron wavelength range for $\mu = 4$, $r = 0.1, 0.5, 1, 2, 4, 6$ and 10 microns, and $S_m = 0.2$ km.

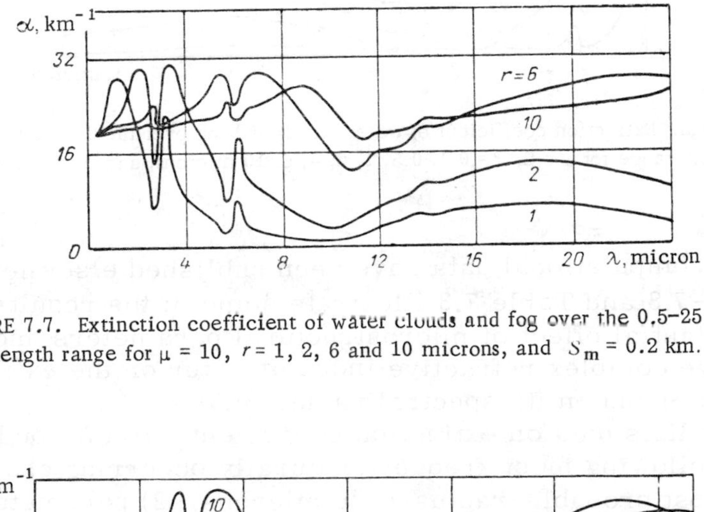

FIGURE 7.7. Extinction coefficient of water clouds and fog over the 0.5–25 micron wavelength range for $\mu = 10$, $r = 1, 2, 6$ and 10 microns, and $S_m = 0.2$ km.

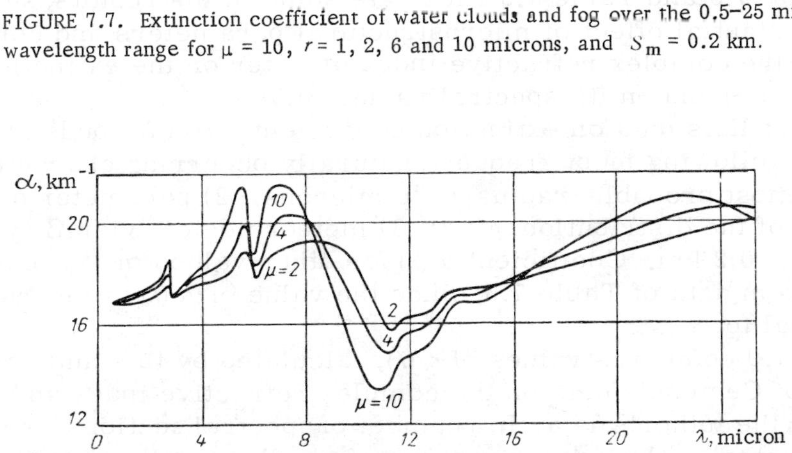

FIGURE 7.8. Extinction coefficient of water clouds and fog over the 0.5–25 micron wavelength range for $r = 6$ microns, $\mu = 2, 4$ and 10, and $S_m = 0.2$ km.

TABLE 7.3. Coefficient $\alpha(\lambda)$ for the most probable cloud and fog microstructural parameters at meteorological visibility $S_m = 0.2$ km or particle concentration $N = 28$ cm^{-3}

λ	α (λ)	λ	α (λ)	λ	α (λ)	λ	α (λ)
0.31	19.5	1.18	19.9	2.94	20.1	6.15	21.6
0.51	19.6	1.24	19.9	3.12	20.3	6.21	21.9
0.53	19.6	1.26	20.0	3.34	20.6	6.27	22.1
0.56	19.6	1.39	20.0	3.39	20.7	6.44	22.9
0.59	19.7	1.43	20.0	3.51	20.8	6.66	23.1
0.625	19.7	1.45	20.0	3.58	20.9	7.69	24.1
0.63	19.7	1.47	20.1	3.85	21.0	8.36	24.6
0.67	19.7	1.56	20.1	4.17	21.4	10.04	22.0
0.69	19.7	1.67	20.1	4.55	21.6	10.6	19.6
0.71	19.7	1.79	20.2	4.77	21.8	11.29	17.3
0.77	19.7	1.82	20.2	5.00	22.0	11.42	17.5
0.81	19.7	1.89	20.2	5.13	22.1	11.56	18.0
0.84	19.7	1.94	20.2	5.27	22.2	12.42	18.6
0.91	19.8	2.00	20.3	5.41	22.6	13.79	20.5
0.97	19.8	2.22	20.4	5.56	23.5	14.18	20.6
1.01	19.8	2.36	20.5	5.82	23.7	14.38	20.6
1.05	19.9	2.50	20.6	5.87	22.7	16.81	21.8
1.06	19.9	2.63	20.8	5.93	22.0	17.39	22.0
1.11	19.9	2.71	21.0	5.98	21.6	22.47	24.5
1.13	19.9	2.78	20.0	6.03	21.4	25.31	25.8
1.15	19.9	2.91	20.0	6.09	21.4		

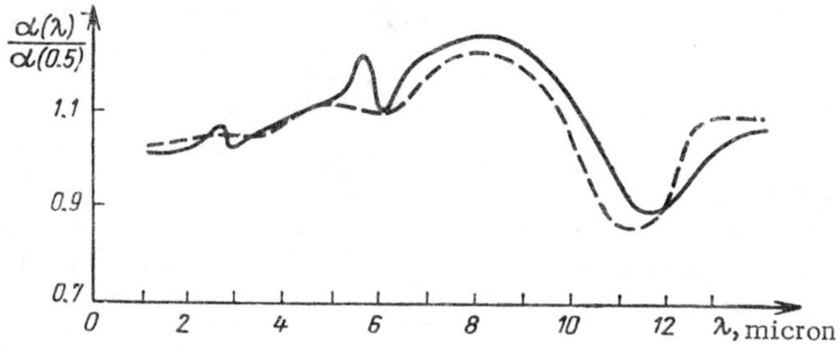

FIGURE 7.9. Comparison of calculated extinction coefficient of water clouds and fog over the wavelength range 0.5–14 microns, carried out on the basis of data of /518/ (dashed curve), and data of the author /520/ (solid curve) on the components of the complex refractive index of water for microstructural parameters $\mu = 2$ and $r = 6$ microns.

Calculations /525/ were carried out for virtually all possible values of the microstructural parameters of water clouds and fog, with the exception of formations with the largest droplet sizes for which, as follows from Figures 7.5–7.8 and from qualitative considerations, $\alpha(\lambda)$ can be assumed to be independent of wavelength with accuracy sufficient for practical purposes. In calculations, the particle size distribution function was the gamma distribution, which describes single-peak curves of particle size spectra. The results of these calculations cannot be applied to other particle size distributions. However, as was shown in Chapter 5, most data on the particle size spectra of clouds and fog are quite satisfactorily approximated by the gamma distribution.

Finally, it should be reemphasized that coefficient $\alpha(\lambda)$ describes attenuation of radiation propagating along a straight line from the source of the receiver. In experimental studies of the aerosol component of the transparency of clouds and fog to visible and infrared waves, the receiving system records not only direct attenuated radiation, but also scattered radiation. The question of the limits of applicability of the data obtained on $\alpha(\lambda)$ for determining the transparency of clouds and fog for different specific sources and receivers will be examined later in Sec. 9.

3. Experimental studies of extinction coefficients. Very complete data on coefficient $\alpha(\lambda)$ for different microstructural parameters of artificial fogs have been obtained /513, 526–529, 702/. These data are discussed elsewhere /1/, where the main results of others are also examined /379, 520–537/.

Reference /1/ contains an analysis of available experimental data and a comparison with theoretical values of coefficient $\alpha(\lambda)$. The experimental data of different investigators are found to be in satisfactory agreement with one another. In the majority of cases it is impossible to make a quantitative comparison due to the absence in the respective publications of necessary information on the microstructure of the clouds and fogs in which the measurements were carried out.

Comparison of experimental and theoretical data obtained by this author on the values of $\alpha(\lambda)$ showed that these are not in satisfactory agreement. This discrepancy was observed also by a number of other investigators. Here the measured values of $\alpha(\lambda)$ are found, as a rule, to be larger than those calculated from data of individual microstructural measurements. It is very important to clarify the reasons for these discrepancies.

It was suggested in analyzing the results of the first series of experiments that the traps used for determining the absolute particulate concentrations did not provide correct results. On the assumption that the traps for some reason do not collect all the droplets from the corresponding volume, we immediately conclude that the coefficients calculated from the microstructural data of traps will be smaller than the experimental values.

To confirm this hypothesis we carried out, during the second series of experiments, special simultaneous measurements of coefficient α (0.42) by means of a photometer, and of the water content by two methods using filters /529/, and used data of microstructural measurements.

It was found that the water content obtained from filters exceeded in all cases that calculated from data of simultaneous microstructural measurements by the trap. Coefficient $\alpha(0.42)$ measured by a photometer was found to be directly proportional to the water content determined by the filtration method. The obtained relationship is approximated by the expression

$$\alpha(0.42)(\mathrm{cm}^{-1}) = 2.04 \cdot 10^{-3} q(\mathrm{g/m}^3). \qquad (7.27)$$

A similar expression is obtained from the equations

$$\alpha(0.42) = 2Q = 2N \frac{\pi d_2^2}{4} 10^{-8} (\mathrm{cm}^{-1}), \qquad (7.28)$$

$$q = N \frac{\pi d_3^3}{6} 10^{-6} (\mathrm{g/m}^3). \qquad (7.29)$$

Here d_2 and d_3 are the root-mean-square and root-mean-cube diameters of the size distribution of particles (in microns).

Relationships (7.28) and (7.29) yield

$$\alpha(0.42) = 3 \frac{d_2^2}{d_3^3} 10^{-2} q (\mathrm{cm}^{-1}). \qquad (7.30)$$

To compare the empirical (7.27) and theoretical (7.30) expressions for $\alpha(0.42)$ as a function of q from data of microstructural measurements we calculated d_2^2/d_3^3 for each value of the water content. The average value of this ratio was found to be 0.0714 which, upon substitution in (7.30), yields

$$\alpha(0.42) = 2.14 \cdot 10^{-3} q. \qquad (7.31)$$

This expression is quite close to (7.27).

A more detailed comparison of water contents, measured by the method of filters and calculated from equation (7.30), is presented in Figure 7.10 for each value. To calculate q from equation (7.30), values of $\alpha(0.42)$ were found from photometer measurements and ratio d_2^2/d_3^3 was determined from data obtained from traps. The relationship between the calculated (q_t) and measured (q_f) water-content values is found to be approximated by the linear equation

$$q_t = 0.983 q_f - 0.01, \qquad (7.32)$$

which implies satisfactory agreement between the calculated and measured values of the water content.

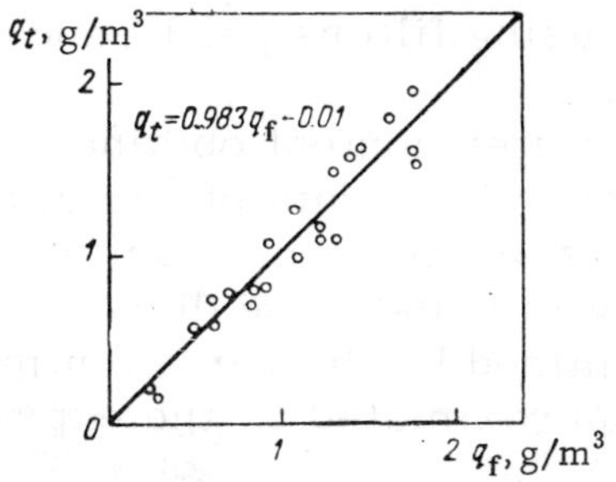

FIGURE 7.10. Calculated q_t and experimentally determined q_f water content of artificial fog.

The measurements carried out have thus demonstrated the validity of the assumption that large differences between theoretical and measured absolute values of bulk extinction coefficients are due to incorrect data on the absolute particle concentrations obtained by means of throughflow traps.

4. Frequency of different values of meteorological visibility in clouds and fog. To solve many problems, it is necessary to know how frequently one encounters given clouds or fog with different values of either meteorological visibility S_m, or particle concentration N, or water content Q. These quantities are related by the expressions

$$N = \frac{3{,}912\mu^2}{S_m F(0.5)\,\pi r^2 (\mu+2)(\mu+1)}, \tag{7.33}$$

$$Q = \frac{3.912 \cdot 4r(\mu+3)\,d}{3 \cdot S_m F(0.5)\,\mu} \tag{7.34}$$

and formula (7.11), where μ and r are the parameters of the particle size gamma distribution; $F(0.5)$ is the efficiency factor for extinction at wavelength $\lambda = 0.5$ micron; d is the density of water.

This author is aware of only one study /535/ in which the frequency of different values of meteorological visibility S_m in clouds and fogs was tabulated on the basis of direct measurements in the Arctic during the summer season (203 measurements in fog, 1457 in *St*, 433 in *Sc* and 148 in *Ns — As*) (see Table 7.4).

If we use data on the frequency of the water content of clouds, obtained /391/ by processing 5000 aircraft measurements of the water content in different parts of the USSR, and formula (7.33), then the most probable values of meteorological visibility are found to be S_m= 388−130 m, obtained by using the most probable values of water content between 0.1 and 0.3 g/m³ /391/ and data on microstructural parameters r = 6 micron and μ = 2.

TABLE 7.4. Frequency (in %) of the meteorological visibility S_m in different clouds and fogs during summer in the Arctic /535/

Cloud form	S_m, m					
	< 26	27—50	51—100	101—200	201—300	301—400
St	0.1	2.0	21.7	36.1	21.4	5.6
Sc	0.5	2.1	18	35.6	21.1	7.1
Ns—As	8.1	5.4	10.1	33.1	20.3	12.2
Fog	1.0	5.4	10.3	32.9	21.7	11.3

TABLE 7.4. (contd.)

Cloud form	S_m, m						
	401—500	501—600	601—700	701—800	801—900	901—1000	>1000
St	4.4	4.2	2.7	1.5	0.7	0.1	—
Sc	5.3	4.9	0.2	1.8	1.1	0.7	0.5
Ns—As	2.7	2.0	1.3	2.7	1.4	0.7	—
Fog	6.9	6.4	1.0	0.6	1.0	0.5	1.0

TABLE 7.5. Water content Q and particle concentration N of some water clouds and fogs

Nature of cloud	Microstructural parameters		S_m, km	Q, g/m³	N, cm⁻³
	r, microns	μ			
Small-droplet	1	2	0.2	0.031	971
	1	10		0.015	2050
Average-droplet	6	2	0.2	0.194	28
	6	10		0.101	65
Large-droplet	10	2	0.2	0.324	10
	10	10		0.168	2.3

In conclusion, Table 7.5 presents values of water content Q and particle concentration N of clouds and fogs for various characteristic ensembles of microstructural parameters when meteorological visibility $S_m = 0.2$ km. The data were calculated from (7.33) and (7.34). For the one value $S_m = 0.2$ km, the water content and particle concentration of different clouds and fogs can range from 0.015 to 0.324 g/m^3 and from 2.3 to 2050 particles per cm^3.

5. Determination of the aerosol component of transmittance in the infrared from transmittance data in the visible spectral region. The possibility of determining the spectral transmittance of clouds and fog in the infrared from corresponding data in the visible wavelength range is very attractive, since the theoretical and experimental solution of the latter problem is much simpler. It suffices to state that the transmittance in the visible region T_a (0.5) can be determined from visual measurements of the meteorological visibility S_m using the expression /5, 539/ (it is assumed that the threshold of the contrast sensitivity of the eye is 0.02)

$$T_a(0.5) = e^{-3.912\frac{l}{S_m}}, \qquad (7.35)$$

where l is the range and

$$\alpha(0.5) = \frac{3,912}{S_m}. \qquad (7.36)$$

The spectral transmittance at wavelength λ is

$$T_a(\lambda) = e^{-\alpha(\lambda)l}$$

This, together with (7.36), yields

$$T_a(\lambda) = e^{-\omega(\lambda)3.912\frac{l}{S_m}} = [T_a(0.5)]^{\omega(\lambda)}, \qquad (7.37)$$

where

$$\omega(\lambda) = \frac{\alpha(\lambda)}{\alpha(0.5)} = \frac{F[x(\lambda), \mu]}{F[x(0.5), \mu]}. \qquad (7.38)$$

Ratio $\alpha(\lambda)/\alpha(0.5)$ for different microstructural parameters of droplet clouds and fogs were obtained by us for the 0.5–25 micron wavelength range using new data on the complex refractive index of water (see part 2 of this section). These data together with a specified value of S_m or of α (0.5) enables the problem at hand to be solved.

It can be seen from Figures 7.5–7.8 that the same value of S_m or, which is the same thing, the transmittance $T_a(0.5)$ in the visible range correspond to different values of transmittance $T_a(\lambda)$ for the given wavelength in the infrared as a function of r and μ. For specific values of these two parameters a unique relationship exists between $T(\lambda)$ and $T(0.5)$ or S_m.

In practical utilization of this method for solving many problems, the range of variation in r and μ can be substantially limited to their most probable values, listed in Table 5.2 for different clouds and fogs. The most frequently encountered values of S_m in clouds of different types and in fogs are listed in Table 7.4.

A great advantage of the proposed method of determining the spectral transmittance of clouds and fogs in the infrared from corresponding data in the visible spectral region is the fact that it eliminates the need for data on aerosol-particle concentration, which is difficult to determine experimentally.

5. ENERGY CHARACTERISTICS OF SCATTERING OF VISIBLE AND INFRARED RADIATION BY HAZE

1. Extinction coefficients. 1. Theoretical investigations. Shifrin and Raskin /540–542/ studied theoretically the extinction coefficients of polydisperse aerosols, consisting of "soft" transparent particles and possessing size distributions described by the Junge formula, the gamma distribution with $\mu = 2$ and generalized gamma distributions.

Extinction coefficients of polydisperse aerosols, consisting of transparent particles with refractive index $m = 1.5$ and 1.33, were calculated by Ivlev and Yanchenko /821/. The particle size distribution functions were the Junge formula with $\beta = 2$, 2.5, 3, 3.5 and 4, lower and upper size limits being $a_1 = 0.08$ micron and $a_2 = 1$, 3, and 10 microns, and other distributions /810, 811/. It has been shown /821/ that in a real atmosphere the extinction coefficients in the visible part of the spectrum can differ substantially in spectrum shape and in absolute value. Deirmendjian /489, 517/ computed extinction coefficients $\alpha(\lambda)$ for several haze models with consideration of the complex nature of the refractive index of water in the spectral region up to 13 microns.

Calculations of $\alpha(\lambda)$ for haze, the particle size spectrum of which is described by the Junge formula for $\beta = 3, 4$ and 5 with allowance for the complex nature of the refractive index of water, were conducted elsewhere /543/. Minimum and maximum particle radii were taken as 0.1 and 1.0 micron. Calculations were also carried out at 0.5–14 micron wavelengths, for which the values of $\alpha(\lambda)$ are best expressed in the form of ratios $\alpha(\lambda)/\alpha(0.5) = \omega(\lambda)$.

If $\omega(\lambda)$ is known, one can derive the value of $\alpha(\lambda)$ for atmospheric haze at different S_m :

$$\alpha(\lambda) = \frac{3.912}{S_m}\,\omega(\lambda)\,(\text{km}^{-1}). \tag{7.39}$$

A new series of data on $\alpha(\lambda)$ was obtained by this author from refined data on the complex refractive index of water (see Sec. 4 and Appendix 2). Calculations were carried out for many wavelengths in the visible and infrared at wavelengths 0.3–25 microns. The minimum and maximum particle sizes were $a_1 = 0.01$, 0.05 and 0.1 micron and $a_2 = 1.0$, 5.0 and 10 microns, so enabling clarification of the contributions of particle fractions with different dimensions to the value of coefficient $\alpha(\lambda)$. The results of new calculations of $\alpha(\lambda)$ are presented in detail in /544/; some of these data are presented in Figures 7.11–7.13. The most substantial effect on the spectral variation of coefficient $\alpha(\lambda)$ in hazes is exerted by parameter β, which characterizes the steepness of the Junge particle size distribution. The contribution of the small-droplet fraction of haze particles ($a_1 = 0.1$–0.01 micron) is negligible even though the concentration of these particles, according to Junge's formula, is very large.

Figure 7.14 compares the first and second series of calculations of $\alpha(\lambda)$ by this author, and shows the marked difference between the two sets of data in individual spectral regions. Figures 7.15–7.17 depict curves of the aerosol component of the spectral transmittance of haze $T_a(\lambda)$, calculated from new data on $\alpha(\lambda)$.

Table 7.6 lists values of $\alpha(\lambda)$ obtained by this author for the most probable values of haze microstructural parameters $\beta = 4$, $a_1 = 0.5\mu$ and $a_2 = 5.0\mu$ for $S_m = 10$ km in the surface layer. Coefficients $\alpha(\lambda)$ for other values of S_m are obtained immediately from Table 7.6, since their values are inversely proportional to S_m.

Consider now the limits of applicability of the above data on $\alpha(\lambda)$. We first study two assumptions used in the calculations: 1) the haze particles were regarded as spherical and consisting of water; 2) their size spectra was described by Junge's empirical formula.

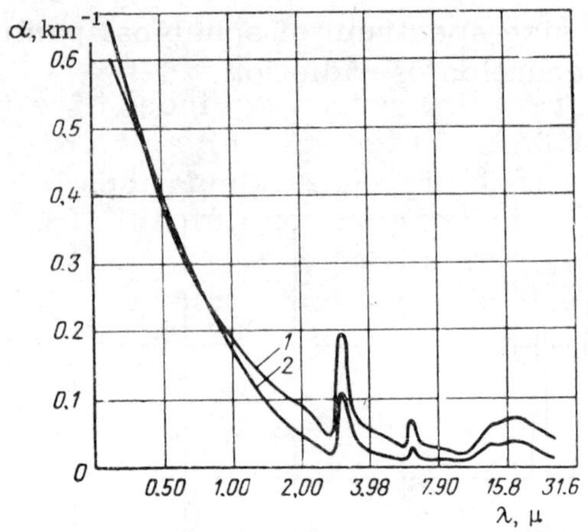

FIGURE 7.11. Extinction coefficient of haze at wavelengths 0.3–25 μ for S_m = 10 km, $\beta = 4$, $a_1 = 0.1\,\mu$, $a_2 = 1.0\,\mu$ (curve 1); $\beta = 4$, $a_1 = 0.01\,\mu$, $a_2 = 10\,\mu$ (curve 2).

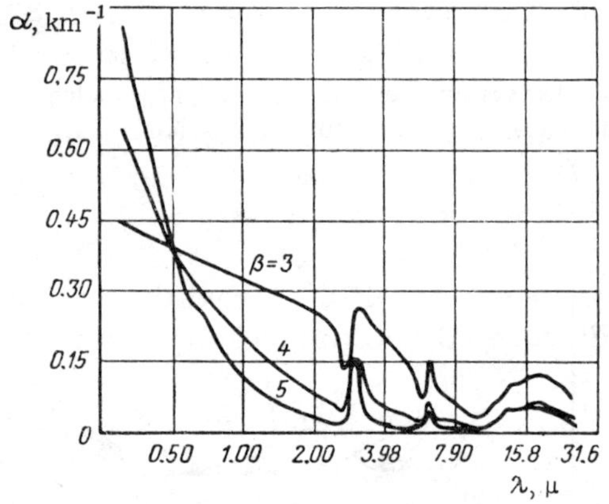

FIGURE 7.12. Extinction coefficient of haze at wavelengths 0.3–25 μ for S_m = 10 km, $a_1 = 0.05\,\mu$, $a_2 = 5.0\,\mu$.

These results are, strictly speaking, inapplicable to haze consisting not only of drops, but also of two-layered particles with solid cores. However, the error incurred by neglecting the two-layered nature of the particles is small when the dimensions of the solid cores are much smaller than those of the water shells /484, 486/. If the haze particles are nonspherical, then the above results should be treated as some approximation of reality, bearing in mind that random orientation of the particles makes it possible, in principle, to find a size spectrum of spherical particles equivalent with respect to attenuation of radiation.

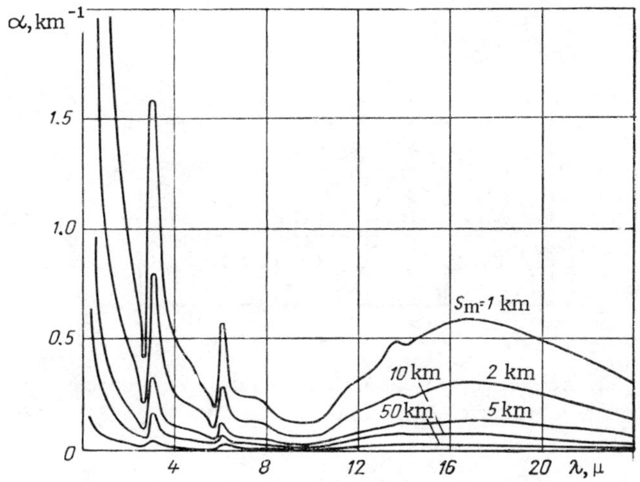

FIGURE 7.13. Extinction coefficient of haze at wavelengths 0.3–25 μ for different S_m with $\beta = 4$, $a_1 = 0.05\,\mu$, $a_2 = 5.0\,\mu$.

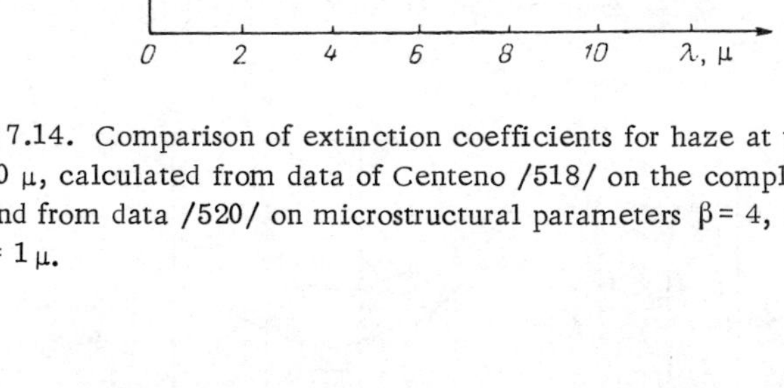

FIGURE 7.14. Comparison of extinction coefficients for haze at wavelengths 0.5–14.0 μ, calculated from data of Centeno /518/ on the complex refractive index and from data /520/ on microstructural parameters $\beta = 4$, $a_1 = 0.1\,\mu$ and $a_2 = 1\,\mu$.

The Junge formula gives an approximate idea of the real particle size spectra of atmospheric aerosol which is usually obtained after averaging a large number of measured particle size distributions (see Chapter 5). Consequently, data obtained on polydisperse extinction coefficients in hazes should be treated as some average characteristics. In each specific case coefficients $\alpha(\lambda)$ will differ from those calculated by Junge's formula. However, to calculate values of $\alpha(\lambda)$ for real particle size distributions one requires more data on these distributions than are presently available.

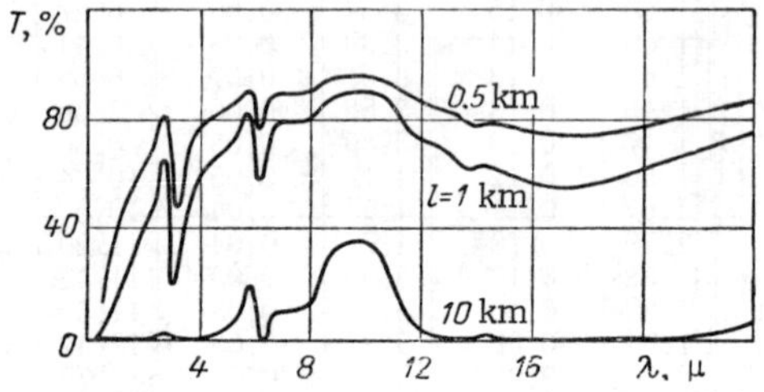

FIGURE 7.15. Aerosol component of the spectral transmittance of haze at wavelengths 0.3–25 μ for $S_m = 1$ km, $\beta = 4$, $a_1 = 0.05\,\mu$ and $a_2 = 5.0\,\mu$.

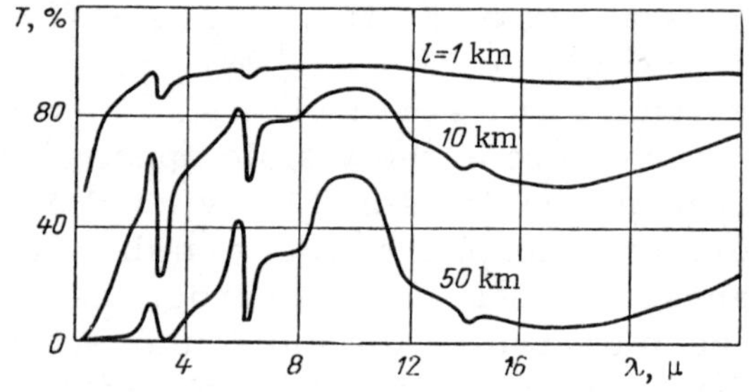

FIGURE 7.16. Aerosol component of the spectral transmittance of haze at wavelengths 0.3–25 μ for $S_m = 10$ km, $\beta = 4$, $a_1 = 0.05\,\mu$ and $a_2 = 5.0\,\mu$.

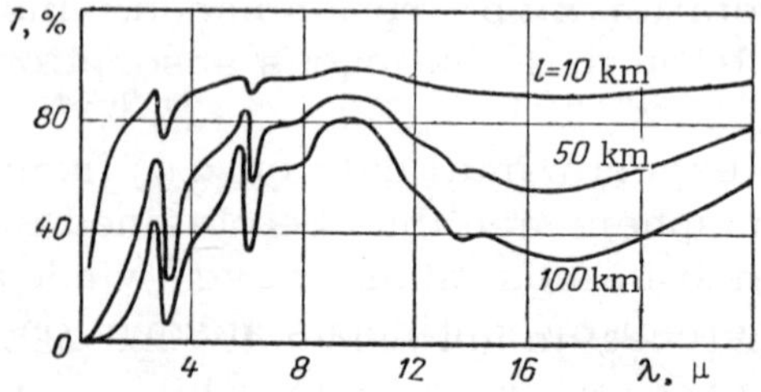

FIGURE 7.17. Aerosol component of the spectral transmittance of haze at wavelengths 0.3–25 μ for $S_m = 50$ km, $\beta = 4$, $a_1 = 0.05\,\mu$ and $a_2 = 5.0\,\mu$.

TABLE 7.6. Coefficients $\alpha(\lambda)$ for haze at most probable values of microstructural parameters and meteorological visibility $S_m = 10$ km in the surface layer of the atmosphere

λ μ	$\alpha(\lambda)$ km^{-1}	λ μ	$\alpha(\lambda)$ km^{-1}	λ μ	$\alpha(\lambda)$ km^{-1}	λ μ	$\alpha(\lambda)$ km^{-1}
0.31	0.65	1.18	0.17	2.94	0.16	6.15	0.05
0.50	0.40	1.24	0.15	3.12	0.13	6.21	0.04
0.53	0.38	1.26	0.15	3.34	0.09	6.27	0.04
0.56	0.36	1.39	0.14	3.39	0.08	6.44	0.03
0.59	0.34	1.43	0.13	3.51	0.07	6.66	0.02
0.625	0.32	1.45	0.13	3.58	0.06	7.69	0.02
0.63	0.32	1.47	0.13	3.85	0.05	8.36	0.01
0.67	0.30	1.56	0.12	4.17	0.05	10.04	0.01
0.69	0.29	1.67	0.11	4.55	0.04	10.6	0.01
0.71	0.26	1.79	0.10	4.77	0.04	11.29	0.02
0.77	0.26	1.82	0.10	5.00	0.03	11.42	0.02
0.81	0.25	1.89	0.10	5.13	0.03	11.56	0.03
0.84	0.24	1.94	0.09	5.27	0.03	12.42	0.03
0.91	0.22	2.00	0.09	5.41	0.03	13.79	0.05
0.97	0.21	2.22	0.08	5.56	0.02	14.18	0.05
1.01	0.20	2.36	0.07	5.82	0.02	14.38	0.05
1.05	0.19	2.50	0.06	5.87	0.02	16.81	0.06
1.06	0.18	2.63	0.05	5.93	0.03	17.39	0.06
1.11	0.18	2.71	0.04	5.98	0.04	22.47	0.04
1.13	0.17	2.78	0.06	6.03	0.05	25.31	0.02
1.15	0.17	2.91	0.15	6.09	0.06		

2. Experimental investigations. Most experimental studies have been carried out in the visible spectral region, due to the relative ease of measurements and the virtual absence of absorption of radiation by atmospheric gases in this region. Hence the quantity of interest is easily obtained from results of measurements which have been corrected for molecular scattering, sufficiently accurate characteristics of which are quite well known.

There exists a chief difficulty in experimental determination of $\alpha(\lambda)$ for the infrared spectral region. This difficulty is related not only to the need to allow for selective absorption of the gaseous component of the atmosphere, but also to the presence of continuous absorption by far strong lines located in the corresponding vibration-rotation and purely rotational absorption bands. Insufficient data are presently available on this continuous absorption (see Chapter 3, Sec. 6).

The most reliable experimental data on $\alpha(\lambda)$ from direct measurements in the atmosphere can thus be obtained in the visible spectral region and also in most short-wave "windows" in the infrared range, where the effect of continuous absorption is not yet of significance.

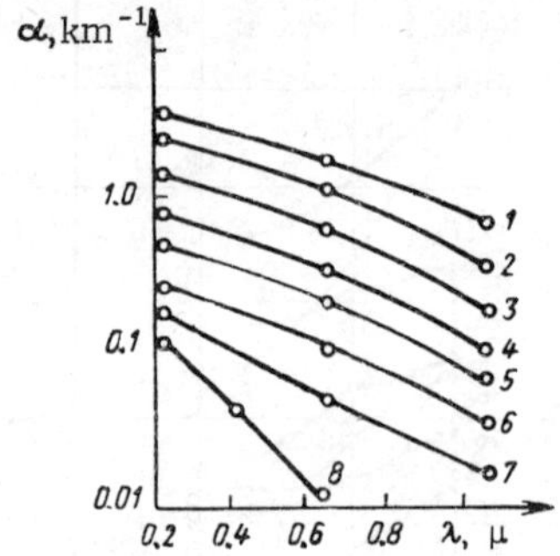

FIGURE 7.18. Scattering coefficients in haze for different values of S_m from data of Popov /545/:

1) S_m = 1.6 km; 2) S_m = 2.5 km; 3) S_m = 4.3 km; 4) S_m = 8.4 km; 5) S_m = 13.4 km; 6) S_m = 29.6 km; 7) S_m = 58.5 km; 8) Rayleigh scattering coefficient.

Popov /545/ measured haze transmittance at 0.313, 0.546 and 1.024μ for different S_m (1.6–58 km) in the surface layer of the atmosphere. He extrapolated these results to other wavelengths within the range under study and constructed curves of the spectral variation of $\alpha(\lambda)$ in haze for different S_m (Figure 7.18).

Coefficients $\alpha(\lambda)$ in ten narrow spectra (0.4–2.27μ) have also been measured /546, 547/. The values of $\alpha(\lambda)$ at 1.65 and 2.27μ were free from the effect of water vapor and methane absorption. Data on the absorption of water vapor and methane were taken from another paper /548/. All the measurements were conducted in the surface layer of the atmosphere over distances of 5.5 and 16.3 km at different meteorological visibilities. Figure 7.19 depicts the extinction coefficient as a function of wavelength for different S_m.

For $\lambda = 0.56\mu$, there exists a correlation between α and the relative humidity when the latter is above 70% /547/. When the relative humidity is below this value, α remains constant. This result is assumed to agree with data of older /549, 550/ and more recent /551, 552/ studies, and confirms the claim made by Middleton /553/ that the radii of condensation nuclei vary little in size at relative humidities up to 70%.

This conclusion disagrees with other results /554, 899/, that α is a function of increasing relative humidity starting with 30–35%. This, together with comprehensive analysis of other optical studies of atmospheric aerosol, gave Rozenberg /538/ the idea that at relative humidities $r \leqslant 30\%$ under measurement conditions the attenuation of light in the atmosphere is due to the dry (dust)

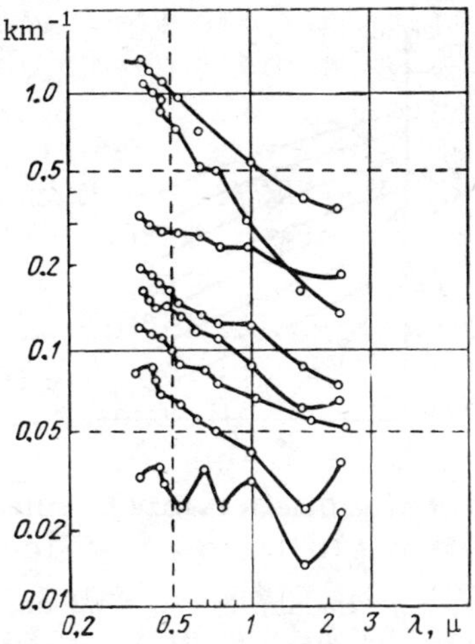

FIGURE 7.19. Aerosol extinction coefficient of hazes for different values of S_m (from S_m = 4 km to S_m = 160 km) /546/.

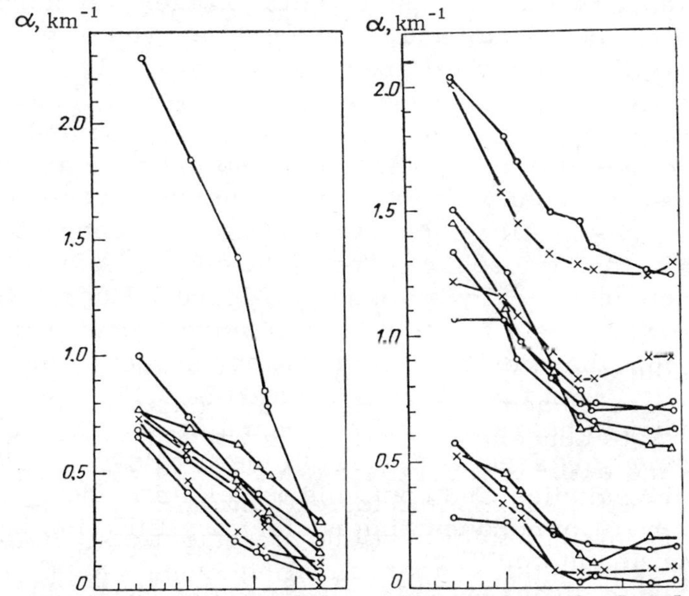

FIGURE 7.20. Aerosol extinction coefficient of hazes in the 0.5–5.0 μ range from measurements in winter (a) and autumn (b) /872/.

fraction of the aerosol, which produces weak turbidity. As the humidity is increased, the air becomes perceptibly turbid due to the formation of the wet haze.

The extinction coefficients of hazes in winter and autumn were measured by Golubitskii et al. /872/. Their results (Figure 7.20) differ in the spectral variation of extinction coefficients at 0.5–5.0μ during these seasons.

An original study of aerosol extinction coefficients in the infrared region for the 2, 4, 8 and 10μ wavelengths was carried out by Dennis /555/. The technique used eliminates the effect of selective as well as continuous absorption of the gas component of the atmosphere; however, the accuracy of measurements is rather poor.

In conclusion, we note that all these experimental investigations have a substantial shortcoming: optical measurements are not accompanied by simultaneous study of the microstructure of the aerosol producing the turbidity. As a result, comparison of experimental and theoretical data is not practical.

2. The aerosol component of haze transmittance at different altitudes and along slant paths in the atmosphere. 1. Theoretical investigations. The following general formulas pertain to the aerosol component of the spectral transmittance of haze for slant directions in the atmosphere:

$$T_a(\lambda) = e^{-\tau}, \tag{7.40}$$

$$\tau = \int_0^l \alpha(\lambda, l)\, dl, \tag{7.41}$$

$$\alpha(\lambda, l) = N(l) \int_{a_1}^{a_2} \sigma(\lambda, a) f(a, \beta)\, da, \tag{7.42}$$

$$\sigma(\lambda, \alpha) = \pi a^2 K(\rho, m), \tag{7.43}$$

where $T_a(\lambda)$ is the spectral haze transmittance component due to aerosol attenuation alone, l is the path traversed by the radiation, N is the particulate concentration which is variable along the beam path, $\sigma(\lambda, a)$ is the extinction coefficient for a particle of radius a, K is the particle efficiency factor for extinction.

The vertical distribution of particle concentration satisfies formulas (5.8), selected after approximating Junge's data on the concentration of optically active aerosol particles at different altitudes.

Expressions (7.40)–(7.43) and (5.8), together with formulas and tabulated data on function $K(\rho, m)$, considered in Chapter 6, enable the aerosol component of the spectral transmittance of haze to be computed along slant paths in the atmosphere. Calculation of $T_a(\lambda)$ in line with this scheme showed that, for the model atmosphere described by (5.8), aerosol attenuation can be neglected in atmospheric layers above 5 km for radiation propagation at all possible angles and up to 1000 km. Consequently, only aerosol attenuation in the lower 5-km layer, where the particulate concentration decreases exponentially with altitude, can be of practical interest (see formulas (5.8)). In this case it is possible to obtain a relatively simple expression for $T_a(\lambda)$.

With the aid of the notation $C' = \int_{a_1}^{a_2} \pi a^2 f(a, \beta)\, da$, we can set

$$\alpha(\lambda) = QF = N(z)\, C'F, \tag{7.44}$$

where Q is the geometric cross-section of particles within unit volume (see (7.11));

$$F = \frac{\int_{a_1}^{a_2} \pi a^2 K\left(\frac{2\pi a}{\lambda}, m\right) f(a, \beta)\, da}{\int_{a_1}^{a_2} \pi a^2 f(a, \beta)\, da}. \tag{7.45}$$

Equation (7.40), with consideration of (7.44), can now be expressed in the differential form

$$dT_a(\lambda) = -N(z)\, C'F\, dl. \tag{7.46}$$

The corresponding initial condition is $T_a(\lambda) = 1$ at $l = 0$.

Suppose the beam passes some distance between altitudes z_1 and z_2 at angle θ to the vertical. Then $dl = \sec\theta\, dz$. We shall assume that only the particulate concentration varies with altitude, while parameter β in (5.6) for the particle size distribution function does not depend on altitude. In this case C' and F are also independent of altitude. Hence integration of (7.46) over the altitude and subsequent rearrangement yields the following expression for quantity τ contained in the expression for $T_a(\lambda)$ in (7.40):

$$\tau(\lambda, z_1, z_2, \theta) = \sec\theta\, \frac{3.912}{S_m}\, \omega(\lambda)\, \frac{1}{b}\left(e^{-bz_1} - e^{-bz_2}\right), \tag{7.47}$$

or

$$\tau(\lambda, z_1, z_2, \theta) = \sec\theta\, [\tau(z_1) - \tau(z_2)], \tag{7.48}$$

where

$$\tau(z) = \frac{3.912}{S_m}\,\omega(\lambda)\,\frac{e^{-bz}}{b}, \qquad (7.49)$$

$$\omega(\lambda) = \frac{\int\limits_{a_1}^{a_2} \pi a^2 K\left(\frac{2\pi a}{\lambda}, m\right) f(a, \beta)\,da}{\int\limits_{a_1}^{a_2} \pi a^2 K\left(\frac{2\pi a}{0.5}, m\right) f(a, \beta)\,da} = \frac{\alpha(\lambda)}{\alpha(0.5)}. \qquad (7.50)$$

Since $N(5) = 0.03$ (see the second of formulas (5.8)), relationships (7.36), (7.42) and (7.50) give

$$b = 2.78 - 0.460 \log S_m(0), \qquad (7.51)$$

where $S_m(0)$ is the meteorological visibility at the earth's surface.

Expressions (7.47) and (7.51) and Table 7.6 enable one to solve the problem of the aerosol component of the spectral transmittance of haze for different source and receiver geometries in the lower 5-km atmospheric layer. Figure 7.21 presents results of computing $T_a(\lambda)$ in the vertical direction at $\beta = 4$, $a_1 = 0.05\mu$ and $a_2 = 5\mu$. The role of various layers in the attenuation of radiation by haze in the vertical direction is demonstrated by Figure 7.22.

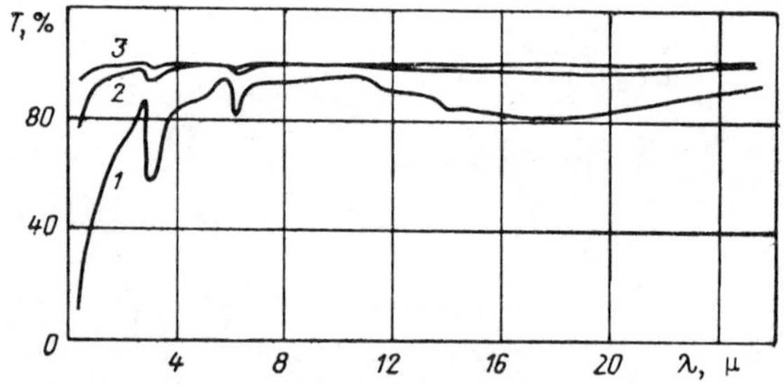

FIGURE 7.21. Aerosol component of the spectral transmittance of haze in the vertical direction for the lower 5 km layer of the atmosphere:

1) $S_m(0) = 1$ km; 2) $S_m(0) = 10$ km; 3) $S_m(0) = 50$ km.

It should be remembered that the above values of $T_a(\lambda)$ were derived for an atmospheric model with some average vertical statistical distribution of atmospheric aerosol particles. The particulate size spectrum was assumed independent of altitude. In each specific case the real distribution of particulate concentration and size can differ to a greater or lesser degree from that described by the assumed model. In addition, one must also remember the limitations on the applicability of data on coefficients $\alpha(\lambda)$ made in the preceding part of this section.

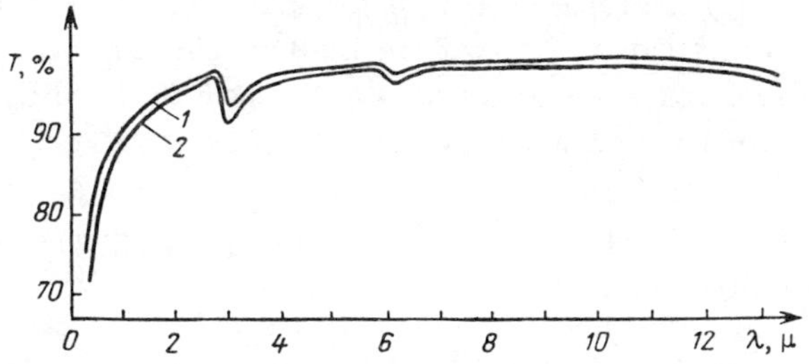

FIGURE 7.22. Aerosol component of the spectral transmittance of haze in the vertical direction for the 0–0.5 km (1) and 0–5.0 km (2) atmospheric layers when $s_m(0) = 5$ km.

The quantitative data on the scattering of radiation by haze particles along slant paths in the atmosphere /544/ and examined above pertain primarily to the infrared spectral region. Analytic expressions for the scattering coefficients and optical thicknesses in the visible spectral region for slant paths in the atmosphere were obtained by Shifrin and Minin /555/, who examined the theory of nonhorizontal visibility.

2. Experimental investigations. In recent years, substantial progress has been attained in the experimental study of aerosol scattering at different altitudes over a wide wavelength range. Only several years ago, primarily data in the visible spectral region, obtained from ground measurements and from a limited number of aircraft observations, were available. At present, quite a large number of investigations have been conducted at various altitudes in the visible and infrared regions. In every measurement the radiation source is the sun. When the atmosphere is sounded by means of searchlights and lasers, twilight, and various other methods, the receiving equipment is carried by aircraft, aerostats, artificial satellites, or placed on the ground.

The most important results of investigations into vertical profiles of aerosol extinction coefficients, associated with describing the atmosphere as a scattering medium, are examined in Chapter 5. We now briefly survey studies not included in that chapter. Very detailed aircraft measurements were obtained in /556, 436, 558, 559, 435, 561, 562/. The measurements were carried out for a number of wavelengths in the visible and very near infrared regions up to 6–7 km over different regions of the USSR. The results can be summed up as follows. The aerosol extinction coefficients at different altitudes, and the aerosol optical thickness of the atmosphere above a given altitude change by 1–2 orders from case to case. The dependence of the average coefficients of aerosol extinction on altitude is exponential to within 15–20%. The bottom 1-km layer plays a dominating role in the overall attenuation of radiation by the entire atmosphere. The same conclusion was reached elsewhere /563–565/.

Exponential reduction in aerosol extinction coefficients with altitude results in their decrease by 1–2 orders at altitudes of 3–5 km, compared with their surface values. Nevertheless, at these altitudes, only in the violet and blue parts of the spectrum are the Rayleigh coefficients found to be equal or to somewhat exceed the aerosol coefficients. In the red, and especially in the infrared parts of the spectrum the aerosol coefficients are greater than the Rayleigh coefficients at every altitude investigated. Aerosol coefficients were found to be greater than Rayleigh coefficients also at high altitude, up to 50 km /433/ and up to 100 km /434/.

Measurements carried out by Faraponova led to the discovery of the virtually moderate variation of aerosol extinction coefficients, equivalent to zero value of n in Angstrom's empirical formula describing the dependence of scattering coefficient α on wavelength λ:

$$\alpha = C\lambda^{-n}, \tag{7.52}$$

where C is an experimentally determined constant.

A similar result was obtained by Rabinovich for atmospheric layers above 1 km; in the bottom 1-km layer and for the entire thickness of the atmosphere, exponent n in (7.52) was found to be approximately equal to unity. Gushchkin conducted measurements from the ground and found that $n \approx 1.32$ on the average, with minima and maxima values of 1.03 and 1.48.

This moderate spectral variation of the extinction coefficient can be understood if we postulate the presence of large particles in the corresponding atmospheric layers. This assumption is in agreement with other data /432, 563, 566, 567/.

Note that Angstrom's formula has been used quite extensively to describe experimental data on the spectral variation of the aerosol extinction coefficient. Different investigators have obtained different values of n: $n = 0.92 \pm 0.25$ /568/, $n = 1.3 \pm 0.6$ /546/, $n = 1.1$, 2.1 and 2.3 /565/. In /540/, where the gamma distribution with $\mu = 2$ was used for the particle size, it was found theoretically for average particle radii of 0.05, 0.1, 0.2 and 0.5 micron that n equals respectively 3.3, 3.0, 2.5 and 2.1. Measurements for $S_m >$ > 7.5 km indicate that $n = 1.2$–1.5, while for $S_m \approx 2$ km $n \approx 0.9$ /569/. Popov /545/ found for transmittances of 49–60% that when $\Delta\lambda = 0.313-0.546\mu$, $n \approx 2$, and that when $\Delta\lambda = 1.014-0.546\mu$, $n \approx 1.2-1.3$. Ivanov et al. /820/ found that n ranged from 0.30 to 2.15, with an average value of 0.81; they note, correctly, that expression (7.52) is by far not always valid.

Vertical profiles of 1-km layer-by-layer attenuation of solar radiation by atmospheric aerosol up to 25–30 km were obtained by Kondrat'ev et al. /445/. These profiles differ somewhat from those determined elsewhere /822/ from spaceship measurements. In the latter work two maxima were detected in the vertical variation of the extinction coefficient (Figure 7.23). In the former work, these maxima are not as clearly expressed and are somewhat shifted in the direction of lower altitudes; in addition, other maxima were found in the vertical variation of the extinction coefficient, especially near the 6-km level /445/.

The existence of an aerosol layer with elevated extinction coefficient at about 10 km was detected in measurements at twilight and by day carried out from the "Vostok" and "Voskhod" spaceships /823/.

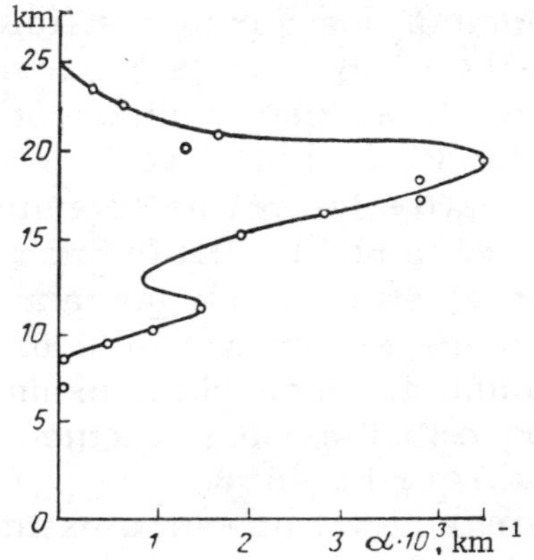

FIGURE 7.23. Vertical profile of aerosol extinction coefficient /822/.

Newkirk et al. /440, 446/ found very thin layers with elevated scattering coefficient from aerostat measurements. The thickness of these layers is approximately 30 m. The interlayer space is occupied by relatively clean air about 300 m thick.

The presence of maxima in the vertical profile of aerosol extinction coefficients is observed up to ionospheric layers. The upper boundary of atmospheric aerosol lies at 100 to 150 km /824/.

A summary graph of the vertical distribution of the aerosol extinction coefficient was constructed by Rozenberg /895/ from optical measurements /432, 559, 822, 901–905/. These data were obtained by different methods and were found to match satisfactorily; they formed an overall pattern of approximately exponential decrease in extinction coefficient for a number of wavelengths of the visible part of the spectrum up to 100 km. The extinction coefficient decreases by approximately six orders of magnitude at all wavelengths as the altitude increases from 0 to 100 km. For all altitudes the aerosol extinction coefficient was found to be at least severalfold greater than the coefficient of molecular scattering. An exception to this are two narrow altitude intervals (approximately from 5 to 8–9 and from 25 to 30 km) with very low aerosol extinction coefficients.

In summing up experimental data on aerosol scattering at different altitudes, one must first emphasize the large variability of the vertical profiles of extinction coefficients and their stratification over the entire atmospheric aerosol thickness. Maxima in the vertical variation of scattering coefficients are due to the presence of corresponding aerosol layers, some of which are stable in time (such as the layer at about 20 km), while the appearance of others is random. In spite of the many studies concerned with the experimental investigation into the aerosol component of atmospheric transparency to visible and infrared waves, the data obtained so far do not suffice for definite conclusions as to average statistical vertical distributions of extinction coefficients and their variability.

As in the case of experimental studies examined in Part One, all the measurements of aerosol extinction coefficients at different altitudes were not accompanied by simultaneous investigations of the aerosol microstructure. It is consequently impractical to compare experimental and theoretical data.

3. **The Elterman atmospheric attenuation model.** Elterman /447/ summarized Western experimental data on the spectra of dimensions, particulate concentrations and extinction coefficients in the surface layer and constructed an optical model of the cloudless atmosphere for $S_m(0) = 25$ km. The model was constructed for the following wavelengths: 0.27, 0.28, 0.30, 0.32, 0.34, 0.36, 0.38, 0.40, 0.45, 0.50, 0.55, 0.60, 0.65, 0.70, 0.80, 0.90, 1.06, 1.26, 1.67, 2.17, 3.50 and 4.00 μ. At all these wavelengths he tabulated coefficients of molecular scattering, aerosol extinction and molecular absorption by ozone. The last coefficients are given for wavelengths subjected to ozone absorption: 0.27, 0.28, 0.30, 0.32, 0.34, 0.36, 0.45, 0.50, 0.55, 0.60, 0.65, 0.70, and 0.80 μ. In addition, the model contains the optical thickness of molecular scattering and the resulting optical thickness due to molecular scattering, aerosol

attenuation and ozone absorption for all wavelengths. All the data are listed for altitudes to 50 km at 1-km intervals. Aerosol coefficients above 30 km are taken as zero.

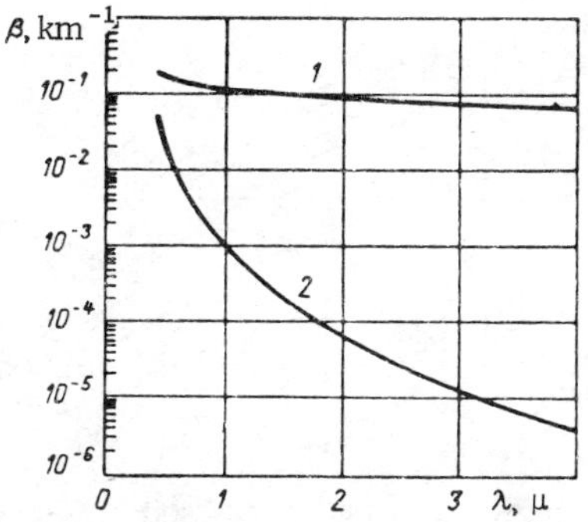

FIGURE 7.24. Coefficients of aerosol extinction (1) and molecular scattering (2) in the surface layer of the atmosphere as a function of wavelength /447/.

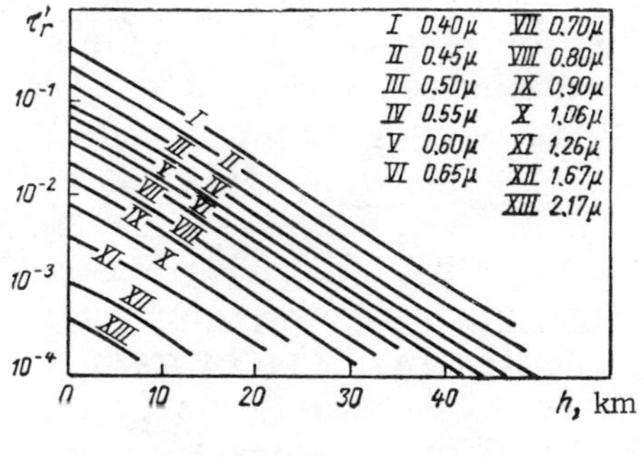

FIGURE 7.25. Optical thickness of the Rayleigh scattering of a vertical column of the atmosphere from different altitudes for different wavelengths /447/.

Data of the Elterman model were used to construct a number of graphs demonstrating the role of aerosol attenuation and molecular scattering in the atmosphere (Figures 7.24–7.26). It should be remembered that data on aerosol attenuation in the model at hand were obtained for $S_m(0) = 25$ km. For other values of S_m coefficients $\alpha(\lambda)$ are easily obtained directly from the tables, since they are inversely proportional to $S_m(0)$.

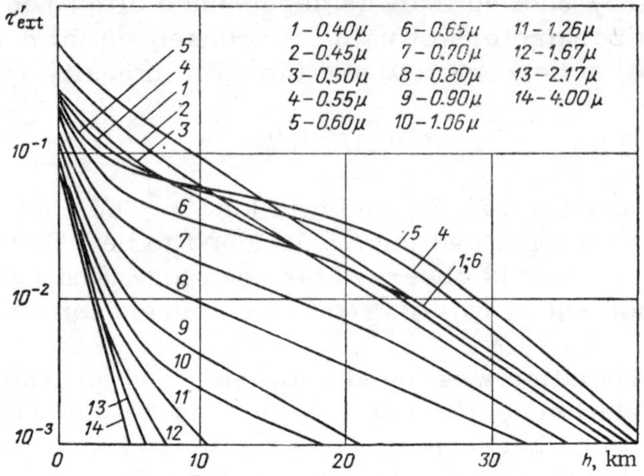

FIGURE 7.26. Vertical variation of the total optical thickness for different wavelengths /447/.

The Elterman model describes quite approximately aerosol scattering at different altitudes. It does not incorporate the abrupt change in the vertical variation of aerosol particle concentration at 5 km and the presence of aerosol layers. In addition, this model assumes zero aerosol attenuation above 30 km. However, no better model is presently available.

6. ENERGY CHARACTERISTICS OF SCATTERING OF VISIBLE AND INFRARED RADIATION BY PRECIPITATION

Precipitation particles for visible and infrared radiation can be regarded as large, in which case $K(\rho, m) = 2$. If the particles are assumed to be spherical, the extinction coefficient is given by

$$\alpha(\lambda) = N \int_0^\infty \pi a^2 K(\rho, m) f(a)\, da = 2Q,$$

where Q is the geometric cross-section of particles in unit volume, defined by (7.11).

Rain droplets can be regarded as spherical. The extinction coefficient of rain is thus independent of wavelength and is determined uniquely by the geometric cross-section Q which, for given microstructural characteristics, is related uniquely to the water content.

Polyakova /453/ showed that the dependence of $\alpha(\lambda)$ on the microstructural parameters of rain is inconsequential compared with the correlation between $\alpha(\lambda)$ and the rain intensity I:

$$\alpha = 0.21 I^{0.74}, \tag{7.53}$$

where I is measured in mm/hr and α in km^{-1}. The coefficient of correlation between $\log \alpha$ and $\log I$ was found to be 0.95 ± 0.01. An even higher correlation is observed between $\log \alpha$ and $\log \omega$, where ω is the water content of rain in g/m^3 (the correlation coefficient is 0.97 ± 0.01).

A similar correlation was found between α and snowfall intensity /570/; the correlation coefficient was found to be 0.91 ± 0.02. The obtained correlations between α and rain intensity and snowfall /453, 570/ were later confirmed by Jefferson /571/.

The extinction coefficients of visible and infrared radiation for precipitation particles can be treated as virtually independent of wavelength. This conclusion is valid at least for emission wavelengths $\lambda < 15$–20μ. The numerical values of these coefficients are easily determined for known precipitation intensity.

7. SCATTERING DIAGRAMS OF POLYDISPERSE AEROSOLS

Shifrin and Raskin /540–542/ obtained analytic expressions which make it possible to calculate scattering diagrams of polydisperse aerosols consisting of "soft" transparent particles possessing different particle size distributions. The distribution functions used were the gamma distribution with $\mu = 2$, the generalized gamma distribution, and the Junge formula for $\beta = 4$, 5 and 6. Data on scattering diagrams with Junge and Junge-type distributions for $\beta = 3$, 4, 5, 6, 8 and with minimum sizes $a_1 = 0.0175$, 0.052, and 0.082 micron, refractive index $n = 1.33$, 1.44, and 1.50 were obtained by Shifrin and Chayanova /572/. Tables for calculating scattering diagrams in the region of small scattering angles have also been obtained /573/.

Deirmendjian /517/ computed the scattering diagram for two haze models and one cloud model with allowance for the complex nature of the refractive index of particles at different wavelengths in the 0.45–16.6μ range. Polydisperse scattering diagrams for media with Junge particle size distribution at $n = 1.25$ and 1.50 for five wavelengths in the visible spectral region were calculated elsewhere /825, 826/. Parameter β and the upper limit of particle sizes assumed various values.

Quite a large number of investigations are concerned with experimental studies into scattering diagrams of visible radiation under different conditions in the surface layer of the atmosphere and with measurements of solar radiation scattering under clear sky. The results of these studies have been summarized /583, 820, 827, 828/.

Detailed experimental studies of the scattering diagram of visible radiation in the surface layer of the atmosphere in different geographic regions were carried out by Barteneva /574/. All the diagrams were grouped into 10 classes on the basis of the value of the asymmetry factor K, which equals the ratio of light fluxes scattered forward and backward. Each class contained diagrams for which K differed by 50%. In one series of observations at El'brus, there was first discovered a near-Rayleigh scattering diagram with $K = 1$ for $\alpha = 0.0136$ km^{-1} or $S_m = 220$ km. The maximum value of K was 35.

Measurements show that there is no unique relationship between the shape of the diagram and the transparency of the atmosphere. Barteneva /574/ notes that this conclusion is in agreement with Belov's measurements /575/, while elsewhere /569, 576–579/ a fixed transparency interval is associated with a certain value of K. Barteneva /574/ pointed out that this conclusion was reached on the basis of an insufficient number of measurements.

Barteneva /580/ elsewhere examines his results /574/ and finds that one can use the nephelometric method of transparency measurement. A theoretical substantiation of this method is given by Shifrin and Chayanova /829/.

Measurements of scattering diagrams of visible light in the surface layer /565, 581, 582/ also indicate that there is no clearly defined relationship between K and atmospheric transparency. Toropova /581/ determined separately the values of K for the total diagram and its aerosol part. The measurement intervals were, respectively, 1.57–8.75 and 1.69–15.1 with average values of 3.29 and 6.23. Toropova /581/ noted that in other measurements /577, 569, 579/the average values were found to be 6.41, 8.17 and 11.5.

Boiko /582/ found the ratio of the intensities of light scattered at angles of 10 and 140°, which characterizes the elongation of the diagram and varies from 8.8 to 63.8. The same study also reports ratios of the intensity of light scattered at different angles, to the corresponding intensities for Rayleigh scattering. In particular, for the minimum and maximum angles of measurements (10 and 160°) these ratios were 8–314 and 1.1–9.6, respectively.

The diagrams were measured for two narrow spectral regions with effective wavelength $\lambda_{ef} = 0.428$ and 0.560μ /565/. It was found that the elongation of the diagram increases with wavelength. The same effect for the entire thickness of the atmosphere was detected elsewhere /583/. A detailed explanation of this phenomenon, which is at first sight incomprehensible ($\rho = \frac{2\pi a}{\lambda}$ decreases with increasing λ; in the case of monodisperse aerosol this should result in a decrease in the elongation of the diagram), is given by Kastrov /584/ when analyzing the contributions of small and large particles to the diagram of polydisperse aerosol.

A detailed study of the aureole part of the scattering diagram in the surface layer of the atmosphere for narrow spectral regions approximately 50 Å wide, with centers near 0.47, 0.585 and 0.888μ, has been carried out /585, 906/. These results led to the conclusion that there exist three qualitatively different optical states of the surface layer of the atmosphere: haze, foggy haze, and fog. Each of these states corresponds to a characteristic variation in the angular dependence of the intensity of light scattered at small angles, as well as fluctuations in the aureole part of the diagram. For fogs, one observes an extremely sharp reduction in intensity $I(\theta)$. When the scattering angle varies from 10' to 7°, $I(\theta)$ changes by 3–7 orders of magnitude. Coronas were observed in the case of foggy haze, which points to the presence of particle fractions of a certain size.

In haze and stable foggy haze, fluctuations in the aureole scattering diagram are small. In the course of transformation from foggy haze to haze, the brightness of the scattered light at angles 10'–3° can change over several seconds by 2–3 orders of magnitude! Lyubovtseva and Rozenberg /585/ noted that in the majority of their aureole diagrams there are virtually no recurrent cases.

A large number of scattering diagrams in the surface layer of the atmosphere were obtained /506, 507/ from experimental studies of the components of the scattering matrix (see Chapter 6, Sec. 3). Gorchakov /507/ presents graphs of the angular dependence of the scattering matrix component M_{11} (scattering diagram) for haze, dense haze, fog and drizzle.

Scattering diagrams in the free atmosphere were measured at different altitudes /586–588/. The results show that the degree of elongation of the diagram varies from one measurement to another. Here the asymmetry of diagrams prevails at all altitudes under study up to 20 km, increasing in the region of the tropopause.

Measurements of the atmospheric scattering diagram for some wavelengths of visible and ultraviolet radiation for scattering angles close to 0° and 180° were carried out by Pavlov /830/. In the region of small scattering angles the elongation of the diagram increased with increasing wavelength and absolute size of the aureole around the sun. In the ultraviolet region the scattering diagram approaches the Rayleigh diagram.

In conclusion, we note that in the studies known to this author the experimental studies of scattering diagrams of polydisperse aerosols under natural and laboratory conditions were carried out virtually for only certain wavelengths in the visible and very near infrared regions (not further than the 0.9μ wavelength). In the majority of cases no simultaneous measurements were carried out of the microstructure of aerosols, which makes it impossible to compare quantitatively experimental data with other such data or with corresponding theoretical data. The latter are available in a more or less sufficient quantity as regards the visible and very near infrared regions. Calculations of polydisperse scattering diagrams with allowance for the complex nature of the refractive indices of particles were conducted for only several specific models of size distribution functions of water spheres. It is noteworthy that at present there exist insufficient data on the chemical composition of particles and especially on the components of their complex refractive indices, which are needed for calculating polydisperse scattering diagrams of nonaqueous aerosols /589/.

8. SCATTERING OF LASER RADIATION IN THE ATMOSPHERE

The coherence, high degree of monochromaticity, high directionality and power of laser radiation generally give rise to corresponding features in its propagation in the atmosphere. These features may prevail also in the absorption of radiation, in its scattering in the region of linear optics where the Bouguer law is valid and in the region of nonlinear optics.

The absorption features of laser radiation in the region of linear optics were presented in Chapter 3, Sec. 14. The nonlinear effects accompanying the absorption and scattering of powerful laser radiation in the atmosphere will be described in Chapter 10. Here we consider only the energetics of laser radiation propagating in a scattering medium when the Bouguer law holds.

Purely qualitative deliberations show that the extinction coefficients for laser radiation and for noncoherent radiation from thermal sources of identical spectral composition should not differ substantially. In fact, the coherence of a wave incident on a layer of scattering particles can result in interference of waves scattered by these particles. This will bring about an intensity distribution of light scattered in different directions which differs from the corresponding pattern obtained on irradiating the layer of incoherent radiation. Experimental proof of this was obtained by Kabanov /590/. This effect will lead to different values of the effective field for particles situated in different parts of the laser beam cross-section. However, if the particles are uniformly distributed over the medium and the growth in the effective field under study does not bring about nonlinear effects, the attenuation of the beam averaged over its cross-section will probably differ unsubstantially from the attenuation obtained when the same layer of this medium is irradiated by incoherent radiation of identical spectral composition.

These purely qualitative considerations require experimental verification, especially as Carrier and Nugent /591/ found that the extinction coefficient for laser radiation and for incoherent radiation of the same composition differ from one another.

The extinction coefficients of water fogs for laser and thermal-source radiations were measured using corresponding interference filters /592/. The lasers used and some of their characteristics are listed in Table 7.7.

TABLE 7.7. Some lasers and their characteristics

Laser	Emitted wavelength λ, μ	Operating regime
He–Ne	0.63	Continuous, multimode
Ruby	0.69	Pulsed, multimode
Nd activated glass	1.06	Pulsed, multimode
He–Ne	1.15	Continuous, multimode
He–Ne	3.39	Continuous, multimode

Figure 7.27 presents a comparison of the extinction coefficients of He–Ne lasers (wavelength 0.63μ), of a ruby laser ($\lambda = 0.69\mu$) and of a thermal source with corresponding interference filters. All three extinction coefficients are close to one another. The equation of the straight line passing through the experimental points is

$$\alpha(0.63) = \alpha(0.69) = 1.06\alpha_f, \tag{7.54}$$

where $\alpha(0.63)$, $\alpha(0.69)$ and α_f are respectively the extinction coefficients for lasers with 0.63 and 0.69μ wavelengths and for thermal sources with interference filters. Similar conclusions to the effect that the extinction coefficients for lasers and thermal sources are close can be also drawn for lasers with wavelengths 1.06, 1.15 and 3.39μ, even though in these cases the extinction coefficient for the thermal radiation were measured for the 0.4μ wavelength. Woodward /593/ measured the extinction coefficients for a laser and thermal source with different wavelengths and detected a difference in the obtained values; he did not interpret his results in great detail.

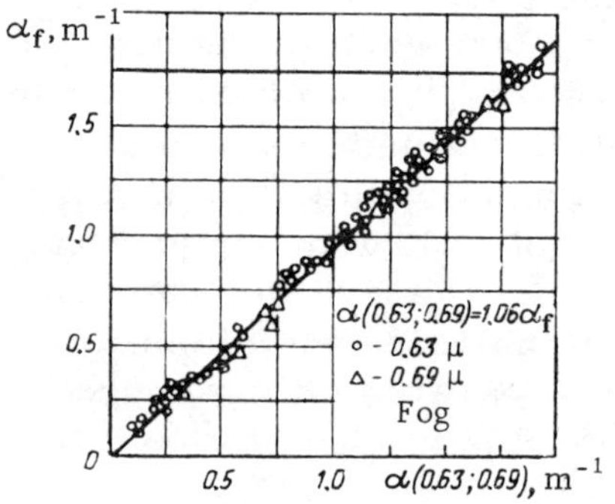

FIGURE 7.27. Comparison of extinction coefficients for laser radiation with 0.63 and 0.69 μ wavelength and of the extinction coefficient pertaining to a thermal source with the same wavelengths in an artificial water fog.

Results of careful laboratory investigations into the angular dependence of radiation scattered by latex substances were published recently /594/. The radiation sources were an He-Ne laser with wavelength of 6328 Å and a high-pressure xenon arc lamp emitting a band about 100 Å wide with center near 6328 Å. The diameters of the latex particles ranged from 0.088 to 3.49μ. It was found that, within the limits of measurement error, the scattering diagrams of the laser and the xenon lamp coincide. Coincidence of diagrams is equivalent to coincidence of extinction coefficients.

Note in conclusion that the results obtained in /592, 594/ should not be interpreted to imply that no difference exists in the propagation of coherent and incoherent radiation of identical spectral composition in scattering media. Reference is had only to approximate equality of extinction coefficients of these emissions.

9. EFFECT OF MEASURING APPARATUS ON EXPERIMENTAL RESULTS IN SCATTERING OF VISIBLE AND INFRARED RADIATION IN THE ATMOSPHERE

The theoretical data on the extinction coefficients of clouds, fogs and hazes examined in this chapter characterize the attenuation of direct monochromatic radiation propagating through media. It is impossible to create experimental conditions in which only direct attenuated light is recorded. Any receiving system records not only direct radiation attenuated by the scattering medium, but also scattered light, the intensity of which should depend on the viewing angle of this system, the properties of the scattering medium and the directivity of the incident radiation. The measured extinction coefficient and the measured transmittance of the layer should thus differ from the theoretical values. A second cause for the difference between measured and predicted values is that in experiments, unlike theory, one does not deal with strictly monochromatic radiation.

Our task consists in analyzing the effect of experimental conditions and of measuring apparatus on the results of corresponding measurements. This problem is considered in detail elsewhere /1/. Below we present a summary of results obtained by various investigators. It is assumed that the effect of source polychromaticity on measurement results can be neglected. This assumption is not of great importance, since the extensively used methods of monochromatization of emissions from thermal sources enables one to single out sufficiently narrow spectral emission regions within which the aerosol extinction coefficient can be regarded as independent of wavelength.

1. Dependence of the measured extinction coefficient on the extent of the atmosphere over which measurements are made. This dependence was first detected experimentally by Sinclair /595/, who measured the scattering coefficients of lycopodium spores (size $15 \pm 1\mu$) located at various distances from the receiver. When this distance varied from 5.5 m to 15 cm, the measured scattering coefficient of a parallel beam of visible light assumed values from $2\pi a^2$ to πa^2. Mie theory for this case yields the value $2\pi a^2$.

The problem was solved theoretically for large particles possessing arbitrary electrical properties /467/. The case of a nontransparent sphere was examined theoretically by Brillouin /596/. A similar result was obtained for a cylinder /597/. Analogous conclusions were subsequently drawn in studies of other nonspherical particles /598/. The cause for the dependence of the scattering coefficient on distance lies in the quite peculiar form of scattering diagrams for large particles. The receiver may receive a larger or smaller part of the energy scattered by the particle, depending on the distance between the particle and the receiver. The measured scattering coefficient is thus found to depend on the experimental arrangement. It was shown by Shifrin /467/ that the measured scattering coefficient can be expressed in the form

$$\alpha_{\text{mea}} = \pi a^2 [2 - \varphi(z)] = 2\pi a^2 \left[1 - \frac{\varphi(z)}{2}\right], \tag{7.55}$$

where $z = \frac{2\pi a}{\lambda} \frac{d}{2l}$, a is the droplet radius, λ the wavelength of the scattered light, d the diameter of the receiver aperture, l the distance from the droplet to the receiver, and $\varphi(z)$ is a tabulated function /467/ which makes allowance for the shape of the diagram. An experimental check of equation (7.55) has been carried out for rain droplets /599/.

The measured coefficients of a localized layer was expressed /597/ by the equation

$$\alpha_{\text{mea}} = \pi a^2 R(z), \tag{7.56}$$

where

$$R(z) = 1 + J_0^2(z) + J_1^2(z); \tag{7.57}$$

$J_0(z)$ and $J_1(z)$ are zero- and first-order Bessel functions of first kind.

The dependence of the measured scattering coefficient on the receiver aperture was detected during studies in a milk medium /600, 601/.

Kabanov /602/ derived an expression for function $B(z)$ appearing in the equation for α_{mea} in the case of large particles:

$$\alpha_{\text{mea}} = \frac{1}{2} \alpha B(z), \tag{7.58}$$

where z has the same value as in (7.55).

Results of calculations of $B(z)$ for z from 0 to 15 are presented elsewhere /1, 602/. When z varies from 0 to ∞, function $B(z)$ assumes values from 2 to 1. Consequently, the measured scattering coefficient of large particles can differ from its true value by at most a factor of 2. Kabanov /602/ shows further that in the case of small particles α_{mea} under atmospheric conditions depends only very weakly on distance and differs little from α. Experimental results of the dependence of the measured scattering coefficient on the parameters of the receiver and of the atmospheric extent being explored are presented in /1/. The measurements were conducted in artificial fogs and the results were found to be in satisfactory agreement with theoretical data.

2. Effect of scattered light on the measured transmittance of the atmosphere. If the emission of a source is concentrated in a given solid angle, then the receiver records not only the radiation propagating from the source along a straight path, but also the radiation scattered by particles situated outside the source-receiver paths. These fractions arriving at the radiation receiver are termed, respectively, direct and scattered-in radiation. In this case the illuminance E produced by these radiations can be expressed in the form

$$E = E_d + E_D = \frac{I_0}{l^2} e^{-\frac{1}{2}\alpha B(z)l} + E_D, \tag{7.59}$$

where E_d and E_D are respectively the illuminances produced at the receiving area of the receiver by the direct and scattered-in radiation; I_0 is the intensity of radiation incident on the scattered layer of thickness l.

Thus the signal at the receiver, usually taken proportional to the atmospheric transmittance, is due to illuminances produced by both direct and scattered-in radiation. In the majority of studies the role of the second factor is disregarded when measuring atmospheric transmittance and, as will be shown below, can result in large errors.

First estimates of the roles of once scattered-in radiation in the atmosphere were made by Middleton /603/, who calculated the illuminances produced by direct E_d and once scattered-in E_D radiation for the following conditions: $a = 2.5\mu$, $N = 10^8$ m^{-3}, $l = 0.2$, 0.5 and 1.0 km; $a = 10\mu$, $N = 6.25 \cdot 10^6$ m^{-3} and $l = 0$ and 5 km; $\theta = 0.02$, 0.10, 0.20 and 1.0 radian, $0 \leq \Psi \leq 0.05$ radian. Here a, N, l, θ and Ψ are respectively the particle radius, concentration, distance from source to receiver, cone half-angle of the source emission and half-angle of the viewing field of the receiver (see Figure 7.28). Ratio E_D/E_d under the above conditions attained its maximum value of 1.6. These theoretical values were confirmed by Middleton in a fog, where it was found that $E_D/E_d \approx 2$ for a range of 114 m at $\theta = 0.30$ radian, $\Psi = 0.008$ radian, and meteorological visibility equal to 125 m.

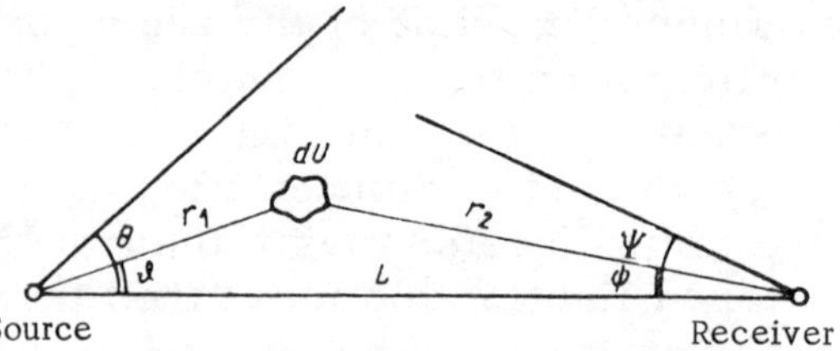

FIGURE 7.28. Schematic diagram of radiation propagation from source to receiver.

Stewart and Curcio /604/ conducted experiments to obtain data on the role of once scattered-in radiation in the case of a slightly turbid atmosphere. They dealt with large paths (3.7 and 16.7 km) over a wide range of Ψ (from 1.5 to 13°) for individual spectral regions in the 0.36–0.61 μ range, which were isolated by filters. Processing of the measured results yielded the following empirical formula relating the true transmittance T to the measured value T_{Ψ}:

$$T_{\Psi} = T + 0.5(1 - T)(1 - e^{-2\Psi}). \tag{7.60}$$

This formula was obtained for $T > 0.70$; the coefficient 0.5 arises due to the albedo of the underlying surface.

Extensive studies of the role of once scattered-in radiation have been carried out /605–607/; the illuminance E_D is given by the expression

$$E_D = gI_0 \int\limits_v \frac{F(\varphi)\, e^{-\alpha(r_1+r_2)} \cos \Psi}{r_1^2 r_2^2}\, dv, \tag{7.61}$$

where g varies from 0.5 to 1, depending on the nature of the underlying surface; v is the scattering volume; $F(\varphi)$ is the scattering diagram; $\varphi = \psi + \vartheta$; the remaining notation follows Figure 7.27. Formula (7.61) was obtained previously by Meshkov /608/ for both single and double scattering.

The value of E_D was calculated in /605–607/ by numerical integration of (7.61) when: $\theta = \pi/2$, $\Psi = 8°20'$, 25, 50, 75 and 90°, $\tau = \alpha l = 0.1$, 0.3, 0.9, 1.2, and 3.6.

The experimental studies /605–607/ were carried out for long paths, high Ψ and low atmospheric turbidity in San Francisco Bay and the Nevada Desert. The paths in the first case were from 1.18 to 14.37 miles long and in the second, 0.51 to 13.17 miles.

The investigations were carried out for wavelengths of 0.40, 0.50, 0.70, 0.83 and 0.90μ.

Similar measurements were conducted /609/ for slant directions in the atmosphere for $\Psi = 1.5$, 15 and 75° in the visible region and at 0.35–2μ and 0.75–2μ for atmospheric thicknesses from 0.5 to 9 km under a variety of meteorological conditions. More detailed data for fogs and hazes were obtained for an isotropic source and receiver with $\Psi = \pi$ /610/. The empirical equation for the resulting transmission function differs from the Bouguer law, and is attributed to the marked effect of the scattered-in radiation /610/.

Theoretical and experimental studies of the effect of scattered radiation on the measured transmittance of the atmosphere were carried out elsewhere /611–615/. An attempt was first made to simplify (7.61) so as to be able to calculate E_D for a wide range of atmospheric conditions. The essence of this simplification is as follows. Variables r and ψ are replaced in equation (7.61) by ϑ and φ according to standard rules /616/; we then set $g = 1$ and $F(\varphi) = \alpha/4\pi\, f(\varphi)$, where $f(\varphi)$ is a normalized scattering diagram. Making use of the fact that $\varphi = \psi + \vartheta$, we obtain

$$E_D = \frac{\alpha I_0}{2l} \int_0^{\Psi} d\psi \int_0^{0} e^{-kl \frac{\sin\psi + \sin\vartheta}{\sin(\vartheta+\psi)}} f(\psi + \vartheta) \cos\psi \, d\vartheta. \tag{7.62}$$

At small ψ or ϑ the exponent in equation (7.62) can be removed from under the integral sign:

$$E_D = \frac{\alpha I_0 e^{-\alpha l}}{2l} \int_0^{\Psi} d\psi \int_0^{0} f(\psi + \vartheta) \cos\psi \, d\vartheta. \tag{7.63}$$

The double integral in (7.63) for the given experimental conditions does not depend on the distance between the source and receiver. If we introduce the notation

$$D = D(\rho, \Psi, \theta) = \frac{1}{2} \int_0^{\Psi} \int_0^{\theta} f(\psi + \vartheta) \cos\psi \, d\psi \, d\vartheta, \tag{7.64}$$

then

$$E_D = \frac{\alpha I_0 e^{-\alpha l}}{l} D. \tag{7.65}$$

In (7.64) the dependence of D on ρ is introduced via the scattering diagram $f(\psi, \vartheta)$.

Estimates show that equation (7.63) describes E_D with an error known a priori to be less than 5% when $\Psi \leqslant 0.1$ and atmospheric transmissivity $T \geqslant 0.1$; these values are attained in the majority of cases during measurements of atmospheric transmittance.

Expressions (7.61), (7.63) and (7.64) yield an expression for the total illuminance at the receiving area of the receiver:

$$E = E_{\text{d}} + E_D = \frac{I_0}{l^2} e^{-\alpha l} [1 + \alpha l D(\rho, \Psi, \theta)], \tag{7.66}$$

from which follows the simple expression

$$\frac{E_D}{E_{\text{d}}} = \tau D, \tag{7.67}$$

where $\tau = \alpha l$ is the optical thickness of the medium.

Expressions (7.63)–(7.66) for $\Psi \leqslant 0.1$ are valid for any wavelengths and any homogeneous scattering medium with known scattering diagram.

The techniques and results of computing D for different ρ, Ψ and θ have been described in detail /1/. Here we present only one figure, which illustrates the dependence of D on Ψ at different ρ for two values, $\theta = \pi/2$ and $\pi/6$ (Figure 7.29).

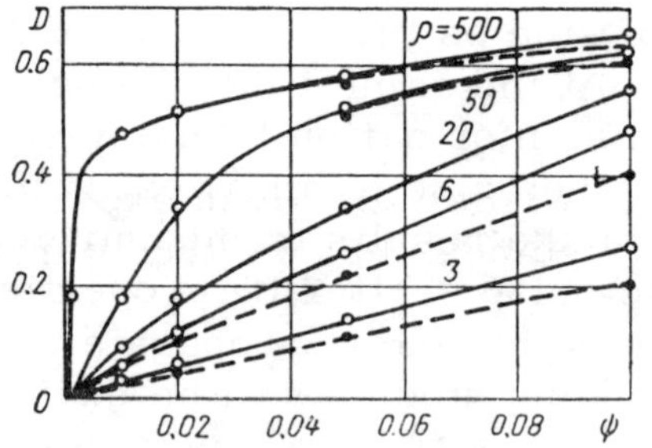

FIGURE 7.29. Curves of D as a function of receiver aperture for various ρ and half-cone angles θ of the source emissions. Solid curves pertain to $\theta = \pi/2$ and dashed curves to $\theta = \pi/6$.

The numerical values obtained for D at different ρ, θ and Ψ make possible a quantitative estimate of the role of once scattered-in light under different experimental conditions in the atmosphere. It follows from these data, in particular, that at relatively small Ψ (0.1) the value of D for large particles may be as high as 0.5–0.7.

At the frequently encountered values of τ (several units) this value is equivalent to the contribution to the signal of scattered-in light, increasing severalfold the contribution of direct attenuated light. Consequently, neglect of this circumstance in experimental studies of atmospheric transmittance can lead to overestimates of the measured transmittance which are several factors more than its true values.

A description of experimental studies of D in a fog chamber and in natural fogs and snowfalls has been provided /1/. Comparison of results of measurements in artificial fogs with controlled micro-structural with theoretical values of D shows satisfactory agreement between experiment and theory.

A detailed experimental study into the effect of the cone angle of source emission and the angular aperture of the receiver on the results of measurements of the extinction coefficient for different optical thicknesses of model scattering media was carried out in /617/. The model media were selected in such a manner that their microphysical and optical properties (especially the probable photon lifetime) would correspond to natural media. Tables of $(\alpha_{mea} - \alpha)\,\alpha^{-1}$ for different Ψ, θ and τ (the notation is the present author's) are presented there /617/.

The scattering of white light by polydisperse aerosols with allowance for the measuring apparatus parameters was calculated in /618/. The calculations were conducted on the basis of Mie theory for refractive index $n = 1.2\,(0.2)\,2.4$ and the Mie parameter satisfying $0.1 < \rho < 100$.

The above results of studies into the value of E_D pertain to quite wide radiation beams. At the same time, the use of lasers in communications systems, information transmission, laser radar, telemetering, etc., has given rise to an investigation into the role of scattered radiation in the results of measurements of various quantities which characterize scattering properties of media when narrow collimated light beams propagate through them. The results of the corresponding studies will be examined in Chapter 9.

Part Three

PROPAGATION OF TIGHT LIGHT BEAMS IN THE ATMOSPHERE

8. *RADIATION FROM VARIOUS TYPES OF LASERS*

INTRODUCTION

This chapter briefly examines those characteristics of laser radiation which are of significance in the treatment of laser radiation propagation in the atmosphere. These characteristics primarily include: 1) spectrum, 2) energy and power, 3) coherence, 4) directivity, 5) operating regime, 6) polarization of radiation. This chapter does not aim at an up-to-date comprehensive survey of the radiation properties of individual lasers, but deals primarily with those laser parameters of importance in laser beam propagation in the atmosphere.

The overwhelming majority of known lasers can be subdivided into four classes: 1) solid crystalline lasers; 2) glass lasers; 3) gas lasers; 4) semiconductor lasers. Their principal characteristics are described below, primarily on the basis of reviews published during the past few years /619–643, 701, 831/.

1. SOLID CRYSTALLINE LASERS

The first demonstration of laser action was achieved using ruby, a single crystal of sapphire (Al_2O_3), activated by the trivalent chromium ion Cr^{3+}. At present this type of laser remains most prevalent, although laser operation was obtained with many other ionic crystals.

Solid crystalline lasers have a high density of active particles ($10^{17} - 10^{20}$ cm^{-3}), more or less uniformly distributed over the volume of the crystal and a long lifetime of the active levels of the ions (approximately 10^{-3} sec). The above properties of the active media of the lasers under study results in high emission energies in the free laser operation regimes and gigantic outputs under Q-switching conditions, while simultaneously allowing easy solution of the cooling problem of the active media.

TABLE 8.1. Crystals and activating ions inducing laser operation at different wavelengths /624, 628/

Crystal	Activating ion	Activating ion concentration, %	Emitted wavelength, Å
CaF_2	H_0^{3+}		5 512
LaF_3	Pr^{3+}	1	5 985
Y_2O_3	Eu^{3+}	5	6 113
Al_2O_3	Cr^{3+}	0.05	6943, 6 934, 6 929
SrF_2	Sm^{2+}	0.01	6 969
Al_2O_3	Cr^{3+}	0.5	7 009, 7 041, 7 670
CaF_2	Sm^{2+}	0.01	7 083
$CaWO_4$	Nd^{3+}	1	9 145, 10 580, 13 392
$Y_3Al_5O_{12}$	Yb^{3+}		10 296
CaF_2	Nd^{3+}	1	10 460
$CaWO_4$	Pr^{3+}	0,5	10 468
$CaMoO_4$	Nd^{3+}	1,8	10 610
$Y_3Al_5O_{12}$	Nd^{3+}		10 648
CaF_2	Tm^{2+}	0.01	11 160
Er_2O_3	Tm^{3+}		13 340
$Ca(NbO_3)_2$	Er^{3+}		16 100
$CaWO_4$	Er^{3+}	1	16 120
MgF_2	Ni^{2+}	1	16 220
$Y_3Al_5O_{12}$	Er^{3+}		16 602
MgF_2	Co^{2+}	1	17 500, 18 030
$CaWO_4$	Tm^{3+}		19 110
$Y_3Al_5O_{12}$	Tm^{3+}		20 132
$CaWO_4$	Ho^{3+}	0.5	20 460
$Y_3Al_5O_{12}$	Ho^{3+}		20 975
SrF_2	U^{3+}		24 070
CaF_2	Dy^{2+}	0.01	25 588
ZnF_2	Co^{2+}	1	26 113
CaF_2	U^{3+}	0.05	26 130

The principal active media in solid crystalline lasers are: Al_2O_3, CaF_2, $CaWO_4$, SrF_2, BaF_2, $SrCl_2$, LaF_3, CeF_3, $SrWO_4$, $CaMoO_4$, $PbMoO_4$, Y_2O_3, Er_2O_3, $Y_3Ga_5O_{12}$, $Gd_3Ga_5O_{12}$, MgF_2, ZnF_2, $Ca(NbO_3)_2$ (Table 8.1). The crystals used most extensively are Al_2O_3, CaF_2, $CaWO_4$ and synthetic garnets.

Depending on the operating regime, all solid crystalline lasers can be subdivided into three groups: 1) continuous operation; 2) pulsed in the free-generation regime; 3) pulsed with Q-switching.

To obtain continuous laser emission it is necessary to have available excess excited ions per unit active medium volume, which is obtained on continuous pumping. This excess of ions is by far not attained for all active media of lasers using known optical pumping sources. Continuous laser operation becomes possible when this condition is satisfied. Parameters of this radiation (see, for example, Table 8.2) depend on many factors which are not discussed here.

TABLE 8.2. Emission characteristics of the most effective continuous solid crystalline lasers /628/

Material of active system	Emitted wavelength, μ	Efficiency, %	Power, W	Operating temperature, °K
$Al_2O_3+Cr^{3+}$	0.69	0.1	1	300
$Y_3Al_5O_{12}+Nd^{3+}$	1.06	0.2	200	300
$Y_3Al_5O_{12}+Ho^{3+}$	2.12	5	15	77
CaF_2+Dy^{2+}	2.36	1—1.5	100	77

Pulsed emission from solid crystalline lasers is attained more easily. Pulsed pumping ensures generation in a much greater number of ionic crystals than continuous pumping. Here, under pulsed laser operation conditions, a very high output is obtained and energies of thousands of joules are released over short periods of time.

When operating in the Q-switched mode the outputs are fantastic. For example, when operating a commercially produced ruby laser in the Q-switch mode an output of 10^9 watts was obtained for a pulse duration of 20 nsec with a divergence of 1 milliradian. Special mode synchronization measures (coupled mode operation) make it possible to obtain outputs of $10^{10}-10^{12}$ watts in $10^{-11}-10^{-13}$ sec. These can be estimated only by indirect methods, since the best rapid-response radiation receivers are not capable of recording pulses of such duration.

Consider briefly the spectral characteristics of the emission of solid crystalline lasers. Table 8.1 lists the emitted wavelengths, which characterize the center of the narrow spectral interval emitted by the laser. Unless special measures are taken, the laser emission is multimode and occupies a section with width of the order of several tenths of an angstrom with the characteristic peak structure of the frequency and time distribution of the intensity.

Special measures enable laser emission bands to be narrowed substantially to values of the order of 10^{-3} Å. The divergence angles of solid crystalline laser emissions without collimating systems depends on the geometric dimensions of the active media, emitted wavelength, material of the active medium, and the number of modes generated. The divergence angle usually amounts to several milliradians. The use of special reflectors with selective distribution of reflection coefficient as a function of the angle of incidence makes it possible to obtain near-diffraction divergence. The energy distribution in the cross-section of the emitted beam is generally inhomogeneous.

The coherence of the emission of solid crystalline lasers depends on the number of modes. Each mode emits light with a certain degree of coherence. The same should also be said of the degree of polarization of the laser beam.

2. GLASS LASERS

A distinguishing feature of glass lasers activated by trivalent ions of neodymium, ytterbium, holmium, and erbium is related to the properties of glass, which can be varied over quite wide limits by modifying its chemical properties (Table 8.3). In addition, glass lasers can be made in different shapes and sizes, such as micron diameter fibers which are 10 m long, or large rods up to 60 mm in diameter and 2 m long. The most extensively used lasers are of the glass-neodymium type.

TABLE 8.3. Some characteristics of the emitted radiation of glass lasers activated by trivalent neodymium, ytterbium, holmium and erbium ions /629/

Glass composition	Activating ion	Emitted wavelength, μ
Na—La—Si	Nd^{3+}	0.92
Li—Mg—Al—Si	Yb^{3+}	1.015
K—Ba—Si	Nd^{3+}	1.06
K—Ba—Si	Yb^{3+}	1.06
Li—Mg—Al—Si	Ho^{3+}	1.35
La—Ba—Th—B	Nd^{3+}	1.37
Yb—Na—K—Ba—Si	Er^{3+}	1.54
Li—Mg—Al—Si	Er^{3+}	1.55

The most powerful laser so far constructed is a glass-neodymium laser with an output of 5000 joules, in the form of a rod 90 cm long and 30 mm in diameter. Glass-neodymium lasers in Q-switched applications have an output of several tens of gigawatts during 10–50 nsec.

The emission of glass lasers has a more or less uniform energy distribution over the beam cross-section. The beam spread is approximately the same as for solid crystalline lasers.

Glass lasers have a much wider emitted frequency band than solid crystalline lasers. Unless special measures are taken, glass lasers usually possess an emitted bandwidth of several tens of angstroms. The singling-out of individual emitted lines ensures substantial narrowing of the laser emission spectrum. Glass-neodymium lasers are now operated so as to produce a spectral width of the order of 10^{-3} Å.

The coherence of glass lasers depends markedly on the number of modes, as in the case of solid crystalline lasers.

In glass lasers it is possible to attain laser operation such that the rod may be completely saturated at one end by the spontaneous emission from the other end (superradiance). In this case the emission spectrum has a uniform intensity distribution over the spontaneous emission line profile, while the radiation proper is nonpolarized. Its output attains 10^9 W for a pulse duration of 70 nsec.

3. GAS LASERS

The most prevalent type of laser is the gas laser. Its principal qualitative feature, distinguishing it from other lasers, consists in the high monochromaticity of the emitted beam and high frequency stability, which depend little on external conditions. These qualities of gas lasers result in a much higher coherence than in other lasers. The emissions of gas lasers occupy a very wide, and continually widening, range of wavelengths. At present this range extends from 2315.5 Å (emitted length of Xe II) to 1 mm. The highest continuous mode output was obtained with gas lasers.

Gas lasers are classified by the following four types: 1) neutral atom lasers, 2) ion lasers, 3) molecular lasers, and 4) chemical lasers. The emitted line finds itself in one or other spectral region, depending on the particles at which the respective energy transition occurs. High-energy electronic transitions in ions pertain to the ultraviolet and visible regions; electronic transitions in atoms yield emission lines in the visible and very near infrared regions.

Vibration-rotation and purely rotational molecular transitions generate lines filling a wide range of wavelengths, from the near infrared to the microwave. The emitted wavelengths of chemical lasers depend on the energy levels of particles forming as a result of chemical reaction at which the corresponding transition occurs.

The most prevalent neutral atom gas lasers are those using He-Ne mixtures. Of these, the most extensively used are lasers emitting at 6328 Å, 1.15 and 3.39μ. The most powerful atom laser, emitting at 6328 Å, has a continuous rating of 1 watt with a discharge tube length of 5 m. Commercial specimens of this laser have a rating not exceeding 100 milliwatts.

Recent years have seen the development of pulsed atom lasers operating on self-terminating transitions with outputs of the order of several kilowatts and emission times of the order of several nanoseconds. A laser operating on copper vapor was constructed recently. It is expected that the efficiency of this device will be as high as 30%.

Ion lasers have the highest continuous output (up to 50–100 watts) in the visible spectral region. The most extensively used ion laser is the ionized argon laser with a series of emission lines from 0.48 to 0.52μ.

The most extensively used molecular laser consists of molecular nitrogen and carbon dioxide devices. In the first case use is made of self-terminating electronic-vibration transitions, emitting at wavelength $\lambda = 3371$ Å and 20 nsec duration with maximum output of 0.3–2.5 milliwatts with repetition frequency of 300 Hz. Commercial specimens are rated at 100 kilowatts, and possess 10 nsec pulse duration, pulse repetition frequency of 1 to 100 Hz, and average maximum output of 100 milliwatts.

The CO_2 laser is the most powerful of all continuous output devices known over any wavelength range. Recently a laser with continuous output of several kilowatts was built to operate at 10.6μ. The advantage of CO_2 as operating medium consists also in the fact that laser operation can occur for an entire series of transitions over a spectral interval of the order of 100 cm^{-1} in the region of the long-wave window of the atmosphere. In addition, the large lifetime of the upper level of CO_2 (more than 1 millisec) makes it possible to operate this laser in the Q-switched mode.

The first among chemical gas lasers is a device in which laser operation occurs due to transition between the metastable and ground state of the iodide. The iodide atoms are obtained on photodissociation of the CH_3I molecule due to ultraviolet radiation. The emitted wavelength of this laser is 1.315μ. The use of photodissociation of CF_3I molecules ensured a high efficiency and an output of 60 joules with 100 microsec pulse duration. A very promising chemical laser is a device in which the chemical reaction between chlorine and hydrogen is utilized. Its emission spectrum is in the 3μ region.

We shall now examine the spectral emission characteristics of gas lasers. The width of the spectral emission region of multimode gas lasers is due to Doppler profiles. The width of the latter at 6328 Å is sufficiently well known for emissions of He-Ne lasers (1700 MHz). The Doppler width is inversely proportional to wavelength. For He-Ne lasers at 1.15 and 3.39μ the widths are 920 and 310 MHz, respectively; for CO_2 lasers it is approximately 60 MHz.

The Doppler width of ion lasers is much greater than for atom and molecular lasers, because laser operation in ion devices occurs at high temperatures. For argon lasers the Doppler width is 3500 MHz, while for krypton and xenon lasers it is 2500 MHz.

Gas lasers make is possible to obtain single-mode and single-frequency continuous laser operation much easier than with other types of lasers and, in addition, the former possess exceedingly high emitted-frequency stability. The short duration frequency stability may be as high as 10^{-9} or more. Long-term stability is equal to several units per 10^{-8} and is virtually independent of environmental conditions.

A continuously emitting single-frequency gas laser is the best type of laser with respect to time and space emission coherence. Gas lasers also possess the highest emission directivity. The angle of emission spread of gas lasers depends on the geometric dimensions of the device and the emitted wavelength. The viewing angle of commercial gas laser specimens without collimating systems amounts to fractions of a milliradian.

4. SEMICONDUCTOR LASERS

The exclusively high concentration of active particles in semiconductor lasers (of the order of 10^{22} cm^{-3}) makes it possible to obtain laser operation using very small volumes of the active medium. This in turn results in a high radiation spread (of the order of 10°).

The most prevalent semiconductor laser is a device operating on gallium arsenide, which emits a band several tens of angstroms wide in the 0.84μ region. This laser provides a maximum continuous output of 3 watts at an efficiency of 15%.

Semiconductor lasers (Table 8.4) are capable of both continuous and pulsed operation. In the latter case the pulse repetition rate can range over very wide limits.

TABLE 8.4. Some characteristics of semiconductor laser emissions /631/

Material	Emitted wavelength, μ	Method of excitation
ZnS	0.32	Electron beam
ZnO	0.38	Electron beam
CdS	0.50	Electron beam, optical
CdS_xSe_{1-x}	0.50–0.69	Electron beam
GaScSe	0.59	Electron beam
$Ga(As_xP_{1-x})$	0.64–0.88	*p-n* junction
CdSe	0.68	Electron beam
CdTe	0.79	Electron beam
CdS_xSe_{1-x}	0.83	*p-n* junction
GaAs	0.84	*p-n* junction, optical, avalanche, electron beam
InP_xAs_{1-x}	0.89	*p-n* junction
InP	0.90	*p-n* junction
GaSb	1.51	*p-n* junction, electron beam
InAs	3.13	*p-n* junction, electron beam, optical
Te	3.65	Electron beam
$Hg_xCd_{1-x}Te$	3.76–4.13	Optical
PbS	4.27	*p-n* junction, electron beam
InSb	5.38	*p-n* junction, electron beam, optical
PbTe	6.54	*p-n* junction, electron beam, optical
$Pb_xSn_{1-x}Te$	6.54–16.67	Optical
PbSe	8.55	*p-n* junction, electron beam

5. VARIATION IN LASER RADIATION PARAMETERS DUE TO VARIOUS CAUSES

Strictly speaking, the radiation parameters of all types of lasers can vary due to a variety of causes. These causes are sometimes associated with laser operation conditions, in others they are created intentionally.

An example of the first kind of cause is the effect of the temperature of the active substance of the laser on its emission spectrum. This effect is manifested most clearly in ruby lasers. Different direct experiments showed that when the ruby temperature is altered by 13°C the emitted wavelength changes on the average by 1 Å. This shift changes little from one specimen to another. We recall that a shift in laser emission in the 0.69μ region by as little as one tenth of an angstrom can markedly alter its absorption in the atmosphere (see Figure 3.10). The temperature-dependence of the radiation parameters of other lasers has not yet been investigated sufficiently.

The temperature at which laser operation occurs can have a substantial effect on the emission of not only solid state lasers, but also of gas lasers. For example, depending on the temperature, operation of the CO_2 laser can occur at different lines.

Alongside temperature variations, laser emission characteristics can be affected by the concentration of active additives, their distribution over the volume, quality of resonators, ambient temperature, total and partial pressure of gases in gas lasers, and so on.

In addition to the above causes, cases exist when investigators intentionally alter some parameters in order to control the laser radiation characteristics, especially the spectral characteristics of the emitted beam.

At present, a number of methods have been developed for altering the spectral characteristics of laser beams: 1) method for separating individual modes in laser emissions; 2) method of parametric generation; 3) method of harmonic generation; 4) method of induced Raman scattering, etc.

The simplest method of separating individual laser emission modes consists in adding mirrors to the laser resonator. The addition of one mirror results in the formation of an additional Fabry-Perot interferometer, the reflection coefficients of which depend on the emitted frequency. As a result the quality of certain modes is improved and undesirable forms of vibration are suppressed. Here the energy of the suppressed modes can be pumped over to the generating modes. Selection of generated modes increases if two mirrors are introduced in the laser resonator, or if a specially selected mirror system is used. Rotation of the mirror placed in the resonator alters the quality of the respective modes, thus making it possible to select a given mode or to scan the generated modes on the radiance profile /643/. At present smooth scanning of laser emission on the radiance profile has been attained under experimental conditions /644/.

By selecting individual modes it was possible to obtain the following data on ruby laser radiation. Single-mode operation of the ruby laser was obtained with a spectral emission line width of less than 0.001 Å, a single-pulse output of 5 milliwatts and wavelength stability of ± 0.005 Å without special control of the active medium temperature /645/. Outputs of 14 and 25 milliwatts were obtained at emitted line widths of 0.005 Å and 0.008 Å. Single-mode laser operation was also obtained for glass-neodymium lasers.

The obtained monochromatization and output of ruby and glass-neodymium lasers make it possible to use methods of parametric generation, second-harmonic generation, and induced Raman scattering. Here the width of the emitted band can be the same, or even smaller under certain conditions, compared with its values at the master oscillator. Combination of the second-harmonic generator and induced Raman scattering methods yields coherent emission at different wavelengths. In the 0.4–1.2μ spectral region these methods provided for generation at more than 40 wavelengths.

The use of parametric generation ensures scanning of laser beams over relatively wide spectral regions. At present prototypes have been constructed of frequency-tuned lasers employing KH_2PO_4 (KDP) and $LiNbO_3$ crystals. The following data were obtained. On passing the emission of the second-harmonic glass-neodymium laser (emitted wavelength 0.53μ, pulsed rating 30–35 Megawatts/cm^2, pulse duration $2.5 \cdot 10^{-8}$ sec, pumping spread 7'–8') through the KDP crystal the emitted wavelength at 9571–11, 775 Å becomes reconstructed. Here, the output power was 0.5–1 kW and the spectral half-width 0.03–0.07 cm^{-1} /644/. Reconstruction of the emitted frequency, when using the same second-harmonic glass-neodymium laser and an $LiNbO_3$ crystal, was obtained in the 0.684–2.335μ spectral region. For pump powers of $5 \cdot 10^4$ watts the total output powers were 50 watts /645/. It is assumed that reconstruction of the emitted frequency using this crystal can be carried out in the 0.59–5μ region, if one also "tunes" the crystal temperature.

In conclusion, note that one can scan laser radiation frequency by means of a magnetic field in which the laser is placed. A relatively low magnetic field intensity can ensure scanning of the emitted frequency of gas lasers of several tenths of cm^{-1}. Such a frequency shift can bring about a substantial change in the absorption coefficient of a laser beam propagating through the atmosphere.

9. *PROPAGATION OF TIGHT LIGHT BEAMS IN ATMOSPHERIC AEROSOLS*

When tight light beams propagate in atmospheric aerosols the radiation is extinguished, giving rise to a light field which is scattered by the particles. When the direct and scattered radiations are superposed, the radiation brightness distribution over the beam cross-section depends in a complicated manner on the optical thickness of the layer, the geometric dimensions of the beam, the properties of the scattering medium, and the polarization state of the incident radiation. This pattern can be distorted by instruments used in experimental studies.

The purpose of this chapter is to examine the propagation of tight light beams (especially laser beams) on passing through various aerosols. Particular attention is paid to the light structure in the direction of small scattering angles (forward scattering) and scattering angles close to 180° (back scattering), since information on forward and back scattering is especially required when using lasers in communications, information transmission, radar, telemetry and other systems.

1. BRIGHTNESS OF FORWARD-SCATTERED RADIATION

1. Theoretical investigations. The light field produced by the scattering of radiation on aerosol particles is described by a transport equation, the general solution of which involves great mathematical difficulties /646/. For small scattering angles the brightness of forward-scattered radiation can be calculated by the small-angle approximation of the transport equation /647–649/. This method, however, has rather severe limitations, since it disregards polarization effects and is suitable only for beams of unlimited width and large particles (highly elongated scattering diagrams). Calculations of the brightness of forward-scattered radiation by the small-angle approximation in various scattering media were conducted by

Kabanov and Savel'ev /650/, who also established the applicability limits of Gershun's formula describing the brightness of forward-scattered light with allowance for only single scattering /651/.

Theoretical determination of the brightness of forward-scattered light is extremely difficult in the case of narrow collimated light beams. This problem has so far been solved for only some special cases. An approximate formula was derived /652/ for the brightness B of light scattered at small angles in a searchlight:

$$B = B_0 \frac{f(0)}{4\Gamma_2} e^{-\tau}\varphi(\tau)\,\theta^2, \tag{9.1}$$

where

$$\varphi(t) = \begin{cases} \left[\tau + \ln\left(\frac{\Gamma_2}{2} + \frac{\Gamma_1}{2} e^{-\tau}\right)\right], & \tau \leqslant \tau_0 (z \leqslant z_0), \\ \left[\tau + \ln\left(\frac{\Gamma_2}{2} + \frac{\Gamma_1}{2} e^{-\tau_0}\right)\right] + \tau_0\left(1 - \frac{z_0}{z}\right), & \\ \tau \geqslant \tau_0 (z \geqslant z_0); & \end{cases} \tag{9.2}$$

Γ_1 and Γ_2 are the radiation scattered in the forward and back hemisphere, respectively; quantity z_0 corresponds to the searchlight focus and is measured from the z axis, which corresponds to the direction of beam propagation; θ is the aperture angle of the beam, related to the beam radius a by the expression $\theta = a/z_0$; $f(0)$ is the forward-scattering diagram; τ is the optical thickness. As expected, the formula contains parameters characterizing the dimensions of the light beam.

More general formulas were obtained for tight beams /653, 654/; however, it is difficult to use them in quantitative calculations.

Kabanov /656/ obtained expressions which make it possible to calculate the brightness of once-scattered light for different geometric parameters of a narrow beam, the spread of which is equal to the viewing angle of the receiving system:

$$V = V_0 e^{-\tau}\left[1 + \tau\frac{1}{2}\int_0^{\theta}\int_0^{\Psi} f(m, \rho, \psi + \vartheta)\, d\psi\, d\vartheta\right] = V_0 e^{-\tau}[1 + \tau D], \tag{9.3}$$

where V_0 is the signal magnitude without extinction in the medium, τ is the optical thickness of the layer, θ and Ψ are the viewing angles of the radiation source and of the receiving system, $f(m, \rho, \psi + \vartheta)$ is the normalized scattering diagram and is a function of the scattering angle $\varphi = \psi + \vartheta$, parameter ρ and the complex refractive index of the scattering particles. Quantity D defines the once-scattered radiation for different Ψ and θ and in various scattering media.

If the distance between the source and receiver is such that $l \gg d/\theta$, where d is the diameter of the receiving objective, formula (9.3) can be used for describing the damping of a narrow collimated beam.

The expression for D was obtained by Kabanov /656/ when θ and Ψ are equal to one another, and the scattering diagram is expressed in terms of Legendre polynomials:

$$D=\frac{1}{2}\left[\Psi\theta+\sum_{n=1}^{\infty} a_n \int_0^{\Psi}\int_0^{\theta} P_n\left[\cos(\psi+\vartheta)\right] d\psi\, d\vartheta\right] \cong$$

$$\cong \frac{1}{2}\varphi^2 \sum_{n=0}^{\infty} a_n = \frac{1}{2}\varphi^2 A_n, \qquad (9.4)$$

where $\varphi=\Psi=\theta$; a_n are coefficients of Legendre polynomials and depend on the optical properties of the scattering medium.

Kabanov /656/ showed that when $\rho \geq 5$, (9.4) can be replaced by a simpler expression derived with the aid of Penndorf's approximation formula /657/:

$$D=\frac{\rho^2\varphi^2}{2}\,\frac{K(\rho, m)}{2}, \qquad (9.5)$$

where $K(\rho, m)$ is the efficiency factor for particle scattering.

Expressions (9.4) and (9.5) are inapplicable for very high ρ, since they were obtained without making allowance for the angular dependence of the Legendre polynomials at low φ. In this connection, Bessel functions were involved in an expression /656/ for D representing the scattering diagram:

$$D=\frac{1}{2}\int_0^{\varphi}\int_0^{\varphi} f(\psi+\vartheta)\, d\psi\, d\vartheta=\rho\varphi\int_{\rho\varphi}^{2\rho\varphi}\left[2\frac{J_1(z)}{z}\right]^2 dz+$$

$$+2\left[1-J_0^2(\rho\varphi)-J_1^2(\rho\varphi)\right]-\left[1-J_0^2(2\rho\varphi)-J_1^2(2\rho\varphi)\right], \qquad (9.6)$$

where J_0 and J_1 are Bessel functions. Note that the representation of the scattering diagram in terms of Bessel functions describes the angular distribution of the scattered light intensity distribution over a range of scattering angles φ of several degrees to within 3% for particles having $\rho \geq 60$.

Figure 9.1 presents the results of computing D from (9.5) and (9.6). The results show that when the viewing angles of the receiving system and the source coincide for narrow collimated radiation beams with a spread of several thousandths of a radian, one may neglect the once, forward-scattered light compared with direct radiation of the ray up to optical thicknesses of several tens in fogs, and especially in haze. In the case of rains the value of D under the same conditions is close to unity, and so already at $\tau = 1$ once-scattered light in the zone of the beam will possess the same brightness as that of the direct beam.

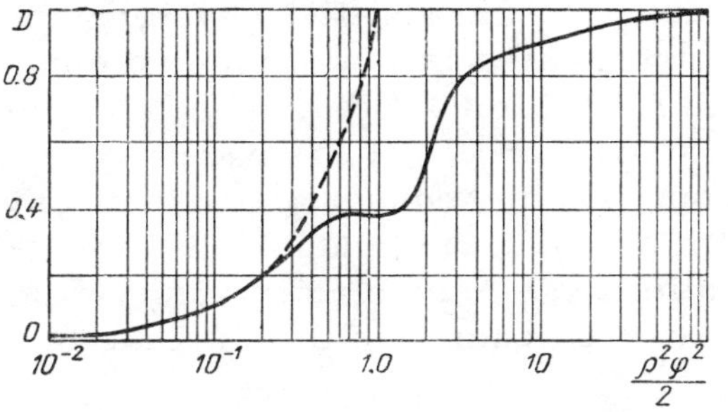

FIGURE 9.1. Curves of D as a function of ρ, obtained from formula (9.5) (dashed curve) and formula (9.6) (solid curve).

In the opinion of this author a highly promising method is to apply the Monte-Carlo method for solving problems pertaining to the transport of radiation of tight light beams in scattering and absorbing media. The Monte-Carlo method is very effective in solving multidimensional problems of radiation transport in media with different optical characteristics for radiation sources and receivers of finite geometry. The relatively slow convergence of the method depends solely on the number of statistical tests and the quality of the simulation. Development of various modifications of the method in conjunction with modern computers have led to the solution of many problems of atmospheric physics.

The theory of the Monte-Carlo method as applied to the question of radiation transport has been developed quite extensively (see, for example, /832/). The application of this method to problems of atmospheric optics was treated elsewhere /655/. Some modifications of the method were obtained, algorithms were formulated, and a number of problems of the propagation of tight light beams in various scattering media were solved /658, 659, 783, 833–837, 843/. Some of these results will be analyzed below. We present briefly the results of investigations into multiply forward-scattered radiation during the propagation of various tight light beams and recorded by receivers with different areal dimensions. The computation program makes it possible to vary the optical properties of the medium and the geometric parameters of the source and receiver over very wide limits. Only some characteristic cases will be discussed here.

Krekov /837/ computed the angular distribution of the intensity of multiply scattered light in the plane of the receiver for the following conditions: the emitting and receiving surfaces are circular; the optical axes of the source and receiver coincide; the intensity of source emission is distributed uniformly over a cone with angles θ equal to 6', 1°, 5° and 10°; the emitted wavelengths are 0.45 and 0.7μ; the radiation propagates in a water cloud corresponding to the diagram and extinction coefficient obtained by Deirmendjian /517/; diameters D_1 of the source and D_2 of the receiver are, respectively, 4 and 150 mm; the optical thicknesses τ traversed by the direct radiation of the beam are 3.0, 5.0 and 10.0. Figure 9.2 illustrates the computation results for the angular intensity distribution of multiply forward-scattered radiation in the plane of the receiver.

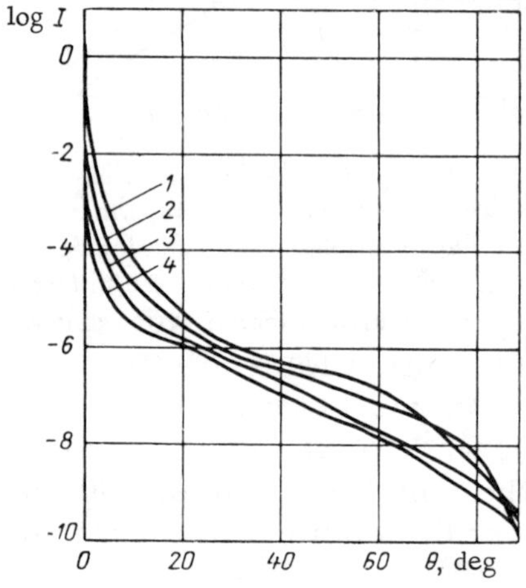

FIGURE 9.2. Results of computing the angular intensity distribution of multiply forward-scattered radiation (λ = 0.45μ) in a water cloud with optical thickness τ = 5.0 for various source divergence angles: θ = 6' (curve 1); 1° (curve 2); 5° (curve 3); 10° (curve 4) (diameters of the source and receiver are 4 and 150 mm, respectively).

For all values of the divergence angle of the source under study, the intensity of radiation incident on the receiving area of the indicator at different angles Ψ varies by 9–12 orders as Ψ goes from 0 to 90°. Note that figures such as Figure 9.2 were constructed from data for angles Ψ = 0, 30', 1°, 2°, 5°, 10°, 15°, 20°, 30°, 40°, 50°,

60°, 70°, 80°, and 90°. As expected, the critical angular dependence of the forward-scattered radiation incident at the receiver is observed at low Ψ and θ.

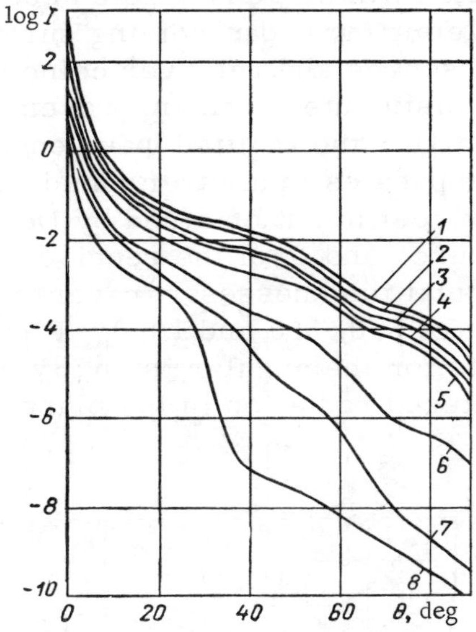

FIGURE 9.3. Results of computing the angular intensity distribution of multiply forward-scattered radiation ($\lambda = 0.45\mu$) in a water cloud with optical thickness $\tau = 4.0$ for various receiver diameters: D (r) = 4.0 m (curve 1); 3.0 m (curve 2); 1.5 m (curve 3); 1.0 m (curve 4); 0.5 m (curve 5); 0.2 m (curve 6); 0.15 m (curve 7) (the diameter and divergence angle of the source are 4 mm and 30', respectively).

Figure 9.3 shows the results of computing the angular intensity distribution for different dimensions of the receiving area of the indicator.

The obtained computational results make possible quantitative analysis of the dependence of the intensity of multiply forward-scattered radiation recorded by the receiver for different geometric parameters of the source and receiver. Comparison of computational and experimental results shows that they are in quite satisfactory agreement.

2. Experimental investigations. Detailed experimental studies on the brightness of forward-scattered light during propagation of laser radiation through different media were carried out in the author's laboratory /660—665, 781/. The apparatus, specially constructed for these investigations, consisted of the following principal elements: 1) lasers with active wavelengths of 6328 Å (continuous output gas laser, beam diameter about 0.4 cm, divergence angle 6′) and 6943 Å (pulsed ruby laser, beam diameter 20–100 mm, divergence angle 30″–40″); 2) collimated thermal source, equipped with narrow-band interference filters with transmission maxima near 0.63–0.69 μ; 3) receiving system, ensuring photometric scanning of individual sections of the focal pattern of the source and background image in the direction of small scattering angles. The radiation brightness distribution in the focal plane of the receiving system was recorded by photographic or photoelectric techniques, depending on the method of diaphragming of the focal pattern.

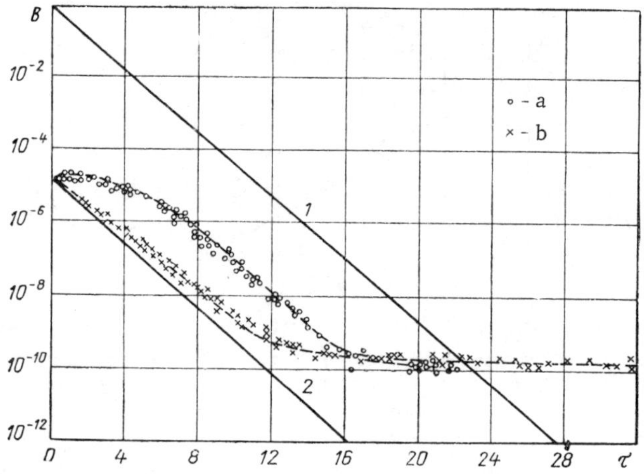

FIGURE 9.4. Total brightness of forward- and back-scattered radiation as a function of optical thickness τ in fog (circles) and smoke (crosses).

Lines 1 and 2 characterize the reduction in brightness of the direct radiation of the beam and the initial source background with τ increasing according to Bouguer's law.

The measurements were conducted in a chamber 30 m^3 in volume, in which were produced wood smoke (particulate diameter 0.8–1.0 μ) and artificial water fog (rms droplet size 8–15 μ). This selection of scattering media is attributed to the need to obtain data suitable for studying the propagation of narrow collimated beams in the earth's atmosphere. The brightness of the direct and scattered radiation were measured by means of two measuring schemes. In one version the radiation source and the receiving system were situated on a single straight line (direct-beam scheme). The beam path length in this case was 4 m. In the second version the beam emitted at the source was rotated by a flat mirror and directed to a receiving device situated alongside the source (reflected-beam scheme). Then the beam pathlength became 8.3 m; the angle of mirror rotation varied from 1°30' to 8°30'.

The specific properties of the attenuation of coherent as compared with incoherent radiation were explored by simultaneously varying the brightness of direct laser and thermal-source radiations in the scattering medium.

Figure 9.4 illustrates the results of measuring the brightness of forward-scattered radiation in artificial water fog and smoke, obtained by the reflected-beam scheme with a gas laser. All the measurements were reduced relative to an initial source brightness of $B = 1$. The figure indicates that the background brightness at $\tau = 0$ does not vanish. To find the background brightness due to light scattering alone it is necessary to subtract the initial source background brightness from the measured total background brightness. It can be assumed that the extinction of the initial source background obeys Bouguer's law (line 2). The contribution of the latter to the total background is found to be substantial for fog when $\tau < 4$ and for smoke when $\tau < 14$. Bearing the above remarks in mind it would appear that the background brightness becomes equal to the brightness of direct radiation when $\tau \cong 22$—24 for both fogs and smokes, which are media with very different scattering particle size. For optical thicknesses of approximately 2 to 13, the background brightness in fog is by 1—1.5 orders greater than in smoke. Starting with some value of τ in smoke and fog, a deflection appears in the background brightness curves. Here it occurs in smoke for lower values of τ than in fog. When $\tau > 20$, the background brightness curves of smoke and fog virtually coincide and are independent of the value of τ.

The results shown in Figure 9.4 were obtained by the reflected-beam scheme. This means that the instruments recorded not only the background due to forward scattering at small angles, but also that induced by back scattering at small angles when the beam traveled from the source to the mirror. It was of interest to obtain the brightness values of both background components. For this reason the mirror was covered, and the background brightness due to back scattering alone was measured. These measurements showed that the background brightness at large τ is determined primarily by back-scattered radiation.

Figure 9.5 depicts the measured values of the brightness of forward-scattered radiation in fog and smoke from the direct-beam scheme. Over the entire range of values of τ under study the background brightness is much smaller than that of direct radiation.

The dependence of the background brightness is described satisfactorily by data of calculations using the single-scattering formula.

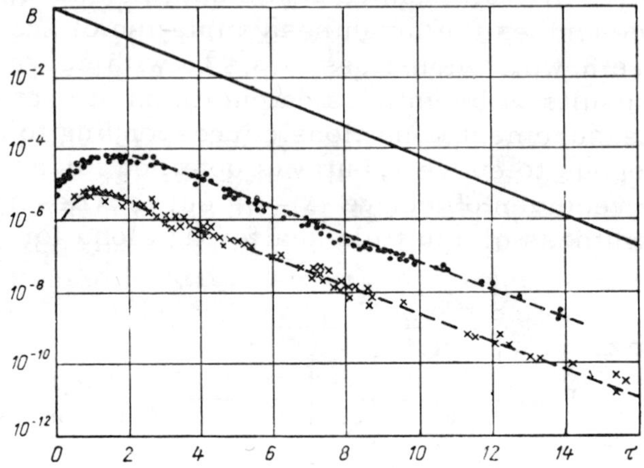

FIGURE 9.5. Brightness of forward-scattered radiation as a function of optical thickness in fog (points) and smoke (crosses).

Dashed curves were constructed from calculations of the theory of single scattering; the straight line describes the Bouguer extinction of direct radiation.

In addition to the above measurements in the fog chamber, the author also investigated the brightness of forward-scattered radiation in model media (slightly turbid water solution of milk, rms particle diameter $3.5 \pm 1.5\mu$, and lycopodium spore suspension in a water-alcohol solution, particle radius $15 \pm 0.5\mu$). The beam width was given by the optical diameter

$$\Delta = \alpha d, \tag{9.7}$$

where α is the extinction coefficient and d the geometric diameter of the beam. The value of Δ in measurements was 0.34. The angular aperture of the receiver was equal to the divergence angle of the source radiation.

Figure 9.6 shows the results of measurements for a slightly turbid milk solution. In these figures the optical thickness of the medium is laid off on the abscissa with measured brightnesses of direct and scattered light on the ordinate axis. The brightness of direct laser radiation in the absence of scattering is taken as unity.

The curves were drawn along experimental points, the scatter of which with respect to brightness did not exceed 5%. Curve 1 characterizes the variation in brightness of direct radiation and of radiation scattered in the zone of the beam at an optical diameter of the receiving objective $\Delta = 0.5$; curve 2 was obtained under the same conditions and describes the brightness variation of scattered radiation alone; curve 3, which coincides at $\tau < 15$ with curve 2, was constructed from results of brightness calculations of once-scattered radiation for measurement conditions corresponding to curve 2; curve 4 is analogous to curve 1, but was obtained for an optical diameter of the receiver objective $\Delta = 9$; curve 5 describes the variation in brightness of multiple scattering alone for $\Delta = 9$.

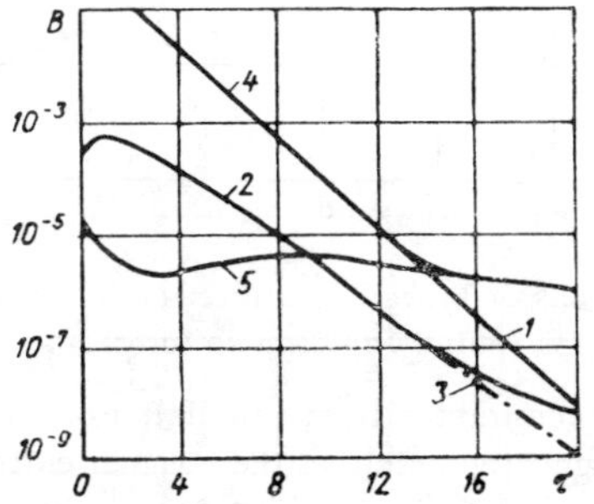

FIGURE 9.6. Total brightness of direct radiation and of radiation scattered in the zone of the beam as a function of the optical thickness of a slightly turbid milk solution.

It can be seen from Figure 9.6 that the scattered light brightness in the zone of a light beam with divergence 6' and optical diameter $\Delta = 0.34$ when the optical diameter of the receiving objective is 0.50, is described by single-scattering formulas /651/ up to $\tau = 15$, and its absolute value is at least an order of magnitude smaller than the brightness of direct radiation extinguished according to Bouguer's law over the range $\tau = 0-18$. Increasing the optical diameter of the receiving objective (which was equivalent in the above experiments to increasing its geometric diameter, since the beam width and density of the scattering media did not change) results in a substantial increase in the role of multiple scattering in the formation of forward-scattered radiation for different τ. For example, when $\tau > 15$ (optical diameter of receiving objective 9.0) the brightness of multiple scattering is found to be greater than that of direct scattering. The absolute magnitude of the brightness of multiple scattering exceeds that of single scattering when $\tau > 9$, being here little dependent on the value of τ.

To determine the dependence of the brightness distribution of multiply forward-scattered light over the beam cross-section on the diameter of the receiving objective, the brightness was measured in individual annular zones separated by diaphragms in front of the receiving objective. The beam zone (zone 1) was, as before, screened by a diaphragm so that $\Delta = 0.5$; zones 2–6 were separated so that each would correspond to an increase in the optical diameter of the receiving objective of 1.70. The measurement results are plotted in Figure 9.7. Curves 6–10 characterize the variation in brightness of multiply scattered radiation as a function of τ in zones 2–6 for a photon survival probability of $\lambda = 0.99$; curves 6a–10a were obtained for a medium with $\lambda = 0.92$ for the same zones. Curves 1–5 are the same as in Figure 9.4. When obtaining curves 6 and 6a, it was impossible to eliminate the effect of the source background in zone 2.

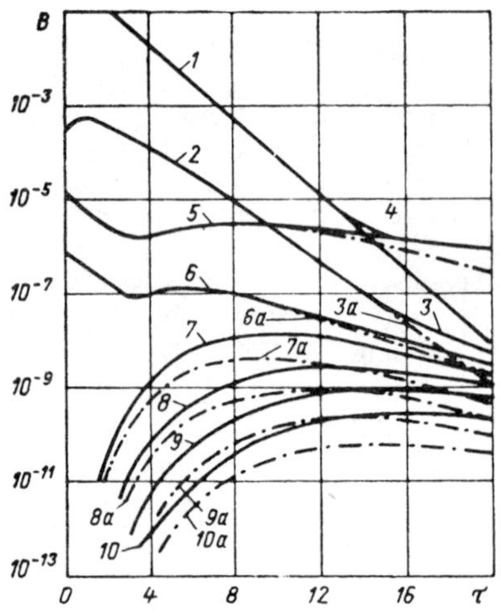

FIGURE 9.7. Brightness variation of multiply forward-scattered radiation in different zones of the cross-section of the receiving objective for two photon survival probabilities: $\lambda = 0.99$ (curves 6–10) and $\lambda = 0.92$ (curves 6a–10a). Curves 1 through 5 are the same as in Figure 9.4.

The results were similar to the above for suspensions of lycopodium spores in a water-alcohol solution. The role of multiple scattering of light in the beam zone was found to be greater than for the slightly turbid milk solution, due to a substantial difference in the elongation of the scattering diagram of these media.

Krekov /837/ reproduced the conditions of these measurements and conducted corresponding calculations, in order to check the accuracy of the Monte-Carlo method in calculating the brightness of multiply forward-scattered radiation in different scattering media and with different geometric parameters of the source and receiver. Figure 9.8 compares the calculated results with experimental data of Figure 9.7. The theoretical results are plotted as points through which pass vertical segments characterizing the rms deviation. The results of calculations are in quite satisfactory agreement with experimental data. Some discrepancy between these two sets of results in the region of low τ is attributable to the effect of the initial source background, which was disregarded in the experimental studies.

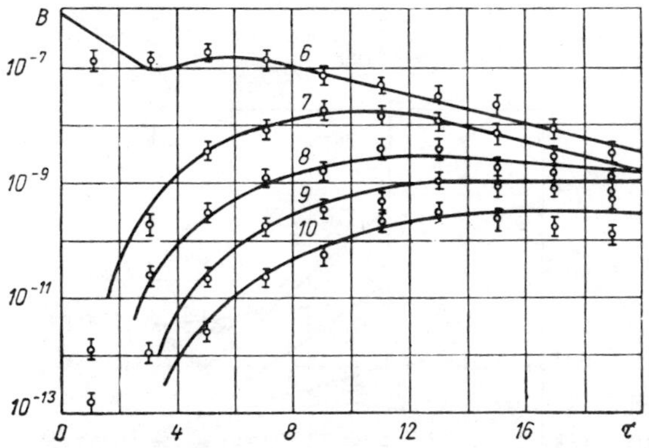

FIGURE 9.8. Comparison of experimental data of Figure 9.7 (curves 6–10) with theoretical results obtained by the Monte-Carlo method /837/.

It was of interest to clarify the limiting values of the optical thickness for which the dominant role in forward scattering in different scattering media is played by single scattering. The calculation is rather simple.

The intensity of once forward-scattered radiation is given by Gershun /651/ in the form

$$I_{sc} = I_0 \tau e^{-\tau} \frac{\omega}{4\pi} f(0), \tag{9.8}$$

where I_0 is the intensity of radiation incident on the scattering layer, τ is the optical thickness of the layer, ω is the solid aperture angle of the beam, $f(0)$ is the value of the scattering diagram at $\theta = 0$

(forward scattering). For small collimation angles one can set $\omega = \pi\alpha^2$, where α is the aperture angle of the beam. If we use Penndorf's approximating formula /656/

$$\frac{\omega}{4\pi} f(\theta) = \frac{\alpha^2\rho^2}{4} K(\rho, m), \tag{9.9}$$

then equation (9.8) yields

$$I_{sc} = I_0 \tau e^{-\tau} \frac{\alpha^2\rho^2 K(\rho, m)}{16}, \tag{9.10}$$

where $K(\rho, m)$ is the efficiency factor for the extinction of a particle with radius a; $\rho = 2\pi a/\lambda$, λ being the wavelength.

Equation (9.10) can also assume the form

$$I_{sc} = I_0 \tau e^{-\tau} \eta, \tag{9.11}$$

where

$$\eta = \frac{\alpha^2\rho^2 K(\rho, m)}{16} \tag{9.12}$$

is a parameter characterizing the forward-scattered fraction of light. The values of this parameter can be found by using experimentally measured values of I_{sc}.

The experimentally determined limiting values of η made it possible to obtain the limiting values of τ at which the effects of multiple scattering can be neglected in determining forward-scattered radiation. It was found that for a beam spread within the limits of $\alpha = 30'-10''$ the formulas describing the intensity of forward scattering light alone due to single scattering are valid up to $\tau < 18$ for particles with $\rho = 1-3000$. This range of ρ virtually overlaps for the visible and infrared spectral regions with particulate dimensions of atmospheric aerosols (haze, cloud, fog, drizzle, smoke, dust).

Experimental studies of the brightness of forward-scattered light propagating through different model media at $\tau \leq 20$ have been carried out /617, 838, 839/. The parameters characterizing the radiation source and receiver and the scattering medium varied over a wide range (source divergence angle from 7' to 1°44', angular aperture of receiver from 16'' to 7°, optical diameter of emitted beam from 0 to 0.30 and optical diameter of radiation receiver from 0 to 0.40). These results led to estimates of the systematic error in measuring the extinction coefficient of the scattering medium, associated with the recording of forward-scattering radiation by the receiver. This error is determined uniquely for known optical characteristics of the scattering medium and geometric parameters of the source and receiver.

In conclusion we examine briefly the results of experimental studies into the brightness of forward-scattered light at very high τ during the onset of the "depth regime." A series of measurements of light scattering in highly turbid media was carried out by Timofeeva /666–668/. According to these measurements the brightness of forward scattered radiation varies exponentially with increasing τ of the medium; however, the extinction coefficient is lower than in the case of small optical thicknesses. The brightness of light scattered at different angles at very high $\dot{\tau}$ decreases according to the same exponential law as the brightness of the forward-scattered light. The range of values of τ at which this dependence prevails is usually termed the region of the depth (or steady) regime. The brightness of radiation in the direction of angle θ in this case has the form

$$B(\theta) = B_0'\varphi(\theta)\, e^{-k'(z-z_0)};\quad B_0' = B_0 C, \tag{9.13}$$

where B_0 is the initial beam brightness in the forward direction; C is a constant which depends on the boundary conditions and geometric parameters of the light beam; $\varphi(\theta)$ is the angular brightness distribution in the depth regime; z is the geometric thickness of the scattering layer; z_0 is the geometric thickness of the scattering layer at which the depth regime sets in; k' is an extinction coefficient, related to the extinction coefficient α of the medium in the region of small τ by the expression

$$k' = \sqrt{A\alpha}; \tag{9.14}$$

A is a parameter, which is a function of the optical properties of the scattering medium and is related to absorption coefficient α_{abs} of the medium by the linear relationship

$$A = b\alpha_{\text{abs}}. \tag{9.15}$$

Constant b in (9.15) depends on the scattering diagram of the medium.

For constant distance between the source and receiver, expressions (9.13)–(9.15) yield the following relationships for the brightness of forward-scattered radiation at a depth with optical thickness τ:

$$B(0) = B_0 C e^{-\sqrt{b\alpha_{\text{abs}} z}\left(\sqrt{\tau} - \sqrt{\frac{z_0 \tau_0}{z}}\right)} = B_0 C' e^{-k\sqrt{\tau}}, \tag{9.16}$$

$$C' = Ce^{-k\sqrt{\frac{z_0\tau_0}{z}}};\quad k = \sqrt{b\alpha_{abs} z}. \tag{9.17}$$

For a purely scattering medium ($\alpha_{abs} = 0$), the brightness of forward-scattered light is seen from (9.16) and (9.17) not to depend on τ. The presence of absorption in the medium ($\alpha_{abs} \neq 0$) results in damping of $B(0)$ with increasing τ.

The diagram of the angular distribution of scattered-light brightness in the presence of the depth regime is a circle. In the case of a purely scattering medium the pole of the diagram is at the center. For absorption in the medium this pole moves backward relative to the direction of propagation of the light beam, and the brightness diagram proper elongates somewhat. The shape of the diagram remains unchanged with increasing τ and is characterized by the quantity

$$R = \frac{B(0)}{B(\theta)} = 1 + \mu\frac{\alpha_{abs}}{\alpha_{sc}} \tag{9.18}$$

where μ is a parameter depending on the scattering diagram of the medium. An expression analogous to (9.18) was obtained by Rozenberg /669/ from theoretical considerations.

Many measurements have been conducted of the brightness of scattered radiation in the presence of the depth regime in different model scattering media /670–677/. The geometric parameters of the light beam were the experimental variables. At low light-beam diameters the depth regime sets in when $\tau > 500$. By studying experimental data the following empirical formula was derived for the illumination of the medium at large depths beneath a parallel light beam:

$$E = E_0 e^{-(\tau-\tau_0)\sqrt{\frac{\alpha_{abs}}{\alpha_{sc} q}}}\left(1 - e^{-p\frac{S}{\tau}}\right), \tag{9.19}$$

where τ_0 is the optical thickness at which the depth regime sets in; q is a parameter depending on the scattering diagram; $S = k^2 Q$ is the optical cross-sectional area of the beam; k is the extinction coefficient; Q is the geometric cross-section of the beam; p is a parameter depending on the absorption and scattering coefficients and on the scattering diagram.

Equation (9.19) can be expressed in the form

$$B = B_0'\left(1 - e^{-p'\frac{d}{\tau}}\right), \tag{9.20}$$

where d is the beam diameter, τ the optical thickness, and p' a parameter depending on the optical properties of the medium.

Ivanov et al. /670–677/ found that the brightness of scattered light is proportional to the angle of spread of a narrow collimated beam in the depth regime.

2. EXTINCTION OF DIRECT RADIATION IN SCATTERING MEDIA. LIMITS OF APPLICABILITY OF THE BOUGUER LAW

Direct monochromatic radiation of a collimated beam, propagating in a scattering medium not emitting natural radiation, is extinguished according to the Bouguer law, providing the particles of the medium scatter the radiation independently of one another. Direct radiation usually carries some kind of useful information. Hence from a practical standpoint it is very important to know those limiting conditions under which direct radiation can be separated from the background of forward-scattered radiation. This can evidently be done when the brightness of direct radiation is greater than that of the background brightness. If, however, the background brightness is greater than that of direct radiation, the extraction of useful information carried by the beam can become hopelessly difficult, because the forward-scattered radiation in produced, in the final count, by the beam itself. In this case, in order to select the direct-radiation signal against the background of scattered radiation, it is necessary that the signal brightness be greater than that of forward-scattered light. It is clear that the most convenient conditions for measuring the useful signal are such when the angular aperture of the receiver and the angle of divergence of the source are identical. In this case the receiving system will receive the entire beam and the minimum radiation of the scattered background, since at high input values the background interference signal is larger for the same useful signal (see Sec. 1).

When the useful signal is much greater than the signal of the forward-scattered light, one should check the applicability of the Bouguer law to describe the measured extinction of a narrow collimated beam propagating in a scattering medium and completely intercepted by the receiving system. Such a study enables one to solve the problem of the operating range of a given system employing narrow collimated light beams to transmit useful information through different scattering media.

A detailed investigation of the applicability limits of the Bouguer law for various specific measurement schemes in different scattering media was carried out in the author's laboratory. Measurements were conducted of the brightness of forward-scattered radiation alongside a determination of the brightness of the direct beam. Here we shall summarize these investigations from the viewpoint of the limits of applicability of the Bouguer law. It should be again emphasized that reference is had to the applicability limits of the Bouguer law to describe the experimentally measured extinction of a light beam propagating through a scattering medium. It is clear that these limits should depend not only on the optical properties of the scattering medium, but also on the geometric parameters of the beam and the receiving system. We shall assume that the beam spread and the angular aperture of the receiver are equal.

Figure 9.9 presents the results of measuring brightness B of a laser beam as a function of the optical thickness of milk diluted with water. The measurements were conducted with a gas laser possessing a 6' spread and beam diameter $d = 0.4$ cm. Each curve in the figure pertains to some value of the optical diameter of the beam $\Delta = \alpha d$. The value of Δ was varied from 0.09 to 12.0 by increasing the density of the scattering media. All the results were reduced to unit brightness of a beam incident on the scattering layer.

The criterion of applicability of the Bouguer law to describe the damping of the measured beam brightness with some value of τ is the existence of a linear relationship between log B and τ. Figure 9.9, in which B is plotted on a logarithmic scale, shows that the linear segments of the curves depend markedly on the optical diameter of the beam. A particularly rapid change in the limiting values of τ occurs at small Δ ($\Delta < 0.74$). When Δ is increased the limiting values of τ decrease increasingly slowly, and when $\Delta > 1.48$ they virtually do not depend on Δ and τ is approximately equal to 12.

The dependence of the applicability limits of the Bouguer law to describe the measured brightness of a beam with some value of τ on the optical diameter of the beam can be attributed to the difference in the role of multiple scattering of light in the zone of the beam at different Δ. On increasing Δ by changing the particle concentration for constant geometric diameter of the beam, or on increasing the latter while the concentration of the scattering particles remains constant, the role of multiple scattering in the formation of light scattered in the beam increases compared with once-scattered radiation. At sufficiently high Δ the contribution of multiple scattering becomes "saturated." At small Δ a more substantial contribution to the forward-scattered field is made by once-scattered radiation. This results in a shift of the applicability limits of the Bouguer law to describe measured damping of the beam in the direction of large τ. It follows from Sec. 1 that analogous variations in the applicability limits of the Bouguer law arise on increasing the optical diameter of the objective of the receiving collimator while maintaining constant the optical diameter of the light beam.

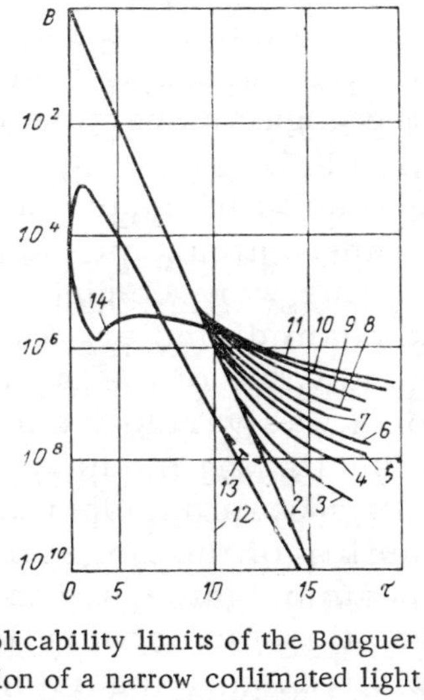

FIGURE 9.9. Applicability limits of the Bouguer law to describe the measured extinction of a narrow collimated light beam in a scattering medium as a function of the optical diameter Δ of the beam:

1) Δ = 0.09; 2) 0.34; 3) 0.74; 4) 1.11; 5) 1.48; 6) 2.22; 7) 2.96; 8) 3.56; 9) 7.12; 10) 9.0; 11) 12.0.

The results of measurements in a model medium (Figure 9.9) allow one to draw a number of conclusions on the applicability of the Bouguer law to describe the extinction of narrow collimated light beams in different atmospheric aerosols. The optical diameter of these beams in aerosols with greatest optical density (clouds and fogs at meteorological visibility $S_m = 50$ m and geometric beam diameter $d = 0.4$ cm in the visible region of the spectrum) does not exceed the minimum of the investigated values of Δ in Figure 9.9. In fact, the extinction coefficient in the case under study (see Chapter 7) is approximately equal to $8 \cdot 10^{-4}$ cm^{-1}, hence at $d =$ $= 0.4$ cm, $\Delta = 0.0003$. The difference in the microphysical and optical properties of clouds and fogs on the one hand, and of a weak solution of milk in distilled water on the other, are not substantially different. Therefore the results plotted in Figure 9.9 can be used for approximate analysis of the applicability limits of the Bouguer law to describe damping of a collimated beam in clouds and fogs. Bearing in mind that in a dense cloud or fog ($S_m = 50$ m) the value of Δ at $d = 0.4$ cm is substantially smaller than its minimum value in Figure 9.6, we can reliably claim that the applicability limits of the Bouguer law in this case lie beyond the limits corresponding to $\tau = 25$. This assertion is in complete agreement with the author's experimental data /660/ on measurements in artificial water fogs. It can be expected that this limit will apply also when the geometric beam diameter is 100 cm, since then $\Delta = 0.08$.

The optical diameters of beams propagating through haze, precipitation and dry aerosols are, as a rule, of lesser significance than in clouds and fogs. Hence the Bouguer law can describe the measured damping of collimated light beams when the divergence and diameter of the beam do not exceed 6' and 100 cm and when the angular aperture of the receiver is equal to the beam spread in all atmospheric aerosols for τ not smaller than 25 in the visible part of the spectrum. It is noteworthy that the most sensitive radiation sensors in the visible spectral region record radiation which passed through a layer of a medium with $\tau \cong 30-35$, if the energy density incident on the layer is equal to 10^3 watts. This conclusion can be also extended to the infrared spectral region, which differs from the visible by relatively smaller values of ρ and higher values of the absorption coefficient of the medium. Both these factors only increase the limiting values of τ for which the Bouguer law can still be employed to describe the measured damping of the beam in the scattering medium.

3. BRIGHTNESS OF BACK-SCATTERED RADIATION

Knowledge of the brightness of back-scattered radiation is of great importance in solving many applied problems, such as the use of narrow collimated light beams for radar operations and telemetering in scattering media, determining the aerosol components of the transmittance of these media, finding the albedo of various surfaces. In this section we consider briefly recent theoretical and experimental studies of back-scattering in the atmosphere.

1. Theoretical investigations. Feigel'son /652/ computed the albedo of clouds at different solar zeniths by numerical solution of the transport equation. Because the albedo of clouds depends little on the zenith distance of the sun, he concluded that the back-scattered radiation depends little on the scattering angle. The dependence of the albedo of clouds on their optical thickness τ is found to be more substantial; here the increase in albedo with τ slows down at large optical thicknesses, tending to "saturation." The brightness of the back-scattered radiation is thus determined by some finite scattering layer. The optical thickness of this effective layer decreases in the presence of absorption in the atmosphere /678/. Calculations of cloud albedo for different zenith distances of the sun in the 1.9–4.3μ infrared region were carried out by Bartky /679/.

Theoretical data on the intensity of back-scattered radiation, which must be available in determining the aerosol component of atmospheric transmittance, can be found elsewhere /680–682/. The amount of light back-scattered by the aerosol layer is found to be proportional to the degree of turbidity of the atmosphere. The calculations were conducted for different haze and fog models. Quantities entering the expression for the back-scattering coefficient have been calculated in detail /683/. The intensity of once back-scattered radiation during fog illumination by a narrow collimated beam was calculated by Kozlov and Fedorova /684/.

Munick /685/ derived an expression for the intensity of light back-scattered by turbulent inhomogeneities in the atmosphere. Calculations with this formula showed that the intensity of light from a ruby laser (emitted wavelength $\lambda = 0.69\mu$) back-scattered on turbulent inhomogeneities amounts to $5.9-10^{-7}$ of the intensity of back-scattered radiation by molecular scattering.

Detailed calculations of the deformation of an instantaneous light flash reflected from scattering media were carried out by the Monte-Carlo method /837/. Some of the data obtained there will be discussed in Sec. 5, which presents results of studies on the propagation of light flashes in scattering and absorbing media.

2. Experimental investigations. Experimental studies of the brightness of back-scattered light for different scattering media illuminated by narrow collimated radiation beams were carried out in the author's laboratory /660, 686/. The need for these studies arose due to a lack of theoretical as well as experimental data on back-scattering by atmospheric aerosols. The measurements were conducted with lasers with emitted wavelengths of 0.63μ (continuous-output He-Ne laser), 0.69μ (pulsed ruby laser) and 1.06μ (pulsed glass-neodymium laser). Measurements at 0.63 and 0.69μ were carried out in artificial fogs and haze, and at 1.06μ under natural conditions in the presence of haze and snowfall.

The results of measuring the brightness of light scattered at an angle of 171°30' emitted by a gas laser ($\lambda = 0.63\mu$, beam diameter 1 cm, spread 6') in artificial fogs for different optical thicknesses are plotted in Figure 9.10 (crosses). The same figure shows data of measurements of the total brightness of forward- and back-scattered radiation (circles) obtained with a source — mirror — receiver arrangement (beam rotation angle 8°30'). The mirror was covered when measuring the brightness of light scattered at an angle of 171°30'.

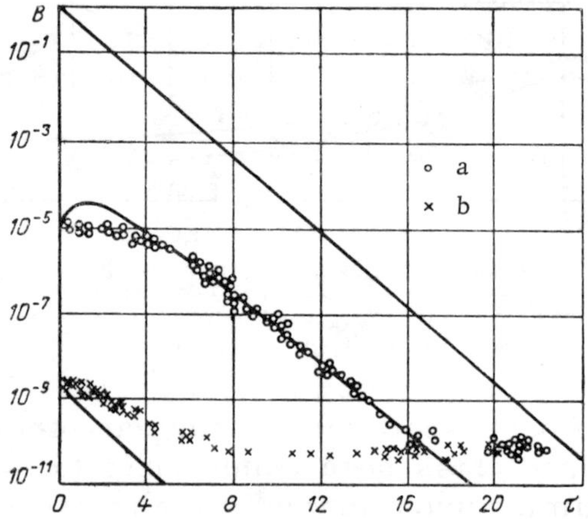

FIGURE 9.10. Total brightness of forward- and back-scattered radiation as a function of the optical thickness of fogs (circles) and brightness of back-scattered radiation (crosses). The straight lines describe the reduction in brightness of direct radiation of the beam and the initial radiation background of the source according to the Bouguer law.

If one eliminates from the measurement results the initial background of the source, it can be seen that when τ increases from 0 to 1 the brightness of the back-scattered light increases, attains a maximum, and then decreases with increasing τ first relatively rapidly and then slowly. When $\tau > 8$ it remains virtually constant, attaining a value equal to the brightness of the direct beam at $\tau \simeq 23$. Measurements of the brightness of radiation of a gas laser scattered at an angle of 171°30' by wood smoke showed that its values coincide with those obtained in fogs at $\tau > 20$. For lower values of τ the brightness of the back-scattered light is smaller for fog than for haze.

Figure 9.11 depicts the measured brightness of laser emission at 0.63μ, scattered at 172° by fog and haze. The measurements were made without using the mirror, with a beam 0.4 cm in diameter and a spread of 6′. The optical axes of the receiver and source intersected at an angle of 8°. The values of τ plotted in the figure pertain to the scattering layer between the source and receiver. The values of brightness are reduced to the initial brightness of the direct beam. Each point on the figure was obtained as an average of 5–10 measurements. The absolute values of the brightness of light scattered at 172° in fog and smoke over the range of τ under study amount to approximately $10^{-8} - 10^{-9}$ of the initial brightness of the direct beam.

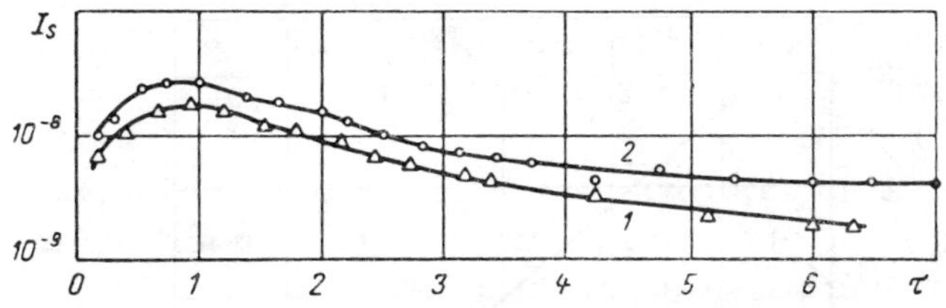

FIGURE 9.11. Variation in the brightness of laser radiation ($\lambda = 0.63\,\mu$) scattered at 172° by fog (1) and wood smoke (2).

Measurements of back-scattering in haze and snowfall were carried out by a pulsed glass-neodymium laser ($\lambda = 1.06\mu$) and provided some preliminary estimates of the magnitude of back-scattering. In the presence of haze, within which S_m = 3–7 km, the brightness of back-scattering amounts to 10^{-10} of the initial brightness of the direct beam. The appearance of snowfall hardly alters the brightness of back-scattered laser radiation at 1.06μ.

4. POLARIZATION CHARACTERISTICS OF SCATTERED RADIATION

The polarization state of radiation scattered at different angles is of importance in solving many applied problems associated with transmission of information by means of a narrow radiation beam. If a scattering layer of the atmosphere is illuminated by a linearly polarized beam, the particle layer may depolarize this radiation on scattering. The use of polarizing devices in receiving direct and scattered radiation may reduce the contribution of the scattered field to the magnitude of the signal, compared with the contribution of the direct radiation of the beam to this quantity. The latter circumstance may increase the range of the equipment in question.

Thus, Gilbert and Pernicka /687/ note that the use of polaroids in receiving systems increases underwater visibility by at least a factor of two. A similar effect should be expected in the atmosphere.

Theoretical consideration of the polarization state of scattered radiation by solving the general transport equation is very difficult. No solution of the transport equation without allowance for the polarization characteristics of the incident and scattered fields has been obtained for narrow collimated beams. Information on the polarization state of scattered radiation, required for solving many problems, is obtained experimentally. Further, experimental studies pertain primarily to investigation of polarization characteristics of scattered radiation for an elementary volume /506, 507, 688/ or for wide beams and large optical thicknesses of the scattering media /672, 689–691/.

New possibilities for experimental study of polarization characteristics of the scattered field pertaining to propagation of narrow collimated light beams in scattering media are becoming available by using lasers with linearly polarized radiation as radiation source. Of primary importance here is the study of the polarization state of radiation scattered forward and back-scattered in different scattering media. An attempt at this kind of study was made in the author's laboratory /692/.

Measurements of the polarization characteristics of scattered radiation were carried out for angles of 1°30' (forward) and 172° (back-scattering). The quantities measured were the light intensity I_1 with vector $\vec{E}$ lying in the scattering plane, and light intensity I_2 with vector $\vec{E}$ perpendicular to the scattering plane. Ratio $\Delta = I_1/I_2$ characterizes the depolarization of radiation. Knowing the value of Δ it is possible to determine the degree of polarization of scattered light:

$$p = \frac{1-\Delta}{1+\Delta}. \tag{9.21}$$

Figures 9.12 and 9.13 present the results of measuring I_1 and I_2, and the degree of polarization p of back-scattered radiation in fog and wood smoke calculated on the basis of the former measurements. The values of I_1 and I_2 were reduced to the initial intensity of the direct radiation of a beam incident on a scattering layer. It is seen from the figures that the dependence of the degree of polarization of back-scattered light on the value of τ over the range of τ under study can be approximated by the straight line

$$p(\tau) = p_0 - k\tau, \tag{9.22}$$

where p_0 and k are empirical constants, the values of which depend on the properties of the scattering medium and change little when the divergence angle of the source varies from 40'' to 6'.

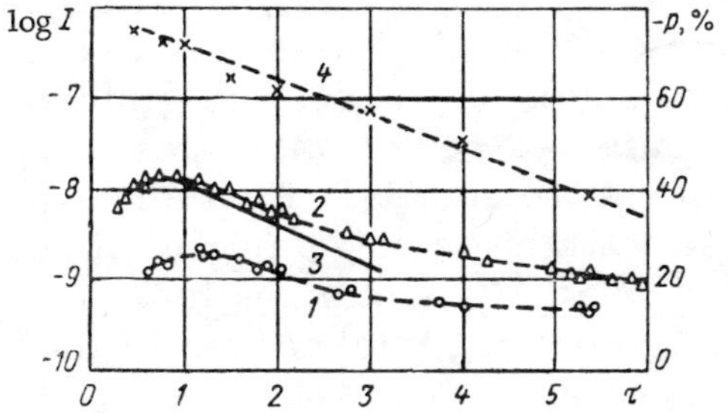

FIGURE 9.12. Intensity of back-scattered light with vector $\vec{E}$ lying in the scattering plane (1) and in a plane perpendicular to the scattering plane (2); intensity of back-scattered light calculated from the single-scattering theory (3); degree of polarization of back-scattered light as a function of the optical thickness of a fog (4).

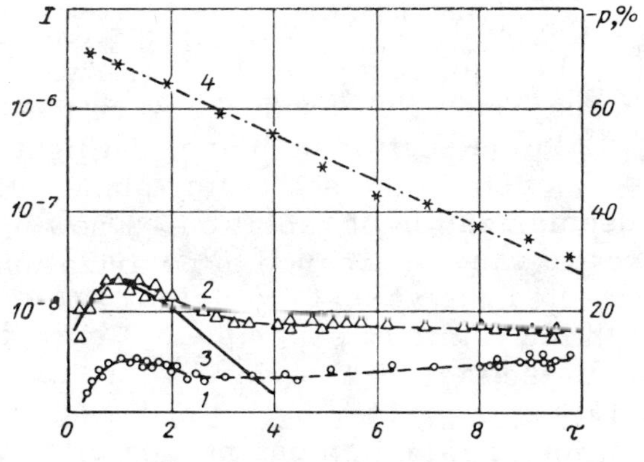

FIGURE 9.13. Intensity of back-scattered light with vector $\vec{E}$ lying in the scattering plane (1) and in a plane perpendicular to the scattering plane (2); intensity of back-scattered light calculated from the single-scattering theory (3); degree of polarization of back-scattered radiation as a function of the optical thickness of wood smoke (4).

The degree of polarization of back-scattered radiation differs from the value of p for the source even at the smallest values of τ under study ($\tau \simeq 0.5$). Extrapolation of linear equation (9.22) to the zero value of τ yields a value of p smaller than at the source.

(The degree of polarization of the radiation at the source, according to measurements, was more than 0.99.) This means that in the region of small τ the dependence of p on τ cannot be linear.

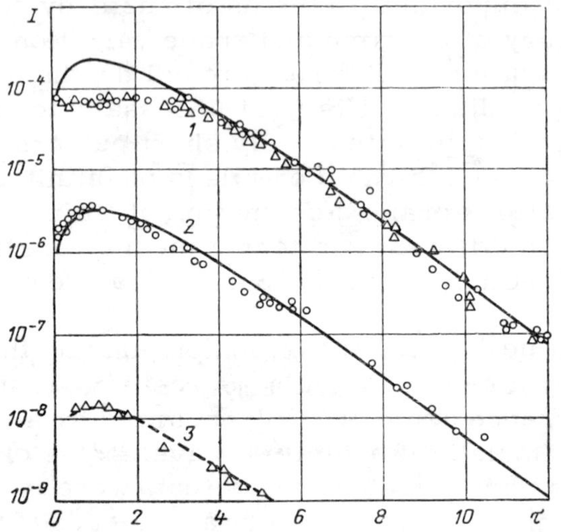

FIGURE 9.14. Total intensity $I_1 + I_2$ (triangles) and intensity I_1 (circles) of forward-scattered radiation in fog (curve 1) and of intensities I_1 (curve 2) and I_2 (curve 3) as a function of τ in wood smoke.

The solid and dashed curves were calculated from the formula of single-scattering theory.

Figure 9.14 presents results of measuring intensities I_1 and I_2 of light scattered forward at an angle of 1°30' in fog and smoke at different τ. As expected, the value of $I_1 + I_2$ in fog measurements did not differ (within the limits of experimental error) from the value of I_1, pointing to the fact that forward-scattered radiation is not depolarized. Quantity I_1 is greater by 2.5 orders of magnitude than I_2. Depolarization of radiation during forward scattering at small angles is thus practically nonexistent also in the case of wood smoke. The solid and dashed curves in the figure were obtained from an expression describing single scattering. They describe satisfactorily the results of measurements for $\tau \leq 12$.

5. PROPAGATION OF A PULSE OF DIRECTED RADIATION IN A SCATTERING AND ABSORBING MEDIUM

The design and practical employment of pulsed lasers with small pulse duration in radar and telemetry, and other devices intended for operation through the atmosphere require a study of the unsteady scattering of light in media with different photon survival probabilities. The latter quantity can be highly variable in the atmosphere, depending on the emitted wavelength.

One of the most important problems of unsteady scattering of radiation is the study of the time and space variation of the pulse shape during propagation to different optical depths of the absorbing and scattering medium. It is first necessary to know the variation in shape of an infinitesimally short pulse, since any pulse can be expressed as an ensemble of pulses of infinitesimal duration.

Detailed calculations of the deformation of an instantaneous pulse propagating in different scattering media were carried out /836/ by the Monte-Carlo method. Some of the results are considered below.

The free-path time distribution of photons during the illumination of a water cloud by a δ-pulse from a source with diameter $D_1 =$ $= 8$ mm and divergence angle $\theta = 30'$, recorded by a receiver with $D_2 = 15$ cm, was calculated for emitted wavelengths of 0.45 and 0.7μ. These wavelengths were selected because data were available on the scattering coefficients and diagrams of water clouds /517/. The assumed measure of pulse blurring was the dimensionless photon free-path time $t = t'c\sigma$. Here t' is the true free-path time, given by $t' = l/c$, where l is the mean free path and c is the speed of light; $\sigma = 0.016525\ \mathrm{m}^{-1}$ is the extinction coefficient. Quantity t can be interpreted as the optical thickness overcome by the photon on moving from the source to the receiver. The calculations were carried out for optical thicknesses $\tau = 1.0$, 2.0, 5.0 and 10.0. In constructing the dependence of the intensity of multiply forward-scattered radiation on the angle at which this radiation is incident on the receiver, t varied at an interval of 0.05 for $\tau = 1.0$, 2.0 and 5.0 and at an interval of 0.1 for $\tau = 10$. For example, when $\tau = 5.0$ the values of t were 5.0, 5.05, 5.1, etc. It should be noted that the incremental step $\Delta t = 0.05$ corresponds approximately for $\sigma = 0.016525\ \mathrm{m}^{-1}$ to a time of 10 nsec. The values of the angle of arrival of the photons at the receiving area of the radiation sensor were selected as 0°, 0.5°, 1°, 2°, 5°, 10°, 15°, 20°(10°), 90°.

The computation results were plotted as a family of curves and produce the complete pattern of the angular and time distribution of the photon flux for the given values of the source and receiver parameters, optical thickness and emitted wavelength. Figure 9.15 illustrates the calculated data for $\tau = 5.0$ and $\lambda = 0.7\mu$. The angle at which the photon is incident on the receiver area (the angle is measured from the normal to this area) is laid off on the abscissa, while the radiation intensity on a logarithmic scale is laid off on the ordinate axis. Curve 1 describes the angular distribution of the photon flux moving through path-lengths equivalent to values of τ

in the interval from 5.0 to 5.05. In other words, this curve determines the role of photons arriving at the receiving area of the indicator with a time lag not exceeding 10 nsec. Curves 2, 3 and 4 pertain to photons with a time lag from 10 to 20, from 20 to 30 and from 30 to 40 nsec.

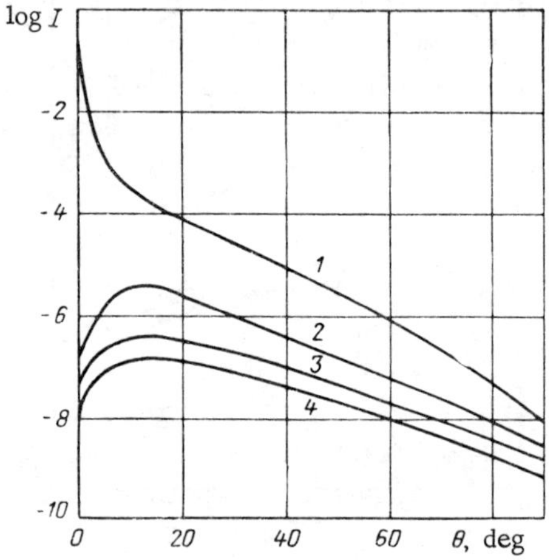

FIGURE 9.15. Results of computing the angular distribution of the photon flux in the plane of a receiver 150 mm in diameter during propagation of a light beam 8 mm in diameter with divergence angle of 30' and wavelength of 0.70 μ in a water cloud with optical thickness $\tau = 5.0$.

Curves 1 through 4 were constructed for values of t differing from one another by $\Delta t = 0.05 \approx 10$ nsec.

Figure 9.16 depicts some of the results of computing the angular distribution of photon fluxes when the scattering medium is a haze with a Junge particle-size distribution ($\beta = 4$) and $\tau = 2.0$. Data on the scattering diagram for refractive index $n = 1.50$ were used /572/. The extinction coefficient was $\sigma = 0.00010555\ m^{-1}$.

Krekov /843/ studied also the angular and time characteristics of photon distributions with different scattering frequency when the optical thickness of a water cloud between the source and receiver assumes the values $\tau = 1.0$, 3.0, 5.0 and 10.0. Deirmendjian's /517/ scattering diagram and coefficient were used. The calculations were made for source and receiver diameters of respectively 4 and 150 mm, and a source divergence angle of 30'.

Figures 9.17 and 9.18 depict curves of the time-dependence of the radiation intensity produced by photons of different scattering multiplicity, when $\tau = 1.0$ and 10.0 and $\lambda = 0.7\mu$. The figures show the redistribution of the role of different scattering multiplicities in the deformation of an instantaneous light pulse as τ varies from 1.0 to 10.0. Figure 9.19 depicts some of the results of computing the back-scattering of a δ-radiation pulse of a narrow light beam propagating in a cloud medium.

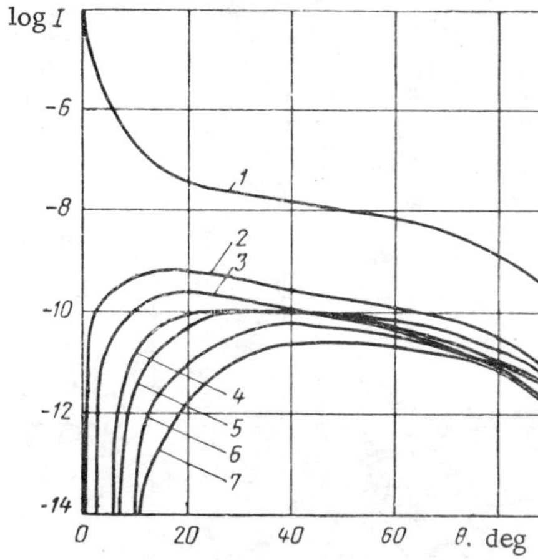

FIGURE 9.16. Results of computing the angular distribution of the photon flux in the plane of a receiver 150 mm in diameter during propagation of a light beam 8 mm in diameter with divergence angle of 30' in haze and $\tau = 2.0$ at particle refractive index $n = 1.5$.

Curves 1 through 7 were constructed for values of t differing from one another by 0.05, or approximately by 1.56 microsec.

Romanova /693–695/ obtained an approximate solution of the unsteady transport equation for an infinite beam and hence studied the distribution of photons by free-path times in reflected and incident light during isotropic scattering of an infinitesimally short radiation pulse. Calculations show that the main role in the formation of the signal reflected from the scattering medium when the latter was illuminated by a δ-pulse is played by a limited layer of medium adjoining the radiation source.

The unsteady transport equation for a narrow collimated beam was formulated and solved /842/ in the single- and double-scattering approximation. The calculations were conducted with a spherical scattering diagram. The solutions of the equation are valid with a photon survival probability smaller than 0.8. (the absorption coefficient of the medium amounts to one quarter of the scattering coefficient). Calculations showed that distortion of a rectangular pulse on passing through a layer of a medium is due primarily to double scattering and depends substantially on the relationship between the duration of the pulse incident on the pulse and the receiver aperture. The shape of the front of the reflected pulse is determined by both once- and twice-scattered radiation.

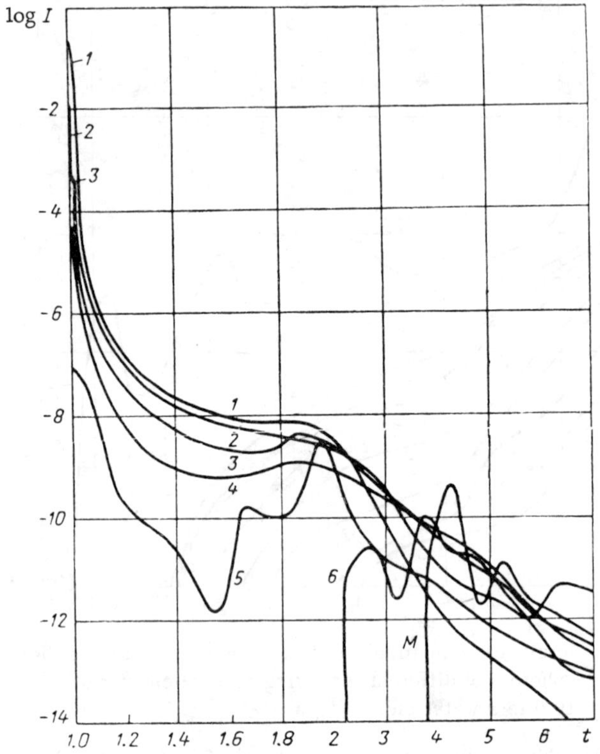

FIGURE 9.17. Calculated time-dependence of the radiation intensity due to photons of different scattering multiplicity for a wavelength of 0.7 μ and τ = 1.0 in a water cloud.

The source and receiver diameters were 4 and 150 mm. Curves 1 through 6 correspond to scattering multiplicities of 1 through 6, and curve M corresponds to all scattering multiplicities above 6.

The reflection of a short pulse from an infinite layer of medium with a very forward-elongated scattering diagram (characteristic of atmospheric conditions to a higher degree than in the preceding case) has been examined elsewhere /840, 841/. The scattering diagram is specified in the form of the ratio of the radiant flux scattered forward by an elementary volume, to the sum of forward- and back-scattered fluxes. The computation results show that, the greater the photon survival probability and the more the scattering diagram is elongated forward, the longer the duration of the reflected pulse. At the same time the duration of the reflected pulse is inversely proportional to the extinction index of the medium.

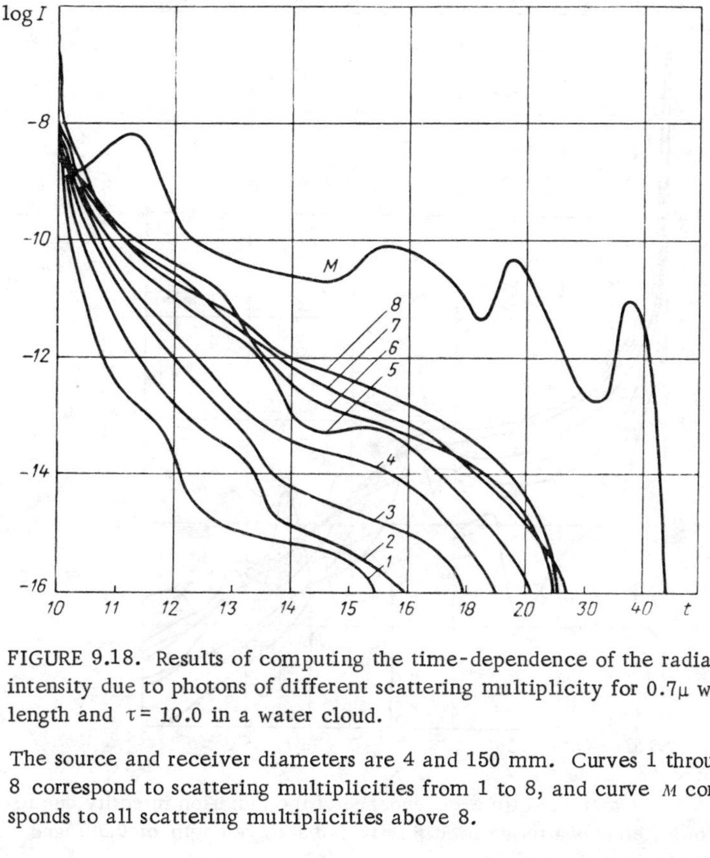

FIGURE 9.18. Results of computing the time-dependence of the radiation intensity due to photons of different scattering multiplicity for 0.7μ wavelength and $\tau = 10.0$ in a water cloud.

The source and receiver diameters are 4 and 150 mm. Curves 1 through 8 correspond to scattering multiplicities from 1 to 8, and curve *M* corresponds to all scattering multiplicities above 8.

Experimental studies on the propagation of short duration pulses in various scattering media are as yet in the initial stage.

An attempt was made /852/ to find a relationship between atmospheric transmittance and the parameters of the pulse reflected from it. It was found in natural fog measurements that the duration and delay of the reflected pulse in the medium depends on the value of extinction coefficient α. As α is increased from 0.001 to 0.01 m^{-1}, the duration of the reflected pulse changed from 2–3 to 0.5 microsec.

An experimental investigation into the deformation of the pulse of a gallium arsenide laser ($\lambda = 0.84\mu$) on reflection from artificial water fog and haze was carried out in the author's laboratory /853/.

The duration of the starting pulse at the 0.5 amplitude level was 8 nsec. The extinction coefficients of the medium ranged from 0.02 to 1.0 m^{-1}. The receiver and source were arranged in such a manner that their optical axes intersect within the medium under study at an angle of 1°, at a distance of 7 m from the forward boundary.

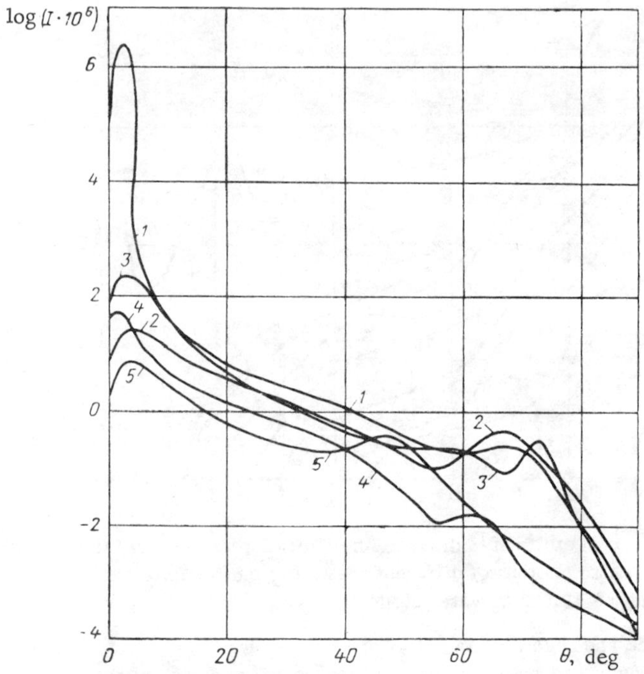

FIGURE 9.19. Results of computing back-scattering of a δ-pulse of radiation by a cloud medium.

Curve 1 corresponds to the source and receiver in direct contact. Curves 2 through 5 correspond to different distances between the source and receiver.

Some of the results of investigating the deformation of a pulse reflected from fogs with different density are plotted in Figure 9.20. The vertical lines in this figure are removed from one another by 10 nsec. The source pulse is shifted to the right by 50 nsec from its actual position and its amplitude is reduced by a factor of 10^5. When the extinction coefficient is reduced or, which is the same thing, on reduction of the optical thickness of the medium, the length of the forward front of the reflected pulse increases, while its curvature decreases. At the same time the length of the rear

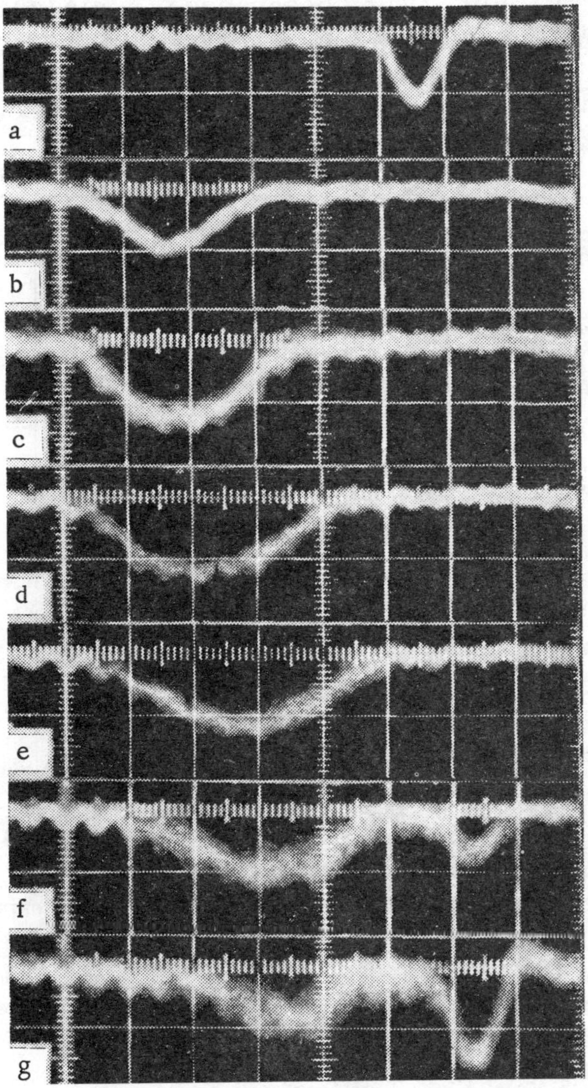

FIGURE 9.20. Oscillograms of emitted pulses reflected from fog layers with different values of the extinction coefficient:

a) emitted source pulse; b) $\alpha = 1.5\ m^{-1}$; c) $\alpha = 0.84\ m^{-1}$; d) $\alpha = 0.43\ m^{-1}$; e) $\alpha = 0.25\ m^{-1}$; f) $\alpha = 0.13\ m^{-1}$; g) $\alpha = 0.08\ m^{-1}$.

front of the reflected pulse first also increases and then starts decreasing. The latter is due to the effect of the rear boundary of the scattering medium. The appearance of a clear pulse in the right-hand side of the bottom oscillogram is due to reflection of radiation from the back wall of the chamber in which the measurements were made.

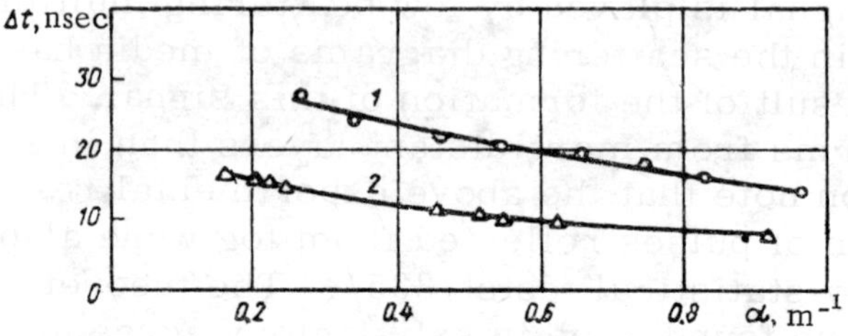

FIGURE 9.21. Duration of pulses reflected from fog and smoke as a function of the extinction coefficient.

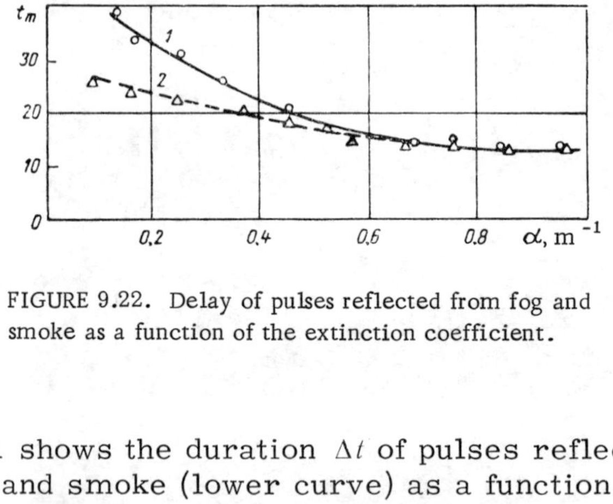

FIGURE 9.22. Delay of pulses reflected from fog and smoke as a function of the extinction coefficient.

Figure 9.21 shows the duration Δt of pulses reflected from fog (upper curve) and smoke (lower curve) as a function of the extinction coefficient. In both cases the duration of reflected pulses decreases with increasing extinction coefficient and tends to the duration of the source pulse, because the reflected signal is formed primarily in layers immediately adjoining the source. The size of these layers decreases with increasing medium density. The different duration of pulses reflected from fog and smoke for the same values of α are apparently due to respective differences in the scattering diagrams of these media.

The dependence of delays t_m of pulses reflected by fog (1) and smoke (2) on the values of the extinction coefficient is shown in

Figure 9.22. Note that the term delay here denotes the time interval between the sending of the pulse and the maximum value of the received signal. It is seen from the figure that for fog the value of t_m decreases more rapidly with increasing α than for smoke. When $\alpha > 0.5$ m^{-1}, the difference in t_m for fog and smoke disappears. It is possible to give the following qualitative explanation of the different behavior of t_m as a function of small α in fog and smoke. At low optical densities of the medium the principal role in forming the reflected signal is played by low scattering multiplicities, when the difference in the scattering diagrams of media has a substantial effect on the result of the formation of this signal. The reflected pulse in fog forms from more distant layers than in smoke.

In conclusion note that the above experimental results concerning the deformation of pulses reflected from fog were also obtained theoretically by statistical tests /835/. The theoretical and experimental data were found to be in satisfactory agreement, which again emphasizes the effectiveness of the Monte-Carlo method when used to study the propagation of tight light beams in scattering and absorbing media.

10. NONLINEAR EFFECTS IN THE PROPAGATION OF HIGH-POWER LIGHT RADIATION IN THE ATMOSPHERE

Propagation through the atmosphere of very high-power optical radiation, which is attained with individual Q-switched pulsed lasers, a number of lasers operating with free-generated pulses as well as in the continuous mode, is accompanied by a large number of nonlinear effects associated with the propagation of these emissions in the atmosphere.

The presence of these effects means in the first place that the Bouguer law is inapplicable for describing radiation damping due to molecular absorption by the gaseous constituents of the atmosphere or by aerosol scattering. Other regularities of optical wave propagation, described in the preceding chapters of this book, were also found to be inapplicable at the high powers and energies of optical radiation.

This chapter deals with various effects for which interference phenomena may be disregarded. Topics discussed include some multiple-photon effects (multiple-photon ionization, dissociation resulting in breakdown of the layer of pure air), spectroscopic saturation effect resulting in partial brightening of the absorbing gas layer, self-focusing of the beam brought about by the nonlinear dependence of the refractive index of the medium illuminated by intense radiation fluxes, and effects of aerosol particle evaporation due to pulsed and continuous radiation sources of high power and energy.

1. PROBABILITY OF MULTIPLE-PHOTON PROCESSES

Theoretical investigations of multiple-photon processes are the subject of many studies /857–860/. Bonch-Bruevich and Khodovoi /857/ examined studies in which multiple-photon processes, caused by interaction between the radiation field and the medium, are described by ordinary perturbation theory. Processes examined in greatest detail /857/ are two-photon processes, in particular resonance scattering of light at atoms. Multiple-photon processes are described by Wallace /858/ using formal scattering theory. The probability of these processes was calculated by means of diagrams and perturbation theory. Multiple-photon transitions in the optical range of wavelengths are surveyed elsewhere /859/. Studies on multiple-photon ionization in inert gases are considered in detail. Breakdown of gases, multiple-photon effect and a number of other related problems are treated by Raizer /860/.

In this chapter the question of multiple-photon processes during the propagation of intense optical radiation in gaseous media is presented on the basis of the model developed by Nesmelova and Tvorogov /696/. In the course of presentation reference is made only to studies which have a direct bearing on the material discussed. Consider first the calculation of the probability of transition of some quantum system from one given state to another.

The Hamiltonian of this system is designated by H. Then the wave function ψ of the system is described by the Schrödinger equation

$$i\hbar \frac{\partial\psi}{\partial t} = H\psi. \tag{10.1}$$

The formal solution of this equation can be expressed in the form

$$\psi(t) = S(t)\,\psi(0), \tag{10.2}$$

where $\psi(0)$ is the initial value of the wave function, while $S(t)$ is a unitary operator describing the time development of the system. It follows from (10.1) and (10.2) that

$$i\hbar \frac{\partial S}{\partial t} = HS, \quad S(0) = 1. \tag{10.3}$$

Relationships (10.3) completely define the mathematical problem of computing operator S.

Consider some operator M with eigenfunctions φ_n. If our system is in a state described by some function φ_n, then the quantity corresponding to operator M assumes some fixed value M_n.

We now select some function φ_n as initial condition for ψ. By virtue of the fact that φ_n is orthonormalized, function $\psi(t)$ can be written in the form

$$\psi(t) = \sum C_m(t)\,\varphi_m. \tag{10.4}$$

According to the general rule of quantum mechanics quantity $|C_m(t)|^2$ defines the probability that the system under study will be in the state described by function φ_m at time t.

If we substitute equation (10.4) into (10.2) and set (by virtue of selection of the initial condition) $\psi(0) = \varphi_n$, then

$$\sum_{m'} C_{m'}(t)\,\varphi_{m'} = S(t)\,\varphi_n. \tag{10.5}$$

It we premultiply both parts of equation (10.5) by φ_m^*, integrate functions φ with respect to variables ξ, and use the fact that $\int \varphi_m^* \varphi_{m'} d\xi = \delta_{mm'}$, then $C_m = < \varphi_m \mid S(t) \mid \varphi_n >$. Here $< \quad >$ denotes a scalar product, defined in the usual manner by

$$\langle f \mid S \mid g \rangle = \int f^*(\xi)\, Sg(\xi)\, d\xi.$$

The probability of transition of our system from state φ_n to φ_m over time t is denoted by W_{mn}. It follows from the meaning of C_m that

$$W_{mn} = |\langle \varphi_m \mid S(t)\,\varphi_n \rangle|^2. \tag{10.6}$$

Expression (10.6) is now applied to analysis of a system consisting of some molecule (atom) and an electromagnetic field. The general Hamiltonian of the system in this case is

$$H = H_0 + H_R + H_{0R}, \tag{10.7}$$

where H_0 is the Hamiltonian of the molecule, H_R is the Hamiltonian of the radiation field, and H_{0R} is the Hamiltonian of interaction between the field and the molecule.

We denote by $\varkappa_n(x)$ and $\psi_{n_k}(\xi)$ the eigenfunctions of H_0 and H_R. Function $\varkappa_n(x)$ depends solely on molecular variables x and describes some state of the molecule. Function $\psi_{n_k}(\xi)$ defines the state of a field with n_k photons and depends on the field variables. We shall subsequently consider a monochromatic field only.

We set $M = H_0 + H_R$. Then $\varphi_n = \varkappa_n(x)\,\psi_{n_k}(\xi)$ by virtue of commutation of H_0 and H_R, since these operators act on different variables. These functions define a situation in which the molecule is in state $\varkappa_n$ and the field contains n_k photons. Suppose the molecule has passed into some state $\varkappa_m$ and this transition resulted in the absorption of N photons. Then the final state of the "molecule plus field" system is defined by the function $\varphi_m = \varkappa_m \psi_{n_k - N}$. According to (10.6), the probability of this transition can be calculated from the expression

$$W = |\langle \varkappa_m \psi_{n_k - N} \mid S \mid \varkappa_n \psi_n \rangle|^2. \tag{10.8}$$

Operator S can be found from equation (10.2) upon substituting (10.7) into (10.3). Operator H_0 is an ordinary operator of a molecule or atom, incorporating all interactions between the particles of the atom or molecule. The two other operators in (10.7) possess long-wave approximations of the form

$$H_R = \hbar\omega a^* a,$$
$$H_{0R} = \beta P(a + a^*) + \alpha(a + a^*)^2,$$

where

$$\beta = -(e/m)(2\pi\hbar/\omega)^{1/2}, \quad \alpha = e^2\pi\hbar/m\omega c;$$

$\hbar$ is Planck's constant, c the speed of light and ω the field frequency; a and a^* are photon creation and annihilation operators; P is the component of vector $m\vec{v}$ along the direction of photon polarization; m, e and $\vec{v}$ are the particle mass, charge and velocity operator. In addition, the eigenfunctions of H_R describe the eigenfunctions of a harmonic oscillator.

Formula (10.8) allows one to calculate the probability of N-photon absorption over time t. Usually matrix elements represent integrals of rapidly fluctuating functions. Hence the probability of transition F in unit time is given by

$$F = \lim_{t\to\infty} \frac{d}{dt} |\langle \varkappa_m \psi_{n_k - N} | S | \varkappa_n \psi_{n_k} \rangle|^2. \tag{10.9}$$

The energy spectrum of operator H_0 consists of discrete levels and of a continuous part. The wave functions corresponding to energies of the continuous spectrum describe the dissociated or ionized state of the atom or molecule. If these functions are substituted in place of $\varkappa_m$, then (10.9) will yield the probability of N-photon dissociation or ionization.

2. COMPUTATION OF THE TRANSITION PROBABILITY BY SUCCESSIVE APPROXIMATIONS

Substitution of (10.7) into (10.3) yields an expression for operator S:

$$i\hbar \frac{\partial S}{\partial t} = (H_0 + H_R + H_{0R}) S. \tag{10.10}$$

Suppose the interaction is represented by setting $S = g(t)\exp[(t/i\hbar)(H_0 + H_R)]$. Then it follows from equation (10.10) that
that

$$i\hbar\frac{\partial g}{\partial t} = H'_{0R}(t)g(t), \quad g(0) = 1, \tag{10.11}$$

and

$$H'_{0R}(t) = \exp[-(t/i\hbar)(H_0 + H_R)]\, H_{0R} \exp[(t/i\hbar)(H_0 + H_R)].$$

If $g(t) = \sum_{n=0}^{\infty} g_n(t)$ and we use the usual iteration scheme, we derive the recurrence equation $i\hbar(\partial g_n/\partial t) = H'_{0R} g_{n-1}(t)$, $g_0 = 1$. Successive solution of these equations yields F.

The final form of F is /696/

$$F = \sum_{n=n'}^{\infty} F_n \sum_{k=k'}^{2n-1} n_k^k \beta^{2(2n-k)} \alpha^{2(k-n)} \times$$
$$\times \left(C_k^{\frac{k+N}{2}}\right)^2 \cdot |\{P_{2n-k},\ I_{k-n}\}|^2. \tag{10.12}$$

Here $C_{j'}^{j}$ is a binomial coefficient; $n' = (N+1)/2$ when N is odd and $n' = N/2$ when N is even; $k' = N$ for $n \leqslant N$ and $k' = n$ for $n > N$; the prime in the second summation denotes that in summing over k one should retain those terms for which $(k+N)/2$ is an integral number;

$$F_n = \frac{2\pi}{\hbar^2} \times$$
$$\times \left\{ \frac{\delta(E_A/\hbar - E_B/\hbar - n\omega)}{(E_B - E_{A_1} + \hbar\omega)^2 (E_B - E_{A_2} + 2\hbar\omega)^2 \ldots (E_B - E_{A_{n-1}} + (n-1)\hbar\omega)^2} + \right.$$
$$+ \frac{\delta[E_A/\hbar - E_{A_1}/\hbar - (n-1)\omega]}{(E_{A_1} - E_B - \hbar\omega)^2 (E_{A_1} - E_{A_2} + \hbar\omega)^2 \ldots (E_{A_1} - E_{A_{n-1}} + (n-2)\hbar\omega)^2} + \ldots$$
$$\left. \ldots + \frac{\delta(E_A/\hbar - E_{A_{n-1}}/\hbar - \omega)}{(E_{A_{n-1}} - E_B - (n-1)\hbar\omega)^2 (E_{A_{n-1}} - E_{A_1} - (n-2)\hbar\omega)^2 \ldots (E_{A_{n-1}} - E_{A_{n-2}} - \omega)^2} \right\},$$

where E_B and E_A are respectively the energy of the initial and final states of the system, while E_C are the eigenvalues of operator H_0.

The expression defined by symbol $\{P_{2n-k},\ I_{k-n}\}$ is constructed as follows. One writes the matrix element $\langle \varphi_{A_n} | \check{N}(PI) | \varphi_B \rangle$, where φ_{A_l} are the eigenfunctions of operator H_0, while $N(PI)$ denotes the sum of all possible permutations of the $(2n-k)$ operators $\check{P}$ and $(k-n)$ unit operators I. The intermediate subscripts are arranged in the sequence $A_n, A_{n-1}, A_{n-2}, \ldots, B$, in which case summation over these subscripts is removed. Summation in the expression for F_n is carried out over all the subscripts $A_1, A_2, \ldots, A_l, \ldots$

Index n in (10.12) registers different orders of the perturbation theory (with respect to operator H_{0R}). If only the first term is taken into account in operator H_{0R}, then summation over n takes place with $n' = N$.

Further, it follows from corresponding estimates that, in analyzing F, consideration must be given to both terms of interaction operator H_{0R}. We note that calculation of multiple-photon processes using only the first term of the interaction operator has been carried out elsewhere /697, 861/. Cohan and Hameka /698/ studied processes in which several photons with different frequencies participate, and which are also termed multiple-photon. However, they only treat cases which can be termed single-photon in the sense of equation (10.8) ($N = 1$) /698/. Clearly, in examining such situations it suffices to take into account only the first term of operator H_{0R}, which is stated by the authors of /698/ in the form of a special theorem.

In the case of transitions to the continuous spectrum (multiple-photon dissociation and ionization) relationship (10.12) must be integrated with respect to the energy of the final state, having first multiplied it by the density of these states. The presence of δ-functions in this expression reduces the integration to the replacement of δ-functions by $\hbar\rho$, where ρ is the aforementioned density of states:

$$\rho = (m/\hbar^2)(2\pi)^{-3} k\, d\Omega;$$

$d\Omega$ is a solid-angle element; $k = \hbar^{-1}\sqrt{E/m}$, $E = E_{A_l} + (n - l)\hbar\omega - E'$, E' is the dissociation or ionization energy; n and l are defined in line with the scheme for constructing F_n.

3. ESTIMATES FOR THE PROBABILITY OF MULTIPLE-PHOTON PHOTOIONIZATION AND PHOTODISSOCIATION

Specific calculations of the probability of multiple-photon processes can be made if the matrix elements of the dipole moment and the energy levels of the molecule or atom are known. Using available data on the energy spectrum and matrix elements of dipole transitions we calculated the probability of ionization of a hydrogen atom for two-, three- and four-photon absorption. The resulting values of F for selected wavelengths are listed in Table 10.1.

TABLE 10.1. Values of F for selected λ

$N=2$		$N=3$		$N=4$	
λ, μ	F, sec^{-1}	λ, μ	F, sec^{-1}	λ, μ	F, sec^{-1}
0.135	$7\cdot10^{-30}n_k^2$	0.258	$10^{-50}n_k^3$	0.344	$2\cdot10^{-72}n_k^4$
0.128	$2\cdot10^{-29}n_k^2$	0.253	$10^{-50}n_k^3$	0.335	$10^{-72}n_k^4$
0.124	$9\cdot10^{-29}n_k^2$	0.248	$10^{-50}n_k^3$	0.326	$6\cdot10^{-73}n_k^4$
0.122	$2\cdot10^{-25}n_k^2$	0.243	$2\cdot10^{-48}n_k^3$	0.317	$6\cdot10^{-73}n_k^4$
0.118	$3\cdot10^{-29}n_k^2$	0.238	$10^{-51}n_k^3$	0.310	$5\cdot10^{-73}n_k^4$
0.113	$2\cdot10^{-30}n_k^2$	0.234	$10^{-51}n_k^3$	0.302	$5\cdot10^{-73}n_k^4$

Note: n_k is the number of photons per unit volume (cm^{-3}); N is the number of absorbed photons.

In the case of molecules of atmospheric gases, when analyzing the processes at hand one must deal with the electron state of the molecules. Calculation of matrix elements and energy levels in this case is extremely difficult. At present pertinent data are available only for the simplest atoms and molecules. The situation is similar with respect to analysis of the wave function of a dissociated polyatomic molecule, knowledge of which is necessary in calculating the matrix elements of the dipole moment, contained in (10.12).

By virtue of the above circumstances we present below estimates of only some quantities characterizing multiple-photon processes. It is assumed in these estimates that the matrix elements of the dipole moment are quantities of the order of 10^{-18} CGSE units. The characteristics of the N-photon process are estimated irrespective of the specific process which occurs (photoionization, photodissociation, alteration of the electronic or vibration-rotation state).

Table 10.2 lists terms of (10.12) for N from 1 to 12, calculated for the above conditions. Table 10.3 presents data on the probability of multiple-photon processes ($N = 1\,(1)\,12$) for different n_k. The table presents quantity b appearing in the expression $F = 10^b$ for different a in the expression $n_k = 10^a$.

When the probability of the respective transitions is known one can estimate the coefficient of absorption τ of the multiple-photon process. Quantities τ and F are related by the expression /71/

TABLE 10.2. Values of individual terms of sum (10.12)

N	E, sec^{-1}					
	1	2	3	4	5	6
1	$10^{-9.82}n_k$					
2	$10^{-32.40}n_k^2$					
3	$10^{-52.6}n_k^3$	$10^{-94.0}n_k^5$				
4	$10^{-74.9}n_k^4$	$10^{-116.2}n_k^6$				
5	$10^{-95.4}n_k^5$	$10^{-136.7}n_k^7$	$10^{-178.2}n_k^9$			
6	$10^{-117.7}n_k^6$	$10^{-158.7}n_k^8$	$10^{-199.9}n_k^{10}$			
7	$10^{-133.3}n_k^7$	$10^{-179.5}n_k^9$	$10^{-220.6}n_k^{11}$			
8	$10^{-160.5}n_k^8$	$10^{-201.2}n_k^{10}$	$10^{-242.6}n_k^{12}$	$10^{-284.1}n_k^{14}$		
9	$10^{-181.2}n_k^9$	$10^{-222.0}n_k^{11}$	$10^{-263.5}n_k^{13}$	$10^{-304.8}n_k^{15}$	$10^{-346.3}n_k^{17}$	
10	$10^{-203.2}n_k^{10}$	$10^{-244.0}n_k^{12}$	$10^{-285.3}n_k^{14}$	$10^{-326.5}n_k^{16}$	$10^{-367.9}n_k^{18}$	
11	$10^{-224.1}n_k^{11}$	$10^{-265}n_k^{13}$	$10^{-306.1}n_k^{15}$	$10^{-347.4}n_k^{17}$	$10^{-388.8}n_k^{19}$	$10^{-430.2}n_k^{21}$
12	$10^{-246.2}n_k^{12}$	$10^{-287}n_k^{14}$	$10^{-327.8}n_k^{16}$	$10^{-369.1}n_k^{18}$	$10^{-412.3}n_k^{20}$	$10^{-451.3}n_k^{22}$

$$\tau(\mathrm{cm}^{-1}) = \frac{F(I)\, n_0}{Ic}, \tag{10.13}$$

where n_0 is the number of absorbing molecules per unit volume and I is the emission intensity; here, in the case of a plane wave

$$I = c\hbar\omega_k n_k. \tag{10.14}$$

The differential equation for the absorbed radiation now has the form

$$dI = -\tau(I)\, I\, dx, \tag{10.15}$$

x being the distance traversed by the light beam in the medium. In the case of a homogeneous medium equations (10.15) and (10.13) yield

$$x = -\frac{c}{n_0}\int \frac{dI}{F(I)} \tag{10.16}$$

The integral here can be always evaluated, since in our case F represents (according to equations (10.12) and (10.14)) some polynomial with respect to I. However, in order to derive specific expressions it is necessary to know the pertinent coefficients exactly. Once the integral is evaluated it is possible to find $I(x)$ by solving the resulting equation. The entire problem can be solved in general form for any N.

For single-photon processes equation (10.16) yields the known exponential reduction in intensity. For two-photon process

$$T = \frac{I_0 x \gamma_2 n_0}{c+1},$$

where $T = I/I_0$, I_0 is the intensity of radiation incident on a layer of the medium, γ_2 is the coefficient of I^2 in the expression for F in the two-photon process. For the three-photon process we obtain

$$x = \Gamma_1 I^{-2} + \Gamma_2 \ln I + \Gamma_3 \arctan \Gamma_4 I + \mathrm{const},$$

where Γ_j are some constants. This equation, however, cannot be used to express T in explicit form. Equation (10.16) for $N > 3$ is even more complicated.

Analysis of data in Tables 10.2 and 10.3 shows that for some N and n_k it is possible to retain in sum (10.12) only a certain term.

TABLE 10.3. Values of quantities a and b in expressions $n_k = 10^a$ and $F = 10^b$

N	a						
	18	19	20	21	22	20.5	21.5
1	8,2	9.2	10,2	11,2	12,2	10.7	11.7
2	3.6	5.6	7.6	9.6	11.6	8.6	10,6
3	1,4	4.4	7,4	11,1	16.0	9.5	13.6
4	−2.9	1,1	5.1	9,8	15.8	7.3	12,8
5	−5,4	−0,1	4,6	11.1	19.8	7.5	15,3
6	−9.7	−3.7	2.3	10,3	20.1	5,8	15,0
7	−12.1	−5.1	1,9	10,9	23.7	6.0	17.2
8	−16,5	−8,5	−0.5	10,1	23.9	4.0	17,0
9	−19.3	−10.3	−1,3	10,9	27.7	3,9	19.2
10	−23.2	−13.2	−3.2	10.8	28,1	2,5	19.2
11	−26.1	−15.2	−4,1	11,5	31.8	2,0	21.3
12	−30.2	−18.2	−6.2	10.7	32.7	0.3	21.7

TABLE 10.4. Distances traveled by light for different N and n_k
$n_0 = 2.4 \cdot 10^{19}$ and $T = 0.1$

N	n_k, cm^{-3}				
	10^{17}	10^{18}	$10^{18.5}$	10^{19}	10^{20}
2	$3 \cdot 10^4$	$3 \cdot 10^3$	$3 \cdot 10^{2.5}$	$3 \cdot 10^2$	$3 \cdot 10^1$
3	$2 \cdot 10^6$	$2 \cdot 10^4$	$2 \cdot 10^2$	2	$2 \cdot 10^{-4}$
4	∞	$3 \cdot 10^7$	$2 \cdot 10^{3.5}$	$2 \cdot 10^1$	$2 \cdot 10^{-4}$
5	∞	$5 \cdot 10^5$	$5 \cdot 10^4$	$8 \cdot 10$	0
6	∞	$7 \cdot 10^7$	$4 \cdot 10^{4.5}$	4	0
7	∞	∞	$2 \cdot 10^5$	$2 \cdot 10^{-1}$	0
8	∞	∞	$3 \cdot 10^{6.5}$	3	0
9	∞	∞	10^7	2	0
10	∞	∞	$8 \cdot 10^{7.5}$	$8 \cdot 10^{-1}$	0
11	∞	∞	∞	$8 \cdot 10^{-2}$	0
12	∞	∞	∞	$9 \cdot 10^{-2}$	0

Then $F(I) = \gamma_j I^{b_j}$, and

$$\frac{I}{I_0} = \frac{1}{\sqrt[b_j]{\frac{I_0^{b_j-1}x}{\sigma}+1}}, \tag{10.17}$$

$$\sigma = \frac{c}{n_0 k\,(b_j-1)\,(\hbar\omega_k c)^{b_j-1}}. \tag{10.18}$$

For sufficiently high I_0 the unity in equation (10.17) can be neglected, in which case

$$x = \sigma\left(\frac{I/I_0}{I_0}\right)^{b_j-1}. \tag{10.19}$$

This can be used for calculating distance x corresponding to a given transmittance $T = I/I_0$ (Table 10.4).

It is noteworthy that according to estimates the results of such a calculation do not significantly depend on the selected value of T. The zero in the table means that the required absorption occurs at the very start of radiation propagation in the medium (the phenomenon should be explosionlike); the sign ∞ denotes that the corresponding distance exceeds 10^7 cm.

Quantity F can be employed for determining the time needed for the decay of some part of the molecules or atoms. The corresponding expression follows from the known decay equation

$$n = n_0 e^{-Ft}, \tag{10.20}$$

where n and n_0 are the number of "survived" molecules (atoms) at time t and $t = 0$. If the time of partial decay τ^0 is defined as the interval during which ratio n/n_0 attains a certain value, then

$$\tau^0 = \frac{\ln n_0/n}{F}. \tag{10.21}$$

Table 10.5 lists values of τ^0 for $n/n_0 = 0.1$ and different N and n_k. The infinity sign in this table means that τ^0 exceed 10^4 sec.

This process can occur during the decay of molecules or atoms of a medium in the focus of some lens. The relationship between photon density n_{k_f} in the lens focus and density n_k in the incident beam can be established from the correspondence principle. If R is the radius of the lens and f is its focal distance, then

$$n_{k_f} = n_k \frac{R^5\omega_k^2}{8f^3c^2}. \tag{10.22}$$

TABLE 10.5. Values of τ° (in second) for different N and n_h when $n/n_0 = 0.1$

N	n_h		
	10^{18}	10^{19}	10^{20}
2	$10^{-3.2}$	$10^{-5.2}$	$10^{-7.2}$
3	10^{-1}	10^{-4}	10^{-7}
4	$10^{3.3}$	$10^{0.7}$	$10^{-4.7}$
5	∞	$10^{0.8}$	$10^{-4.2}$
6	∞	$10^{4.1}$	$10^{-1.9}$
7	∞	∞	$10^{-1.5}$
8	∞	∞	$10^{0.9}$
9	∞	∞	$10^{1.7}$
10	∞	∞	$10^{3.6}$
11	∞	∞	$10^{4.5}$
12	∞	∞	$10^{6.6}$

The order of an N-photon process can be easily estimated from the expression

$$N = \left[\frac{E_1 - E_2}{\omega_k}\right],$$

where the symbol [ ] denotes the integral part of the number, E_1 is the ionization or dissociation energy, and E_2 is the energy of the lower (as a rule, ground) state.

TABLE 10.6. Dissociation energy E_D of O_2, N_2, CO_2 and H_2O molecules, ionization energy E_I of N, O, C and H atoms, and orders of multiple-photon processes for a glass-neodymium laser with $\lambda = 1.06\mu$ (N_1) and a ruby laser with $\lambda = 0.69\mu$ (N_2)

Molecule	O_2	N_2	CO_2	H_2O	Atom	N	O	C	H
E_D, eV	5.08	9.76	2.8	5.1	E_I, eV	14.5	13.6	11.27	13.6
N_1	5	9	3	5	N_1	13	12	10	12
N_2	3	6	2	3	N_2	9	8	7	8

Table 10.6 presents values of the dissociation energy of molecules of principal atmospheric gases and the ionization energy of their atoms.

4. SPECTROSCOPIC SATURATION EFFECT

The spectroscopic saturation effect for the case of elastic molecular collisions has been explored in sufficient detail, for example, in /699, 700/. We shall analyze completely the quantum version of this effect, in which it is possible to examine inelastic molecular collisions during single-photon light absorption. The question of saturation in multiple-photon processes has been described elsewhere /862/.

The Hamiltonian of the problem at hand can be expressed in the form

$$H = H_1 + H_2 + H_c(t) + H_R + H_{1R}, \tag{10.23}$$

where H_1, H_2 and H_R are the Hamiltonians of the first and second molecules and of the radiation field. Operator $H_c(t)$ makes allowance for the time-dependent interaction between two molecules (collisions), and H_{1R} is the interaction between the field and the first (light-absorbing) molecule. The second molecule does not absorb the radiation (at least in single-photon interactions), but only deforms the state of the first molecule.

The problem under study deals with bound states, and so the initial condition must be formulated for some time $t = 0$, which can be treated as coinciding with the onset of some collision. The solution of problem (10.23) should subsequently be sought at time $t \gg \tau$, where τ is some critical (correlation) time, which depends on the nature of interaction $H_c(t)$. Some approximation τ should be thought of as designating the average time spent by one molecule next to the other, or the molecular collision time.

Deliberations which are almost identical with those presented in Sec. 1 make it possible to obtain the result

$$P_{BA} = |\langle \psi_{n_{k-1}}(\tau)\, \varphi_m(\tau) \,|\, S \,|\, \psi_{n_k}(0)\, \varphi_n(0) \rangle|^2. \tag{10.24}$$

Here P_{BA} is the probability of transition of the system from state A to state B during time τ; state A is defined as some state of a field with n_k photons and the n-state of interacting molecules; state B is the m-state of the molecules and the state of the field with n_{k-1} photons. The notation in equation (10.24) is: ψ_{n_k} – eigenfunctions of operator H_R; $\varphi_n(0)$ — eigenfunctions of operator H_1+H_2, $\varphi_n(t) = S_c(t)\varphi_n(0)$. Operators S and S_c satisfy the equations

$$i\hbar\frac{\partial S}{\partial t}=HS, \tag{10.25}$$

$$i\hbar\frac{\partial S_c}{\partial t}=(H_1+H_2+H_c(t))S_c. \tag{10.26}$$

Further $\psi_{n_k-1}(\tau)=S_R\psi_{n_k-1}$ and $S_R=\exp\left(\frac{t}{i\hbar}H_R\right)$. Hence in (10.24) $\psi_{n_k-1}(\tau)$ can be replaced by $\psi_{n_k-1}(0)$.

Expression (10.24) should then be summed over m in order to take into account all possible transitions accompanying absorption of a photon, from one given state to another. After this operation

$$\sum_{(B)}P_{BA}=\langle P_n(0)\,|\,F\,|\,\varphi_n(0)\rangle, \tag{10.27}$$

$$F=\int\bar{\psi}_{n_k}(p)\,U^{-1}(p,q)\,\psi_{n_k-1}(p)\,\psi_{n_k-1}(p')\times \times U(p',q)\,\psi_{n_k}(p')\,dp\,dp', \tag{10.28}$$

where P are the variables of the field functions and q are the variables of the molecular functions. Here, $U=SS'^{-1}$ and $S'=S_cS_R$. It follows from the fact that U is a unitary operator that F is a Hermitian operator.

Now equation (10.27) must be averaged over the initial molecular states. According to the general rule, averaging of a quantity of this type yields the expression

$$P=S_p\rho_0F, \tag{10.29}$$

where P denotes sum $\sum_{(A)}P_{BA}$, ρ_0 averaged over the initial molecular states, and ρ_0 is the Boltzmann density matrix for Hamiltonian H_1+H_2.

Note that result (10.29) can be obtained by starting with an expression for the density matrix ρ satisfying the equation

$$i\hbar\frac{\partial\rho}{\partial t}=H_\rho-\rho H \tag{10.30}$$

with Hamiltonian given by (10.23) and initial conditions

$$\rho^{(0)}_{nn_k,\,nn_k}=1,\quad \rho^{(0)}_{mm_k,\,nn_k}=0. \tag{10.31}$$

Then equation (10.24) for the probability of transition $A \to B$ is given by the diagonal element of density matrix $\rho_{mn_k-1,\ mn_k-1}$; after summation over final states and averaging over initial states of system $H_1 + H_2$ we arrive at equation (10.29). The expression for P should be averaged over the collisions. Technically speaking this is carried out according to the scheme of Tsao and Curnutte /71/.

Equation (10.25) is then solved to the first order of perturbation theory with respect to interaction operator H_{1R}. Integral (10.28) is evaluated by means of set rules for operating with photon creation and annihilation operators. The method of evaluating (10.29) is based on the theory of representation of rotation groups /71, 56/. Due to its cumbersomeness we omit the intermediate manipulation and present the final expression for

$$\omega = \frac{P}{n_k c\tau} \tag{10.32}$$

which has the meaning of absorption coefficient calculated for a single absorbing molecule:

$$\omega = 2\gamma \sum_{\substack{a,\ b\\ \alpha,\ \beta}} \frac{\rho_{aa}^{(0)}}{2a+1} \times$$

$$\times \frac{|\langle a\alpha \mid \mu \mid b\beta\rangle|^2 (\sigma_{ba}^{(r)})^2}{(\omega_k - \omega_{ba} + \sigma_{ba}^{(i)})^2 + (\sigma_{ba}^{(r)})^2 + 2\eta\tau n_k(\sigma_{ba}^{(2)}) \mid \langle a\alpha \mid \mu \mid b\beta\rangle|^2}. \tag{10.33}$$

Here, $\gamma = 2\pi\omega_k/\hbar c$, $\eta = 2\pi\omega_k\hbar^{-1}$, $\sigma_{ba}^{(r)}$ is the half-width, $\sigma_{ba}^{(i)}$ is the shift, ω_{ba} is the center of the spectral line, $\rho_{aa}^{(0)}$ are the diagonal matrix elements of the Boltzmann density matrix for operator H_1, $\langle a\alpha \mid \mu \mid b\beta\rangle$ are matrix elements of the dipole moment in the representation of degenerate eigenfunctions of operator H_1, and c is the speed of light. It should also be noted that (10.33) was evaluated by incorporating only those terms which contain a frequency difference.

The term in the denominator which is a function of n_k formally represents the saturation effect. The above calculations emphasize the long-known idea that the saturation effect arises because the density matrix of the molecule in the presence of a strong field differs from the Boltzmann matrix.

5. ABSORPTION OF LIGHT WITH ALLOWANCE FOR SATURATION EFFECTS

Expression (10.33) can be used to derive the following expression for the absorption coefficient of light with frequency ν:

$$k(\nu,\ I)=\sum_j \frac{a_j}{b_j(\nu)+c_j(\nu)\,I}\ . \qquad (10.34)$$

Summation is carried out over all lines contributing to the absorption of light of given frequency, $a_j = S_j\gamma_j/\pi$, $b_j = (\nu - \nu_j)^2 + \gamma_j^2$, $c_j = \tau\gamma_j\,|\,\mu_j\,|^{-2}c^{-2}$, S_j, γ_j, ν_j and μ_j are the strength, half-width, center, and square of the matrix element of the dipole moment of the j th line, I is the radiation intensity. The dimensions of ν_j and γ_j are cm^{-1}.

The variation in radiation intensity in the medium satisfies

$$dI = -k(I)\,I\,d\omega, \qquad (10.35)$$

where ω is the precipitable layer of absorbing gas. The initial condition for equation (10.35) is standard: $I=I_0$ when $\omega = 0$.

Equation (10.35) follows in essence from the fact that (10.32) defines ω as the ratio of the absorbed energy to the intensity of incident radiation.

For a homogeneous medium equation (10.35) can be integrated exactly. However, this involves cumbersome numerical calculations. It is therefore simpler to solve equation (10.35) by successive approximations, utilizing the smallness of c_jI. For this purpose equation (10.35) should be written in the form $dI/I = -k(I)\,d\omega$ with approximation $I^{(0)} = I_0\cdot\exp(-k_0\omega)$, where $k_0 = \Sigma a_j/b_j$. Simple computation yields /703/:

$$T=T_0\prod_j\left(\frac{1+I_0\varkappa_j}{1+I_0\varkappa_jT_0}\right)^{\frac{a_j}{b_jk_0}}, \qquad (10.36)$$

where $T=I/I_0$, $T_0=\exp(-k_0\omega)$, $\varkappa_j=c_j/b_j$.

It follows from the expressions for a_j and b_j that k_0 has the meaning of an extinction coefficient calculated without allowance for the dependence of the molecular density matrix on field strength. Then T_0 is the usually determined transmittance for the given radiation frequency. It follows from equation (10.36) that when $I_0\varkappa_j \ll 1$, $T = T_0$.

Equation (10.36) makes it possible to estimate the effect of saturation on the transparency of the absorbing layer. If

$$I_0 \varkappa_j T_0 \gg 1, \text{ then } T = 1. \tag{10.37}$$

Quantity τ can be estimated from the following considerations. The collision time can be treated as the time of transition of colliding molecules from one rectilinear path of their motion to another. The corresponding solution of this problem /704/ yields

$$\tau = \tau_0 P^{-1/2} T^{-1/6}. \tag{10.38}$$

Here T is the absolute temperature, P is the pressure and τ_0 is some, almost constant quantity. If P is expressed in atmospheres, then τ_0 is of the order of 10^{-11}.

The last remark enables one to use (10.37) for estimating the intensity at which the saturation effect appears. In line with the estimates of Sec. 3, and setting additionally $\alpha_j = 0.1\ \text{cm}^{-1}$ and $T_0 = 0.1$, we obtain for I_0 a value of the order of 10^{14} ergs/cm$^2 \cdot$ sec.

6. SELF-FOCUSING OF A LASER BEAM

The propagation of a high-power laser beam in a medium can result in altering the refractive index in the beam channel. The difference between the refractive index of the medium in and outside of the beam channel can be such for which the beam spread is eliminated. This phenomenon has been termed self-focusing of a laser beam and has been examined elsewhere in detail /705–719, 844, etc./.

It is found that in order to eliminate beam spread φ it is necessary to alter the refractive index in the beam channel compared with its value outside the channel by the amount Δn, which is approximately equal to φ^2.

If the first, nonlinear term of the expansion of the polarization vector with respect to the field in the medium is proportional to the third power of electric field strength E, then the refractive index of such a medium can be represented in the form

$$n = n_0 + n_2 E^2, \tag{10.39}$$

where n_0 is the ordinary refractive index, while n_2 is a coefficient characterizing the nonlinear properties of the medium.

Equation (10.39) was written on the assumption that subsequent nonlinear terms of the expansion are small compared with the first two.

The nonlinear dependence of the refractive index of the medium on the electric field strength is primarily due to the Kerr effect and electrostriction. In addition, in the presence of absorption in the medium the respective nonlinear effects can be caused by thermal expansion of the gas while being heated by the absorbed wave energy.

The Kerr effect is associated with orientation of the dipole moments of the molecules in the direction of a strong electric field, which results in the virtually inertialess appearance of anisotropy in the properties of the refracting medium. Electrostriction consists in the fact that in the region of action of radiation there arise the additional volume pressure

$$P = \frac{1}{8\pi} \rho \frac{\partial \varepsilon}{\partial \rho} E^2 \tag{10.40}$$

and force

$$F = \text{grad}\, P, \tag{10.41}$$

where ρ is the density and ε the dielectric constant of the medium.

Solution of the equations of electrodynamics and the linearized perturbation equations of the dielectric constant yields expressions for the threshold self-focusing power P_{th} induced by the Kerr effect and electrostriction.

The magnitude of the threshold self-focusing power can be approximately estimated from the simple expression

$$P_{\text{th}} \approx \frac{\lambda c}{n_2}, \tag{10.42}$$

where λ is the emitted wavelength and c the speed of light.

The threshold self-focusing power decreases with decreasing wavelength and increasing nonlinear properties of the medium. The lowest self-focusing thresholds, of the order of tens of kilowatts, are those of some liquids. The threshold powers for gases amount to tens and hundreds of thousands of kilowatts. This means that self-focusing conditions should be expected during propagation of laser beams in the atmosphere. Recall that the pulsed output of a number of lasers exceeds substantially the above threshold self-focusing power in gases.

Individual measurements in the atmosphere show that self-focusing of a laser beam actually occurs, and also that alongside focusing of the beam as a whole there are formed fine filaments inside the beam. It should be noted that the self-focusing theory predicts the breakup of the light beams into individual components, when the output power substantially exceeds P_{th}. However, under experimental conditions the fine filaments were observed at powers not exceeding P_{th}. It is presently difficult to give a comprehensive qualitative interpretation of the origin of these filaments. One possible mechanism can be associated with fluctuations in the refractive index of air, induced by turbulent motions in the atmosphere. Apparently, these fluctuations can aid in bringing about the self-focusing effect at sub-threshold powers for a homogeneous medium. Detailed studies on the self-focusing of laser beams in the atmosphere have not yet been conducted.

7. ACTION OF A HIGH-POWER RADIATION FIELD ON ATMOSPHERIC AEROSOL PARTICLES

The illumination of atmospheric aerosols by powerful directional radiation is accompanied by a number of effects, which result in altering the conditions of radiation propagation in the medium. Of greatest practical interest among these effects is particle evaporation due to absorption of the field energy. The change in particulate size results in corresponding modification of the bulk polydisperse extinction coefficients. The transparency of the scattering-medium layer thus becomes dependent on the intensity of incident radiation.

Slow evaporation of particles in a thermal field has been studied extensively. A very comprehensive survey of pertinent experimental and theoretical investigations was given by Fuchs /720/. The results obtained in these studies cannot be used for estimating the evaporation of particles in a powerful radiation field. Studies on this phenomenon started only recently and some of the results are presented below.

The evaporation of solid and liquid spherical particles in a radiation field in an infinite gaseous medium was examined by Williams /721/. Using the Knudsen kinetic equation, he found the rate of particle evaporation. However, the evaporation coefficient in the Knudsen equation remains indeterminate and is only estimated approximately.

The kinetics of droplet evaporation due to solar radiation has also been studied /722/. Assuming steady-state evaporation, simple analytic expressions were derived for the variation in droplet radius with time:

$$\left.\begin{aligned} r &= r_0 e^{-\eta t}, & \frac{8\pi\varkappa}{\lambda} r \ll 1, \\ r &= r_0 - \eta' t, & \frac{8\pi\varkappa}{\lambda} r \gg 1, \end{aligned}\right\} \tag{10.43}$$

where r_0 and r are the droplet radii before and after incidence of radiation, t is time, η and η' are parameters, $\varkappa$ is the absorption index of the particle material, and λ is wavelength.

The latter equation shows that, for given wavelength, large droplets change their size linearly due to solar radiation, while the variation in size of small particles is exponential.

Theoretical studies of the evaporation of atmospheric aerosol particles due to strong radiation were conducted recently in the author's laboratory /723–725/. The principal results of these studies are presented below.

1. General equations. The heat and mass transfer on obtaining the general equations is described on the basis of the Enskog-Chapman mathematical theory of nonuniform gases /726/. The final equations obtained for the distribution of partial water vapor pressure and temperature outside the particles in a spherical coordinate system have the form

$$\frac{\partial p}{\partial t} - \frac{p}{T^l}\frac{\partial T^l}{\partial T} = D\left(\frac{2}{\rho}\frac{\partial p}{\partial \rho} + \frac{\partial^2 p}{\partial \rho^2}\right)\frac{P}{P-p} + D\frac{P}{(P-p)^2}\left(\frac{\partial p}{\partial \rho}\right)^2, \tag{10.44}$$

$$\begin{aligned} \frac{\partial T^l}{\partial t} = {} & \frac{\lambda^l R_n}{C_p}\,\frac{T^l}{\mu^a P - (\mu^a - \mu^v) P}\left(\frac{\partial^2 T^l}{\partial \rho^2} + \frac{2}{\rho}\frac{\partial T^l}{\partial \rho}\right) + \\ & + D\frac{T^l}{P-p}\left(\frac{\partial^2 p}{\partial \rho^2} + \frac{2\partial p}{\rho\,\partial \rho}\right) + \\ & + D\mu^v P\frac{T^l}{(P-p)^2\,[\mu^a P - (\mu^a - \mu^v) p]}\left(\frac{\partial p}{\partial \rho}\right)^2, \end{aligned} \tag{10.45}$$

where p is the partial water vapor pressure, P is the total pressure in the medium, T^l is the temperature of the medium, D is the coefficient of water vapor diffusion in air, ρ is the polar radius of the particle, λ^l is the thermal conductivity of the medium, C_p is the specific heat of the medium at constant pressure, R_n is the universal gas constant, μ^a is the molecular weight of air, and μ^v is the molecular weight of water vapor.

The temperature field inside the particle is described by the inhomogeneous thermal diffusion equation

$$\frac{\partial T^i}{\partial t} = \varkappa^i\left(\frac{\partial^2 T^i}{\partial \rho^2} + \frac{2}{\rho}\frac{\partial T^i}{\partial \rho}\right) + I, \tag{10.46}$$

where T^i is the temperature inside the particle, $\varkappa^i$ is the thermal diffusivity of the material of the particle, I is the source function

$$I = \frac{3\mathcal{J}\omega(t)\,K_{ab}(r,\lambda,m)}{4C_L\gamma_L r^3}. \qquad (10.47)$$

Here $\mathcal{J}$ is the thermal equivalent of work, ω is the radiation source strength, C_L and γ_L are the specific heat and density of the particle material, and K_{ab} is the optical absorption cross-section of the particle.

Equations (10.44)–(10.46) define the evaporation kinetics of a single particle in the radiation field. The initial and boundary conditions for these equations have the following form:

Initial conditions:

$$t=0;\; p(\rho,0)=p_\infty;\; T^l(\rho,0)=T^i(\rho,0)=T_\infty, \qquad (10.48)$$

Boundary conditions:

$$\begin{gathered}\rho=r,\; T^l(r,t)=T^i(r,t)\equiv T_0(r,t),\; \rho(r,t)=p_0(T_0),\\ Lj_m^0+j_Q^0=\lambda^i\left(\frac{\partial T^i}{\partial\rho}\right)_{\rho=r},\\ \rho\to\infty,\; T^l\to T_\infty,\; p\to p_\infty,\end{gathered} \qquad (10.49)$$

where T_0 is the particle surface temperature; j_m and j_Q are the vapor and heat fluxes.

The initial conditions are formulated on the assumption of thermodynamic equilibrium between the particle and the ambient medium before the incidence of radiation. Function $p_0(T_0)$ in (10.49) characterizes the dependence of the saturation vapor pressure on the particle surface temperature. For particles with radii $r > 1\,\mu$ this function has the form

$$p_0(T_0)=p_r e^{L\mu^V/R_nT_r}e^{-L\mu^V/R_nT_0}, \qquad (10.50)$$

where L is the specific heat of evaporation. For particles with $r \leqslant 1\,\mu$ equation (10.50) should be corrected appropriately.

The limits of applicability of equations (10.44)–(10.46) can be obtained by analyzing the assumptions on which they were obtained. These equations are applicable for describing mass and heat transport in the "water droplet in air" system when $p=p_{cr}<P,\; T^l=T_b$; $v=v_s$, where T_b is the boiling temperature of water, v is the rate of mass flow, and v_s is the speed of sound.

When quantities p, T^l and v attain values exceeding p_{cr}, T_b and v_s, the vapor and heat fluxes can be described by approximate equations if it is assumed that

$$j_m = \frac{\mu^v p}{R_n T^l} v_s , \tag{10.51}$$

$$j_Q = C_p \gamma_g T^l v_s , \tag{10.52}$$

where γ_g is the density of the vapor-gas mixture. The boundary conditions in this case take the form

$$L \frac{\mu^v p_0}{R_n T_0} v_s + C_p^v \frac{\mu^v p_0}{R_n} v_s = \lambda^i \left(\frac{\partial T^i}{\partial \rho} \right)_{\rho=r} , \tag{10.53}$$

while the droplet evaporation kinetics are described by the equation

$$\frac{dr}{dt} = -\frac{1}{\gamma_L} \frac{\mu^v p_0}{R_n T_0} v_s . \tag{10.54}$$

Equations (10.44) and (10.45) can only be solved numerically. Accordingly appropriate algorithms were obtained, and programs for solution using ALGOL-60 language were formulated.

2. Quasi-steady fields. In the case of quasi-steady fields of the temperature and partial water vapor pressure, when they are determined solely by the particle surface temperature, equations (10.44) and (10.45) yield expressions for partial pressure p and temperature T^l in the form

$$p = P - \frac{(P - p_0) \frac{T^l}{T_0} \frac{r}{\rho}}{(P - p_\infty) \frac{T^l}{T_\infty} (\frac{r}{\rho} - 1)} , \tag{10.55}$$

$$T^l = \frac{r}{\rho} (T_0 - T_\infty) + \frac{C_p D}{\lambda^l R_n} \times$$
$$\times \left[\mu^v P \ln \frac{P - p}{P - p_\infty} - (\mu^a - \mu^v)(p - p_\infty) \right] +$$
$$+ \frac{r}{\rho} \frac{C_p D}{\lambda^l R_n} \left[\mu^v P \ln \frac{P - p_\infty}{P - p_0} - (\mu^a - \mu^v)(p_\infty - p_0) \right] + T_\infty . \tag{10.56}$$

From these equations it is possible to obtain asymptotic expressions for p which do not contain T^l, and expressions for T^l which do not contain p when $(\rho - r) \ll 1$ (points near the particle surface). Due to the cumbersomeness of these equations they are not cited here.

In calculating the droplet evaporation rate it is necessary to know the temperature of its surface, which depends on the fields outside the particle, as well as on the temperature distribution within the particle. To determine the temperature inside the droplet it is necessary to solve a boundary value problem, known as the third boundary layer problem for a parabolic equation. The solution yields the following transcendental equation for the droplet surface temperature T_0:

$$T_0 + \frac{2r}{\varkappa^{i2}} \sum_{n=1}^{\infty} \frac{(-1)^n \pi n}{\pi^3 n^3 (1+n^2\pi^2)} \left\{ I_r + \frac{2\varkappa^{i2}}{r^2} \times \right.$$
$$\left. \times \theta [1-(-1)^n] \right\} \left(1 - e^{-\left(\frac{\varkappa^i n\pi}{r}\right)^2 t}\right) + \frac{2}{r} \sum_{n=1}^{\infty} \frac{(-1)^n}{1+n^2\pi^2} \times$$
$$\times \left\{ T_\infty r - \theta \frac{(-1)^n [n^2\pi^2+2]-2}{n^2\pi^2} \right\} e^{\left(\frac{\varkappa^i n\pi}{r}\right)^2} + \frac{\theta}{r} = 0, \qquad (10.57)$$

where

$$\theta = \frac{\lambda^l}{\lambda^i} T_0 r - \frac{D\mu^{\mathrm{v}} P r}{\lambda^i R_n} \ln (P - p_0) \left(C_g + \frac{L}{T_0} \right) +$$
$$+ \frac{C_p D_r}{\lambda^i R_n} (\mu^{\mathrm{a}} - \mu^{\mathrm{v}}) p_0 - \frac{\lambda^l T_\infty}{\lambda^i} r + \frac{D\mu^{\mathrm{v}} P r}{\lambda^i R_n} \times$$
$$\times \ln (P - p_\infty) \left(C_p + \frac{L}{T_\infty} \right) - \frac{C_p D}{\lambda^i R_n} (\mu^{\mathrm{a}} - \mu^{\mathrm{v}}). \qquad (10.58)$$

If the solution of equation (10.57) is known, then the rate of droplet evaporation can be calculated easily from the expressions

$$\frac{dr}{dt} = -\frac{1}{\gamma L} j_m^0, \qquad (10.59)$$

$$j_m^0 = \frac{\mu^{\mathrm{v}} D p}{r R_n} \left[\frac{1}{T_\infty} \ln (P - p_\infty) - \frac{1}{T_0} \ln (P - p_0) \right]. \qquad (10.60)$$

3. Steady-state droplet evaporation in the presence of continuous irradiation. When a particle is subjected to a radiation flux constant in time, it can be assumed that the heat fluxes absorbed by the droplet and carried away together with the evaporating vapor are identical. This circumstance makes it possible to write the balance equation in the form

$$4\pi\lambda' (T_0 - T_\infty) + 4\pi r L (T) D [c_{\mathrm{v}} (T_0) - c_{\mathrm{v}\infty}] =$$
$$= \pi \mathcal{J} \omega K_{\mathrm{ab}}(r) r^2, \qquad (10.61)$$

where c_{v} is the water vapor concentration in the medium. Equation (10.61) can be used when determining T_0. The graphical solution of equation (10.61) is plotted in Figure 10.1.

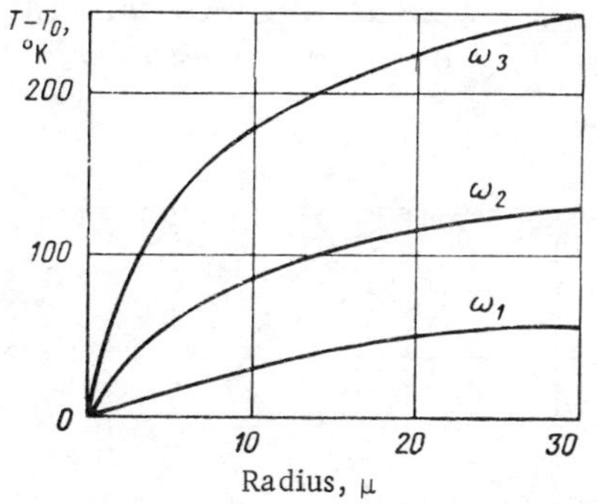

FIGURE 10.1 Computed variation in the temperature of droplets of different size continually irradiated by a radiant flux with $\omega_1 = 10^2$ W/cm², $\omega_2 = 10^3$ W/cm², $\omega_3 = 10^4$ W/cm², $T_0 = 273$°K, $\lambda = 10.6\,\mu$.

The rate of droplet evaporation can be estimated by integrating the equations of steady-state evaporation kinetics:

$$r\frac{dr}{dt} = -\frac{D}{\gamma_L}[c_v(T) - c_{v\infty}]. \qquad (10.62)$$

Approximate graphical integration of this equation using tabulated values of $c_v(T)$ for the same values of the parameters as in Figure 10.1 made it possible to obtain curves of the droplet radii as a function of time (Figure 10.2). These curves are approximated satisfactorily by the equation

$$r(t) = \begin{cases} r_0 - 0.494\omega^{1.102}t, & pr \gg 1, \\ r_0 e^{-0.494\omega^{1.102}}t, & pr \ll 1, \end{cases} \qquad (10.63)$$

where $p = 8\pi\varkappa/\lambda$; $q = \exp[-0.2(|m|-1)]$, and m is the complex refractive index of water.

4. Evaporation of droplets irradiated by single radiation pulses. If the droplet is acted upon by a single pulse with a duration for which the energy absorption by the particle can be treated as adiabatic, then the temperature of the droplet after being acted upon by the pulse is

$$T_i = T_\infty + \frac{1.8 \cdot 10^3 K_{ab}(r_0)\,E}{r_0}, \qquad (10.64)$$

where E is the energy of the pulse in joules per cm² and r_0 is the initial particle droplet in microns. Estimates show that, when

$T_i \leq 400°K$ (brought about by an individual pulse), the droplet evaporation does not substantially change its radius.

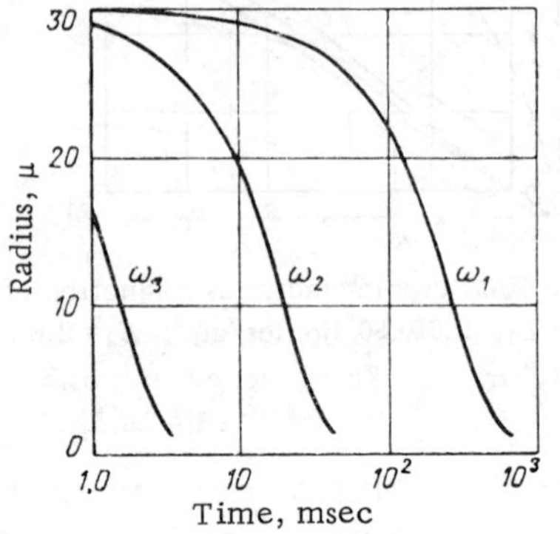

FIGURE 10.2. Computed droplet radius as a function of time during continuous irradiation of the droplet at 10.6μ and powers $\omega_1 = 10^2$ W/cm², $\omega_2 = 10^3$ W/cm², $\omega_3 = 10^4$ W/cm². Initial droplet radius $r_0 = 30\mu$.

There is no sense in considering values of $T_i \gg T_{cr} = 647°K$ for the liquid phase. Thus the effect of a single pulse on the droplet of interest is that for which temperature T_i satisfies

$$400° K \leqslant T_i \leqslant T_{cr}. \tag{10.65}$$

If the droplet temperature is assumed to be distributed uniformly over the volume, it can be found from the simple expression

$$C_L m\, dT = L\, dm, \tag{10.66}$$

from which it is easy to derive the dependence of the droplet temperature on its radius:

$$T = T_i + \beta \ln \frac{r}{r_0}, \tag{10.67}$$

where

$$\beta = \frac{6L(T_i)\, L(T_\infty)}{C_L [L(T_i) + L(T_\infty)]} \tag{10.68}$$

Equation (10.67) yields an expression for the final droplet radius after being acted upon by a single pulse:

$$r_f = r_0 e^{-(T_i - T_\infty)/\beta}. \tag{10.69}$$

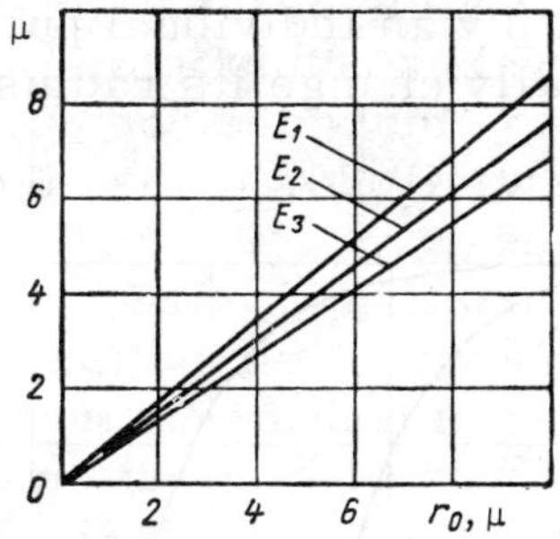

FIGURE 10.3. Computed final droplet radius as a function of the initial radius for pulse energies $E_1 = 2.00 \cdot 10^3$ joules/cm^2, $E_2 = 2.66 \cdot 10^3$ joules/cm^2, $E_3 = 2.93 \cdot 10^3$ joules/cm^2.

Figure 10.3 depicts the results of computing the final radius of a droplet irradiated by a single pulse at 10.6μ. It should be mentioned that, in general, the dependence of the final particle radius on its initial radius is not linear, since the difference $T_i - T_\infty$ and quantity β depend on r_0. It follows from equation (10.69) that the final droplet radius vanishes only for $T_i = T_{cr} = 647°K$, when $\beta = 0$, since $L(T_{cr}) = 0$.

The above expressions are valid when the pulse duration is smaller than the time during which the droplet size and temperature can change perceptibly. Estimates show that this time should be of the order of 10^{-4} sec.

When $T_i > T_{cr}$ one should expect explosion of the droplet, in which water vapor at high pressure impacts on the air and forms a compression wave. Estimates show that this wave equalizes the vapor pressure in the medium surrounding the droplet; at a distance of about $10\, r_0$ from the droplet the products of explosion give up their energy to the air. It is found that the time during which this phenomenon occurs is expressed by the simple formula

$$t = \frac{r_0}{U_x} \frac{10^3}{3} \text{ (sec)}, \tag{10.70}$$

where U_x is the propagation speed of the compression wave. For water droplets 10μ in radius the [droplet] breakup time is found to be of the order of 10^{-6} sec.

5. Irradiation of polydisperse aerosols. To determine the effect of radiation on polydisperse aerosols it is necessary to know the variation in the particle size distribution function which this effect induces. It can be assumed in the first

approximation that each particle of the polydisperse aerosol is affected independently. This assumption is justified by the fact that, in the majority of cases, the moisture content of aerosol particles per unit volume is substantially lower than the water content of air in which the aerosol resides.

Bearing the above assumption in mind it is easy to derive an expression for the particle size distribution function $f(r,t)$, which contains time. In the case of continuous irradiation of a polydisperse aerosol with initial particle size distribution function described by a gamma-distribution (see (5.4)), it is possible to obtain the following expressions for $f(r, t)$:

$$f(r, t)=\begin{cases} \frac{1}{\Gamma(\mu+1)}\mu^{\mu+1}\frac{1}{\bar{r}_0}\left(\frac{r+S}{\bar{r}_0}\right)^{\mu} e^{-\mu\left(\frac{r+S}{\bar{r}_0}\right)}, & (10.71) \\ \frac{e^{pS}}{\Gamma(\mu+1)}\mu^{\mu+1}\frac{1}{\bar{r}_0}\left(\frac{re^{pS}}{\bar{r}_0}\right) e^{-\mu\left(\frac{re^{pS}}{\bar{r}_0}\right)}, & (10.72) \\ pr \ll 1, \end{cases}$$

where μ is a parameter characterizing the half-width of the distribution, $S=0.494q\omega^{1.102}t$; $\bar{r}_0$ is the most probable particulate radius. The bulk extinction coefficient can be calculated by substituting formulas (10.71) and (10.72) into (7.3).

In the case of large particles, when $K(\rho, m) = 2$, the bulk extinction coefficient for $\mu = 2$ is given by

$$\alpha(\omega, t)=2\pi N_0(S^2+3\bar{r}_0S+3\bar{r}_0^2)\,e^{-2S/\bar{r}_0}, \tag{10.73}$$

where N_0 is the number of particles per unit volume. For small and "soft" particles

$$\alpha(\omega, t)=\frac{\pi N_0}{\Gamma(\mu+1)}\mu^{\mu+1}\times$$
$$\times\left[\frac{32\pi(n-1)(\tan\beta)\,r^{-3}\Gamma(\mu+4)}{3\lambda\mu^{\mu+1}}e^{-3pS}+\right.$$
$$\left.+\frac{8\pi^2(1-\tan^2\beta)\,\bar{r}_0^4\Gamma(\mu+5)}{\lambda^2\mu^{\mu+5}}e^{-4pS}\right], \tag{10.74}$$

where $\tan\beta=\frac{\varkappa}{n-1}$; n is the refractive index of water.

When a polydisperse aerosol is subjected to pulsed radiation, and when the relationship between the final and initial particle radii can be treated as linear, the overall particle size distribution function is altered, thus changing the value of the most probable particulate radius. In this case the bulk coefficient can be calculated from standard formulas, except that the most probable radius should be represented by quantity $\bar{r}_0 k$, where k is a parameter depending on the energy of the pulse and the optical cross-section of the particle.

This approximate theory of the evaporation of spherical particles subjected to powerful continuous, quasi-continuous and pulsed optical radiation represents a first attempt at a quantitative description of this quite complicated phenomenon. The degree of its reliability can be determined from pertinent experimental studies, the initial results of which were obtained by Bukatyi et al. /727, 728/. However, the light sources used in these studies were too weak and so definite conclusions could not be reached on agreement between theoretical and experimental data.

It should also be noted that droplet evaporation is by far not the only effect accompanying the propagation of powerful optical radiation in a polydisperse water aerosol. When a water aerosol is subjected to high-power and short-duration light pulses, the aerosol particles acquire directional motion, induced by their nonuniform heating. This motion can result in droplet coalescence.

When the radiation incident on the aerosol is powerful, it may become necessary to make allowance for the radiation pressure effect, which can be computed rather easily. Unsteady evaporation of water aerosol particles can apparently result in the appearance of conditions conducive to defocusing and anomalous refraction of the directional radiation in the beam channel.

Study of the above effects and their influence on droplet evaporation is a task for the future. At present it can only be stated that, from the standpoint of energy losses of high-power optical radiation in a water aerosol, the droplet evaporation effect is predominant if no consideration is given to losses of the directional radiation flux associated with scattering at the particles.

11. *PROPAGATION OF TIGHT LIGHT BEAMS IN A TURBULENT ATMOSPHERE*

A light wave propagating in a real atmosphere experiences not only energy losses due to absorption and scattering of radiation, but also amplitude and phase fluctuations due to the random space-time distribution of the refractive index of air. The latter is due to turbulent motion in the atmosphere which brings about random variations in the temperature and in the refractive index. Estimates show that a 1°C variation in air temperature alters the refractive index by the order of 10^{-6} /729/. The amplitude of observed air temperature fluctuations at a given point may be as high as tenths of a degree; the fluctuation period varies from milliseconds to seconds. The amplitude of temperature fluctuations along horizontal paths in the atmosphere can reach several degrees for points located at distances of the order of 10^2–10^3 m /730/.

The effect of space-time fluctuations in the refractive index of the atmosphere on the characteristics of infinite plane and spherical waves has been studied quite extensively both theoretically and experimentally. The books by Tatarskii /463, 731/ contain a systematic presentation of the theory and an analysis of experimental results. However, the theoretical and experimental data obtained for unrestricted beams cannot, in general, be employed directly for quantitative estimates of the corresponding parameters of tight beams. This requires new theoretical and experimental studies, the results of which are of great applied significance in connection with the employment of tight laser beams in communications, information transmission, telemetry, and other systems intended for operation through the atmosphere.

This chapter examines the principal results of investigations into the effect of a turbulent atmosphere on the parameters of tight light beams under different propagation conditions.

1. GENERAL REMARKS

As a result of turbulent motion, inhomogeneities are produced in the atmosphere the length of which varies from hundreds of meters (macroinhomogeneities) to dimensions l_0 of order given by

$$l_0 \sim \sqrt[4]{\nu^3/\varepsilon}; \tag{11.1}$$

quantity l_0 is the inner turbulence scale, ν is the kinematic viscosity, ε is the rate of thermal dissipation of energy by the inhomogeneity. Dimension l_0 near the surface is of the order of 1 mm.

The refractive index n of a medium with random inhomogeneities is usually characterized by the structure function /463/

$$D_n(r) = \overline{[n(r_1 + r) - n(r_1)]^2}, \tag{11.2}$$

where r is the distance between the relevant points. We similarly introduce a structure function for the temperature field:

$$D_T(r) = \overline{[T(r_1 + r) - T(r_1)]^2}. \tag{11.3}$$

The form of functions $D_n(r)$ and $D_T(r)$ in the inertial interval /731/ on the assumption of locally isotropic and homogeneous turbulence was derived by Obukhov /732/. The corresponding equations are termed the Kolmogorov-Obukhov two-thirds law:

$$D_n(r) = C_n^2 r^{2/3}, \tag{11.4}$$
$$D_T(r) = C_T^2 r^{2/3}, \tag{11.5}$$

where $l_0 < r < L_0$, L_0 being the outer turbulence scale.

Quantities C_n and C_T are the structure constants of the refractive index and of the index of air temperature fluctuations, and are related for the 0.5μ wavelength by the expression

$$C_n(\text{cm}^{-1/3}) = \frac{79.2 \cdot 10^{-6} P\,(\text{mb})}{T^2\,(^\circ\text{K})} C_T(\text{deg} \cdot \text{cm}^{-1/3}). \tag{11.6}$$

The values of C_n in the surface layer of the atmosphere lie between 10^{-7} and 10^{-9} cm$^{-1/3}$ /741/. Methods for determining the structure constant of the refractive index employ data on the average values of meteorological characteristics /733, 846/ and data on temperature fluctuations /847/. Optical methods for determining C_n have recently been developed /848–850/.

The rms fluctuations of the logarithm of intensity of an infinite plane monochromatic wave propagating in a turbulent atmosphere is given by

$$\sigma^2 = 2.24k^{7/6} \int_0^L C_n^2 (L-x)^{5/6}\, dx, \tag{11.7}$$

where k is the wave number and L the distance traversed by the wave in the atmosphere.

In the case of an atmosphere which is homogeneous from the standpoint of the turbulent regime, C_n^2 does not vary over the entire path L and so expression (11.7) reduces to

$$\sigma_0^2 = 1.23 C_n^2 k^{7/6} L^{11/6}. \tag{11.8}$$

The value of C_n depends on the altitude of the atmospheric layer:

$$C_n^2(z) = C_n^2(z_0)\left(\frac{z}{z_0}\right)^{-\alpha}, \tag{11.9}$$

where z_0 is a fixed altitude ($z > z_0$); $\alpha = 2/3$ in the presence of dynamic turbulence in the atmosphere and $\alpha = 4/3$ in the case of free convection /463/.

Formulas for the variance of fluctuations in the logarithm of the intensity describe correctly the case of so-called weak fluctuations, defined by the condition $\sigma \ll 1$. Tatarskii /851/ solved the wave equation in the approximation of geometrical optics and obtained an expression for the rms fluctuations in the logarithm of the amplitude σ_x^2 in the case of strong fluctuations:

$$\sigma_x^2 = 1 - [1 + 6\sigma_0^2]^{-1/6},$$

which is in satisfactory agreement with experimental data /751/, although the applicability conditions of geometrical optics were not satisfied during the experiment.

The rms fluctuation in the angle of incidence of an infinite plane wave in the focal plane of the objective is /463/

$$\sigma_\varphi^2 = 2.8b^{-1/3} \sec\theta \int_{z_0}^{\infty} C_n^2(z)\, dz, \tag{11.10}$$

where θ is the zenith distance of the source, z_0 the altitude at which the receiving objective is located, and b the objective diameter.

Aircraft measurements /734/ show that equation (11.9) with $\alpha = 4/3$ is valid up to altitudes at which cumulus forms. If this altitude is taken as infinite, then equations (11.10) and (11.9) yield a simple equation for σ_φ^2 /735/:

$$\sigma_\varphi^2 = 2.8 b^{-1/3} \sec\theta\, z_0 3 C_n^2(z_0). \qquad (11.11)$$

Quantity σ_φ^2 is related to the frequency spectrum of fluctuations in angles of arrival by the expression

$$\sigma_\varphi^2 = \int \omega_\varphi(f)\, df. \qquad (11.12)$$

Golitsyn et al. /736/ derived an expression for the frequency spectrum of fluctuations in angles of arrival in the case of a homogeneous turbulent atmosphere:

$$\omega_\varphi(f) = 0.045 \sigma_\varphi^2 \sin^2\left(\frac{\pi b f}{v_1}\right)\left(\frac{b}{v_1}\right)^{-5/3} f^{-8/3}, \qquad (11.13)$$

where f is the frequency, and v_1 is the wind velocity component perpendicular to the direction of wave propagation for the frozen-turbulence model.

2. THEORETICAL STUDIES OF FLUCTUATIONS IN THE PARAMETERS OF TIGHT BEAMS PROPAGATING IN THE ATMOSPHERE

Kon and Tatarskii /737/ obtained an expression for calculating the fluctuations in phase and the effective spreading of a tight beam propagating in a turbulent atmosphere. The problem was solved in the Rytov-Obukhov approximation of smooth perturbations. This consists in the following. The expression for the logarithm of the perturbation of a small field ψ satisfies the wave equation and is expressed in terms of the characteristics of the incident and perturbed fields in the following manner:

$$\psi_1 = \psi - \psi_0 = \ln\frac{A}{A_0} + i(s - s_0) \qquad (11.14)$$

(A, A_0 and s, s_0 are, respectively, the amplitude and phase of the perturbed and incident fields). In this expression we substitute the value of the unperturbed field, the amplitude of which is distributed over the beam cross-section according to the law

$$A_0 = A_0' \exp\left(-\frac{y^2+z^2}{a^2}\right), \tag{11.15}$$

where A_0' is the amplitude of the incident wave on the beam axis, a is the beam dimension, y and z are coordinates in the plane of the cross-section of a beam propagating in the x direction. Then one writes an expression for the structure functions of phase fluctuation D_s, which contains the quantity $s_1 = \mathrm{Im}\psi_1$. The integrals contained in the expression for D_s are calculated approximately using the conditions

$$g = 2L/ka^2 \ll 1 \text{ (or } a^2 \gg \lambda L), \tag{11.16}$$
$$a^4 \gg \lambda^3 L^3 l_0^2. \tag{11.17}$$

Only fluctuations in the direction of propagation are considered here, in connection with which one uses the value of D_s for $|\vec{\rho}| \ll l_0$, where $\vec{\rho} = \vec{\rho}_1 - \vec{\rho}_2$, $\vec{\rho}_1$ and $\vec{\rho}_2$ being the radius vectors of points in the plane $x = L$. For D_s one obtains in the approximation of geometrical optics ($\sqrt{\lambda L} \ll l_0$) a formula differing little from the corresponding expression for the case of an infinite plane wave. After deriving an expression for D_s when $\sqrt{\lambda L} \gg l_0$ (a case of practical importance), one introduces the parameter θ given by

$$\theta^2 = 2L^2\varkappa_m^2/k^2a^2, \tag{11.18}$$

where $\varkappa_m$ is a wave number characterizing the inner turbulence scale. Parameter θ characterizes the ratio of the width of the diffracted beam to the geometrical width of the beam.

In the asymptotic case $\theta \gg 1$ the following expression is obtained for structure function D_s:

$$D_s(\rho, R) = 0.46k^3aC_n^2k_m^{-2/3}\rho^2 \ln 2\theta^2 \times \\ \times\left[2\,{}_1F_1\left(\frac{1}{6}, 1, \frac{2R^2}{a^2}\right) - {}_1F_1\left(\frac{1}{6}, 2, \frac{2R^2}{a^2}\right)\right], \tag{11.19}$$

where ${}_1F_1$ is a degenerate hypergeometric function; R is the numerical value of vector $\vec{R} = \frac{1}{2}(\vec{\rho}_1 + \vec{\rho}_2)$ (vectors $\vec{\rho}$ and $\vec{R}$ are assumed to be parallel; the line connecting observation points $\vec{\rho}_1$ and $\vec{\rho}_2$ passes through the beam center). Equation (11.19) differs substantially from the corresponding expression for D_s in the case of an infinite wave.

With the aid of (11.19), an approximate expression was found /737/ for the effective rms fluctuation in the angles of incidence:

$$\sigma_{\varphi\,\mathrm{eff}}^2 \simeq 0.1 C_n^2 k a l_0^{2/3} \ln 2\theta^2, \tag{11.20}$$

from which follows the conclusion that at large θ (large L) the growth of $\sigma_{\varphi\,\mathrm{eff}}^2$ slows down sharply with the range (becomes proportional to $\ln L$ rather than to L, as in the case of an infinite wave). This situation arises because, in the case of a tight beam, the scattered waves leave the beam at L and no longer contribute to the fluctuations. At $\theta = 10$ (beam width 50 cm, $l_0 = 1$ mm, $L = 5$ km, $\lambda = 0.5\mu$) σ_φ^2 differs fourfold from the value calculated from (11.10) for an infinite wave. Note that $\sigma_{\varphi\mathrm{eff}}^2$ is taken to express the value of σ_φ^2 averaged over the beam cross-section with allowance for amplitude distribution (11.15).

An expression for the linear spread of the beam as a function of the distance it has traversed in the turbulent medium has also been obtained /737/. The random spread of the beam over path L is equal to

$$R = \int_0^L \Delta\varphi(x)\,dx. \tag{11.21}$$

Assuming that the longitudinal correlation coefficient for $\Delta\varphi(x)$ is close to unity /738/ and employing (11.20), Kon and Tatarskii /737/ derived R_{eff} in the form

$$R_{\mathrm{eff}} = \sqrt{\langle R^2 \rangle} = \mathrm{const}\, C_n l_0^{1/3} (ka \ln \theta)^{1/2} L, \tag{11.22}$$

which shows that at high L ($\theta \gg 1$) R_{eff} is proportional to L, rather than to $L^{3/2}$, as in the case $\theta \ll 1$.

Calculations of amplitude fluctuations of tight beams showed that the corresponding expression for σ^2 differs little (of the order of g^2) from (11.17) in the case of an infinite wave.

Equation (11.15) describing the distribution of wave amplitude over the beam cross-section does not always correspond to experimental conditions. However, Kon and Tatarskii /737/ emphasize that the main qualitative conclusions of their study remain valid also for other forms of function $A_0(y, z)$.

The spread of a tight beam propagating in a turbulent atmosphere was examined by Feizulin and Kravtsov /739/, who treated the diffusion of the beam, as well as the diffraction divergence and oscillation as a whole about the direction of propagation. The total spread of the beam is

$$\overline{\langle\rho^2\rangle} = \overline{\Delta\rho^2} + \overline{\rho_c^2}, \tag{11.23}$$

where

$$\langle\overline{\rho^2}\rangle = \frac{\int_S \rho^2 \overline{I_s(\vec{\rho})}\, d\vec{\rho}}{\int_S I_s(\vec{\rho})\, d\vec{\rho}}; \tag{11.24}$$

$$\overline{\rho_c^2} = \frac{\int_S \int (\vec{\rho}_1, \vec{\rho}_2)\, I_s(\vec{\rho}_1)\, I_s(\vec{\rho}_2)\, d\vec{\rho}_1\, d\vec{\rho}_2}{\left(\int_S I_s(\vec{\rho})\, d\vec{\rho}\right)^2}. \tag{11.25}$$

The term $\overline{\rho_c^2}$ characterizes the oscillation of the beam as a whole about the direction of propagation and $\overline{\Delta\rho^2}$ defines the beam spread relative to the perturbed location of its center. Expressions were found /739/ for the average field intensity $I_s(\rho)$ at a point with radius vector $\vec{\rho}$ and the intensity correlation function $\overline{I_s(\vec{\rho}_1)\, I_s(\vec{\rho}_2)}$ in plane S perpendicular to the direction of beam propagation. The propagating wave is regarded as plane. The statistical characteristics contained in the sought expressions are calculated by the method of smooth perturbations /463/.

Consider now one of the terms of the equation obtained in /739/ for $\overline{\langle\rho^2\rangle}$ characterizing beam propagation due to scattering during fluctuations in the refractive index of air in the case of isotropic turbulence:

$$\frac{L^2}{2k^2}\Delta D_\psi(0) = 2.19 C_n^2 l_0^{-1/3} L^3, \tag{11.26}$$

where D_ψ is the structure function of the complex phase.

The expression for $\langle\overline{\rho^2}\rangle$ simplifies markedly in the case of a Gaussian field distribution in the beam cross-section:

$$\overline{\langle\rho^2\rangle} = a^2 + \frac{L^2}{k^2 a^2} + 2.19 C_n^2 l_0^{-1/3} L^3. \tag{11.27}$$

It does not follow from this equation that the rate of beam-width growth slows down with range at high L, as concluded in /737/. The authors of /739/ attribute the difference in the results to the different measures of beam width used in the two cases.

Feizulin and Kravtsov /739/ then study two specific cases of oscillation of the "center of gravity" of the beam: large-scale inhomogeneities, when the beam dimensions are much smaller than the dimensions of the inhomogeneities, and small-scale inhomogeneities (dimensions of inhomogeneities much smaller than the beam width). It is shown that in the first case the beam does not widen but only shifts as a whole about the direction of propagation. In the second case quantitative analysis is difficult. It follows from qualitative considerations that the shift of the "center of gravity" of the beam in this case should be insignificant.

Values of the total beam spread were determined /739/ for the following data: $a = 35.7$ cm, $l_0 = 0.1$ cm, $\lambda = 0.5\mu$, $C_n = 0.02 \times \times 10^{-6}$ cm$^{-1/3}$, and ranges of $L_1 = 15$ km and $L_2 = 145$ km. The determined values were found to be, respectively, about 91 cm and 25.5 m.

We now mention briefly other theoretical studies. Qualitative data concerning the theory of fluctuations in the parameters of a laser beam propagating in a turbulent atmosphere were analyzed by Davis /741/ with reference to the operation of laser systems. A detailed calculation of the effect of atmospheric calculations on optical communications lines was carried out elsewhere /744/, while the variance of the logarithm of the amplitude of a laser beam propagating in the atmosphere along a horizontal path was estimated by Fried and Seidman /745/.

A statistical theory governing the transfer of optical images through a turbulent atmosphere has been constructed /750/, and an expression obtained for the optical transfer function characterizing the coefficient of transfer of spatial frequencies through the atmosphere:

$$M(\lambda R f) = e^{-\frac{1}{2}k^2\langle[S(p)]^2\rangle},$$

where $f = p/\lambda R$ is the spatial frequency, p are the linear dimensions of the object under observation, R is the focal length of the receiving lens, and

$$\langle S^2\rangle = 2.91 p^{5/3} \int_0^z C_n^2(z')\,dz'.$$

The angle brackets denote sufficiently extensive time-averaging.

Fried /742, 743/ studied theoretically the effect of the turbulent state of the atmosphere both on the optical resolution (determined by the optical transfer function) for long and short exposures, and on the limiting resolution of images obtained (by photography) looking down through the atmosphere.

A number of quantitative estimates of the effect of atmospheric turbulence on the parameters of tight light beams was also obtained elsewhere /740, 746–749, 752/. Estimates of the limiting dimensions of the radiation source and collector for which no departure occurs from spatial coherence of the light beam due to atmospheric turbulence were obtained by Fried /740/. Consortini et al. /749/ estimated the limiting dimensions of the diameter of the light beam cross-section outside of which turbulent motion in the atmosphere substantially reduces the spatial coherence of the beam. Fried /740/ derived an expression for the normalized variance of intensity fluctuations on averaging over the aperture, which is termed the aperture-averaging factor. The dependence of the source scintillation intensity on its angular dimensions has also been investigated /752/.

3. EXPERIMENTAL STUDIES OF FLUCTUATIONS IN THE PARAMETERS OF TIGHT BEAMS PROPAGATING IN A TURBULENT ATMOSPHERE

1. Intensity fluctuations. Measured fluctuations in the intensity of tight laser beams propagating through a turbulent atmosphere depend on meteorological conditions, optical length of the beam path, diameter of the source and receiving apertures, time constant of the receiver, measuring time (exposure), emitted wavelength, and beam focusing.

The quantitative measure of fluctuations in the intensity I of a beam is usually represented by the quantity

$$\sigma^2 = \overline{(\ln I - \overline{\ln I})^2}, \tag{11.28}$$

or /759/

$$M^2 = \overline{\overline{I^2(t)}/\overline{I(t)}^2}, \tag{11.29}$$

$$\overline{I(t)} = \lim_{T\to\infty} \frac{1}{2T} \int_{-T}^{T} I(t)\,dt;$$

$$\overline{I^2(t)} = \lim_{T\to\infty} \frac{1}{2T} \int_{-T}^{T} [I(t) - \overline{I(t)}]^2\,dt, \tag{11.30}$$

and the standard deviation of the measured signal, referred to its average value,

$$\sigma_n = \sqrt{M^2}. \tag{11.31}$$

The experimentally measured quantities are

$$\bar{I}=\frac{1}{N}\sum_{i=1}^{N} I_i, \tag{11.32}$$

$$\overline{I^2}=\frac{1}{N}\sum_{i=1}^{N} I_i^2, \tag{11.33}$$

$$\sigma_k^2=\frac{\overline{I^2}-\bar{I}^2}{\bar{I}^2}. \tag{11.34}$$

When the fluctuations in intensity I follow a lognormal distribution, σ_k^2 and σ^2 are related by the expression

$$\sigma^2=\ln(1+\sigma_k^2). \tag{11.35}$$

A general idea about the nature and amplitude of signal intensity fluctuations is provided by Figures 11.1 and 11.2, which show recordings of signals /751, 766/.

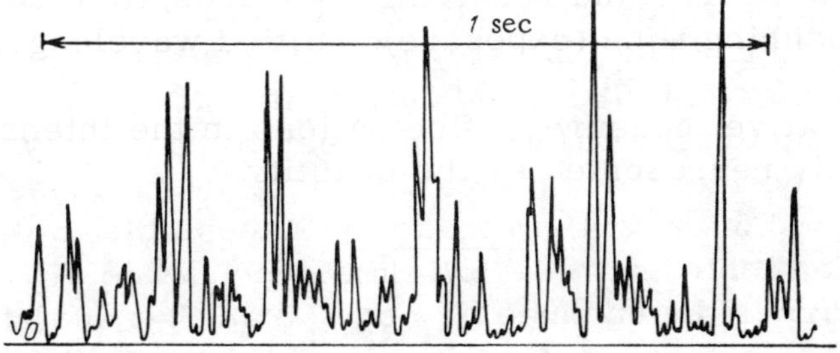

FIGURE 11.1. Example of a recording of fluctuations in a signal propagating through the atmosphere /751/.

Figures 11.3 through 11.5 present relationships between σ and the distance between the source and receiver, the divergence angle of the laser beam and the diameter of the receiving aperture /755/. All the curves are drawn through points corresponding to arithmetic mean values of σ; vertical line segments designate the range of values of σ in individual measurements, due to measuring errors and variations in prevailing meteorological conditions. All the curves were obtained using a He–Ne laser with $\lambda = 6328$ Å, operating in the continuous mode. The measurements were made at night.

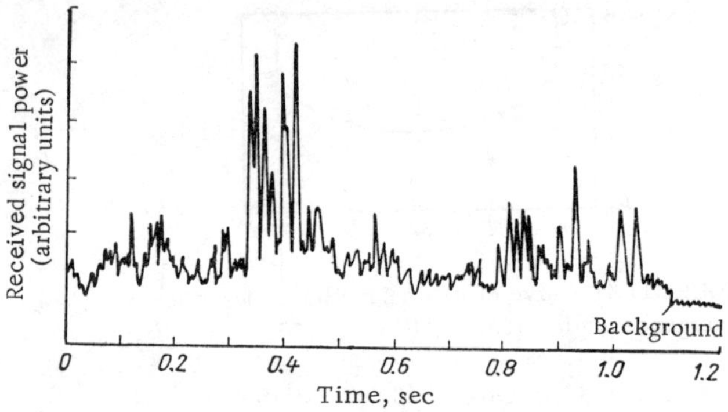

FIGURE 11.2. Example of the time-dependence of a signal propagating in the atmosphere /766/.

Figure 11.3 was constructed on the basis of measurements in the atmosphere over paths 0.1, 0.25 and 1.2 km long. A flat rotating mirror was placed at the ends of each of the paths. The light beam during these measurements thus traversed distances of 0.2, 0.5 and 2.4 km. The beam diameter at the output of the collimating system was 14.4 cm; the divergence angle was 10'' for $L = 0.2$ and 0.5 km and 17'' for 2.4 km. The diameter of the receiving aperture was 4 mm for $L = 0.2$ and 0.5 km, and 14 mm for $L = 2.4$ km. Over this range of path lengths the value of σ is seen to change by approximately 50%. The fact that no measurements were made for path lengths between 0.5 and 2.4 km makes it impossible to definitely resolve the question of saturation in the dependence of σ on L. Note that a sharper dependence of σ on L than that depicted in Figure 11.3 was detected by Gracheva and Gurvich /751/. Buck /766/ obtained saturation of σ at a range of 0.55 km.

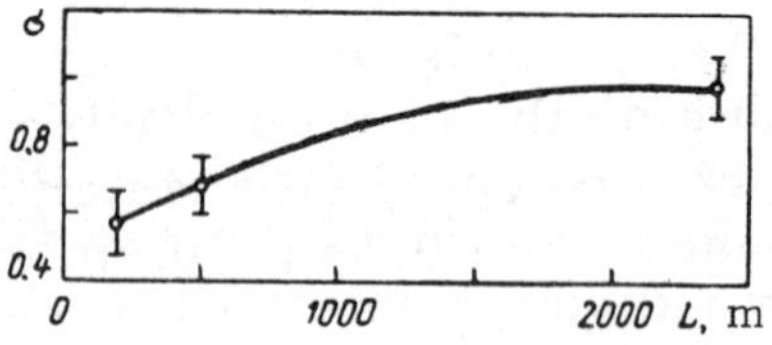

FIGURE 11.3. Curve of σ as a function of distance L between the source and receiver /755/.

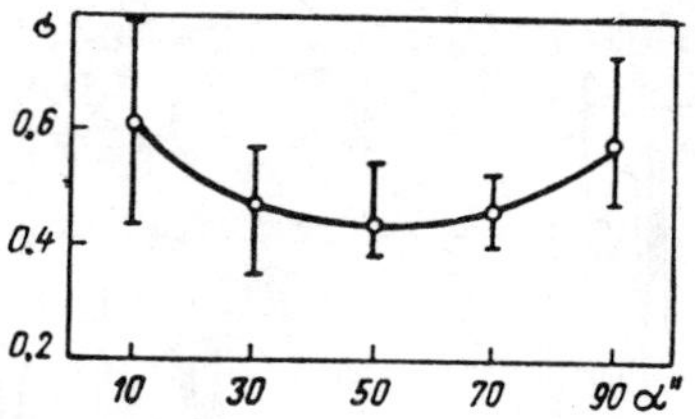

FIGURE 11.4. Curve of σ as a function of laser beam divergence angle /755/.

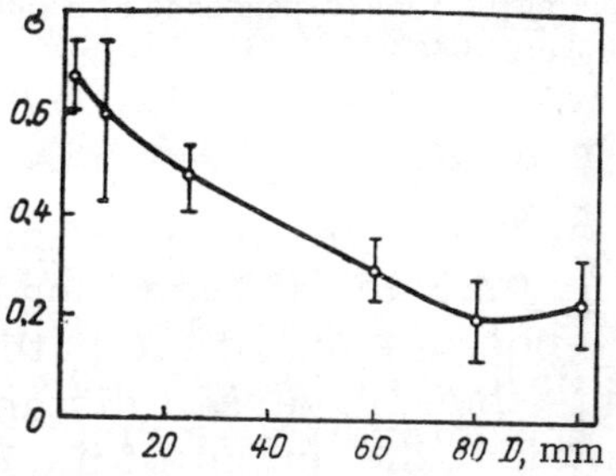

FIGURE 11.5 Curve of σ as a function of the diameter of receiving aperture /755/.

In measuring the dependence of σ on the angular beam divergence, the latter ranged from 10'' to 90'' for a 12.5 cm beam diameter leaving the collimating system, 500 m measurement distance, and 1.6 mm diameter of the receiving aperture. It is seen from Figure 11.4 that when the beam opening angle α is increased from 10'' to 50'' the value of σ decreases; as α is increased further, σ increases.

When measuring the dependence of σ on the receiving aperture diameter *D* (see Figure 11.5), the latter was set at 0.2, 0.8, 2.4, 6.0, 8.0 and 10.0 cm, $L = 0.5$ km, the beam diameter leaving the collimating system was 12.5 cm, and the angular divergence was 70''. When *D* was increased from 0.2 to 8.0 cm the value of σ decreased approximately threefold; when $D = 10$ cm, the measured value was found to be somewhat greater than when $D = 8$ cm.

Detailed investigations on the dependence of a gas laser beam intensity fluctuations at wavelength 0.63μ (beam diameter of 3.5 mm at the 0.7 level) on the receiving aperture diameter over an 8-km path were conducted by Fried et al. /768/. The results show that σ^2 decreases smoothly as *D* is increased from 0.1 to 10 cm.

Subsequent increases in diameter to 1 m have virtually no effect on the variation in σ^2. The lognormal intensity distribution was found to apply to all values of D. Calculations of the variance of intensity fluctuations for different values of D /854/ employing formulas of the existing theory /731/ were found to contradict the experimental data obtained by Fried /854/ at high D. Fried et al. /768/ offer a hypothesis, that there exist two independent effects which determine the magnitude of the variance of fluctuations in the intensity of a beam propagating in a turbulent atmosphere. One of these effects produces a normal intensity distribution at high D, while the other is responsible for a lognormal distribution. However, here the dominating contribution is that of the latter effect.

It is noteworthy that the nonconformance between the theory of weak fluctuations and experimental results at high D was also detected in experimental investigations carried out in the author's laboratory /754/. The measurements were carried out with a receiving aperture of 70 cm. The exit diameter of the light beam and its divergence were 25 cm and 30', respectively. The value of the fluctuation variance in these measurements was found to be greater than predicted; the intensity distribution corresponded to a lognormal law. In addition, within the limits of measurement error, the variance of the fluctuations was not found to depend on the emitted wavelength.

Buck /766/ found (in measurements with large D, when the entire beam is recorded by the receiver) that σ is smaller than 0.02, at which value the signal can be regarded as virtually stable and independent of time. This circumstance pointed to the smallness of signal fluctuations brought about by a corresponding variation in molecular absorption and aerosol scattering during measurements.

The intensity distribution of laser beams propagating at 6328 Å over different distances in the atmosphere was investigated by Höhn /767/ and Khmelevtsov and Tsvyk /755/. In the first study measurements were made over path lengths of 4.5 and 14.5 km with a 0.35 cm beam diameter entering the atmosphere. The beam used in the second study had an initial diameter of 1.4 cm and was employed over lengths of 0.2, 0.5, 2.4 and 7.0 km. Sixty-eight intensity distributions were processed in /767/ and 28 in /755/, of which respectively 51 and 23 were found to be lognormal.

An investigation of fluctuations in the intensity of three lasers with different wavelengths (0.63, 3.5 and 10.6μ) was carried out by Chu /855/. The measurements were made over a 2.6 km path at D equal to 5, 12.5 and 15 cm, respectively, for the above three wavelengths. The frequency spectrum of intensity fluctuations was investigated by a spectrum analyzer. Exponential reduction of the signal with frequency was found for frequencies up to 100 Hz.

At higher frequencies this reduction is slowed down, and is attributed to the possible effect of terrestrial noise. The dependence of intensity fluctuations on emitted wavelength was found to be slower than $\lambda^{-1/2}$, which follows from the thoery of the frozen turbulence model. This result cannot be explained by differences in the value of *D* when measuring at different wavelengths.

Höhn /769/ studied frequency spectra of intensity fluctuations of a laser beam ($\lambda = 6328$ Å) with an initial diameter of 0.35 cm for path lengths of 4.5 and 14.5 km. He found that the normalized spectrum fluctuation of light intensity *U* (*f*), where *f* is the frequency, depends little on the value of *D*, which ranged in the experiments from 0.5 to 8.0 cm. He also found data on the characteristic frequency of fluctuations or on the frequency corresponding to the maximum of curve *U* (*f*). The dependence of this frequency on the distance and on the average wind speed component perpendicular to the direction of beam propagation differs from that predicted by theory, as well as from the relationship following from the empirical formula obtained by Tatarskii /463/. Satisfactory agreement of this relationship was obtained with pertinent experimental data of Ryznar /761/.

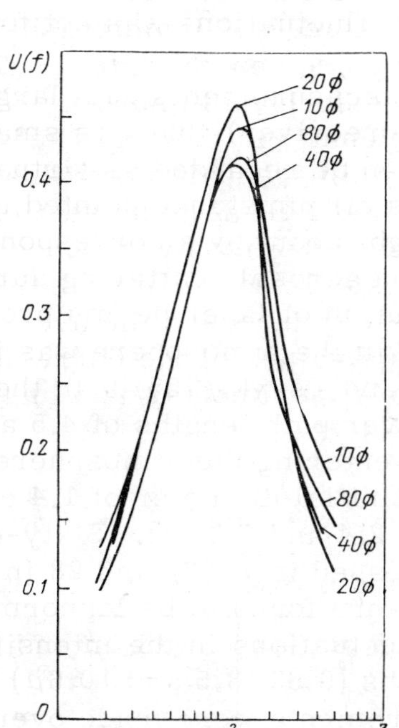

FIGURE 11.6. Normalized spectra *U* (*f*) for different receiving aperture diameters /769/.

Figure 11.6 displays spectra of $U(f)$ for different values of D obtained from one series of measurements /769/. The values of frequency f_m corresponding to the maximum of $U(f)$ were varied from one measurement to another over a range of approximately 40–140 Hz. The half-width of the distribution ranged from 4 to 18, the latter being defined as

$$\gamma_{1/2} = f_2/f_1, \tag{11.36}$$

where f_1 and f_2 are frequencies satisfying the condition

$$U(f_2) = U(f_1) = \frac{1}{2} U(f_m). \tag{11.37}$$

An investigation into the spatial structure of a pulsed laser beam in a turbulent atmosphere was carried out in the author's laboratory /770/. The measurements were made with a ruby laser ($\lambda = 0.69\mu$) operating in the free-generation mode (single pulses of approximately 0.5 μsec duration, output of several hundred watts, beam spread 6'). The structure of laser emission was studied at distances of 0.25, 1.2, 3.5 and 10 km from the source. The corresponding patterns of this structure were photographed from screens placed at these distances. Photographs with normal density were obtained by using neutral filters. The sections of the images thus obtained were scanned in such a manner so as to have at least 10 readings of intensity in the region of the theoretical value of the correlation radius, equal approximately to $\sqrt{\lambda L}$ /731/. From 2 to 6 longitudinal and 1 transverse section of each frame were scanned. The scanned sections were processed on an electronic computer, which calculated the correlation functions of intensity fluctuations, the variance, and the average value of the intensity. Table 11.1 lists values of the correlation radius determined from processing of experimental data and from calculations using the expression $\rho = \sqrt{\lambda L}$.

TABLE 11.1. Values of correlation radius ρ determined experimentally and theoretically

Distance L, km	Experimental values of ρ, cm			Theoretical value of ρ, cm	Number of scanned sections
	minimum	maximum	average		
0.25	4.7	17	9.1	1.3	37
1.2	4	12.8	10	2.9	23
3.5	3	14	9.5	4.9	15
10.0	–	–	8.1	8.3	52

Analysis of the above experimental data from the standpoint of the intensity distribution shows that for distances of 0.25, 1.2 and 3.5 km, in all cases, no deviations were observed from the normal distribution of the value of the natural logarithm of intensity. In this case equation (11.35) can be employed for determining σ^2 from experimental data. For measurements along a 10-km path the intensity distribution was not found to be lognormal.

In the case of a lognormal distribution we can determine σ^2 as a function of path length and beam spread. It was found that σ^2 increases up to $L = 3.5$ km and then drops slowly. The curve of σ^2 as a function of the beam divergence angle φ has a minimum, corresponding to a value of φ equal to approximately 50''. Note that φ is here identified with the angle of beam divergence, which is determined from the dimensions of the beam spot and length when measuring over short paths.

Experimental data on fluctuations in the intensity of tight light beams were also obtained elsewhere /753, 758, 760, 762–765, 892/. The intensity of fluctuations and their frequency spectrum for a gas laser beam ($\lambda = 0.63\mu$) as a function of receiving aperture diameters from 3.5 to 50 cm was investigated by Mevers et al. /758/. Fluctuations in the polarization of the radiation of the same laser, associated with intensity fluctuations, were investigated for 5-mile paths in the surface layer of the atmosphere /765/. An explanation of the nature of the detected polarization fluctuations requires careful additional studies, in which it would be possible to decompose (with sufficient accuracy) the measured fluctuations into components due to atmospheric aerosol and turbulent inhomogeneities in the atmosphere.

Fluctuations in the intensity of a searchlight at distances from 1 to 30 km were investigated by Siedentopf and Mayer /892/. The distribution of the logarithm of the intensity fluctuations was found to be approximately normal. Also studied were the frequency spectrum of intensity fluctuations and the degree of modulation of radiation by the atmosphere for various values of D between 13 and 4 cm /892/. The degree of modulation is defined as the ratio of the average peak-to-peak amplitude of signal fluctuations to its average value. It was found that the degree of modulation is inversely proportional to D. At low values of the latter saturation was observed in the relationship under study.

2. Variation in beam diameter with distance. Turbulent motion in the atmosphere induces widening of the beam diameter, in addition to the widening induced by geometric and diffraction spreading. This problem has not yet been investigated to the required degree.

Some data were obtained /757, 766, 770/ and can be used to obtain some idea about possible spreading of laser beams under different atmospheric conditions. Spreading of 8'' and 13'' was observed during measurements over paths of lengths 15 and 145 km, respectively /757/. On the basis of measured beam diameters D at different distances up to 145 km, with initial diameter of 10 cm and under night conditions for a slightly turbid atmosphere, the following empirical expression for D was obtained by Buck /766/:

$$D = \alpha L^{\beta}, \tag{11.38}$$

where $\alpha = 4.5 \cdot 10^{-6}$, $\beta = 1.2$. D is in cm, and L is in meters. It is noted /766/ that under hazy conditions and during daylight hours the beam spreads to several times the values following from this equation.

Buck /766/ also studied the dependence of the diameter of the focused beam on the value of its initial diameter. On propagation in a vacuum these quantities are inversely proportional. Measurements in turbulent atmosphere showed that reduction in the diameter of the focused beam compared with the diameter of the source aperture occurs only up to values of the latter equal to 11 cm. As the initial beam diameter is increased further, focusing does not improve. Figure 11.7 shows the experimentally obtained relationship (on the upper curve, the length of the vertical segments characterizes the scatter of experimental data). The lower curve in this figure pertains to the case of propagation of a focused beam in vacuum. The slight upswing in the upper curve at source apertures in excess of 11 cm is attributed to aberration of the optical system.

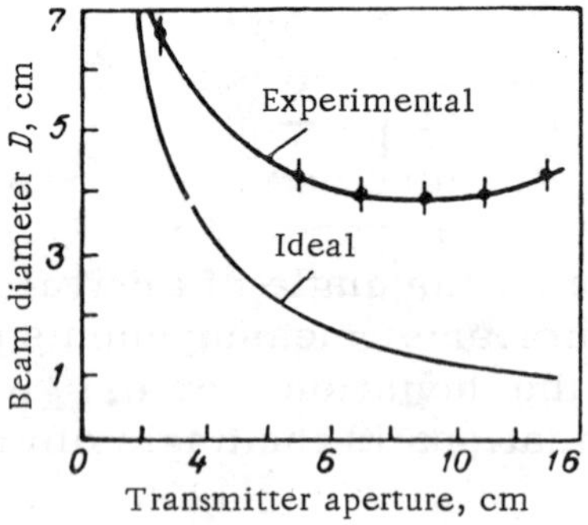

FIGURE 11.7. Diameter of received beam as a function of its initial diameter /766/.

Different linear dimensions of the spot produced by a pulsed ruby laser during measurements at the same distance have been observed /770/. For example, at a distance of 10 km the linear dimensions of the beam cross-section ranged approximately from 0.7 to 2 m in pulses following one another at 45-sec intervals.

3. Fluctuations in the angle of arrival of a tight radiation beam. Fluctuations in the angle of arrival of radiation, by means of which the image is formed in the focal plane of the objective, induce image dancing and blurring, Image dancing occurs when the dimensions of inhomogeneities, induced by turbulent motions in the atmosphere, are greater than the diameter of the objective. Small-scale inhomogeneities shift individual elements of the radiation beam and induce blurring of the image.

Experimental investigations into fluctuations of the angles of arrival of a gas laser beam propagating in the surface layer of the atmosphere were carried out in the author's laboratory /756/. The measurements were conducted at a distance of 1.2 km with He–Ne lasers (emitted wavelength 6328 Å) generating continually axial oscillations, and a thermal source, the luminescent body of which was limited by appropriate diaphragms. The initial beam diameter d_0 and its spread φ were 2.5 mm and 110''.

When using a telescope with objective lens diameter $d = 19$, 27 and 38 mm, the beam divergence angle was near to that induced by diffraction and equal to 20–70''. The receiving objective diameter was 2.4, 4.8, 8.0 and 10.0 cm. The source images were recorded on a highly sensitive moving film. The time during which random fluctuations in the image location on the film occurred was 20–30 sec. A total of 96 angles of arrival were recorded.

The measure of source scintillation was

$$\sigma_{\varphi} = \sqrt{\frac{\sum_i \Delta_i^2}{n}}, \tag{11.39}$$

where Δ_i is the deviation of the angle of arrival measured at 0.2-sec intervals and n is the number of measurements of Δ_i. The summation was taken over all the deviations for the given measurement.

The variance of fluctuations in the angle of arrival was calculated from the expression

$$\sigma_{\varphi}^2 = \frac{\sum_i x_i^2 n_i}{n}, \tag{11.40}$$

where x_i is the deviation of the image track from the average value. For each value of the beam diameter and of the receiving aperture the average value of σ_φ^2 was determined as the arithmetic mean of all the measurements. The error in determining σ_φ^2 was approximately in the 10–20% range.

Distribution functions of fluctuations in the angle of arrival were also determined. It was found that of 64 distributions, 62 are normal with values of σ_φ^2 between 0.8 and 10.8 square seconds of angular measure.

The dependence of σ_φ^2 for a noncollimated beam on diameter D_r of the receiver objective (in cm) has the form

$$\sigma_\varphi^2 \sim (D_r)^{-1/3}. \tag{11.41}$$

The numerical values of σ_φ^2 for each D_r range over wide limits, depending on the prevailing meteorological conditions.

As the initial beam diameter increased from 2.5 to 27 mm, the variance of the angles of arrival increased; this is apparently due to transition from a spherical to a plane wave. A further increase in beam diameter results in a small reduction in the variance of fluctuations in the angle of arrival.

In addition to high-frequency fluctuations in the angle of arrival, laser beams propagating in the atmosphere are slowly shifted. Deviations of a laser beam with $\lambda = 6328$ Å as a whole amounted to 4 m at distances of 10 km over a period of 30 min /766/.

4. Study of the coherence of a laser beam propagating in a turbulent atmosphere. The distortion of the wave front, and the intensity redistribution in the cross-section of a light beam induced by turbulent motion in the atmosphere, result in deterioration in the space and time coherence of laser radiation. The latter circumstance has a substantial effect on the effectiveness of communications systems constructed on the basis of the optical heterodyne schemes.

The quantitative measure of radiation coherence is the coherence function, the normalized value of which coincides with the optical transfer function characterizing the resolving power of optical receivers which record the image of objects through the atmosphere /750, 856/. Measurement of the coherence function can facilitate the solution of the inverse problem, namely, finding the parameters of atmospheric turbulence from these measurements. The study of the coherence function of radiation propagating through the atmosphere thus possesses great practical importance.

In experimental studies of the coherence of radiation propagating through the atmosphere one analyzes the interference pattern forming upon superposition of the fields of a beam which passed through the atmosphere and of a reference beam. In studies on wide-line base interferometry, photography is employed for recording interference bands of finite width. It was shown in these studies that it is possible to observe interference bands quite clearly. An interference method has been developed /774/ which makes it possible to find the characteristics of atmospheric turbulence from measurements of the coherence of optical radiation. Data on coefficient C_n were obtained with the aid of this technique. Also investigated was the effect of atmospheric turbulence on the resolving power of interference receivers /775/, as well as the spectrum of a photomultiplier signal in the optical heterodyne scheme /776/. Investigations into the spatial coherence of laser radiation and the effect of atmospheric turbulence on it were also conducted /777, 778/. The measurements were carried out in the surface layer of the atmosphere over paths of length 100 and 200 m. Two methods were used, the basis for which was developed elsewhere /863, 856, 864, 750/. The results of studies on the dependence of the radiation coherence function on the temperature gradient were found to be in satisfactory agreement with theory. The dependence of the coherence function on the distance between points in the beam cross-section does not agree with theoretical predictions; this is attributed by the above investigators to the limited nature of the approximation of geometrical optics used in the theory.

Experimental data on the coherence function was used to derive the structure constant C_n of the refractive index /777/. At the same time values of C_n were determined from measurements of meteorological elements at altitudes of 20 cm and 2.0 m. Comparison of these values of C_n shows that they are in satisfactory agreement at lapse rates $\Delta T \leq 0.6°C$ (except when $\Delta T = 0$, in which case according to meteorological data $C_n = 0$, while according to the second method $C_n = 0.008 \cdot 10^{-6}$ cm$^{-1/3}$). When $\Delta T > 0.6°C$, the values of C_n determined from measurements of the coherence function and from meteorological data differ, the difference increasing with higher ΔT. For example, at $\Delta T = 1.0°C$ this difference is 100%, while at $\Delta T = 2.2°C$ it is 200%. In every case the value of C_n determined from the measured meteorological elements was found to be smaller than that obtained from measurements of the coherence function. This means that when $\Delta T > 0.6°C$, the spatial coherence of radiation decreases more rapidly than follows from the theory. During all measurements the wind speed did not exceed 0.2–0.3 m/sec.

APPENDIX 1

1. The ARDC model of the atmosphere for altitudes from 0 to 30 km /3/

Altitude, km	Temperature, °K	Pressure, mb	Density, g/cm^3	Number of molecules per unit volume, cm^{-3}	Height of homogeneous atmosphere, km
0	288.16	$1.01325 \cdot 10^3$	$1.2250 \cdot 10^{-3}$	$2.5476 \cdot 10^{19}$	8.4344
1	281.66	$8.9876 \cdot 10^2$	1.1117	2.3118	8.2468
2	275.16	7.9501	1.0066	2.0933	8.0591
3	268.67	7.0121	$9.0926 \cdot 10^{-4}$	1.8909	7.8713
4	262.18	6.1660	8.1935	1.7039	7.6835
5	255.69	5.4048	7.3643	1.5315	7.4957
6	249.20	4.7217	6.6011	1.3728	7.3077
7	242.71	4.1105	5.9002	1.2270	7.1198
8	236.23	3.5651	5.2578	1.0934	6.9317
9	229.74	3.0800	4.6706	$9.7130 \cdot 10^{18}$	6.7436
10	223.26	2.6500	4.1351	8.5993	6.5554
11	216.78	2.2700	3.6480	7.5864	6.3672
12	216.66	1.9399	3.1194	6.4870	6.3656
13	216.66	1.6579	2.6659	5.5441	6.3676
14	216.66	1.4170	2.2785	4.7385	6.3696
15	216.66	1.2112	1.9475	4.0501	6.3716
16	216.66	1.0353	1.6647	3.4619	6.3736
17	216.66	$8.8496 \cdot 10$	1.4230	2.9593	6.3756
18	216.66	7.5652	1.2165	2.5298	6.3776
19	216.66	6.4674	1.0399	2.1627	6.3796
20	216.66	5.5293	$8.8909 \cdot 10^{-5}$	1.8490	6.3816
21	216.66	4.7274	7.6015	1.5808	6.3836
22	216.66	4.0420	6.4995	1.3516	6.3856
23	216.66	3.4562	5.5575	1.1557	6.3876
24	216.66	2.9554	4.7522	$9.8828 \cdot 10^{17}$	6.3896
25	216.66	2.5273	4.0639	8.4513	6.3916
26	219.34	2.1632	3.4359	7.1453	6.4728
28	225.29	1.5949	2.4663	5.1290	6.6525
30	231.24	1.1855	1.7861	3.7144	6.8323

2. Altitude of isobaric surfaces in the layer from 25 to 100 km, geopotential kilometers /3/

Atmospheric level, mb	Summer								
	Latitude								
	0°	10°	20°	30°	40°	50°	60°	70°	80°
20	26.5	26.5	26.6	26.9	27.1	27.3	27.4	27.5	27.6
16	27.9	28.0	28.1	28.4	28.6	28.8	28.9	29.0	29.1
12	29.8	29.9	30.0	30.3	30.6	30.8	30.9	31.0	31.1
10	31.1	31.1	31.2	31.6	31.8	32.1	32.2	32.3	32.4
8	32.6	32.6	32.8	33.1	33.4	33.7	33.8	33.9	34.0
6	34.6	34.6	34.8	35.2	35.5	35.7	35.9	36.0	36.1
4	37.5	37.5	37.8	38.2	38.5	38.8	38.9	39.0	39.1
3	39.7	39.7	40.0	40.4	40.7	41.0	41.1	41.2	41.2
2.5	41.1	41.1	41.4	41.8	42.2	42.5	42.5	42.6	42.6
2.0	42.8	42.9	43.2	43.6	44.0	44.2	44.3	44.4	44.3
1.6	44.6	44.6	44.9	45.4	45.8	46.1	46.1	46.1	46.0
1.2	47.0	47.0	47.3	47.7	48.2	48.5	48.5	48.5	48.4
1.0	48.4	48.5	48.8	49.2	49.7	50.1	50.1	50.0	49.9
0.8	50.3	50.3	50.6	51.1	51.6	52.0	52.0	51.9	51.8

Atmospheric level, mb	Latitude								
	0°	10°	20°	30°	40°	50°	60°	70°	80°
0.6	52.7	52.7	53.0	53.4	54.0	54.4	54.5	54.4	54.2
0.4	—	56.0	56.2	56.6	57.2	57.7	57.8	57.8	57.6
0.3	—	—	58.4	58.8	59.4	60.0	60.1	60.1	59.9
0.2	—	—	—	61.8	62.3	63.0	63.2	—	—
0.15	—	—	—	63.9	64.4	65.1	65.4	—	—
0.10	—	—	—	66.7	67.1	67.8	68.2	—	—
0.08	—	—	—	68.3	68.6	69.3	70.0	—	—
0.04	—	—	—	72.7	73.0	73.6	74.0	—	—
0.02	—	—	—	77.0	77.1	77.6	78.0	—	—
0.01	—	—	—	81.1	81.1	81.4	81.8	—	—
0.004	—	—	—	86.6	86.4	86.5	86.7	—	—
0.002	—	—	—	90.9	90.6	90.6	90.5	—	—
0.001	—	—	—	95.4	95.1	94.9	94.7	—	—
0.0005	—	—	—	100.2	99.8	99.5	99.2	—	—

Winter

Atmospheric level, mb	Latitude								
	0°	10°	20°	30°	40°	50°	60°	70°	80°
20	26.1	26.1	26.2	26.2	26.2	26.1	25.7	25.2	25.0
16	27.5	27.5	27.6	27.6	27.6	27.5	27.1	26.6	26.3
12	29.4	29.4	29.5	29.5	29.4	29.3	28.8	28.3	28.0
10	30.6	30.6	30.7	30.6	30.6	30.4	30.0	29.4	29.0
8	32.1	32.1	32.2	32.1	32.0	31.8	31.3	30.7	30.3
6	34.1	34.0	34.1	34.0	33.9	33.7	33.1	32.4	32.0
4	37.0	37.0	37.0	36.8	36.6	36.3	35.7	34.9	34.5
3	39.1	39.0	39.0	38.9	38.7	38.2	37.6	36.7	36.2
2.5	40.4	40.4	40.4	40.2	40.0	39.5	38.8	37.9	37.6
2.0	42.1	42.1	42.1	41.9	41.6	41.1	40.3	39.3	38.8
1.6	43.9	43.8	43.8	43.6	43.2	42.6	41.8	40.8	40.3
1.2	46.2	46.1	46.1	45.8	45.4	44.7	43.8	42.8	42.2
1.0	47.6	47.6	47.5	47.2	46.8	46.1	45.1	44.0	43.5
0.8	49.5	49.4	49.3	49.0	48.5	47.7	46.7	45.6	45.0
0.6	51.8	51.8	51.6	51.3	50.7	49.8	48.7	47.6	47.0
0.4	55.2	55.1	55.0	54.5	53.9	52.9	51.7	50.4	49.7
0.3	—	57.4	57.2	56.8	56.1	55.1	53.8	52.5	51.8
0.2	—	60.5	60.4	60.0	59.3	58.3	56.9	55.5	54.8
0.15	—	—	—	62.1	61.4	60.4	59.1	—	—
0.10	—	—	—	65.0	64.4	63.6	62.2	—	—
0.08				66.5	66.0	65.2	63.9	—	—
0.04	—	—	—	71.1	70.6	70.0	68.9	—	—
0.02	—	—	—	75.4	75.0	74.4	73.5	—	—
0.01	—	—	—	79.6	79.2	78.6	77.8	—	—
0.004	—	—	—	85.0	84.7	84.3	83.7	—	—
0.002	—	—	—	89.3	89.1	88.9	88.4	—	—
0.001	—	—	—	93.7	93.7	93.6	93.3	—	—
0.0005	—	—	—	98.4	98.6	98.6	98.3	—	—

3. Temperature in the atmospheric layer from 25 to 100 km, °K /3/

Summer

Atmospheric level, mb	Latitude								
	0°	10°	20°	30°	40°	50°	60°	70°	80°
20	220	220	222	224	226	230	233	235	236
16	223	223	225	227	229	232	234	236	237

Atmospheric level, mb	Latitude								
	0°	10°	20°	30°	40°	50°	60°	70°	80°
12	227	227	229	231	233	236	237	239	239
10	230	230	232	234	237	239	240	242	242
8	234	234	237	239	242	244	244	245	244
6	240	240	243	246	249	251	250	250	248
4	251	251	254	257	261	261	259	257	254
3	260	260	263	266	269	267	264	261	257
2.5	265	265	267	270	273	271	267	264	260
2.0	271	271	272	274	277	276	273	268	265
1.6	274	274	276	277	281	282	279	274	271
1.2	278	278	279	279	285	290	289	284	279
1.0	279	279	280	279	285	293	293	289	283
0.8	280	280	280	278	283	291	294	292	288
0.6	280	279	277	274	277	285	290	291	290
0.4	—	274	270	265	265	272	276	278	280
0.3	—	—	262	259	254	261	267	269	270
0.2	—	—	—	248	244	247	254	—	—
0.15	—	—	—	242	237	238	245	—	—
0.10	—	—	—	233	227	226	232	—	—
0.08	—	—	—	228	222	220	226	—	—
0.04	—	—	—	214	208	205	203	—	—
0.02	—	—	—	202	197	189	186	—	—
0.01	—	—	—	200	194	186	180	—	—
0.004	—	—	—	208	202	194	185	—	—
0.002	—	—	—	218	212	204	195	—	—
0.001	—	—	—	229	226	217	210	—	—
0.0005	—	—	—	241	243	238	234	—	—

Winter

Atmospheric level, mb	Latitude								
	0°	10°	20°	30°	40°	50°	60°	70°	80°
20	214	213	213	214	214	214	209	204	201
16	218	217	216	216	214	214	208	203	200
12	222	221	220	218	216	213	208	203	199
10	226	225	223	220	217	213	208	203	199
8	230	229	228	224	220	214	209	203	199
6	237	236	234	230	225	218	212	206	202
4	246	245	243	239	234	226	219	212	208
3	253	253	251	247	240	232	224	217	213
2.5	257	257	256	252	245	236	228	220	217
2.0	263	262	260	256	249	241	233	225	221
1.6	268	267	264	260	253	244	236	229	225
1.2	274	273	269	265	257	248	238	232	229
1.0	277	276	272	267	260	250	239	233	231
0.8	279	279	274	270	262	252	241	235	232
0.6	281	280	277	273	265	255	245	237	234
0.4	276	276	275	274	268	259	250	243	239
0.3	—	270	269	269	267	262	255	247	244
0.2	—	257	256	257	259	263	259	254	251
0.15	—	—	—	249	253	261	262	—	—
0.10	—	—	—	236	243	254	262	—	—
0.08	—	—	—	232	237	247	258	—	—
0.04	—	—	—	218	220	224	229	—	—
0.02	—	—	—	207	209	211	216	—	—
0.01	—	—	—	200	204	207	214	—	—

Atmospheric level, mb	Winter								
	0°	10°	20°	30°	40°	50°	60°	70°	80°
0.004	—	—	—	205	210	218	224	—	—
0.002	—	—	—	214	220	228	236	—	—
0,001	—	—	—	224	233	239	244	—	—
0.0005	—	—	—	236	245	250	250	—	—

APPENDIX 2

Values of absorption coefficient α, absorption index $\varkappa$, and refractive index of pure water n according to data of this author and his coworkers for various wavelengths in the 0.55–25.3 μ interval

Serial No.	ν, cm^{-1}	λ, μ	α, cm^{-1}	$\varkappa$	n
1	395	25.31	1130	0.229	1.700
2	405	24.65	1190	0.234	1.702
3	415	24.09	1230	0.236	1.709
4	425	23.52	1210	0.227	1.738
5	435	22.98	1300	0.238	1.784
6	445	22.47	1790	0.320	1.833
7	455	21.97	2300	0.402	1.802
8	465	21.50	2590	0.444	1.763
9	475	21.05	2800	0.470	1.722
10	485	20.61	3010	0.495	1.695
11	495	20.21	3180	0.511	1.665
12	505	19.80	3370	0.531	1.637
13	515	19.41	3480	0.537	1.601
14	525	19.04	3540	0.537	1.575
15	535	18.69	3710	0.552	1.544
16	545	18.34	3780	0.552	1.516
17	555	18.01	3870	0.553	1.496
18	565	17.69	3980	0.560	1.459
19	575	17.39	4070	0.563	1.421
20	585	17.09	4100	0.557	1.394
21	595	16.80	4110	0.550	1.367
22	605	16.52	4150	0.546	1.329
23	615	16.26	4150	0.536	1.324
24	625	16.00	4130	0.526	1.292
25	635	15.74	4020	0.504	1.280
26	645	15.50	3990	0.492	1.248
27	655	15.26	3970	0.483	1.242
28	665	15.03	3960	0.474	1.222
29	675	14.81	3850	0.454	1.192
30	685	14.59	3680	0.427	1.177
31	695	14.38	3520	0.403	1.175
32	705	14.18	3380	0.382	1.194
33	715	13.98	3520	0.392	1.192
34	725	13.79	3600	0.396	1.166
35	735	13.61	3510	0.380	1.147
36	745	13.42	3350	0.358	1.126
37	755	13.24	3220	0.339	1.115
38	765	13.07	3100	0.323	1.107
39	775	12.91	2940	0.302	1.102
40	785	12.73	2750	0.279	1.097

Serial No.	ν, cm^{-1}	λ, μ	α, cm^{-1}	$\varkappa$	n
41	795	12.57	2620	0.262	1.096
42	805	12.42	2470	0.245	1.095
43	815	12.26	2360	0.230	1.096
44	825	12.12	2270	0.219	1.100
45	835	11.97	2170	0.207	1.102
46	845	11.83	2020	0.190	1.103
47	855	11.69	1960	0.182	1.107
48	865	11.56	1930	0.177	1.113
49	875	11.42	1690	0.154	1.110
50	885	11.29	1520	0.137	1.112
51	896	11.16	880	0.122	1.118
52	946	10.57	800	0.067	1.144
53	996	10.04	580	0.047	1.105
54	1047	9.55	550	0.042	1.180
55	1098	9.11	530	0.038	1.203
56	1148	8.71	530	0.037	1.228
57	1198	8.35	540	0.036	1.250
58	1249	8.01	540	0.034	1.262
59	1300	7.69	560	0.034	1.277
60	1350	7.41	560	0.033	1.292
61	1401	7.14	560	0.032	1.303
62	1451	6.89	600	0.033	1.316
63	1502	6.66	610	0.032	1.301
64	1553	6.44	720	0.037	1.297
65	1595	6.27	1130	0.056	1.379
66	1611	6.21	1140	0.056	1.401
67	1626	6.15	2180	0.107	1.371
68	1641	6.09	2940	0.143	1.342
69	1656	6.03	2760	0.133	1.275
70	1671	5.98	2390	0.114	1.239
71	1686	5.93	1730	$0.815 \cdot 10^{-1}$	1.224
72	1701	5.87	1090	$0.511 \cdot 10^{-1}$	1.202
73	1716	5.83	550	$0.211 \cdot 10^{-1}$	1.208
74	1731	5.77	445	$0.205 \cdot 10^{-1}$	1.222
75	1746	5.72	467	$0.213 \cdot 10^{-1}$	1.231
76	1761	5.67	650	$0.294 \cdot 10^{-1}$	1.261
77	1797	5.57	271	$0.125 \cdot 10^{-1}$	1.275
78	1848	5.41	232	$0.105 \cdot 10^{-1}$	1.287
79	1898	5.27	232	$0.103 \cdot 10^{-1}$	1.309
80	1949	5.13	249	$0.107 \cdot 10^{-1}$	1.312
81	1999	5.00	309	$0.123 \cdot 10^{-1}$	1.317
82	2046	4.89	362	$0.136 \cdot 10^{-1}$	1.321
83	2096	4.77	386	$0.142 \cdot 10^{-1}$	1.323
84	2146	4.66	386	$0.138 \cdot 10^{-1}$	1.327
85	2196	4.55	348	$0.122 \cdot 10^{-1}$	1.329
86	2246	4.45	296	$0.100 \cdot 10^{-1}$	1.329
87	2296	4.36	252	$0.083 \cdot 10^{-1}$	1.329
88	2396	4.17	181	$0.559 \cdot 10^{-2}$	1.335
89	2446	4.09	160	$0.521 \cdot 10^{-2}$	1.338
90	2496	4.01	134	$0.428 \cdot 10^{-2}$	1.343
91	2546	3.93	127	$0.397 \cdot 10^{-2}$	1.347
92	2596	3.85	116	$0.359 \cdot 10^{-2}$	1.357
93	2646	3.78	122	$0.369 \cdot 10^{-2}$	1.364
94	2696	3.71	134	$0.398 \cdot 10^{-2}$	1.370
95	2746	3.64	166	$0.484 \cdot 10^{-2}$	1.376
96	2796	3.58	218	$0.621 \cdot 10^{-2}$	1.394
97	2846	3.51	290	$0.813 \cdot 10^{-2}$	1.398
98	2896	3.45	418	$0.115 \cdot 10^{-1}$	1.416
99	2946	3.39	630	$0.170 \cdot 10^{-1}$	1.478
100	2996	3.34	900	$0.240 \cdot 10^{-1}$	1.458
101	3006	3.33	313	$0.828 \cdot 10^{-2}$	1.460

Serial No.	ν, cm^{-1}	λ, μ	α, cm^{-1}	ϰ	n
102	3021	3.31	417	$0.111 \cdot 10^{-1}$	1.468
103	3036	3.29	680	$0.179 \cdot 10^{-1}$	1.467
104	3051	3.28	1300	$0.338 \cdot 10^{-1}$	1.458
105	3066	3.26	990	$0.257 \cdot 10^{-1}$	1.452
106	3081	3.25	1010	$0.261 \cdot 10^{-1}$	1.463
107	3096	3.23	1180	$0.303 \cdot 10^{-1}$	1.465
108	3111	3.71	1410	$0.361 \cdot 10^{-1}$	1.478
109	3126	3.20	1740	$0.443 \cdot 10^{-1}$	1.489
110	3141	3.18	2100	$0.532 \cdot 10^{-1}$	1.492
111	3156	3.17	3060	$0.637 \cdot 10^{-1}$	1.495
112	3171	3.15	3070	$0.772 \cdot 10^{-1}$	1.504
113	3186	3.16	3690	$0.923 \cdot 10^{-1}$	1.509
114	3201	3.12	4420	0.110	1.513
115	3216	3.11	5320	0.132	1.514
116	3231	3.10	6660	0.164	1.515
117	3246	3.08	7610	0.187	1.489
118	3261	3.07	6910	0.171	1.467
119	3276	3.05	7930	0.193	1.494
120	3291	3.03	9250	0.224	1.469
121	3306	3.02	9950	0.240	1.451
122	3321	3.01	10390	0.249	1.430
123	3336	2.99	10830	0.259	1.414
124	3351	2.98	11120	0.264	1.396
125	3366	2.97	11280	0.266	1.374
126	3381	2.96	11350	0.267	1.353
127	3396	2.94	11660	0.273	1.337
128	3411	2.93	11750	0.274	1.314
129	3426	2.92	11880	0.276	1.288
130	3441	2.91	12230	0.283	1.260
131	3456	2.89	11670	0.269	1.232
132	3471	2.88	11080	0.254	1.210
133	3486	2.87	10770	0.237	1.190
134	3501	2.86	9940	0.226	1.175
135	3516	2.84	9310	0.210	1.147
136	3531	2.83	8610	0.194	1.131
137	3546	2.82	7530	0.169	1.110
138	3561	2.81	5940	0.133	1.094
139	3576	2.79	4890	0.109	1.088
140	3591	2.78	4480	$0.994 \cdot 10^{-1}$	1.100
141	3606	2.77	3600	$0.794 \cdot 10^{-1}$	1.108
142	3621	2.76	2950	$0.648 \cdot 10^{-1}$	1.112
143	3636	2.75	2360	$0.517 \cdot 10^{-1}$	1.133
144	3651	2.74	1780	$0.384 \cdot 10^{-1}$	1.138
145	3666	2.73	1050	$0.227 \cdot 10^{-1}$	1.141
146	3696	2.71	730	$0.157 \cdot 10^{-1}$	1.181
147	3746	2.67	247	$0.526 \cdot 10^{-2}$	1.221
148	3796	2.63	183	$0.386 \cdot 10^{-2}$	1.234
149	3846	2.60	146	$0.304 \cdot 10^{-2}$	1.245
150	3896	2.57	128	$0.263 \cdot 10^{-2}$	1.255
151	3946	2.53	119	$0.240 \cdot 10^{-2}$	1.261
152	3996	2.50	103	$0.206 \cdot 10^{-2}$	1.264
153	4046	2.47	88.9	$0.175 \cdot 10^{-2}$	1
154	4096	2.44	63.8	$0.124 \cdot 10^{-2}$	1.269
155	4146	2.41	56.3	$0.108 \cdot 10^{-2}$	
156	4196	2.38	48.6	$0.921 \cdot 10^{-3}$	
157	4246	2.36	40.4	$0.757 \cdot 10^{-3}$	1.274
158	4296	2.33	34.0	$0.631 \cdot 10^{-3}$	
159	4346	2.30	29.9	$0.548 \cdot 10^{-3}$	
160	4396	2.27	27.2	$0.492 \cdot 10^{-3}$	1.286
161	4446	2.25	25.3	$0.453 \cdot 10^{-3}$	
162	4496	2.22	24.5	$0.424 \cdot 10^{-3}$	1.289
163	4546	2.19	24.9	$0.436 \cdot 10^{-3}$	
164	4596	2.18	25.9	$0.449 \cdot 10^{-3}$	1.294

Serial No.	ν, cm⁻¹	λ, μ	α, cm⁻¹	ϰ	n
165	4646	2.15	27.5	$0.471\cdot10^{-3}$	
166	4696	2.13	29.6	$0.502\cdot10^{-3}$	1.297
167	4746	2.11	32.7	$0.548\cdot10^{-3}$	
168	4780	2.09	37.0	$0.616\cdot10^{-3}$	1.300
169	4830	2.07	40.6	$0.669\cdot10^{-3}$	
170	4878	2.05	45.5	$0.743\cdot10^{-3}$	1.304
171	4938	2.03	48.3	$0.778\cdot10^{-3}$	
172	4987	2.01	55.8	$0.891\cdot10^{-3}$	1.304
173	5040	1.98	70.3	$0.111\cdot10^{-2}$	
174	5105	1.96	85.2	$0.133\cdot10^{-2}$	1.306
175	5165	1.94	106	$0.163\cdot10^{-2}$	
176	5208	1.92	123	$0.187\cdot10^{-2}$	1.305
177	5277	1.89	119	$0.179\cdot10^{-2}$	
178	5333	1.88	68.4	$0.102\cdot10^{-2}$	1.303
179	5376	1.86	25.0	$0.371\cdot10^{-3}$	
180	5449	1.84	15.5	$0.227\cdot10^{-3}$	1.309
181	5602	1.78	13.6	$0.193\cdot10^{-3}$	
182	5635	1.77	12.6	$0.177\cdot10^{-3}$	
183	5710	1.75	12.4	$0.173\cdot10^{-3}$	1.310
184	5880	1.70	9.71	$0.131\cdot10^{-3}$	
185	6030	1.66	9.71	$0.128\cdot10^{-3}$	1.311
186	6200	1.61	9.71	$0.125\cdot10^{-3}$	
187	6370	1.57	10.7	$0.133\cdot10^{-3}$	1.312
188	6470	1.55	14.6	$0.179\cdot10^{-3}$	
189	6570	1.52	17.6	$0.213\cdot10^{-3}$	1.313
190	6660	1.50	21.8	$0.253\cdot10^{-3}$	
191	6770	1.48	28.5	$0.335\cdot10^{-3}$	1.313
192	6690	1.43	30.9	$0.352\cdot10^{-3}$	1.314
193	7100	1.41	24.0	$0.269\cdot10^{-3}$	
194	7210	1.39	14.5	$0.160\cdot10^{-3}$	1.314
195	7275	1.37	6.91	$0.756\cdot10^{-4}$	
196	7420	1.35	3.33	$0.347\cdot10^{-4}$	1.315
197	7540	1.33	2.21	$0.233\cdot10^{-4}$	
198	7670	1.30	1.62	$0.169\cdot10^{-4}$	1.315
199	7800	1.28	1.33	$0.136\cdot10^{-4}$	
200	7920	1.26	1.33	$0.133\cdot10^{-4}$	1.316
201	8060	1.24	1.38	$0.136\cdot10^{-4}$	
202	8180	1.22	1.44	$0.141\cdot10^{-4}$	1.316
203	8320	1.20	1.49	$0.143\cdot10^{-4}$	
204	8460	1.18	1.48	$0.139\cdot10^{-4}$	1.317
205	8600	1.16	1.41	$0.131\cdot10^{-4}$	
206	8760	1.14	1.13	$0.102\cdot10^{-4}$	1.317
207	8900	1.12	0.423	$0.378\cdot10^{-5}$	
208	9040	1.11	0.242	$0.212\cdot10^{-5}$	1.318
209	9190	1.09	0.190	$0.176\cdot10^{-5}$	
210	9350	1.07	0.161	$0.137\cdot10^{-5}$	1.319
211	9500	1.05	0.187	$0.156\cdot10^{-5}$	
212	9660	1.04	0.243	$0.199\cdot10^{-5}$	1.320
213	9800	1.02	0.321	$0.260\cdot10^{-5}$	
214	9950	1.01	0.402	$0.322\cdot10^{-5}$	1.321
215	10100	0.99	0.508	$0.401\cdot10^{-5}$	
216	10200	0.98	0.522	$0.408\cdot10^{-5}$	0.323
217	10350	0.97	0.456	$0.352\cdot10^{-5}$	
218	10450	0.96	0.363	$0.277\cdot10^{-5}$	1.325
219	10650	0.94	0.221	$0.165\cdot10^{-5}$	
220	10800	0.93	0.171	$0.126\cdot10^{-5}$	1.326
221	10900	0.92	0.121	$0.88\cdot10^{-6}$	
222	11000	0.91	$0.853\cdot10^{-1}$	$0.62\cdot10^{-6}$	1.327
223	11200	0.89	$0.749\cdot10^{-1}$	$0.53\cdot10^{-6}$	
224	11300	0.88	$0.673\cdot10^{-1}$	$0.47\cdot10^{-6}$	1.329
225	11450	0.87	$0.606\cdot10^{-1}$	$0.42\cdot10^{-6}$	

Serial No.	ν, cm^{-1}	λ, μ	α, cm^{-1}	ϰ	n
226	11600	0.86	$0.563\cdot10^{-1}$	$0.39\cdot10^{-6}$	1.330
227	11700	0.85	$0.533\cdot10^{-1}$	$0.36\cdot10^{-6}$	
228	11820	0.85	$0.510\cdot10^{-1}$	$0.34\cdot10^{-6}$	1.331
229	11950	0.84	$0.466\cdot10^{-1}$	$0.31\cdot10^{-6}$	
230	12100	0.83	$0.386\cdot10^{-1}$	$0.25\cdot10^{-6}$	1.331
231	12150	0,82	$0.306\cdot10^{-1}$	$0.20\cdot10^{-6}$	
232	12300	0,82	$0.250\cdot10^{-1}$	$0.16\cdot10^{-6}$	
233	12400	0,81	$0.248\cdot10^{-1}$	$0.16\cdot10^{-6}$	
234	12550	0,80	$0.241\cdot10^{-1}$	$0.15\cdot10^{-6}$	1,333
235	12650	0.79	$0.263\cdot10^{-1}$	$0.17\cdot10^{-6}$	
236	12800	0.78	$0.287\cdot10^{-1}$	$0.18\cdot10^{-6}$	
237	12900	0.78	$0.300\cdot10^{-1}$	$0.19\cdot10^{-6}$	
238	13000	0.77	$0.312\cdot10^{-1}$	$0.19\cdot10^{-6}$	1.333
239	13150	0.76	$0.318\cdot10^{-1}$	$0.19\cdot10^{-6}$	
240	13300	0,75	$0.319\cdot10^{-1}$	$0.19\cdot10^{-6}$	
241	13450	0.74	$0.312\cdot10^{-1}$	$0.18\cdot10^{-6}$	1.333
242	13550	0.74	$0.295\cdot10^{-1}$	$0.17\cdot10^{-6}$	
243	13650	0.73	$0.226\cdot10^{-1}$	$0.13\cdot10^{-6}$	
244	13800	0,72	$0.159\cdot10^{-1}$	$0.92\cdot10^{-7}$	
245	13850	0,72	$0.131\cdot10^{-1}$	$0.75\cdot10^{-7}$	
246	14000	0.71	$0.100\cdot10^{-1}$	$0.57\cdot10^{-7}$	1.333
247	14100	0.71	$0.86\cdot10^{-2}$	$0.49\cdot10^{-7}$	
248	14300	0.70	$0.73\cdot10^{-2}$	$0.47\cdot10^{-7}$	
249	14300	0.70	$0.68\cdot10^{-2}$	$0.38\cdot10^{-7}$	
250	14500	0.70	$0.62\cdot10^{-2}$	$0.34\cdot10^{-7}$	1.333
251	14600	0.68	$0.60\cdot10^{-2}$	$0.33\cdot10^{-7}$	
252	14750	0,68	$0,68\cdot10^{-2}$	$0.37\cdot10^{-7}$	1,332
253	14900	0,67	$0,59\cdot10^{-2}$	$0.32\cdot10^{-7}$	
254	14950	0.67	$0,56\cdot10^{-2}$	$0.30\cdot10^{-7}$	1,332
255	15100	0.662	$0,57\cdot10^{-2}$	$0.30\cdot10^{-7}$	
256	15200	0,644	$0.55\cdot10^{-2}$	$0.29\cdot10^{-7}$	
257	15300	0.653	$0,55\cdot10^{-2}$	$0,29\cdot10^{-7}$	
258	15400	0.649	$0,52\cdot10^{-2}$	$0.27\cdot10^{-7}$	
259	15500	0,645	$0.54\cdot10^{-2}$	$0,28\cdot10^{-7}$	
260	15600	0,641	$0,52\cdot10^{-2}$	$0.27\cdot10^{-7}$	1.3318
261	15700	0.636	$0.52\cdot10^{-2}$	$0.26\cdot10^{-7}$	
262	15800	0.632	$0.49\cdot10^{-2}$	$0,25\cdot10^{-7}$	
263	15850	0,630	$0,50\cdot10^{-2}$	$0.25\cdot10^{-7}$	1.3314
264	16000	0.625	$0,48\cdot10^{-2}$	$0.24\cdot10^{-7}$	
265	16050	0.623	$0,51\cdot10^{-2}$	$0.25\cdot10^{-7}$	
266	16200	0.617	$0.51\cdot10^{-2}$	$0.25\cdot10^{-7}$	
267	16250	0.615	$0.50\cdot10^{-2}$	$0,25\cdot10^{-7}$	1,3314
268	16350	0.611	$0,50\cdot10^{-2}$	$0.24\cdot10^{-7}$	
269	16450	0,607	$0.49\cdot10^{-2}$	$0.24\cdot10^{-7}$	
270	16500	0,606	$0.45\cdot10^{-2}$	$0.22\cdot10^{-7}$	1,3314
271	16700	0,598	$0,43\cdot10^{-2}$	$0.21\cdot10^{-7}$	
272	16900	0,591	$0.39\cdot10^{-2}$	$0,18\cdot10^{-7}$	
273	17050	0.586	$0,37\cdot10^{-2}$	$0.17\cdot10^{-7}$	
274	17200	0.581	$0,36\cdot10^{-2}$	$0.17\cdot10^{-7}$	
275	17400	0,574	$0.39\cdot10^{-2}$	$0,18\cdot10^{-7}$	1.3314
276	17550	0.569	$0.39\cdot10^{-2}$	$0.18\cdot10^{-7}$	
277	17650	0.566	$0.39\cdot10^{-2}$	$0,18\cdot10^{-7}$	
278	17900	0.558	$0.36\cdot10^{-2}$	$0,16\cdot10^{-7}$	
279	18100	0,552	$0.39\cdot10^{-2}$	$0,17\cdot10^{-7}$	
280	18300	0.546	$0.41\cdot10^{-2}$	$0,18\cdot10^{-7}$	1,3314

BIBLIOGRAPHY

1. ZUEV, V.E. Atmospheric Transparency in the Visible and the Infrared. Israel Program for Scientific Translations. 1970.
2. KONDRAT'EV, K.Ya. Actinometry. Gidrometeoizdat. 1965.
3. GOODY, R.M. Atmospheric Radiation, Vol.1, Theoretical Basis. Oxford University Press. 1964.
4. FLEAGLE, R.G. and J.A.BUSINGER. An Introduction to Atmospheric Physics. New York–London, Academic Press. 1963.
5. TVERSKOI, P.N. Physics of the Atmosphere. A Course in Meteorology. Israel Program for Scientific Translations. 1965.
6. GUTNICK, M. – Appl. Optics, Vol.1, No.5, p.670. 1962.
7. MALKEVICH, M.S., Yu.B.SAMSONOV, and L.I.KOPROVA. – Uspekhi Fizicheskikh Nauk, Vol.80, No.1, p.93. 1963.
8. HOUGHTON, J.T. – In: Les problèmes météorologiques de la stratosphère et de la mesosphère. Paris, Presses universitaires de France. 1966.
9. KOMAROV, V.S. – Izvestiya AN SSSR, Fizika Atmosfery i Okeana, Vol.4, No.11, p.1160. 1968; Trudy Nauchno-Issledovatel'skogo Instituta Aeroklimatologii, No.47. 1968.
10. HOWARD, J.N. and J.S.GARING. – Infrared Phys., Vol.2, No.3, p.155. 1962.
11. KOMAROV, V.S. and A.S.MARCHENKO. – Trudy Nauchno-Issledovatel'skogo Instituta Aeroklimatologii, No.58. 1968.
12. KOMAROV, V.S. – Meteorologiya i Gidrologiya, No.1. 1969.
13. KOMAROV, V.S. – Trudy Nauchno-Issledovatel'skogo Instituta Aeroklimatologii, No.57. 1968; No.58. 1969.
14. CARLON, H.R. – Appl. Optics, Vol.4, No.9, p.1089. 1965.
15. BAZHENOV, V.A., R.N.IVANOVA, and M.M.MIROSHNIKOV. – Izvestiya AN SSSR, Fizika Atmosfery i Okeana, Vol.2, No.3, p.248. 1966.
16. ROZENBERG, G.V. Twilight. Fizmatgiz. 1963.
17. FEDYNSKII, A.V., S.P.PEROV, and A.F.CHIZHOV. – Izvestiya AN SSSR, Fizika Atmosfery i Okeana, Vol.3, No.5, p.557. 1967.
18. KHRGIAN, A.Kh. (editor). Atmospheric Ozone. Izdatel'stvo MGU. 1961.
19. KHRGIAN, A.Kh., G.I.KUZNETSOV, and A.M.KONDRAT'EVA. Atmospheric Ozone. Israel Program for Scientific Translations. 1967.
20. KHVOSTIKOV, I.A. Physics of the Ozonosphere and Ionosphere. Izdatel'stvo AN SSSR. 1963.
21. GUSHCHIN, G.P. Ozone and Aerosynoptic Conditions in the Atmosphere. Gidrometeoizdat. 1964.
22. DÜTSCH, H.U. – Archiv Meteorol. Geophys. Biocl., Ser.A, Vol.11. 1959.
23. YAKOVLEVA, A.V., L.A.KUDRYAVTSEVA, A.S. BRITAEV, et al. – In: Iskusstvennye sputniki Zemli, No.14, p.57. Izdatel'stvo AN SSSR. 1962.
24. HALDANE, J.B.S. – Nature, Vol.137, No.3466, p.575. 1936.
25. CALLENDAR, G.S. – Q.J.Roy. Meteorol. Soc., Vol.64, No.275. 1938; Vol.66, No.287. 1940.
26. GLÜCAUF, E. – Nature, Vol.153, No.3890, p.620. 1944.
27. BUCH, K. – Geophys., Vol.3, p.63. 1948.
28. REVELLE, R. and H.E.SUESS. – Tellus, Vol.9, No.1. 1957.
29. SLOCUM, G. – Monthly Weather Review, Vol.83, No.10. 1955.

30. FONSELIUS, S., F. KOROLEFF, and K.E. WÄRME. – Tellus, Vol.8, No.2. 1956.
31. BRAY, J.R. – Tellus, Vol.11, No.2. 1959.
32. BISHOP, W. – Tellus, Vol.112, No.2. 1960.
33. LYUTSAREV, S.V. – Trudy Instituta Okeanologii AN SSSR, Vol.47, p.199. 1961.
34. BRUEVICH, S.V. and S.V. LYUTSAREV. – DAN SSSR, Vol.136, No.2, p.405. 1961.
35. SUESS, H.E. – Science, Vol.122, p.415. 1958.
36. ARNOLD, J.R. and E.C. ANDERSEN. – Tellus, Vol.9, No.1. 1957.
37. KEELING, C.D. – Tellus, Vol.12, No.2. 1960.
38. TAKAHASHI, T. – J. Geophys. Res., Vol.60, No.2. 1961.
39. TOMISLAV, P. – Meteorol. Zpravy. R. VIII, No.5. 1955.
40. MISRA, R.K. – Ind. J. Meteorol. Geoph., No.1. 1950.
41. VOITIKOVA, T.D. – Trudy Glavnoi Geofizicheskoi Observatorii, No.118, p.82. 1961.
42. SEELEY, J.S. and J.T. HOUGHTON. – Infrared Phys., Vol.1, No.2, p.116. 1961.
43. MIGEOTTE, M., L. NEVEN, and J. SWENSON. – Met. Soc. Sci., Liège (spec. issue). 1956.
44. BOWMAN, R.L. and J. SHAW. – Appl. Optics, Vol.2, No.2, p.176. 1963.
45. SHAW, J.H. – Astrophys. J., Vol.128, p.428. 1958.
46. FINK, U., D.H. RANK, and T.A. WIGGINS. – J. Opt. Soc. Amer., Vol.54, No.4, p.472. 1964.
47. GOODY, R.M. The Physics of the Stratosphere. Cambridge University Press. 1954.
48. BIRKELAND, J.W. and J.H. SHAW. – J. Opt. Soc. Amer., Vol.49, No.6, p.637. 1959.
49. McMATH, R.R., A.K. PIERCE, O.C. MOHLER, L. GOLDBERG, and R.A. DONOVAN. – Phys. Rev., Vol.78, p.65. 1950.
50. HERZBERG, G. Spectra of Diatomic Molecules. 2nd ed. Van Nostrand. 1950.
51. EL'YASHEVICH, M.A. Atomic and Molecular Spectroscopy. Fizmatgiz. 1962.
52. ERLANDSSON, J. – Arkiv für Physik, Vol.10, p.65. 1956.
53. GORDY, W., W.V. SMITH, and R.F. TRAMBARULO. Microwave Spectroscopy. New York, Wiley. 1953.
54. KONDRAT'EV, V.N. Atomic and Molecular Structure. Fizmatgiz. 1959.
55. LORENTZ, H.A. – Proc. Amst. Acad. Soc., Vol.8, p.591. 1906.
56. ANDERSON, P.W. – Phys. Rev., Vol.76, p.647. 1949; Vol.80, p.511. 1950.
57. BENEDICT, W.S. and L.D. KAPLAN. – J. Chem. Phys., Vol.30, p.388. 1959.
58. BENEDICT, W.S. and L.D. KAPLAN. – J. Quant. Spectry. Rad. Trans., Vol.4, No.3, p.453. 1964.
59. GALIBINA, L.I. and V.E. ZUEV. – Izvestiya Vuzov SSSR, Seriya Fizika, No.4, p.69. 1962.
60. PLYLER, E.K., N.M. GAILAR, and T.A. WIGGINS. – J. Res. Nat. Bur. Stand., Vol.48, p.221. 1952.
61. MADDEN, R.F. – J. Chem. Phys., Vol.35, p.2083. 1961.
62. DRAYSON, S.R. – Appl. Optics, Vol.5, No.3, p.385. 1966.
63. GOLDBERG, L. – In: The Earth as a Planet, ed. G.P. Kuiper, p.434. Chicago Univ. Press. 1954.
64. DUFAY, J. – Ann. d'astrophys., Vol.5. 1942.
65. DIANOV-KLOKOV, V.I. – Optika i Spektroskopiya, Vol.16, p.409, 1964.
66. RANK, D.H., A.G. St. PIERRE, and T.A. WIGGINS. – J. Mol. Spectry., Vol.18, No.4, p.418. 1965.
67. BREEN, R.G. The Shift and Slope of Spectral Lines. Oxford–London–New York–Paris, Pergamon Press. 1961.
68. TRAVING, G. Über die Theorie der Druckverbreitung von Spektrallinien. Karlsruhe, Braun. 1960.
69. SOBEL'MAN, I.I. – Uspekhi Fizicheskikh Nauk, Vol.54, No.4, p.551. 1954.
70. CHEN, S.Y. and M. TAKEO. – Rev. Mod. Phys., Vol.29, p.20. 1957.
71. TSAO, C.J. and B. CURNUTTE. – J. Quant. Spectry. Rad. Transf., Vol.2, No.1, p.41. 1962.
72. RAUTIAN, S.G. and I.I. SOBEL'MAN. – Uspekhi Fizicheskikh Nauk, Vol.90, No.2, p.209. 1966.
73. PLASS, G.N. and D.I. FIVEL. – Astrophys. J., Vol.117, p.225. 1953.

74. SURTEES, W.J. – J. Opt. Soc. Amer., Vol.55, No.7, p.893. 1965.
75. FRIED, B.D. and S.D. CONTE. The Plasma Dispersion Function. New York, Academic Press. 1961.
76. YOUNG, C. – J. Quant. Spectry. Rad. Trans., Vol.5, No.3, p.549. 1965.
77. SHVED, G.M. and I.V. TSARITSYNA. – In: Problemy fiziki atmosfery, No.1, p.71. Izdatel'stvo Leningradskogo Universiteta. 1963.
78. WINTERS, B.H., S. SILVERMAN, and W.S. BENEDICT. – J. Quant. Spectry. Rad. Trans., Vol.4, p.527. 1964.
79. KYLE, T.G., D.G. MURCRAY, F.H. MURCRAY, and W.J. WILLIAMS. – J. Opt. Soc. Amer., Vol.55, No.11, p.1421. 1965.
80. BURCH, D.E., R.R. PATTY, and D.A. GRYVNAK. – J. Opt. Soc. Amer., Vol.55, No.5, p.606. 1965.
81. BURCH, D.E. and D.A. GRYVNAK. – J. Quant. Spectry. Rad. Trans., Vol.6, No.3, p.229. 1966.
82. LADENBERG, R. and F. REICHE. – Ann. Physik, Vol.42, p.181. 1913.
83. ELSASSER, W.M. Heat Transfer by Infrared Radiation in the Atmosphere. – Harvard Meteor. Studies, No.6. Harvard University Press. 1942.
84. KONDRAT'EV, K.Ya. Transport of Long-Wave Radiation in the Atmosphere. Gostekhizdat. 1950.
85. ZUEV, V.E. and S.D. TVOROGOV. – Izvestiya Vuzov SSSR. Seriya Fizika, No.1, p.185. 1965.
86. ZUEV, V.E. and S.D. TVOROGOV. – Izvestiya Vuzov SSSR, Seriya Fizika, No.2, p.170. 1965.
87. MATOSSI, F., R. MAYER, and E. RAUSCHER. – Phys. Rev., Vol.76, No.6, p.1049. 1949.
88. ZUEV, V.E. and S.D. TVOROGOV. – Izvestiya Vuzov SSSR. Seriya Fizika, No.4, p.26. 1966.
89. ELSASSER, W.M. Heat Transfer by Infrared Radiation in the Atmosphere. Harvard. 1942.
90. KONDRAT'EV, K.Ya. Radiative Heat Exchange in the Atmosphere. Gidrometeoizdat. 1956.
91. GOODY, R.M. – Q. J. Roy. Meteorol. Soc., Vol.78, p.336. 1952.
92. PLASS, G.N. – J. Opt. Soc. Amer., Vol.48, No.10, p.690. 1958.
93. GODSON, W.L. – J. Opt. Soc. Amer., Vol.12, No.3. 1955; Proc. Toronto Meteorol. Conf., Toronto, 1953. 1954.
94. KAPLAN, L. – Proc. Toronto Meteorol. Conf., Toronto, Vol.43. 1953.
95. SOSHNIKOV, V.N. – Optika i Spektroskopiya, Vol.10, No.4, p.448. 1961.
96. WYATT, P.J., V.R. STULL, and G.N. PLASS. – J. Opt. Soc. Amer., Vol.52, No.11, p.1209. 1962.
97. SOSHNIKOV, V.N. – Optika i Spektroskopiya, Vol.12, No.1, p.123. 1962.
98. PLASS, G.N. Advances in Molecular Spectroscopy, Vol.3, p.1326. Oxford–London–New York–Paris, Pergamon Press. 1962.
99. GODSON, W.L. – Archiv Meteorol. Geophys. Bioklim., Vol.12, No.1, p.1. 1962.
100. GODSON, W.L. – Archiv Meteorol. Geophys. Bioklim., Vol.13, No.2, p.196. 1963.
101. GERSHUN, A. A Short History of Photometry. Selected Works on Photometry and Illumination Engineering. Fizmatgiz. 1958.
102. STRONG, F.S. – Annal. Chem., Vol.24, p.338. 1952.
103. VAVILOV, S.D. – Microstructure of Light. Izdatel'stvo AN SSSR. 1950.
104. HOUGHTON, J.T. – Q. J. Roy. Meteorol. Soc., Vol.89, No.381, p.332. 1963.
105. ANGSTRÖM, K. – Oversing of K. Vetensk Akad. Förh., Vol.46, p.546. 1889; Vol.47, p.331. 1890; Phys. Rev., Vol.1, p.597. 1892; Ann. Physik, Vol.6, p.163. 1901; Arkiv Math. astron. Physik, Vol.4, No.30. 1907.
106. SCHAEFER, C. – Ann. Physik, Vol.16, p.93. 1905.
107. VON BAHR, E. – Ann. Physik, Vol.29, p.780. 1909; Vol. 33, p.585. 1910; Phys. Zeit., Vol.12, p.1167. 1911; Verh. deutsche Phys. ges., Vol.15, p.673, 710. 1913.
108. RANK, D.H., A.H. GUENTHER, J.H. SHEARER, and T.A. WIGGINS. – J. Opt. Soc. Amer. Vol.47, p.144. 1957.
109. PLYLER, E.K. and L.K. BLAINE. – J. Res. Nat. Bur. Stand., Vol.62, No.1. 1959.

110. RANK, D.H., D.P. EASTMAN, W.B. BIRTLEY, G. SKORINKO, and T.A. WIGGINS. – J. Opt. Soc. Amer., Vol.50, No.8, p.821. 1960.
111. NIELSEN, J.R., V. THORNTON, and E.B. DALE. – Rev. Mod. Phys., Vol.16, p.307. 1944.
112. HOWARD, J.N. D.E. BURCH, and D. WILLIAMS. – J. Opt. Soc. Amer., Vol.46, pp.186, 237, 242, 334, 452. 1956.
113. ZUEV, V.E., M.E. ELYASHBERG, and G.I. SAFRONOVA. – Izvestiya Vuzov SSSR, Seriya Fizika, No.5, p.77. 1960.
114. ZUEV, V.E. and B.P. KOSHELEV. – Izvestiya Vuzov SSSR, Seriya Fizika, No.6, p.172. 1960.
115. KOZLOV, V.P. and E.O. FEDOROVA. – Optika i Spektroskopiya, Vol.10, No.5. 1961.
116. DMITRIEVSKII, O.D., B.S. NEPORENT, and V.A. NIKITIN. – Uspekhi Fizicheskikh Nauk, Vol.64, p.447. 1958.
117. RAUTIAN, S.G. – Uspekhi Fizicheskikh Nauk, Vol.66, p.475. 1958.
118. PETRASH, G.G. – Optika i Spektroskopiya, Vol.6, No.6, p.792. 1959; Vol.8, No.1, p.122. 1960; Vol.9, No.3, p.423. 1960.
119. PETRASH, G.G. and S.G. RAUTIAN. – Inzhenerno-Fizicheskii Zhurnal, Vol.1, pp.61, 80. 1958.
120. DMITRIEVSKII, O.D. and V.A. NIKITIN. – Zhurnal Optiko-Mekhanicheskaya Promyshlennost', No.4, p.9. 1957; Optika i Spektroskopiya, Vol.8, No.1, p.120. 1960.
121. NIKITIN, V.A. – Optika i Spektroskopiya, Vol.4, p.523. 1958.
122. KYLE, T.G. and J.O. GREEN. – J. Opt. Soc. Amer., Vol.55, No.7, p.895. 1965.
123. RANK, D.H. and T.A. WIGGINS. – J. Quant. Spectry. Rad. Trans., Vol.3, No.4, p.377. 1963.
124. DENNISON, D. – Rev. Mod. Phys., Vol.3, p.280. 1931.
125. BONNER, L.J. – Phys. Rev., Vol.46, p.458. 1934.
126. ESCART, C. – Phys. Rev., Vol.47, p.552. 1959.
127. KHACHKURUZOV, G.A. – Trudy Gosudarstvennogo Instituta Prikladnoi Khimii, Vol.42, p.51. 1959.
128. WANG, S.E. – Phys. Rev., Vol.34, p.243. 1929.
129. RAY, B.S. – Z. Physik, Vol.78, p.74. 1932.
130. HAINER, R.M., P.C. CROSS, and G.W. KING. – J. Chem. Phys., Vol.11, p.27. 1943.
131. HAINER, R.M., P.C. CROSS, and G.W. KING. – J. Chem. Phys., Vol.17, p.826. 1949.
132. WILSON, E.B. and J.B. HOWARD. – J. Chem. Phys., Vol.4, p.260. 1936.
133. WILSON, E.B. – J. Chem. Phys., Vol.5, p.617. 1937.
134. EL'YASHEVICH, M.A. – Trudy Gosudarstvennogo Opticheskogo Instituta, Vol.4, No.106. 1939.
135. SHAFFER, W.H. and H.H NIELSEN. – Phys. Rev., Vol.56, p.188. 1939.
136. MAJUMDAR, D. – Proc. Phys. Soc., Vol.67, p.351. 1954; Vol.72, p.635. 1958.
137. MAJUMDAR, D. and A.K. GOSH. – Spectrochem. Acta, Vol.18, p.615. 1962.
138. BENEDICT, W., N. GAILAR, and E.K. PLYLER. – J. Chem. Phys., Vol.24, p.1139. 1956.
139. ZUEV, V.E., I.I. IPPOLITOV, Yu.S. MAKUSHKIN, A.A. ORLOV, and V.V. FOMIN. – DAN SSSR, Vol.179, No.1, p.51. 1968.
140. ZUEV, V.E., I.I. IPPOLITOV, Yu.S. MAKUSHKIN, A.A. ORLOV, and V.V. FOMIN. – Optika i Spektroskopiya, Vol.25, No.1. 1968.
141. ZUEV, V.E., I.I. IPPOLITOV, Yu.S. MAKUSHKIN, and V.V. FOMIN. – Report to the 14th General Assembly of the IUGG. Switzerland, September. 1967.
142. MAKUSHKIN, Yu.S. – Izvestiya Vuzov SSSR, Seriya Fizika, No.6, p.70. 1966.
143. MAKUSHKIN, Yu.S. – Izvestiya Vuzov SSSR, Seriya Fizika, No.2, p.79. 1967.
144. MAKUSHKIN, Yu.S. – Izvestiya Vuzov SSSR, Seriya Fizika, No.3, p.153. 1967.
145. MAKUSHKIN, Yu.S. – In: Report Conf. Inst. Higher Learning on Spectral Transmission of the Atmosphere (June 1965). Izdatel'stvo Tomskogo Gosudarstvennogo Universiteta. 1968.
146. MAKUSHKIN, Yu.S. – Optika i Spektroskopiya, Vol.25, No.1. [sic.]
147. IPPOLITOV, I.I. and Yu.S. MAKUSHKIN. – Optika i Spektroskopiya, Vol.24, No.4, p.530. 1968.

148. NIELSEN, H.H. – Rev. Mod. Phys., Vol.23, p.90. 1951.
149. GOLDSMITH M., G. AMAT, and H.H. NIELSEN. – J. Chem. Phys., Vol.24, p.1178. 1956.
150. PODOLSKY, B. – Phys. Rev., Vol.32, p.812. 1928.
151. AMAT, G., M. GOLDSMITH, and H.H. NIELSEN. –J.Chem. Phys., Vol.27, p.838. 1957.
152. AMAT, G. and H.H. NIELSEN. – J. Chem. Phys., Vol.27, p.845. 1957.
153. AMAT, G. and H.H. NIELSEN. – J. Chem. Phys., Vol.36, p.1859. 1962.
154. BLOKHINTSEV, D.I. Principles of Quantum Mechanics. Izdatel'stvo Vysshaya Shkola. 1963.
155. SLEATOR, W.W. – Astrophys. J., Vol.48, p.125. 1918.
156. SLEATOR, W.W. and E.R. PHELPS. – Astrophys. J., Vol.62, p.28. 1925.
157. PLYLER, E.K. and W.W. SLEATOR. – Phys. Rev., Vol.37, p.1493. 1931.
158. PLYLER, E.K. – Phys. Rev., Vol.39, p.77. 1932.
159. MECKE, R. – Z.Physik, Vol.81, p.373. 1933.
160. BOUMANN, W. and R. MECKE. – Z. Physik, Vol.81, p.445. 1933.
161. FREUDENBERG, K. and R. MECKE. – Z. Physik, Vol.81, p.465. 1933.
162. NIELSEN, H.H. – Phys. Rev., Vol.59, p.565. 1941; Vol.62, p.422. 1942.
163. NELSON, R.C. and W.S. BENEDICT. – Phys. Rev., Vol.74, p.703. 1948.
164. BENEDICT, W.S. and E.K. PLYLER. – J. Res. Nat. Bur. Stand., Vol.46, p.246. 1951.
165. BENEDICT, W.S. – Bull. Amer. Phys. Soc., Vol.23, p.54. 1948; Phys. Rev., Vol.74, p.1246. 1948.
166. BENEDICT, W.S., H.H. CLAASSEN, and J.H. SHAW. – J. Res. Nat. Bur. Stand., Vol.49, p.91. 1952.
167. DALBY, F.W. and H.H. NIELSEN. – J. Chem. Phys., Vol.25, p.934. 1956.
168. BENEDICT, W.S., A.M. BASS, and E.K. PLYLER. – J. Res. Nat. Bur. Stand., Vol.52, p.64. 1954.
169. BARNES, R.B. and W.S. BENEDICT. – Phys. Rev., Vol.47, p.918. 1935.
170. RANDALL, H.M., D.M. DENNISON, N. GINSBERG, and L.R. WEBER. – Phys. Rev., Vol.52, p.160. 1937.
171. DENNISON, D.M. – Rev. Mod. Phys., Vol.12, p.175. 1940.
172. YAROSLAVSKII, N.G. and A.E. STANEVICH. – Optika i Spektroskopiya, Vol.5, p.384. 1958; Vol.7, p.626. 1959.
173. FURASHOV, N.I. – Optika i Spektroskopiya, Vol.20, No.3, p.427. 1966.
174. MINNARET, M., G.F. MULDERS, and J. HOUTGAST. Photometric Atlas of the Solar Spectrum λ 3612 to λ 8771. Amsterdam. 1940.
175. BABCOCK, H.D. and C.E. MOORE. The Solar Spectrum, λ 6600 to λ 13495. Publ. No.579. Washington, Carnegie Institute of Washington. 1947.
176. MOHLER, O.C., A.K. PIERCE, R.R. McMATH, and L. GOLDBERG. Atlas of the Solar Spectrum from 0.84 to 2.55 Microns. – Ann Arbor Univ. Mich. Press. 1950.
177. MOHLER, O.C. and W.S. BENEDICT. – Phys. Rev., Vol.74, p.702. 1948.
178. MOHLER, O.C. Tables of Solar Spectrum Wavelength 1.20–2.55 Microns. – Ann Arbor, Univ. Mich. Press. 1955.
179. SHAW, J.H., R.M. CHAPMAN, J.N. HOWARD, and M.L. OXHOLM. – Astrophys. J., Vol.113, p.268. 1951.
180. SHAW, J.H., M.L. OXHOLM, and H.H. CLAASSEN. – Astrophys. J., Vol.116, p.354. 1952.
181. MIGEOTTE, M., L. NEVEN, and J. SWENSON. The Solar Spectrum from 2.8 to 23.7 Microns. Astrophysics Institute, University of Liège, Final Report, Pt.I. 1956; Pt.II. 1957.
182. MIGEOTTE, M., L. NEVEN, J. SWENSON, and W.S. BENEDICT. The Solar Spectrum from 2.8 to 23.7 Microns. Astrophysics Institute, University of Liège, Final Report Pt.II. 1957.
183. DELBOUILLE, L. and G. ROLAND. Atlas Photométrique du spectre solaire de λ 7498 à λ 12016. Liège. 1963.
184. FARMER, C.B. and P.J. KEY. – Appl. Optics, Vol.4, No.9, p.1051. 1965.
185. CURCIO, J.A., L.F. DRUMMETER, and G.L. KNESTRICK. – Appl. Optics, Vol.3, No.12, p.1401. 1964.

186. MOORE, C.E., M.G.J.MINNAERT, and J.HOUTGAST. Tables of Spectral Lines of the Atmosphere from λ 2935 to λ 8770 Å. Washington Gov. Print. Off. 1966.

187. OPPENHEIM, U.P. and A.GOLDMAN. Atlas and Wavenumber Tables of Water Vapor Absorption at 1200°K in the 1.9 and 2.7 μ Regions. – Scientific Report No.1 AFEOAR 63 – 111, Haifa, Israel. 1964.

188. SCHWENDENMAN, R.H. and V.W.LAURIE. Tables of Line Strengths for Rotational Transmissions of Asymmetric Rotar Molecules. New York, Pergamon Press. 1958.

189. CALFEE, R.F. – J. Quant. Spectry. Rad. Trans., Vol.6, No.3, p.221. 1966.

190. IPPOLITOV, I.I. – Izvestiya Vuzov SSSR, Seriya Fizika, No.7, p.145. 1967.

191. CROSS, P.C., R.M.HAINER, and G.W.KING. – J. Chem. Phys., Vol.12, p.210. 1944.

192. IPPOLITOV, I.I. – Optika i Spektroskopiya, Vol.27, No.3, p.458. 1969.

193. KHACHKURUZOV, G.A. – Trudy Gosudarstvennogo Instituta Prikladnoi Khimii, Vol.46, p.66. 1960.

194. PAPOUŠEK, D. and J.PLIVA. – Collect Czech. Chem. Comm., Vol.29, p.1973. 1964.

195. VASILEVSKII, K.P. and B.S.NEPORENT. – Optika i Spektroskopiya, Vol.7, p.572. 1959.

196. VASILEVSKII, K.P. and B.S.NEPORENT. – Optika i Spektroskopiya, Vol.4, p.474. 1958.

197. VASILEVSKII, K.P. – Optika i Spektroskopiya, Vol.11, p.207. 1961.

198. SANDERSON, R.B. and N.GINSBURG. – J. Quant. Spectry. Rad. Trans., Vol.3, No.4, p.435. 1963.

199. SAIEDY, F. – Q. J. Roy. Meteorol. Soc., Vol.87, No.384, p.578. 1961.

200. BOLLE, H.J. – J. Opt. Soc. Amer., Vol.56, No.4, p.560. 1966.

201. GOLDMAN, A. and U.P.OPPENHEIM. – Appl. Optics, Vol.5, No.6, p.1073. 1966.

202. JAFFE, J.H. and W.S.BENEDICT. – J. Quant. Spectry. Rad. Trans., Vol.3, No.1, p.87. 1963.

203. HIRSHFELD, M.A., J.H.JAFFE, and G.ROSS. – J. Quant. Spectry. Rad. Trans., Vol.6, No.3, p.311. 1966.

204. LANDAU, L.D. and E.M.LIFSHITS. Field Theory. Fizmatgiz. 1962.

205. DAVYDOV, A.S. Quantum Mechanics. Fizmatgiz. 1963.

206. ZUEV, V.E. – In: Report Conf. Inst. Higher Learning on Spectral Transmission of the Atmosphere (June 1965). Izdatel'stvo Tomskogo Gosudarstvennogo Universiteta. 1968.

207. GRADSHTEIN, I.S. and I.M.RYZHIK. Tables of Integrals, Sums, Series and Products. Fizmatgiz. 1962.

208. BECKER, G.E. and S.H.AUTLER. – Phys. Rev., Vol.70, p.300. 1946.

209. COWLING, T.G. – Phil. Mag., Vol.41, p.109. 1950.

210. BENEDICT, W.S. – J. Opt. Soc. Amer., Vol.47, p.1056. 1957.

211. KONDRAT'EV, K.Ya. and Yu.M.TIMOFEEV. – Izvestiya AN SSSR, Seriya Fizika Atmosfery i Okeana, Vol.2, No.3, p.272. 1966.

212. PLASS, G.N. and V.R.STULL. Theoretical Study of High Temperature Emissivities and Atmospheric Transmission. Contract, Bedford. 1960.

213. HOWARD, J.N., D.E.BURCH, and D.WILLIAMS. Near Infrared Transmission Through Synthetic Atmosphere. Contract, Ohio University. 1955.

214. STULL, V.R., P.J.WYATT, and G.N.PLASS. The Infrared Absorption of Water Vapour. – Report SSD-TDR-62-127, Vol.II. Space Systems Division, Air Force Systems Command, Los Angeles, California. 1962.

215. STULL, V.R., R.J.WYATT, and G.N.PLASS. The Infrared Absorption of Carbon Dioxide. – Report SSD-TDR-62-127, Vol.III. Space Systems Division, Air Force Systems Command, Los Angeles, California. 1962.

216. BURCH, D.E., D.GRYVNAK, E.B.SINGLETON, W.L.FRANCE, and D.W.WILLIAMS. Infrared Absorption by CO_2, Water Vapour and Minor Atmospheric Constituents. – AFCRL-62-698, Bedford, Mass. July. 1962.

217. WYATT, P.J., V.R. STULL, and G.N. PLASS. – Appl. Optics, Vol.3, No.2, p.229. 1964.
218. OPPENHEIM, U.P. and Y. BEN-ARYEH. – J. Quant. Spectry. Rad. Trans., Vol.4, No.4, p.559. 1964.
219. ZUEV, V.E., L.I. NESMELOVA, V.A. SAPOZHNIKOVA, and S.D. TVOROGOV. – In: Actinometry and Atmospheric Optics (Report 5th Interdepartmental Conf. on Actinometry and Atmospheric Optics), edited by G.V. Rozenberg. Izdatel'stvo Nauka. 1964.
220. ZUEV, V.E. – In: Jubilee Volume of Reports on Physics of Tomsk University. Izdatel'stvo Tomskogo Gosudarstvennogo Universiteta. 1970.
221. BURCH, D.E. and D.W. WILLIAMS. – Appl. Optics, Vol.2, No.6. 1963.
222. LARMOR, L. and S. PASSMAN. – Proc. IRIS, Vol.1, No.1, p.1. 1956; Vol.1, No.2, p.15. 1956.
223. LANGER, R.M. – Proc. IRIS, Vol.2. July. 1958.
224. ROACH, W.T. – Q. J. Roy. Meteorol. Soc., Vol.87, No.373, p.346. 1961.
225. ZUEV, V.E. – Izvestiya Vuzov SSSR, Seriya Fizika, Nos.3, 6. 1961; Nos.1, 3, 6. 1962; In: Scattering and Polarization of Light in the Earth's Atmosphere. Izdatel'stvo AN Kazakhskoi SSR. 1962.
226. BURCH, D.E., E.B. SINGLETON, and D. WILLIAMS. – Appl. Optics, Vol.1, No.3, p.359. 1962.
227. PALMER, C.H. – J. Opt. Soc. Amer., Vol.50, p.1232. 1960.
228. IZATT, J. – Ph.D. Dissertation. Johns Hopkins Univ. 1960.
229. GOODY, R.M. and T.W. WORMELL. – Proc. Roy. Soc., A, Vol.209, p.178. 1951.
230. CROSS, P.S. and F. DANIELS. – J. Chem. Phys., Vol.2, p.6. 1934.
231. EDWARDS, D.K. – J. Opt. Soc. Amer., Vol.50, p.617. 1960.
232. STAUFFER, F.R. and T.E. WALSH. – J. Opt. Soc. Amer., Vol.56, No.3, p.401–406. 1966.
233. PALMER, C.H. – J. Opt. Soc. Amer., Vol.47, pp.1024, 1028, 1054. 1957; Vol.49, p.1139. 1959; Vol.50, p.1232. 1960.
234. BURCH, D.E. and E. REISMAN. – J. Opt. Soc. Amer., Vol.56, No.4, p.556. 1966.
235. GOLDSTEIN, R. – J. Quant. Spectry. Rad. Trans., Vol.4, No.2, p.343. 1964.
236. LUDWIG, C.B., C.C. FERRISO, and C.N. ABEYTA. – J. Quant. Spectry. Rad. Trans., Vol.5, No.2, p.281. 1965.
237. GRAY, L.D. and S.S. PENNER. – J. Quant. Spectry. Rad. Trans., Vol.5, No.4, p.611. 1965.
238. FERRISO, C.C., C.B. LUDWIG, and A. THOMSON. – J. Opt. Soc. Amer., Vol.55, No.5, p.617. 1965.
239. FERRISO, C.C. and C.B. LUDWIG. – J. Opt. Soc. Amer., Vol.56, No.4, p.560. 1966.
240. GOLDMAN, A. and U.P. OPPENHEIM. – J. Opt. Soc. Amer., Vol.55, No.7, p.794. 1965.
241. EDWARDS, D.K., B.J. FLORNES, L.K. GLASSEN, and W. SUN. – Appl. Optics, Vol.4, No.6, p.715. 1965.
242. FERRISO, C.C. and C.B. LUDWIG. – J. Quant. Spectry. Rad. Trans., Vol.4, p.215. 1964.
243. FERRISO, C.C. and C.B. LUDWIG. – J. Chem. Phys., Vol.41, p.1668. 1964.
244. MÖLLER, F. – Beit. Geophys., Vol.58, p.27. 1942.
245. GEBBIE, H.A., W.R. HARDING, C. HILSUM, A.W. PRYCE, and V. ROBERTS. – Proc. Roy. Soc., A, Vol.206, No.1084, p.87. 1951.
246. TAYLOR, J.H. and H.W. YATES. – J. Opt. Soc. Amer., Vol.47, No.3, p.223. 1957.
247. ADEL, A. – Astrophys. J., Vol.89, p.1. 1939.
248. ANTHONY, R. – Phys. Rev., Vol.85, p.674. 1952.
249. FEIGEL'SON, E.M. – Izvestiya AN SSSR, Seriya Geofizicheskaya, No.4, p.92. 1951.
250. ROACH, W.T. and R.M. GOODY. – Q. J. Roy. Meteorol. Soc., Vol.84, p.319. 1958.
251. VOLZ, F. – Ann. Meteorol., Vol.8, p.34. 1957.
252. BIGNELL, K., F. SAIEDY, and P.A. SHEPPARD. – J. Opt. Soc. Amer., Vol.53, No.4, p.466. 1963.
253. KONDRAT'EV, K.Ya., I.Ya. BADINOV, S.V. ASHCHEULOV, and S.D. ANDREEV. – Izvestiya AN SSSR, Fizika Atmosfery i Okeana, Vol.1, No.4, p.363. 1965.

254. KAPLAN, L.D. – In: Les problèmes météorologiques de la stratosphère et de la mesosphère, p. 307. Paris, Presses Universitaires de France. 1966.
255. KAPLAN, L.D. – J. Chem. Phys., Vol.18, No.2, p.186. 1950.
256. YAMOMOTO, G. and T. SASAMORY. – Sci. Reports, Tohoku Univ., Series 5, Geophys., Vol.10, No.2, p.37. 1958; Vol.13, p.1. 1961.
257. GORDON, H.R. and T.K. McCUBBIN, Jr. – J. Mol. Spectry., Vol.18, No.1, p.73. 1965.
258. KAPLAN, L.D. and D.F. EGGERS. – J. Chem. Phys., Vol.25, p.876. 1950.
259. NIELSEN, A.H. and Y.T. YAO. – Phys. Rev., Vol.68, p.173. 1945.
260. BENEDICT, W.S., R.C. HERMAN, and S. SILVERMAN. – J. Chem. Phys., Vol.19, No.10, p.1325. 1951.
261. COURTOY, C.P. – Canad. J. Phys., Vol.35, p.608. 1957.
262. GRAY, L.D. and J.E. SELVIDGE. – J. Quant. Spectry. Rad. Trans., Vol.5, No.2, p.291. 1965.
263. CALFEE, R.F. and W.S. BENEDICT. – Appl. Optics, Vol.5, No.10, p.1695. 1966; Carbon Dioxide Spectral Line Positions and Intensities Calculated for 2.05 and 2.7 Microns Region. – Nat. Bur. Stand., Washington, D.C. 1966.
264. WILLIAMS, D. – Appl. Optics, Vol.5, No.10, p.1696. 1966.
265. GORDON, H.R. and T.K. McCUBBIN, Jr. – J. Mol. Spectry., Vol.19, No.2, p.137. 1966.
266. VASILEVSKII, K.P., V.A. KAZBANOV, and T.E. DERVIZ. – Optika i Spektroskopiya, Vol.23, No.6, p.888. 1967.
267. BOESE, R.W., J.H. MILLER, and E.C.Y. INN. – J. Quant. Spectry. Rad. Trans., Vol.6, No.6, p. 717. 1966.
268. GORCHAKOVA, I.A. and M.S. MALKEVICH. – Izvestiya AN SSSR, Seriya Fizika Atmosfery i Okeana, Vol.2, No.6, p.585. 1966.
269. STULL, V.R., P.J. WYATT, and G.N. PLASS. – Appl. Optics, Vol.3, No.2, p.243. 1964.
270. MEGILL, L.R. and P.M. JAMNICK. – J. Opt. Soc. Amer., Vol.51, No.11, p.1294. 1961.
271. GRAY, L.D. – J. Opt. Soc. Amer., Vol.56, No.4, p.555. 1966.
272. GRAY, L.D. and R.A. McCLATCHEY. – J. Opt. Soc. Amer., Vol.55, No.11, p.1567. 1965.
273. BURCH, D.E., D.A. GRYVNAK, and D. WILLIAMS. – Appl. Optics, Vol.1, No.6, p.759. 1962.
274. EDWARDS, D.K. and W.A. MENARD. – Appl. Optics, Vol.3, No.7, p.847. 1964.
275. EDWARDS, D.K. and W. SUN. – Appl. Optics, Vol.3, No.12, p.1501. 1964.
276. VARANASI, P. and J.L. LAUER. – J. Quant. Spectry. Rad. Trans., Vol.6, No.1, p.127. 1966.
277. CLOUGH, S.A. and F.X. KNEIZYS. – J. Chem. Phys., Vol.44, No.5, p.1855. 1966.
278. CLOUGH, S.A. and F.X. KNEIZYS. Ozone Absorption in the 9.0 Micron Region. – AFCRL-65-862, Report, Air Force Cambridge Research Laboratories. 1965.
279. WHITE, J.U., N.L. ALPERT, A.G. DEBELL, and R.M. CHAPMAN. – J. Opt. Soc. Amer., Vol.47, No.5, p.358. 1957.
280. WALSHAW, C.D. An Experimental Investigation of the 9.6 μ Band of Ozone. – Thesis, Cambridge University. 1954.
281. NEXSEN, W.E., Jr. – Sci Report No.1, Contract AF19(604)–1003 with the Ohio State University AFCRL-TN-56-265. 1965.
282. McCAA, D.J. and J.H. SHAW. – Appl. Optics, Vol.2, p.581. 1963.
283. TRAJMAR, S. – Paper Presented at Symposium on Molecular Structure and Spectroscopy. Columbus, Ohio. 1963.
284. TRAJMAR, S., and D.J. McCAA.–J. Mol. Spectry., Vol.14, No.3, p.244. 1964.
285. GORA, E.K. – J. Mol. Spectry., Vol.3, p.78. 1959.
286. GEBBIE, H.A., N.W.B. STONE. G. TOPPING, E.K. GORA, S.A. CLOUGH, and F.X. KNEIZYS. – J. Mol. Spectry., Vol.19, No.1, p.7. 1966.
287. WALSHAW, C.D. – Q. J. Roy. Meteorol. Soc., Vol.83, No.357, p.315. 1957.
288. ALTSHULLER, T.L. Infrared Transmission of Clear Atmosphere. – Document 61SD199, General Electric Company, Philadelphia, Penns., Dec. 1961.

289. ELSASSER, W.M. and M.F. CULBERTSON. – Meteorol. Monogr., Vol.4, No.23, Am. Met. Soc. 1960.
290. ALLEN, H.C., and E.K. PLYLER. – J. Chem. Phys., Vol.26, No.4, p.972. 1957.
291. HERRANZ, J. and G. THYAGARAJAN. – J. Mol. Spectry., Vol.19, No.3, p.247. 1966.
292. HERRANZ, J., J. MORCILLO, and A. GOMES. – J. Mol. Spectry., Vol.19, No.3, p.266. 1966.
293. BURCH, D.E. and D. WILLIAMS. – Appl. Optics, Vol.1, No.4, p.473. 1962.
294. BURCH, D.E. and D. WILLIAMS. – Appl. Optics, Vol. 1, No.5, p.587. 1962.
295. LONG, R.L. Atmospheric Absorption and Laser Radiation. – Ohio State University Engineering Publications, Bulletin 199. Columbus, Ohio. 1967.
296. HOWARD, J.N. and J.S. GARING. – In: U.S. National Report, 14th General Assembly Transactions, American Geophysical Union, Vol.48, No.2, p.471. 1967.
297. GRAY, L.D. – Appl. Optics, Vol.4, No.11, p.1494. 1965.
298. ABELS, L.L. and D.L. FORD. – J. Opt. Soc. Amer., Vol.55, No.5, p.617. 1965.
299. OPPENHEIM, U.P. and A. GOLDMAN. – J. Opt. Soc. Amer., Vol.56, No.5, p.675. 1966.
300. RANK, D.H., A.F. SLOMBA, F.F. GARDNER, and T.A. WIGGINS. – J. Opt. Soc. Amer., Vol.52, No.8, p.858. 1962.
301. DORMAN, F. and C.C. LIN. – J. Mol. Spectry., Vol.12, No.2, p.119. 1964.
302. GORDON, H.R. and T.K. McCUBBIN. – J. Mol. Spectry., Vol.18, No.1, p.73. 1965.
303. ABELS, L.L. and J.H. SHAW. – J. Opt. Soc. Amer., Vol.53, No.7, p.856. 1963.
304. YAMADA, H. and W.S. PERSON. – J. Chem. Phys., Vol.45, No.6, p.1861. 1966.
305. HATHAWAY, C.E., G.M. HOOVER, and D. WILLIAMS. – J. Opt. Soc. Amer., Vol.56, No.4, p. 555. 1966.
306. EDWARDS, D.K. – Appl. Optics, Vol.4, No.10, p.1352. 1965.
307. BENEDICT, W.S., R. HERMAN, G.E. MOORE, and S. SILVERMAN. – Astrophys. J., Vol.135, No.1, p.277. 1962.
308. SHAW, J.H. and W.L. FRANCE. Intensities and Widths of Single Lines of the 4.7 Microns CO Fundamental. – Scientific Rep. 4 on Project 587. Ohio State Univ. 1956.
309. DOWLING, J.M. and R.T. HALL. – J. Mol. Spectry., Vol.19, No.1, p.108. 1966.
310. WEINBERG, J.M., E.S. FISHBURNE, and K.N. RAO. – J. Mol. Spectry., Vol.18, No.4, p.428. 1965.
311. GRYVNAK, D.A. and J.H. SHAW. – J. Opt. Soc. Amer., Vol.52, No.5, p.539. 1962.
312. HATHAWAY, C.E., G.M. HOOVER, L.D. TUBBS, and D. WILLIAMS. – J. Opt. Soc. Amer., Vol.56, No.10, p.1422. 1966.
313. ABU-ROMIA, M.M. and C.L. TIEN. – J. Quant. Spectry. Rad. Trans., Vol.6, No.2, p.143. 1966.
314. SHAW, J.H. and J.T. HOUGHTON. – Appl. Optics, Vol.3, No.6, p.773. 1964.
315. HOOVER, G.M., C.E. HATHAWAY, and D. WILLIAMS. – Appl. Optics, Vol.6, No.3, p.481. 1967.
316. GRAY, L.D. and R.A. McCLATCHEY. – Appl. Optics, Vol.4, p.1624. 1965.
317. PENNER, S.S. and P. VARANASI. – J. Quant. Spectry. Rad. Trans., Vol.6, No.2, p.181. 1966.
318. SAKAI, H. and F.R. STAUFFER. – J. Opt. Soc. Amer., Vol.54, p.759. 1964.
319. SAKAI, H. – J. Opt. Soc. Amer., Vol.56, No.1, p.127. 1966.
320. PLASS, G.N. – J. Opt. Soc. Amer., Vol.55, No.1, p.104. 1965.
321. SAKAI, H. – J. Opt. Soc. Amer., Vol.56, No.4, p.556. 1966.
322. PLASS, G.N. – J. Opt. Soc. Amer., Vol.55, No.5, p.593. 1965.
323. PLASS, G.N. – Appl. Optics, Vol.4, p.69. 1965.
324. BALANDIN, N.I., E.S. KUZNETSOVA, and M.V. PODKLADENKO. – Prikladnaya Spektroskopiya, Vol.1, No.1, p.66. 1964.
325. PODKLADENKO, M.V. – In: Report Conf. Inst. Higher Learning on Spectral Transmission of the Atmosphere in the Visible and Infrared, p.106. Izdatel'stvo Tomskogo Gosudarstvennogo Universiteta. 1968.

326. CURCIO, J.A. and D.V.E.BUTTREY. – Appl. Optics, Vol.5, No.2, p.231. 1966.
327. ZUEV, V.E. – Izvestiya Vuzov SSSR, Seriya Fizika, No.3, p.138. 1967.
328. ZUEV, V.E. – Izvestiya Vuzov SSSR, Seriya Fizika, No.10, p.53. 1967.
329. ZUEV, V.E. V.V.POKASOV, Yu.A.PKHALAGOV, A.V.SOSNIN, and S.S.KHMELEVTSOV. – Izvestiya AN SSSR, Fizika Atmosfery i Okeana, Vol.4, No.1, p.63. 1968.
330. ANTIPOV, B.A., V.E.ZUEV, and V.A.SAPOZHNIKOVA. – Izvestiya Vuzov SSSR, Seriya Fizika, No.7, p.142. 1967.
331. ANTIPOV, B.A., V.E.ZUEV, and V.A.SAPOZHNIKOVA. – Izvestiya Vuzov SSSR, Seriya Fizika, No.6, p.158. 1967.
332. EDWARDS, B.N. and D.E.BURCH. – J. Opt. Soc. Amer., Vol.55, No.2, p.174. 1965.
333. CHU, T.S. and D.C.HOGG. – The Bell System Technical Journal, Vol.45, No.2, p.301. 1966.
334. LONG, R. and T.H.LEWIS. – In: Lasers and Applications, p.209. Columbus, Ohio, U.S.A. 1963.
335. McCUBBIN, T.K., Jr. and R.DARONE. – J. Opt. Soc. Amer., Vol.56, No.4, p.555. 1966.
336. LONG, R.K. and J.H.McCOY. – J. Opt. Soc. Amer., Vol.57, No.4, p.570. 1967.
337. HOWE, J.A. and R.A.McFARLANE. – J. Mol. Spectry., Vol.19, No.2, p.224. 1966.
338. CALFEE, R.F. and D.M.GATES. – Appl. Optics, Vol.5, No.2, p.287. 1966.
339. ZUEV, V.E. and S.D.TVOROGOV. – Izvestiya Vuzov SSSR, Seriya Fizika, No.6, p.84. 1965.
340. PLASS, G.N. – Appl. Optics, Vol.2, No.5, p.515. 1963.
341. PLASS, G.N. Transmittance Tables for Slant Path in the Stratosphere. – SSD-TDR-62-127, Vol.V. Space System Division, Air Force Systems Command, Los Angeles, California. 22 May. 1963.
342. PLASS, G.N. – Appl. Optics, Vol.3, No.4, p.479. 1964.
343. MAKUSHKIN, Yu.S. – Prikladnaya Spektroskopiya, Vol.8, No.6, p.1068. 1968.
344. KONDRAT'EV, K.Y., I.J.BADINOV, G.N.GAEVSKAYA, G.A.NICOLSKY, and G.M.SHVED. – In: Les problèmes météorologiques de la stratosphère et de la mesosphère, p.271. Paris, Presses Universitaires de France. 1966.
345. MURCRAY, D.G., F.H.MURCRAY, and W.J.WILLIAMS. – J. Opt. Soc. Amer., Vol.55, No.10, p.1239. 1965.
346. MURCRAY, D.G., F.H.MURCRAY, W.J.WILLIAMS, and F.E.LESLIE. – J. Geophys. Res., Vol.65, p.11. 1960; J. Opt. Soc. Amer., Vol.51, No.2, p.186. 1961.
347. MURCRAY, D.G., F.H.MURCRAY, and W.J.WILLIAMS. – J. Geophys. Res., Vol.67, No.2, p.759. 1962.
348. NEPORENT, B.S. and M.S.KISELEVA. – Optika i Spektroskopiya, Vol.16, No.5, p.803. 1964.
349. YATES, D.M. – J. Opt. Soc. Amer., Vol.50, No.12, p.1299. 1960.
350. YATES, D.M. and W.J.HARROP. – Appl. Optics, Vol.2, No.9, p.887. 1963.
351. HOUGHTON, J.T. and J.S.SEELEY. – Q. J. Roy. Meteorol. Soc., Vol.86, No.369, p.358. 1960.
352. HOUGHTON, J.T., N.D.P.HUGHES, T.S.MOSS, and J.S.SEELEY. – Phil. Trans. Roy. Soc., Vol.254, No.1043, p.47. 1962.
353. HOUGHTON, J.T. – Q. J. Roy. Meteorol. Soc., Vol.89, No.381, p.319. 1963.
354. CUMMING, C. Stratospheric Composition from Infrared Solar Spectra. – Thesis, International Symposium on Radiation Processes, Leningrad. 1964.
355. STAUFFER, F. and J.STRONG. – Appl. Optics, Vol.1, No.2, p.129. 1962.
356. BROWN, A.H. and V.ROBERTS. – J. Sci. Instr., Vol.30, No.1, p.5. 1953.
357. JONES, F.H. and V.ROBERTS. – Proc. Roy. Soc., A, Vol.236, No.1205, p.171. 1956.
358. LEVIN, L.M. Studies on Physics of Coarse Aerosols. Izdatel'stvo AN SSSR. 1961.
359. BOROVIKOV, A.M. et al. – In: Cloud Physics, edited by A.Kh.Khrgian. Gidrometeoizdat. 1961.
360. HUANG-MEI-YÜAN. – Izvestiya AN SSSR, Seriya Geofizicheskaya, No.2, p.362. 1963.
361. MASON, B.J. The Physics of Clouds. Oxford. 1957.

362. RITTBERGER, W. – Archiv Meteorol., Geophys. Bioklim., Ser. A., Vol.11, No.3, p.333. 1959.

363. BEST, A.C. – Q. J. Roy. Meteorol. Soc., Vol.77, p.418. 1951.

364. KHRGIAN, A.Kh. and I.P. MAZIN. – Trudy Tsentral'noi Aerologicheskoi Observatorii, No.7, p.56. 1952.

365. BOROVIKOV, A.M. – Trudy Tsentral'noi Aerologicheskoi Observatorii, No.3, p.3. 1948.

366. LEVIN, L.M. – DAN SSSR, Vol.94, No.6, p.1045. 1954.

367. KOLMOGOROV, A.N. – DAN SSSR, Vol.31, No.2, p.99. 1941.

368. LEVIN, L.M. – Izvestiya AN SSSR, Seriya Geofizicheskaya, No.10, p.1211. 1958.

369. SHIFRIN, K.S. – Trudy Glavnoi Geofizicheskoi Observatorii, No.46, p.5. 1955.

370. FUCHS, N.A. The Mechanics of Aerosols. Oxford, Pergamon Press. 1964.

371. D'YACHENKO, P.V. – Trudy Glavnoi Geofizicheskoi Observatorii, No.101, p.3. 1959.

372. KHRGIAN, A.Kh. and I.P. MAZIN. – Trudy Tsentral'noi Aerologicheskoi Observatorii, No.17, p.36. 1956.

373. DERGACH, A.L. – Trudy Arkticheskogo i Antarkticheskogo Nauchno-Issledovatel'skogo Instituta, Vol.228, p.55. 1959.

374. NIKANDROV, V.Ya. – Trudy Arkticheskogo i Antarkticheskogo Nauchno-Issledovatel'skogo Instituta, Vol.228, p.146. 1959.

375. DOLGIN, I.M. – Trudy Arkticheskogo i Antarkticheskogo Nauchno-Issledovatel'skogo Instituta, Vol.239, p.5. 1962.

376. VOSKRESENSKII, A.I. – Trudy Arkticheskogo i Antarkticheskogo Nauchno-Issledovatel'skogo Instituta, Vol.239, p.64. 1962.

377. VOSKRESENSKII, A.I. and A.L.DERGACH. – In: Study on Clouds, Precipitation and Thunderstorm Electricity, p.101. Izdatel'stvo AN SSSR. 1961.

378. KAZAS, V.I. – Izvestiya AN SSSR, Seriya Geofizicheskaya, No.5, p.805. 1963.

379. ARNULF, A., J.BRICARD, E.CURÉ, and C.VÉRET. – J. Opt. Soc. Amer., Vol.47, No.6, p.491. 1957.

380. WIENER, F.M., J.H.BALL, and C.M.GOGOS. – J. Geophys. Res., Vol.66, No.11, p.3974. 1961.

381. ELDRIDGE, R.G. – J. Meteorol., Vol.18, No.5, p.671. 1961.

382. MAY, K.R. – Q. J. Roy. Meteorol. Soc., Vol.87, No.374, p.535. 1961.

383. OKITA, T. – J. Meteorol. Soc. Japan., Vol.40, No.1, p.39. 1962.

384. ELDRIDGE, R.G. – J. Atmos. Sci., Vol.23, No.5, p.605. 1966.

385. MAZIN, I.P. and V.D. SKOSYREVA. – Trudy Tsentral'noi Aerologicheskoi Observatorii, No.36, p.43. 1961.

386. NIKANDROVA, G.T. – Trudy Glavnoi Geofizicheskoi Observatorii, No.126, p.70. 1962.

387. SELEZNEVA, E.S. – Meteorologiya i Gidrologiya, No.2. 1948.

388. PEDERSEN, K. and M. TODSEN. – Geophys. Publ. Vol.21, No.7. 1960.

389. MINERVIN, V.E. – Trudy Tsentral'noi Aerologicheskoi Observatorii, No.36, p.62. 1961.

390. SCHUMACHER, P. – J. Aircraft, Vol.1, No.1. 1964.

391. MINERVIN, V.E. – In: Study on Clouds, Precipitation and Thunderstorm Electricity, p.91. Izdatel'stvo AN SSSR. 1961.

392. BOROVIKOV, A.M., I.P. MAZIN, and A.N. NEVZOROV. – Trudy Tsentral'noi Aerologicheskoi Observatorii, No.36. 1961.

393. BOROVIKOV, A.M., I.P. MAZIN, and A.N. NEVZOROV. – Izvestiya AN SSSR. Fizika Atmosfery i Okeana, Vol.1, No.3, p.291. 1965.

394. OKITA, T. – Tellus, Vol.13, No.4, p.509. 1961.

395. PETRENCHUK, O.P. and V.M. DROZDOVA. – Trudy Glavnoi Geofizicheskoi Observatorii, No.185, p.117. 1966.

396. JUNGE, C.E. – Advances in Geophysics, Vol.4, p.1. 1958.

397. MAKHON'KO, K.P. – Izvestiya AN SSSR, Seriya Geofizicheskaya, No.1, p.183. 1963.
398. TWOMEY, S. and G.T.SEVERYNSE. – J. Atmos. Sci., Vol.21, No.5, p.558. 1964.
399. FEDOROV, M.M. – DAN SSSR, Vol.118, No.4, p.691. 1958.
400. MAKHON'KO, K.P. – Izvestiya AN SSSR, Seriya Geofizicheskaya, No.8, p.1235. 1959.
401. SPURNY, K. and J.PICH. – Studia Geophys. et Geolog., Vol.5, No.11, p.85. 1961.
402. HATCH, T. and S.CHOATE. – J. Franklin Institute, Vol.207. 1929.
403. VOEGTLIN, C. and H.C.HODGE. Pharmacology and Toxicology of Uranium Compounds. New York, McGraw-Hill. 1948.
404. PODZIMEK, J. – Studia Geophys. et Geolog., Vol.3, No.4. 1959.
405. PODZIMEK, J. and I. CERNOCH. – Geophys. Pure et Appl., Vol.50, No.3, p.96. 1961.
406. JUNGE, C.E. – J. Meteorol., Vol.12, p.13. 1955.
407. LAKTIONOV, A.G. – Izvestiya AN SSSR, Seriya Geofizicheskaya, No.3, p.419. 1958.
408. LAKTIONOV, A.G. – Izvestiya AN SSSR, Seriya Geofizicheskaya, No.4, p.566. 1960.
409. PLASS, G.N. – Appl. Optics, Vol.5, No.1, p.149. 1966.
410. OKITA, T. – J. Meteorol. Soc. Japan., Vol.35, No.2. 1957.
411. JUNGE, C.E., C.W.CHAGNON, and J.E.MANSON. – Science, Vol.133, No.3463, p.1478. 1961.
412. JUNGE, C.E. and J.E.MANSON. – J. Geophys. Res., Vol.66, No.7, p.2163. 1961.
413. FRIEDLANDER, C.K. – J. Meteorol., Vol.17, No.3, p.373. 1960; Vol.18, No.6, p.753. 1961.
414. IDLIS, G.M. – Izvestiya Astrofizicheskogo Instituta AN Kazakhskoi SSR, Vol.4, Nos.5–6, p.3. 1957.
415. GOETZ, A., O.PRIENING, and T.KALLAI. – Geophys. Pure et Appl., Vol.50, No.3, pp.67, 79. 1961.
416. KHRGIAN, A.Kh. Physics of the Atmosphere. Gostekhizdat. 1953.
417. LAKTIONOV, A.G. – Izvestiya AN SSSR, Seriya Geofizicheskaya, No.9, p.1397. 1960.
418. GAIVORONSKII, I.I. – Trudy Tsentral'noi Aerologicheskoi Observatorii, No.4, p.46. 1949.
419. PENNDORF, R. – J. Meteorol., Vol.11, No.3, p.245. 1954.
420. SELEZNEVA, E.S. and M.I.YUDIN. – Trudy Glavnoi Geofizicheskoi Observatorii, No.105, p.37. 1960.
421. SELEZNEVA, E.S. – In: Study on Clouds, Precipitation and Thunderstorm Electricity, p.96, 1961; Trudy Glavnoi Geofizicheskoi Observatorii, No.134. 1962.
422. DERGACH, A.L. – Trudy Glavnoi Geofizicheskoi Observatorii, No.105, p.30. 1960.
423. JUNGE, C.E., C.W.CHAGNON, and J.E.MANSON. – J. Meteorol., Vol.18, No.1, p.81. 1961.
424. CHAGNON, C.W. and C.E. JUNGE. – J. Meteorol., Vol.18, No.6, p.746. 1961.
425. MANSON, J.E., C.E.JUNGE, and C.W.CHAGNON. Chemical Reactions in the Lower and Upper Atmosphere, p.139. New York–London, Interscience. 1962.
426. JUNGE, C.E. – J. Meteorol., Vol.18, No.4. 1961.
427. DURBIN, W.S. – Geophys. Pure et Appl., Vol.42, No.1, p.11. 1959.
428. DURBIN, W.S. and G.D.WHITE. – Tellus, Vol.13, No.2, p.260. 1961.
429. SINGLETON, F. and W.S.DURBIN. – Q. J. Roy. Meteorol. Soc., Vol.88, No.377, p.315. 1961.
430. WRIGHT, F.W. and P.W.HODGE. – Nature, Vol.195, No.4838, p.269. 1962.
431. LAKTIONOV, A.G. – Trudy Vsesoyuznogo Nauchnogo Meteorologicheskogo Soveshchaniya, Vol.6, p.60. Gidrometeoizdat. 1963.
432. GEORGIEVSKII, Yu.S., A.Ya.DRIVING et al. – In: Searchlight in the Atmosphere, edited by G.V.Rozenberg. Izdatel'stvo AN SSSR. 1960.
433. ROSSLER, F. and E.VASSY. – C.r. Acad. Sci., Vol.254, No.11, p.2041. 1962.
434. MIKIROV, A.E. – DAN SSSR, Vol.142, No.3, p.587. 1962.
435. GUSHCHIN, G.P. – Trudy Glavnoi Geofizicheskoi Observatorii, No.105, p.43. 1960.
436. RABINOVICH, Yu.I. – Trudy Glavnoi Geofizicheskoi Observatorii, No.125, p.54. 1962.
437. COLLIS, R.T.H. and M.G.H.LIGDA. – J. Atmos. Sci., Vol.23, No.2, p.255. 1966.

438. FIOCCO, G. and G. GRAMS. – J. Atmos. Sci., Vol.21, No.3, p.323. 1964.
439. DEIRMENDJIAN, D. – J. Geophys. Res., Vol.70, No.3. 1965.
440. NEWKIRK, G. and J.L. KROENING. – J. Atmos. Sci., Vol.22, No.5, p.567. 1965.
441. ELTERMAN, L. and A.B. CAMPBELL. – J. Atmos. Sci., Vol.21, No.4, p.457. 1964.
442. ELTERMAN, L. – J. Opt. Soc. Amer., Vol.55, No.11, p.1583. 1965
443. ELTERMAN, L. – Appl. Optics, Vol.5, No.11, p.1769. 1966.
444. ELTERMAN, L. An Atlas of Aerosol Attenuation and Extinction Profiles for the Troposphere and Stratosphere. – Environmental Research Papers, No.241, AFCRL, Mass. 1966.
445. KONDRAT'EV, K.Ya., G.A. NIKOL'SKII, and E.N. ESIPOVA. – Izvestiya AN SSSR, Fizika Atmosfery i Okeana, Vol.2, No.4. 1966.
446. NEWKIRK, G., Jr. and J.A. EDDY. – J. Atmos. Sci., Vol.21, No.1, p.35. 1964.
447. ELTERMAN, L. Atmospheric Attenuation Model, 1964, in the Ultraviolet, Visible and Infrared Regions for Altitudes to 50 km. – Environmental Research Papers, No.46, AFCRL, Mass. 1964.
448. ELTERMAN, L. – Appl. Optics, Vol.3, No.6, p.745. 1964.
449. KELKAR, V.N. – Indian J. Meteorol. Geophys., Vol.13, No.2, p.173. 1962.
450. MASON, B.J. and C.L. ANDREWS. – Q. J. Roy. Meteorol. Soc., Vol.86, p.346. 1960.
451. POLYAKOVA, E.A. and K.S. SHIFRIN. – Trudy Glavnoi Geofizicheskoi Observatorii, No.42, p.84. 1953.
452. POLYAKOVA, E.A. – Trudy Glavnoi Geofizicheskoi Observatorii, No.68, p.92. 1957.
453. POLYAKOVA, E.A. – Trudy Glavnoi Geofizicheskoi Observatorii, No.100, p.45. 1960.
454. MARSHALL, T.S. and W. PALMER. – J. Meteorol., Vol.5, No.4, p.165. 1948.
455. ZEL'MANOVICH, I.L. – Trudy Glavnoi Geofizicheskoi Observatorii, No.100, p.58. 1960.
456. TVOROGOV, S.D. – Izvestiya Vuzov SSSR, Seriya Fizika, No.3, p.147. 1965.
457. GREEN, H.S. and E. WOLF. – Proc. Phys. Soc., Vol.66A, p.1129. 1953.
458. WOLF, E. – Proc. Phys. Soc., Vol.74, p.269. 1959.
459. ROMAN, P. – Acta Phys. Hung., Vol.4, p.209. 1957.
460. TVOROGOV, S.D. – Izvestiya Vuzov SSSR, Seriya Fizika, No.3, p.175. 1965.
461. HULST, H.C. van de. Light Scattering by Small Particles. New York–London, Wiley. 1957.
462. SCHIFF, L.I. – Phys. Rev., Vol.103, p.443. 1956.
463. TATARSKII, V.I. Theory of Fluctuation Phenomena During Propagation of Waves in the Atmosphere. Izdatel'stvo AN SSSR. 1959.
464. BORN, M. and E. WOLF. Principles of Optics. New York, Pergamon Press. 1959.
465. SOKOLOV, V.V. and S.D. TVOROGOV. – Zhurnal Prikladnoi Spektroskopii, Vol.8, No.3, p.493. 1968.
466. LANDAU, L.D. and E.M. LIFSHITS. Electrodynamics of Continuous Media. Oxford, Pergamon Press. 1960.
467. SHIFRIN, K.S. Scattering of Light in a Turbid Medium. Gostekhizdat. 1951.
468. TVOROGOV, S.D. – Izvestiya Vuzov SSSR, Seriya Fizika, No.1, p.87. 1961.
469. SHIFRIN, K.S. – Trudy Glavnoi Geofizicheskoi Observatorii, No.26, p.83. 1951.
470. JOHNSON, J.C. and J.R. TERRELL. – J. Opt. Soc. Amer., Vol.45, No.6, p.451. 1955.
471. PENNDORF, R.B. – J. Opt. Soc. Amer., Vol.46, p.1001. 1956; Vol.47, p.603. 1957; J. Meteorol., Vol.13, p.219. 1956; J. Phys. Chem., Vol.62, p.1537. 1958.
472. DEIRMENDJIAN, D., R. CLASEN, and W. VIEZEE. – J. Opt. Soc. Amer., Vol.51, No.6, p.620. 1961.
473. STEPHENS, J.J. and J.R. GERHARDT. – J. Meteorol., Vol.18, No.6, p.818. 1961.
474. HERMAN, B.M. – Q. J. Roy. Meteorol. Soc., Vol.88, No.376, p.143. 1962.
475. PENNDORF, R. Research in Aerosol Scattering in the Infrared. – Sci. Rep. 5. Tech. Rep. RAD-TR-61-32, U.S. Dept. of Commerce, Office of Technical Services. 1961.

476. IRVINE, W.M. – Bull. Astronom. Inst. Netherlands, Vol.17, No.4, p.266. 1964.
477. IRVINE, W.M. – J. Opt. Soc. Amer., Vol.55, No.1, p.16. 1965.
478. ORCHARD, S.E. – J. Opt. Soc. Amer., Vol.55, No.9, p.1190. 1965.
479. PLASS, G.N. – Appl. Optics, Vol.5, No.2, p.279. 1966.
480. KATTAWAR, G.W. and G.N. PLASS. – Appl. Optics, Vol.6, No.8, p.1377. 1967.
481. SHIFRIN, K.S. and I.L. ZEL'MANOVICH. Light Scattering Tables, Vol.1, Angular Functions. Gidrometeoizdat. 1966.
482. SHIFRIN, K.S. and I.L. ZEL'MANOVICH. Light Scattering Tables, Vol.2. Tables of Scattering Matrices and of Scattered-Field Components. Gidrometeoizdat. 1968.
483. SHIFRIN, K.S. and I.L. ZEL'MANOVICH. Light Scattering Tables, Vol.3. Extinction, Scattering and Radiation Pressure Coefficients. Gidrometeoizdat. 1968.
484. SHIFRIN, K.S. – Izvestiya AN SSSR, Seriya Geofizicheskaya, No.2, p.15. 1952.
485. KERKER, M., J.P. KRATOHVIL, and E. MATIJEVIC. – J. Opt. Soc. Amer., Vol.52, No.5, p.551. 1962.
486. FENN, R.W. and H. OSER. – Appl. Optics, Vol.4, No.11, p.1504. 1965.
487. KERKER, M., L.H. KAUFFMAN, and W.A. FARONE. – J. Opt. Soc. Amer., Vol.56, No.8, p.1053. 1966.
488. PLASS, G.N. – Appl. Optics, Vol.4, No.12. p.1616. 1965.
489. DEIRMENDJIAN, D. – Q. J. Roy. Meteorol. Soc., Vol.86, p.371. 1960.
490. PENNDORF, R. – J. Opt. Soc. Amer., Vol.52, No.4, p.402, 1962.
491. CLARK, G.C., CHIAO-MIN CHU, and S.W. CHURCHILL. – J. Opt. Soc. Amer., Vol.47, No.1, p.81. 1957.
492. VOLZ, F. – Ber. Deut. Wetterdienstes, Vol.2, No.13. 1954.
493. PENNDORF, R. – J. Opt. Soc. Amer., Vol.47, No.2, p.176. 1957.
494. FOLDY, L.L. – Phys. Rev., Vol.67, p.107. 1945.
495. LAX, M. – Rev. Mod. Phys., Vol.23, p.287. 1951; Phys. Rev., Vol.85, p.621. 1952.
496. WATERMAN, P.C. and R. TRUELL. – J. Math. Phys., Vol.2, p.512. 1961.
497. TWERSKY, V. – J. Math. Phys., Vol.3, p.700. 1962.
498. GNEDIN, Yu.N. and A.Z. DOLGINOV. – Zhurnal Eksperimental'noi i Teoreticheskoi Fiziki, Vol.45, p.1136. 1963.
499. GNEDIN, Yu.N. and A.Z. DOLGINOV. – Astronomicheskii Zhurnal, Vol.43, p.800. 1966.
500. BOROVOI, A.G. – Izvestiya Vuzov SSSR, Seriya Fizika, No.2, p.175. 1966.
501. BOROVOI, A.G. – Izvestiya Vuzov SSSR, Seriya Fizika, No.6, p.50. 1966.
502. BOROVOI, A.G. – Izvestiya Vuzov SSSR, Seriya Fizika, No.4, p.97. 1967.
503. BOROVOI, A.G. – Izvestiya Vuzov SSSR, Seriya Fizika, No.5, p.7. 1967.
504. ROZENBERG, G.A. – Uspekhi Fizicheskikh Nauk, Vol.56, No.1, p.77. 1955
505. MALKOVA, V.S. – Izvestiya AN SSSR, Fizika Atmosfery i Okeana, Vol.1, No.1, pp.109–113. 1965.
506. GORCHAKOV, G.I. and G.V. ROZENBERG. – Izvestiya AN SSSR, Fizika Atmosfery i Okeana, Vol.1, No.12, p.1279. 1965.
507. GORCHAKOV, G.I. – Izvestiya AN SSSR, Fizika Atmosfery i Okeana, Vol.2, No.6, p.595. 1966.
508. LEVIN, L.M. In: Studies on Experimental and Theoretical Physics, in Memory of G.S.Lansberg, p.121. Izdatel'stvo AN SSSR. 1959.
509. SHIFRIN, K.S. and V.F. RASKIN. – Optika i Spektroskopiya, Vol.11, No.2, p.268. 1961.
510. TVOROGOV, S.D. – Izvestiya Vuzov SSSR, Seriya Fizika, No.1, p.87. 1961.
511. TVOROGOV, S.D. – Izvestiya Vuzov SSSR, Seriya Fizika, No.4, p.175. 1962.
512. TVOROGOV, S.D. – Izvestiya Vuzov SSSR, Seriya Fizika, No.3, p.174. 1962.
513. ZUEV, V.E., M.V. KABANOV, B.P. KOSHELEV, S.D. TVOROGOV, and S.S. KHMELEVTSOV. – In: Actinometry and Atmospheric Optics (Report 5th Interdepartmental Conf. on Actinometry and Atmospheric Optics), edited by G.V. Rozenberg. Izdatel'stvo Nauka. 1964.

514. ZUEV, V.E., M.V. KABANOV, B.P. KOSHELEV, S.D. TVOROGOV, and S.S. KHMELEVTSOV. – Izvestiya Vuzov SSSR, Seriya Fizika, No.3, p.92. 1964.
515. TVOROGOV, S.D. – Izvestiya Vuzov SSSR, Seriya Fizika, No.3, p.147. 1965.
516. ZUEV, V.E. and S.D. TVOROGOV. – Izvestiya Vuzov SSSR, Seriya Fizika, No.2, p.143. 1966.
517. DEIRMENDJIAN, D. – Appl. Optics, Vol.3, No.2, p.187. 1964.
518. CENTENO, M. – J. Opt. Soc. Amer., Vol.31, p.244. 1941.
519. SONCHIK, V.K., V.P. LOPASOV, and N.A. CHERNYAVSKAYA. – In: Report Conf. Inst. Higher Learning on Spectral Transmission of the Atmosphere in the Visible and Infrared (June, 1965), Izdatel'stvo Tomskogo Gosudarstvennogo Universiteta. 1968.
520. ZUEV, V.E., V.P. LOPASOV, and V.K. SONCHIK. – Izvestiya AN SSSR, Fizika Atmosfery i Okeana, Vol.3, No.1, p.16. 1967.
521. ZUEV, V.E., and V.K. SONCHIK, – Izvestiya AN SSSR, Fizika Atmosfery i Okeana, Vol.5, No.7, p.745. 1969.
522. RUBENS, H. and E. LADENBURG. – Verh. D. Ges., Vol.11, No.16. 1909.
523. KISLOVSKII, L.F. – Optika i Spektroskopiya, Vol.7, No.3, p.311. 1959.
524. KONDRAT'EV, K.Ya., M.P. BURGOVA, I.F. GAINULIN, and G.F. TOGUNOVA. – In: Problems of Atmospheric Physics, No.2. Izdatel'stvo LGU. 1963.
525. ZUEV, V.E., V.V. SOKOLOV, and S.D. TVOROGOV. – In: Report All-Union Interdepartmental Conf. on Scattering of Light in the Atmosphere. Chernovtsy. 1967.
526. ZUEV, V.E., M.V. KABANOV, B.P. KOSHELEV, S.D. TVOROGOV, and S.S. KHMELEVTSOV. – Izvestiya Vuzov SSSR, Seriya Fizika, No.2, p.90. 1964.
527. ZUEV, V.E., B.P. KOSHELEV, S.D. TVOROGOV, and S.S. KHMELEVTSOV. – Izvestiya Vuzov SSSR, Seriya Fizika, No.3, p.121. 1966.
528. ZUEV, V.E., B.P. KOSHELEV, S.D. TVOROGOV, and S.S. KHMELEVTSOV. – Izvestiya AN SSSR, Fizika Atmosfery i Okeana, No.5, p.509. 1965.
529. KOSHELEV, B.P. – Izvestiya Vuzov SSSR, Seriya Fizika, No.3, p.126. 1966.
530. ARNULF, A. and J. BRICARD. – Beit. Phys. Atmos., Vol.33, No.1/2, p.9. 1960.
531. KURNICK, S.W., R.N. ZITTER, and D.W. WILLIAMS. – J. Opt. Soc. Amer., Vol.50, p.578. 1960.
532. GATES, D.M. and C.D. SHAW. – J. Opt. Soc. Amer., Vol.50, p.876. 1960.
533. BOCHAROV, E.I. – Izvestiya AN SSSR, Seriya Geofizicheskaya, No.5, p.678. 1958.
534. ZABRODSKII, G.M. and V.G. MORACHEVSKII. – Trudy Arkticheskogo i Antarkticheskogo Instituta, Voprosy Fiziki Oblakov i Tumanov, Vol.228, No.1, p.68. 1959.
535. ZABRODSKII, G.M. – Trudy Arkticheskogo i Antarkticheskogo Instituta, Voprosy Fiziki Oblakov i Tumanov, Vol.239, No.2, p.48. 1962; Trudy Vsesoyuznogo Nauchnogo Meteorologicheskogo Soveshchaniya, Vol.6, p.102. 1963.
536. TRAPPENBERG, R. – Arch. Meteorol. Geophys. Biokl., A.12, No.1, p.125. 1960.
537. McCUBBIN, T.K. – J. Opt. Soc. Amer., Vol.42, p.876. 1952.
538. ROZENBERG, G.V. – Izvestiya AN SSSR, Fizika Atmosfery i Okeana, Vol.3, No.9, p.836. 1967.
539. SHARONOV, V.V. Measurement and Calculation of the Visibility of Far Objects. Gostekhizdat. 1947.
540. SHIFRIN, K.S. and V.F. RASKIN. – Trudy Glavnoi Geofizicheskoi Observatorii, No.100, p.3. 1960.
541. SHIFRIN, K.S. and V.F. RASKIN. – In: Actinometry and Atmospheric Optics, p.178. Gidrometeoizdat. 1961.
542. SHIFRIN, K.S. and V.F. RASKIN. – Trudy Glavnoi Geofizicheskoi Observatorii, No.109, pp.155, 161. 1961.
543. ZUEV, V.E., V.V. SOKOLOV, and S.D. TVOROGOV. – Izvestiya Vuzov SSSR, Seriya Fizika, No.3, p.7. 1966.
544. ZUEV, V.E., V.V. SOKOLOV, and S.D. TVOROGOV. – Izvestiya Vuzov SSSR, Seriya Fizika, No.1, p.107. 1969.

545. POPOV, O.I. – Optika i Spektroskopiya, Vol.3, No.5, p.504. 1957.
546. CURCIO, J. – J. Opt. Soc. Amer., Vol.51, No.5, p.548. 1961.
547. KNESTRICK, G.L., T.H. COSDEN, and J.A. CURCIO. – J. Opt. Soc. Amer., Vol.52, No.9, p.1010. 1962.
548. PASSMAN, S. and L. LARMORE. – Proc. IRIS, Vol.1, No.2, p.15. 1956.
549. JUNGE, C. Gerlands Beitr. Geophys., Vol.46, p.108. 1935.
550. WRIGHT, H.L. – Q. J. Roy. Meteorol. Soc., Vol.66, p.66. 1940.
551. BOILEAN, A.R. – In: Scripps Institution of Oceanography, University of California, p.1. Feb. 1959.
552. BUMA, T.J. – Bull. Amer. Meteorol. Soc., Vol.41, p.357. 1960.
553. MIDDLETON, W.E.K. Vision Through the Atmosphere, ch.3, p.24. University of Toronto Press. 1952.
554. GEORGIEVSKII, Yu.S. – Izvestiya AN SSSR, Fizika Atmosfery i Okeana, Vol.2, No.5, p.494. 1966.
555. SHIFRIN, K.S. and I.N. MININ. – Trudy Glavnoi Geofizicheskoi Observatorii, No.68, p.5. 1957.
556. FARAPONOVA, G.P. – Trudy Tsentral'noi Aerologicheskoi Observatorii, No.23, p.52. 1957; No.32, p.3. 1959.
557. Handbook of Geophysics and Space Environments AFCRL, Office of Aerospace Research, U.S. Air Force. 1965.
558. RABINOVICH, Yu.I. – In: Actinometry and Atmospheric Optics, p.146. 1961; Trudy Glavnoi Geofizicheskoi Observatorii, No.118, p.18. 1961.
559. RABINOVICH, Yu.I. and L.N. GUSEVA. – Trudy Glavnoi Geofizicheskoi Observatorii, No.118, p.69. 1961.
560. GUSHCHIN, G.P. (editor). Atmospheric Ozone. Proc. 3rd Interdepartmental Conf. on Atmospheric Ozone, 21–23 May, 1963. Gidrometeoizdat. 1965.
561. GUSHCHIN, G.P. – In: Actinometry and Atmospheric Optics, p.218. 1961.
562. GUSHCHIN, G.P. Studies of Atmospheric Ozone. Gidrometeoizdat. 1963.
563. PENNDORF, R. – J. Meteorol., Vol.11, No.3. 1954.
564. PACKER, D. and C. LOCK. – J. Opt. Soc. Amer., Vol.41, No.7. 1951.
565. TOROPOVA, T.P. – Izvestiya Astrofizicheskogo Instituta AN Kazakhskoi SSR, Vol.10, p.71. 1960.
566. ZUEV, V.E., M.V.KABANOV, and Yu.A.PKHALAGOV. – Izvestiya AN SSSR, Fizika Atmosfery i Okeana, Vol.4, No.7, p.778. 1968.
567. DRIVING, A.Ya. and A.I. SMIRNOVA. – Izvestiya AN SSSR, Seriya Geofizicheskaya, No.3, p.337. 1958.
568. SCHMOLINSKY, F. – Meteor. Zeit., Vol.61, No.6. 1944.
569. FOITZIK, L. and H. ZSKHAECK. – Z. Meteorol., Vol.7, No.1, p.1. 1953.
570. POLYAKOVA, E.A. and V.V. TRET'YAKOV. – Trudy Glavnoi Geofizicheskoi Observatorii, No.100, p.53. 1960.
571. JEFFERSON, G.J. – Meteorol. Mag., Vol.90, No.1067, p.163. 1961.
572. SHIFRIN, K.S. and E.A. CHAYANOVA. – Trudy Glavnoi Geofizicheskoi Observatorii, No.170, p.93. 1965.
573. SHIFRIN, K.S., A.Ya. PEREL'MAN, and V.A.PUNINA. – Trudy Glavnoi Geofizicheskoi Observatorii, No.183, p.18. 1966.
574. BARTENEVA, O.D. – Izvestiya AN SSSR, Seriya Geofizicheskaya, No.12, p.1852. 1960.
575. BELOV, V.F. Measurement of the Principal Optical Characteristics of the Ground Air Layer. Gidrometeoizdat. 1956.
576. HULBERT, E.O. – J. Opt. Soc. Amer., Vol.31. 1941.
577. BULLRICH, K. – Meteor. Zeit., Vol.61, p.270. 1944.

578. BULLRICH, K. and F. MÖLLER. - Optik, Vol.2. 1947.
579. REGER, E. and A. SIEDENTOPF. - Optik, Vol.1, p.15. 1946.
580. BARTENEVA, O.D. and G.Ya. BASHILOV. - Izvestiya AN SSSR, Seriya Geofizicheskaya, No.4, p.613. 1961.
581. TOPOROVA, T.P. - Izvestiya Aerofizicheskogo Instituta Kazakhskoi SSR, Vol.9, p.108. 1959.
582. BOIKO, P.N., G.Sh. LIFSHITS, and T.P. TOROPOVA. - Izvestiya Aerofizicheskogo Instituta Kazakhskoi SSR, Vol.8, p.98. 1959.
583. PYASKOVSKAYA-FESENKOVA, E.V. Studies on the Scattering of Light in the Earth's Atmosphere. Izdatel'stvo AN SSSR. 1957.
584. KASTROV, V.G. - Trudy Tsentral'noi Aerologicheskoi Observatorii, No.32, p.84. 1959.
585. LYUBOVTSEVA, Yu.S. and G.V. ROZENBERG. - Izvestiya AN SSSR, Fizika Atmosfery i Okeana, Vol.2, No.3, p.248. 1966.
586. BELOV, V.F. - Trudy Tsentral'noi Aerologicheskoi Observatorii, No.23, p.63. 1957; Trudy Vsesoyuznogo Nauchnogo Meteorologicheskogo Soveshchaniya, Vol.6, p.52. 1963.
587. CHAYANOV, B.A. - Trudy Tsentral'noi Aerologicheskoi Observatorii, No.32, p.17. 1959.
588. KOMPANIETS, A.I. - Izvestiya AN SSSR, Fizika Atmosfery i Okeana, Vol.1, No.3, p.335. 1965.
589. BULLRICH, K. - Adv. Geophys., Vol.10, p.101. 1964.
590. KABANOV, M.V. - In: Report Conf. Inst. Higher Learning on Spectral Transmission of the Atmosphere in the Visible and Infrared (June 1965). Izdatel'stvo Tomskogo Gosudarstvennogo Universiteta. 1968.
591. CARRIER, L.W. and L.J. NUGENT. - Appl. Optics, Vol.4, No.11, p.1457. 1965.
592. BUKATYI, V.I., M.V. KABANOV, B.P. KOSHELEV, B.A. SMIRNOV, B.A. SAVEL'EV, and S.S. KHMELEVTSOV. - Izvestiya Vuzov SSSR, Seriya Fizika, No.8, p.139. 1967.
593. WOODWARD, D.H. - Appl. Optics, Vol.2, No.11, p.1205. 1963.
594. HARRIS, F.S., Jr., G.C. SHERMAN, and E.L. MORSE. - IEEE Transactions on Antennas and Propagation, Vol. AP-15, No.1, p.141. 1967.
595. SINCLAIR, D. - J. Opt. Soc. Amer., Vol.37, p.475. 1947.
596. BRILLOUIN, L. - J. Appl. Phys., Vol.20, No.11, p.1110. 1949.
597. GUMPRECHT, R.O. and C.M. SLIEPCEVICH. - J. Phys. Chem., Vol.57, p.90. 1953.
598. KING, R.W.P. and TAI TSUN WU. The Scattering and Diffraction of Waves. Cambridge, Harvard University Press. 1959.
599. POLYAKOVA, E.A. - Trudy Glavnoi Geofizicheskoi Observatorii, No.68, p.88. 1957.
600. TIMOFEEVA, V.A. - Trudy Morskogo Geofizicheskogo Instituta, Vol.3, p.35. 1953.
601. TIMOFEEVA, V.A. and L.A. KOVESHNIKOVA. - Izvestiya AN SSSR, Fizika Atmosfery i Okeana, Vol.2, No.3, p.320. 1966.
602. KABANOV, M.V. - In: Actinometry and Atmospheric Optics (Report 5th Interdepartmental Conf. on Actinometry and Atmospheric Optics), edited by G.V. Rozenberg. Izdatel'stvo Nauka. 1964.
603. MIDDLETON, W.E.K. - J. Opt. Soc. Amer., Vol.39, No.7, p.576. 1949.
604. STEWART, H.S. and J.A. CURCIO. - J. Opt. Soc. Amer., Vol.42, p.801. 1952.
605. GIBBONS, M.G. - J. Opt. Soc. Amer., Vol.48, p.550. 1958; Vol.49, p.702. 1959.
606. GIBBONS, M.G., J.R. NICHOLS, F.I. LAUGHRIDGE, and R.L. RUDKIN. - J. Opt. Soc. Amer., Vol.51, p.633. 1961.
607. GIBBONS, M.G., F.I. LAUGHRIDGE, J.R. NICHOLS, and N.A. KRAUSE. - J. Opt. Soc. Amer., Vol.52, p.38. 1962.
608. MESHKOV, V.V. Principles of Illumination Engineering. Part 1. Gosenergoizdat. 1957.
609. ELDRIDGE, R.G. and J.C. JOHNSON. - J. Opt. Soc. Amer., Vol.48, p.463. 1958.
610. ELDRIDGE, R.G. and J.C. JOHNSON. - J. Opt. Soc. Amer., Vol.52, p.787. 1962.

611. ZUEV, V.E., M.V.KABANOV, and A.G.BOROVOI. – Izvestiya Vuzov SSSR, Seriya Fizika, No.6, p.162. 1963.
612. ZUEV, V.E., and M.V.KABANOV. – Izvestiya Vuzov SSSR, Seriya Fizika, No.1, p.168. 1964.
613. ZUEV, V.E., M.V.KABANOV, and B.A.SAVEL'EV. – Izvestiya Vuzov SSSR, Seriya Fizika, No.5, p.80. 1964.
614. ZUEV, V.E., and M.V.KABANOV. – In: Actinometry and Atmospheric Optics (Report 5th Interdepartmental Conf. on Actinometry and Atmospheric Optics), edited by G.V.Rozenberg. Izdatel'stvo Nauka. 1964.
615. ZUEV, V.E., and M.V.KABANOV. – Izvestiya Vuzov SSSR, Seriya Fizika, No.6, p.175. 1965.
616. SMIRNOV, V.I. A Course in Higher Mathematics, Vol.2. Gostekhizdat. 1948.
617. IVANOV, A.P. and A.Ya.KHAIRULLINA. – Izvestiya AN SSSR, Fizika Atmosfery i Okeana, Vol.2, No.7, p.721. 1966.
618. HODKINSON, J.R. and J.R.GREENFIELD. – Appl. Optics, Vol.4, No.11, p.1463. 1965.
619. PROKHOROV, A.M. – Uspekhi Fizicheskikh Nauk, Vol.85, p.599. 1965.
620. BASOV, N.G. – Uspekhi Fizicheskikh Nauk, Vol.85, p.585. 1965.
621. AKHMANOV, S.A. and R.V.KHOKHLOV. – Uspekhi Fizicheskikh Nauk, Vol.88, No.3, p.439. 1966.
622. MAYMAN, S. – Elektronika, Vol.38. No.8, p.55. 1965.
623. TOWNES, C.H. – Uspekhi Fizicheskikh Nauk, Vol.88, No.3, p.461. 1966.
624. KALINSKII, A.A. and V.V.OSIKO. – Izvestiya AN SSSR, Neorganicheskie Materialy, Vol.1, No.12, p.2049. 1965.
625. KISS, Z.J. and R.J.PRESSLEY. – Trudy IIER, Vol.54, No.10, p.7. 1966.
626. BRIDGES, W.B. and A.N.CHESTER. – Appl. Optics, Vol.4, No.5, p.573. 1965.
627. JOHNSON, A.R. – Elektronika, Vol.39, No.8, p.3. 1966.
628. KISS, Z.J. and R.J.PRESSLEY. – Appl. Optics, Vol.5, No.10, p.1474. 1966.
629. SNITZER, E. – Appl. Optics, Vol.5, No.10, p.1487. 1966.
630. BLOOM, A.L. – Appl. Optics, Vol.5, No.10, p.1500. 1966.
631. NATHAN, M.I. – Appl. Optics, Vol.5, No.10, p.1514. 1966.
632. KHOKHLOV, R.V. – Uspekhi Fizicheskikh Nauk, Vol.87, p.17. 1965.
633. PIMENTAL, G.C. – Scientific American, Vol.214, No.4, p.32. 1966.
634. BUNKIN, F.V. (editor). Optical Lasers. Modern Developments in Optical Quantum Electronics. Izdatel'stvo Mir. 1966.
635. LABUDA, E.F., E.GORDON, and R.C.MILLER. – IEEE, J.Quant.Electron., Vol.1, No.6, p.273. 1965.
636. POLANYI, I.C. Appl. Optics, Suppl., No.2, p.109. 1965.
637. PAANANEN, R.A. – IEEE, Spectrum, Vol.3, No.6, p.88. 1966.
638. BOGORODITSKII, N.P., A.N.PIKHTIN, and D.A.YAS'KOV. – Izvestiya Leningradskogo Elektrotekhnicheskogo Instituta, Vol.57, No.21, p.3. 1966.
639. PATEL, C.K.N. – Chim. Phys. et Phys. Chim. Bioclim., Vol.64, No.1, p.82. 1967.
640. WALTER, W.T., N.SOLIMENE, M.PILTCH, and G.GOULD. – IEEE, J. Quant. Electron., Vol.2, No.9, p.474. 1966.
641. JOHNSON, A.R.I. – Laser Focus, Vol.2, No.17, p.19. 1966.
642. Survey of Laser Products Microwaves, Vol.5, No.10, pp.56, 58, 62. 1966.
643. MANGER, H. – Z. ang. Phys., Vol.18, No.4, p.265. 1965.
644. AKHMANOV, S.A., A.I.KOVRIGIN, V.A.KOLESOV, A.S.PISKARSKAYA, and V.V.FADEEV.– Zhurnal Eksperimental'noi i Teoreticheskoi Fiziki, Vol.3, No.9, p.372. 1966.
644a. DANILEIKO, M.V., V.I.KRAVCHENKO, S.G.ODULOV, O.N.POGORELYI, and M.S.SOSKIN. – Ukrainskii Fizicheskii Zhurnal, Vol.12, No.4, p.669. 1967.
645. MILLER, R.C. and W.A.NORDLAND. – Appl. Phys. Letters, Vol.10, No.2, p.53. 1967.
646. SOBOLEV, V.V. Radiant Energy Transport in Stellar and Planetary Atmospheres. Gostekhizdat. 1956.
647. WANG, M.C. and E.GUTH. – Phys. Rev., Vol.84, No.6, p.1092. 1951.
648. ROMANOVA, L.M. – Optika i Spektroskopiya, Vol.13, No.3, p.429. 1962.

649. ROMANOVA, L.M. - Optika i Spektroskopiya, Vol.13, No.6, p.819. 1962.
650. KABANOV, M.V. and B.A. SAVEL'EV. - Izvestiya AN SSSR, Fizika Atmosfery i Okeana, Vol.3, No.6, p.658. 1967.
651. GERSHUN, A.A. Selected Works on Photometry and Illumination Engineering. Fizmatgiz. 1958.
652. FEIGEL'SON, E.M. Radiative Processes in Stratus Clouds. Izdatel'stvo Nauka. 1964.
653. DOLIN, L.S. - Izvestiya Vuzov SSSR, Seriya Radiofizika, Vol.9, No.1, p.61. 1966.
654. KOLESOV, A.K. - Vestnik LGU, Seriya Matematika, Mekhanika, Astronomiya, No.1, p.160. 1966.
655. MARCHUK, G.I. and G.A. MIKHAILOV. - Izvestiya AN SSSR, Fizika Atmosfery i Okeana, Vol.3, No.3, p.258. 1967.
656. KABANOV, M.V. - Izvestiya Vuzov SSSR, Seriya Fizika, No.8, p.26. 1967.
657. PENNDORF, R.B. - J. Opt. Soc. Amer., Vol.52, No.7, p.797. 1962.
658. KREKOV, G.M., G.A. MIKHAILOV, and B.A. KARGIN. - Izvestiya Vuzov SSSR, Seriya Fizika, No.4, p.110. 1968.
659. KREKOV, G.M., G.A. MIHKAILOV, and B.A. KARGIN. - Izvestiya Vuzov SSSR, Seriya Fizika, No.5, p.54. 1968.
660. ZUEV, V.E., M.V. KABANOV, and B.A. SAVEL'EV. - DAN SSSR, Vol.175, No.2, p.327. 1967.
661. ZUEV, V.E., M.V. KABANOV, and B.A. SAVEL'EV. - Izvestiya AN SSSR, Fizika Atmosfery i Okeana, Vol.3, No.7, p.724. 1967.
662. KABANOV, M.V. and B.A. SAVEL'EV. - Izvestiya AN SSSR, Fizika Atmosfery i Okeana, Vol.4, No.9, p.960. 1968.
663. KABANOV, M.V., B.A. SAVEL'EV, and V.Ya. FADEEV. - Izvestiya Vuzov SSSR, Seriya Fizika, No.7, p.140. 1967.
664. SAVEL'EV, B.A. - Izvestiya Vuzov SSSR, Seriya Fizika, No.12, p.135. 1967.
665. SAVEL'EV, B.A. - Izvestiya Vuzov SSSR, Seriya Fizika, No.12, p.136. 1967.
666. TIMOFEEVA, V.A. - Trudy Morskogo Gidrofizicheskogo Instituta, Vol.3, p.35. 1953.
667. TIMOFEEVA, V.A. - DAN SSSR, Vol.113, p.556. 1957.
668. TIMOFEEVA, V.A. - Izvestiya AN SSSR, Seriya Geofizicheskaya, No.2, p.265. 1957.
669. ROZENBERG, G.V. - DAN SSSR, Vol.145, No.4, p.775. 1962.
670. IVANOV, A.P. and S.A. MAKAREVICH. - Izvestiya AN SSSR, Seriya Geofizicheskaya, No.11, p.1754. 1963.
671. IVANOV, A.P. and I.D. SHERBAF. - DAN BSSR, Vol.7, No.10, p.673. 1963.
672. IVANOV, A.P. and I.D. SHERBAF. - In: Actinometry and Atmospheric Optics, Izdatel'stvo Nauka. 1964.
673. IVANOV, A.P. and G.K. IL'ICH. - DAN BSSR, Vol.9, No.5, p.301. 1965.
674. IVANOV, A.P. and I.D. SHERBAF. - Optika i Spektroskopiya, Vol.18, No.14, p.698. 1965.
675. IVANOV, A.P. and G.K. IL'ICH. - DAN BSSR, Vol.9, No.7, p.438. 1965.
676. IVANOV, A.P. and I.D. SHERBAF. - Izvestiya AN SSSR, Fizika Atmosfery i Okeana, Vol.2, No.3, p.312. 1966.
677. IVANOV, A.P. and I.D. SHERBAF. - Zhurnal Prikladnoi Spektroskopii, Vol.5, No.2, p.195. 1966.
678. ROMANOVA, L.M. - Optika i Spektroskopiya, Vol.14, No.2, p.262. 1963.
679. BARTKY, C.E. - Appl. Optics, Vol.4, No.7, p.847. 1965.
680. TWOMEY, S. and H.B. HOWELL. - Appl. Optics, Vol.4, No.4, p.501. 1965.
681. FENN, R.W. - Appl. Optics, Vol.5, No.2, p.293. 1966.
682. GAVRILOV, V.A. Atmospheric Visibility. Gidrometeoizdat. 1965.
683. SHIFRIN, K.S. and I.L. ZEL'MANOVICH. - Trudy Glavnoi Geofizicheskoi Observatorii, No.170, p.61. 1965.

684. KOZLOV, V.P. and E.O. FEDOROVA. – Optiko-Mekhanicheskaya Promyshlennost', Vol.1, No.3, p.1. 1967.
685. MUNICK, R.J. – J. Opt. Soc. Amer., Vol.55, No.7, p.893. 1965.
686. DONCHENKO, V.A., M.V. KABANOV, B.A. SAVEL'EV, and I.V. SAMOKHVALOV. – Izvestiya Vuzov SSSR, Seriya Fizika, No.1, p.158. 1968.
687. GILBERT, G.D. and J.C. PERNICKA. – Appl. Optics, Vol.6, No.4, p.741. 1967.
688. EIDEN, R. – Appl. Optics, Vol.5, No.4, p.569. 1966.
689. TIMOFEEVA, V.A. – Izvestiya AN SSSR, Seriya Geofizicheskaya, No.5, p.766. 1961.
690. TIMOFEEVA, V.A. – DAN SSSR, Vol.147, No.3, p.604. 1962.
691. TIMOFEEVA, V.A., A.A. VOSTROKNUTOV, and L.A. KOVESHNIKOVA. – Izvestiya AN SSSR, Fizika Atmosfery i Okeana, Vol.2, No.12, p.1259. 1966.
692. KABANOV, M.V., B.A. SAVEL'EV, and I.V. SAMOKHVALOV. – Izvestiya AN SSSR, Fizika Atmosfery i Okeana, Vol.4, No.10. 1968.
693. ROMANOVA, L.M. – Izvestiya AN SSSR, Fizika Atmosfery i Okeana, Vol.1, No.6, p.599. 1965.
694. ROMANOVA, L.M. – Izvestiya AN SSSR, Fizika i Okeana, Vol.1, No.10, p.1022. 1965.
695. ROMANOVA, L.M. – Izvestiya AN SSSR, Fizika Atmosfery i Okeana, Vol.2, No.8, p.844. 1966.
696. NESMELOVA, L.I. and S.D. TVOROGOV. – Izvestiya Vuzov SSSR, Seriya Fizika, No.5, p.141. 1955.
697. BEBB, H.B. and A. GOLD. – Phys. Rev., Vol.143, No.1, p.1. 1966.
698. COHAN, N.V. and HAMEKA, H.F. – Phys. Rev., Vol.151, No.4, p.1076. 1966.
699. TOWNES, C.H. and A.L. SCHAWLAW. Microwave Microscopy. New York, McGraw-Hill. 1955.
700. KARPLUS, R. and J. SCHWINGER. – Phys. Rev., Vol.73, No.9, p.1020. 1948.
701. AKHMANOV, S.A. and R.V. KHOKHLOV. – Uspekhi Fizicheskikh Nauk, Vol.95, No.1, p.231. 1968.
702. ZUEV, V.E. – In: Jubilee Volume of Reports of Tomsk State University on Physics. Izdatel'sto Tomskogo Gosudarstvennogo Universiteta. 1970.
703. NESMELOVA, L.I. and S.D. TVOROGOV. – In: Report Conf. Inst. Higher Learning on Spectral Transmission of the Atmosphere in the Visible and Infrared. Izdatel'stvo Tomskogo Gosudarstvennogo Universiteta. 1968.
704. LANDAU, L.D. and E.M. LIFSHITS. Mechanics. Fizmatgiz. 1966.
705. ASKAR'YAN, G.A. – Zhurnal Eksperimental'noi i Teoreticheskoi Fiziki, pis'ma, Vol.4, No.10, p.400. 1966.
706. RAIZER, Yu.P. – Zhurnal Eksperimental'noi i Teoreticheskoi Fiziki, pis'ma, Vol.4, No.1, p.3. 1966.
707. RAIZER, Yu.P. – Zhurnal Eksperimental'noi i Teoreticheskoi Fiziki, pis'ma, Vol.4, No.4, p.124. 1966.
708. RAIZER, Yu.P. – Zhurnal Eksperimental'noi i Teoreticheskoi Fiziki, pis'ma, Vol.4, No.8, p.286. 1966.
709. YANKAUSKAS, Z.K. – Izvestiya Vuzov SSSR, Seriya Radiofizika, Vol.9, No.2, p.412. 1966.
710. LITVAK, A.G. – Izvestiya Vuzov SSSR, Seriya Radiofizika, Vol.9, No.3, p.629. 1966.
711. SHEN, Y.R. – J. Opt. Soc. Amer., Vol.56, No.10, p.1425. 1966.
712. CHIAO, R.Y., E. GARMIRE, and C.H. TOWNES. – Phys. Rev. Letters, Vol.13, p.479. 1964; erratum, Phys. Rev. Letters, Vol.14, p.1056. 1965.
713. AKHMANOV, S.A., A.P. SUKHORUKOV, and R.V. KHOKHLOV. – Zhurnal Eksperimental'noi i Teoreticheskoi Fiziki, Vol.50, No.6, p.1537. 1966.
714. BARKHUDAROVA, G.M., G.S. VORONOV, V.M. GORBUNOV, and N.B. DELONE. – Zhurnal Eksperimental'noi i Teoreticheskoi Fiziki, Vol.49, No.2, p.386. 1965.

715. WANG, C.C. – Phys. Rev. Letters, Vol.16, p.344. 1966.
716. AKHMANOV, S.A. and R.V. KHOKHLOV. Problems of Nonlinear Optics. Izdatel'stvo AN SSSR. 1964.
717. AKHMANOV, S.A., A.P. SUKHORUKOV, and R.V. KHOKHLOV. – Uspekhi Fizicheskikh Nauk, Vol.93, No.1, p.19. 1967.
718. ZEL'DOVICH, Ya.B. and Yu.P. RAIZER. – Zhurnal Eksperimental'noi i Teoreticheskoi Fiziki, Vol.3, No.3, p.137. 1966.
719. AKHMANOV, S.A., D.P. KRINDACH, A.P. SUKHORUKOV, and R.V. KHOKHLOV. – Zhurnal Eksperimental'noi i Teoreticheskoi Fiziki, Vol.6, No.2, p.509. 1967.
720. FUCHS, N.A. Evaporation and Droplet Growth in Gaseous Media. Pergamon Press. 1959.
721. WILLIAMS, T.A. – J. Heat Mass Transfer, Vol.8, p.575. 1965.
722. SHIFRIN, K.S. and Zh.K. ZOLOTOVA. – Izvestiya AN SSSR, Fizika Atmosfery i Okeana, Vol.2, No.12, p.1311. 1966.
723. KUZIKOVSKII, A.V. and S.S. KHMELEVTSOV. – Izvestiya AN SSSR, Fizika Atmosfery i Okeana, Vol.4, No.3, p.363. 1968.
724. KUZIKOVSKII, A.V. – Izvestiya Vuzov SSSR, Seriya Fizika, No.5, p.95. 1970.
725. KUZIKOVSKII, A.V. and S.S. KHMELEVTSOV. Proc. 9th All-Union Conf. on Radiowave Propagation, Khar'kov. 1969.
726. CHAPMAN, S. and T.G. COWLING. The Mathematical Theory of Nonuniform Gases. 2nd ed. Cambridge University Press. 1952.
727. BUKATYI, V.I. and V.A. POGODAEV. – Izvestiya Vuzov SSSR, Seriya Fizika, No.1, p.141. 1970.
728. BUKATYI, V.I. and S.S. KHMELEVTSOV. – Izvestiya Vuzov SSSR, Seriya Fizika, No.1, p.91. 1968.
729. HODARA, H. – Trudy IIER, Vol.54, No.3, p.36. 1966.
730. ZAITSEV, V.A.. and A.A. LEDOKHOVICH. – Trudy Glavnoi Geofizicheskoi Observatorii, No.156, p.118. 1964.
731. TATARSKI, V.I. Wave Propagation in a Turbulent Medium. New York, McGraw-Hill. 1961.
732. OBUKHOV, A.M. – Izvestiya AN SSSR, Seriya Geofizicheskaya, Vol.13, p.58. 1949.
733. BOVSHEVEROV, V.M., A.S. GURVICH, and M.A. KALISTRATOVA. – Izvestiya Vuzov SSSR, Seriya Radiofizika, Vol.4, No.5, p.886. 1961.
734. TSVANG, L.R. – Izvestiya AN SSSR, Seriya Geofizicheskaya, Vol.10, p.1594. 1963.
735. KALISTRATOVA, M.A. – Izvestiya Vuzov SSSR, Seriya Radiofizika, Vol.9, No.1, p.50. 1966.
736. GOLITSYN, G.S., A.S. GURVICH, and V.I. TATARSKII. – Akusticheskii Zhurnal, Vol.6, No.2, p.187. 1960.
737. KON, A.I. and V.I. TATARSKII. – Izvestiya Vuzov SSSR, Seriya Radiofizika, Vol.8, No.5, p.870. 1965.
738. CHENROV, L.A. Wave Propagation in a Medium with Random Inhomogeneities. Izdatel'stvo AN SSSR. 1958.
739. FEIZULIN, Z.I. and Yu.A. KRAVTSOV. – Izvestiya Vuzov SSSR, Seriya Radiofizika, Vol.10, No.1, p.68. 1967.
740. FRIED, D.L. – J. Opt. Soc. Amer., Vol.57, No.2, p.169. 1967.
741. DAVIS, J.I. – Appl. Optics, Vol.5, No.1, p.139. 1966.
742. FRIED, D.L. – J. Opt. Soc. Amer., Vol.56, No.10, p.1372. 1966.
743. FRIED, D.L. – J. Opt. Soc. Amer., Vol.56, No.10, p.1380. 1966.
744. FRIED, D.L. and R.A. SCHMELTZER. – Appl. Optics, Vol.6, No.10, p.1729. 1967.
745. FRIED, D.L. and SEIDMAN. – J. Opt. Soc. Amer., Vol.57, No.2, p.181. 1967.
746. MUNICK, R.J. – J. Opt. Soc. Amer., Vol.55, No.5, p.594. 1965.

747. STRAUB, H.W. – Appl. Optics, Vol.4, No.7, p.875. 1965.
748. SHAH, B.R. – J. Opt. Soc. Amer., Vol.56, No.4, p.559. 1966.
749. CONSORTINI, A., L. RONCHI, A.M. SCHEGGI, and G. TORALDO DI FRANCIA. – Radio Sci. J. Res., Vol.1, p.523. 1966.
750. HUFNAGEL, R.E. and N.R. STANLEY. – J. Opt. Soc. Amer., Vol.54, p.52. 1964.
751. GRACHEVA, M.E. and A.S. GURVICH. – Izvestiya Vuzov SSSR, Seriya Radiofizika, Vol.8, No.4, p.717. 1965.
752. KON, A.I. and V.I. TATARSKII. – Izvestiya Vuzov SSSR, Seriya Radiofizika, Vol.7, No.2, p.306. 1964.
753. GURVICH, A.S. and A.I. KON. – Izvestiya Vuzov SSSR, Seriya Radiofizika, Vol.7, No.4, p.790. 1964.
754. KABANOV, M.V., A.A. PERSHIN, and Yu.A. PKHALAGOV. – Izvestiya Vuzov SSSR, Seriya Fizika, No.9, p.121. 1968.
755. KHMELEVTSOV, S.S. and R.Sh. TSVYK. – Izvestiya Vuzov SSSR, Seriya Radiofizika, No.1. 1970.
756. BORISOV, B.D., V.M. SAZANOVICH, and S.S. KHMELEVTSOV. – Izvestiya Vuzov SSSR, Seriya Fizika, No.1. 1969.
757. HINCHMAN, W.R. and A.L. BUCK. – Proc. IEEE, Vol.52, p.305. 1964.
758. MEVERS, G.E., D.L. FRIED, and M.P. KEISTER. – J. Opt. Soc. Amer., Vol.55, No.11, p.1575. 1965.
759. EDWARDS, B.N. and R.R. STEEN. – Appl. Optics, Vol.4, No.3, p.311. 1965.
760. COLLINS, S.A. and M. MONROE. – J. Opt. Soc. Amer., Vol.56, No.4, p.560. 1966.
761. RYZNAR, E. – Appl. Optics, Vol.4, p.1416. 1965.
762. SUBRAMANIAN, M. and J.A. COLLINSON. – Bell. System Tech. J., Vol.46, No.3. 1967.
763. SUBRAMANIAN, M. and J.A. COLLINSON. – Bell. System Tech. J., Vol.44, p.543. 1965.
764. GOLDSTEIN, B.S., P.A. MILES, and A. CHABOT. – Proc. IEEE, Vol.53, p.1172. 1965.
765. FRIED, D.L. and G.E. MEVERS. – J. Opt. Soc. Amer., Vol.55, p.740. 1965.
766. BUCK, A.L. – Appl. Optics, Vol.6, No.4, p.703. 1967.
767. HÖHN, D.H. – Appl. Optics, Vol.5, No.9, p.1427. 1966.
768. FRIED, D.L., G.E. MEVERS, and M.P. KEISTER. – J. Opt. Soc. Amer., Vol.57, No.6, p.787. 1967.
769. HÖHN, D.H. – Appl. Optics, Vol.5, No.9, p.1433. 1966.
770. S"EDIN, V.Ya. and S.S. KHMELEVTSOV. – Izvestiya Vuzov SSSR, Seriya Radiofizika, No.1. 1970.
771. HERRICK, R.B. and J.R. MAYER-ARRENDT. – Appl. Optics, Vol.5, No.6, p.981. 1966.
772. MAROCUMA, T., K.F. NEFFLEN, T.R. LAWRENCE, and T.M. KLUCHER. – J. Opt. Soc. Amer., Vol.53, No.3. 1963.
773. McNISH, A.G. – Science, Vol.146, p.117. 1964.
774. VAITSEL', V.I. and S.S. KHMELEVTSOV. Theses Reports 1st Conf. on Information Transmission by Lasers. Izdatel'stvo Kievskogo Gosudarstvennogo Universiteta. 1968.
775. VAITSEL', V.I. and S.S. KHMELEVTSOV. Theses Report 1st Conf. on Information Transmission by Lasers. Izdatel'stvo Kievskogo Gosudarstvennogo Universiteta. 1968.
776. VAITSEL', V.I. and S.S. KHMELEVTSOV. – Izvestiya Vuzov SSSR, Seriya Fizika, No.6, p.87. 1968.
777. POKASOV, V.V. , and S.S. KHMELEVTSOV. – Izvestiya Vuzov SSSR, Seriya Fizika, No.5, p.82. 1968.
778. POKASOV, V.V. and S.S. KHMELEVTSOV. – Optika i Spektroskopiya, Vol.25, No.6, p.945. 1968.
779. ZUEV, V.E., I.I. IPPOLITOV, and Yu.S. MAKUSHKIN. Proc. Int. Symp. on Radiative Processes in the Atmosphere. Bergen, Norway. August 1968.

780. ZUEV, V.E., G.I.MARCHIK, V.Ya. GALIN, I.I.IPPOLITOV, Yu.S.MAKUSHKIN, S.D.TVOROGOV, and V.V.FOMIN. Proc. Int. Symp. on Radiative Processes in the Atmosphere. Bergen, Norway. August 1968.

781. ZUEV, V.E., M.V.KABANOV, B.A.SAVEL'EV, I.V.SAMOKHVALOV, V.V.SOKOLOV, and S.D.TVOROGOV. Proc. Int. Symp. on Radiative Processes in the Atmosphere. Bergen, Norway, August 1968.

782. ANTIPOV, B.A., V.E.ZUEV, V.A.SAPOZHNIKOVA, A.V.SOSNIN, and S.S. KHMELEVTSOV, Proc. Int. Symp. on Radiative Processes in the Atmosphere. Bergen, Norway. August 1968.

783. KREKOV, G.M. and G.A.MIKHAILOV. Proc. Int. Symp. on Radiative Processes in the Atmosphere. Bergen, Norway. August 1968.

784. KOSTKOWSKI, H.J. and A.M.BASS. – J. Opt. Soc. Amer., Vol.46, p.1060. 1956.

785. SESHADRI, K.S. and R.N.JONES. – Spectrochemica Acta, Vol.19, p.1013. 1963.

786. BASTIN, J.A., A.E.GEAR, G.O. JONES, H.J.T.SMITH, and P.J.WRIGHT. – Proc. Roy. Soc., Vol.278, No.1375, p.543. 1964.

787. RYADOV, V.Ya. and N.I.FURASHOV. – Izvestiya Vuzov SSSR, Seriya Radiofizika, Vol.9, No.5, p.1075. 1966.

788. DRYAGIN, Yu.A., A.G.KISLYAKOV, L.M. KUKIN, A.I.NAUMOV, and L.I.FEDOSEEV. – Izvestiya Vuzov SSSR, Seriya Radiofizika, Vol.9, No.5, p.1078. 1966.

789. RYADOV, V.Ya. and N.I.FURASHOV. – Optika i Spektroskopiya, Vol.24, No.2, p.178. 1968.

790. ZUEV, E.V., L.I.NESMELOVA, and S.D.TVOROGOV. – In: Report Conf. Inst. Higher Learning on Spectral Transmission of the Atmosphere in the Visible and Infrared (June 1965). Izdatel'stvo Tomskogo Gosudarstvennogo Universiteta. 1968.

791. GOLUBITSKII, B.M. and N.I.MOSKALENKO. – Izvestiya AN SSSR, Fizika Atmosfery i Okeana, Vol.4, No.1, p.85. 1968.

792. GOLUBITSKII, B.M., S.O. MIRUMYANTS, and N.I. MOSKALENKO. – Report 7th All-Union Conf. on Actinometry and Atmospheric Physics, Leningrad. 27–31 May 1968.

793. MOSKALENKO, N.I. – Izvestiya AN SSSR, Fizika Atmosfery i Okeana, Vol.4, No.7, p.777. 1968; Report 7th All-Union Conf. on Actinometry and Atmospheric Physics, Leningrad. 27–31 May 1968.

794. KISELEVA, M.S. – Optika i Spektroskopiya, Vol.24, No.3, p.401. 1968.

795. ROSSETTI, C. and P.BARCHEWITZ. – C. r. Acad. Sci., Paris, Ser.B, Vol.262, p.1966. 1966.

796. ROSSETTI, C., F.BOURBONNEUX, R.FARRENG, and P.BARCHEWITZ. – C. r. Acad. Sci., Paris, Ser.B, Vol.262, p.1684. 1966.

797. FARRENG, R., C.ROSSETTI, F.BOURBONNEUX, and P.BARCHEWITZ. – C. r. Acad. Sci., Paris, Ser.B, Vol.263, p.241. 1966.

798. HERZBERG, G. Molecular Spectra and Molecular Structure. Van Nostrand. 1945.

799. BABCOCK, H.D. and L.HERZBERG. – Astrophys. J., Vol.108, p.167. 1948.

800. ADIKS, T.G. and V.I.DIANOV-KLOKOV. – Izvestiya AN SSSR, Fizika Atmosfery i Okeana, Vol.4, No.10, p.1052. 1968.

801. SCHLAPP, R. – Phys. Rev., Vol.51, No.5. 1937.

802. WARK, D.Q. and D.H.MERCER. – Appl. Optics, Vol.4, No.7. 1965.

803. MALKEVICH, M.S. – Trudy Glavnoi Geofizicheskoi Observatorii, No.166, p.282. 1964.

804. SAIEDY, F., D.T.HILLEARY, and W.A.MORGAN. – Appl. Optics, Vol.4, No.4. 1965.

805. MOELLER, G. and J.D.RIGDEN. – Appl. Phys. Letters, Vol.8, p.69. 1966.

806. ZUEV, V.E., A.V.SOSNIN, , and S.S.KHMELEVTSOV. – Izvestiya AN SSSR, Fizika Atmosfery i Okeana, Vol.5, No.2, p.201. 1969.

807. KONDRATYEV, K.J., S.D.ANDREEV, I.J.BADINOV, V.S.GRISHECHKIN, and L.V.POPOVA. – Appl. Optics, Vol.4, No.9, p.1069. 1965.

808. NEPORENT, B.S., M.S. KISELEVA, A.G. MAKOGONENKO, and V.I. SHLYAKHOV. – Appl. Optics, Vol.6, No.11, p.1845. 1967.
809. KISELEVA, M.S., I.N. RESHETNIKOVA, and E.O. FEDOROVA. Report 7th All-Union Conf. on Actinometry and Atmospheric Optics, Leningrad. 27–31 May 1968.
810. IVLEV, L.S., N.S. BURSAKOVA, and O.M. SURIKOV. – In: Problems of Atmospheric Physics, No.6, Izdatel'stvo Leningradskogo Gosudarstvennogo Universiteta. 1968.
811. FENN, R. – Beitr. Physik Atmos., Vol.37, p.69. 1964.
812. BULLRICH, K., R. EIDEN, R. JAENICKE, and W. NOVAK. Final Technical Report, Contract DA-91-591-EUC-3458, Mainz. 1966.
813. IVLEV, L.S. and O.M. SURIKOV. – In: Problems of Atmospheric Physics, No.6. Izdatel'stvo Leningradskogo Gosudarstvennogo Universiteta. 1968.
814. TANAKA, M.– Special Contributions. Geophysical Institute, Kyoto University, No.6, p.47. 1966.
815. TOBA, Y. – Special Contributions. Geophysical Institute, Kyoto University, No.6, p.59. 1966.
816. TOBA, Y. and M. TANAKA. – Special Contributions. Geophysical Institute, Kyoto University, No.5, p.81. 1965.
817. TOBA, Y. and M. TANAKA. – J. Meteorol. Soc., Japan, Ser.II. Vol.41. No.3, p.135. 1963.
818. TOBA, Y. – Tellus, Vol.27, No.1, p.131; No.3, p.366. 1965.
819. TOBA, Y. – Tellus, Vol.28, No.1, p.132. 1966.
820. IVANOV, A.I., G.Sh.LIFSHITS, V.E. PAVLOV, B.T. TASHENOV, and Ya.A. TEIFEL'. Scattering of Light in the Atmosphere. Izdatel'stvo "Nauka" Kazakhskoi SSR. 1968.
821. IVLEV, L.S. and E.L. YANCHENKO. Report All-Union Conf. on Scattering of Light in the Atmosphere. Alma-Ata. November 1969.
822. NIKOLAEVA-TERESHKOVA, V.V. and G.V. ROZENBERG. – Izvestiya AN SSSR, Fizika Atmosfery i Okeana, Vol.1, No.4, p.386. 1965.
823. FEOKISTOV, K.P., G.V. ROZENBERG, A.B. SANDOMIRSKII, V.N. SERGEEVICH, and D.M. SONECHKIN. – In: Reports All-Union Conf. on Physics of Outer Space. Izdatel'stvo Nauka. 1965.
824. BUNAKOVA, A.M. and L.S. IVLEV. – In: Problems of Atmospheric Physics, No.5. Izdatel'stvo Leningradskogo Gosudarstvennogo Universiteta. 1967.
825. TOROPOVA, T.P. – In: Reports of Astrophysical Institute of Kazakh Academy of Sciences. Problems of Astrophysics and Atmospheric Optics, Vol.8. Izdatel'stvo "Nauka" Kazakhskoi SSR. 1967.
826. TOROPOVA, T.P. and S.O. OBASHEVA. – In: Reports of Astrophysical Institute of Kazakh Academy of Sciences. Problems of Astrophysics and Atmospheric Optics, Vol. 8. Izdatel'stvo "Nauka" Kazakhskoi SSR. 1967.
827. LIVSHITS, G.Sh. Scattering of Light in the Atmosphere, Part 1. Reports of Astrophysical Institute of Kazakh Academy of Sciences, Vol. 6. 1965.
828. BARTENEVA, O.D., E.N. DAVGYALLO, and E.A. POLYAKOVA. Experimental Studies on Optical Properties of the Ground Layer of the Atmosphere. – Trudy Glavnoi Geofizicheskoi Observatorii, No.220. Gidrometeoizdat. 1967.
829. SHIFRIN, K.S. and E.A. CHAYANOVA. – Izvestiya AN SSSR, Fizika Atmosfery i Okeana, Vol.3, p.274. 1967.
830. PAVLOV, V.E. – Astronomicheskii Zhurnal, Vol.41, No.1, p.122; No.3, p.546. 1964.
831. TYCHINSKII, V.P. – Uspekhi Fizicheskikh Nauk, Vol.91, No.3, p.389. 1967.
832. MARCHUK, G.I. (editor). Monte-Carlo Method in Radiation Transport Problems. Atomizdat. 1967.
833. KREKOV, G.M., G.A. MIKHAILOV, and B.A. KARGIN. – Izvestiya Vuzov SSSR, Seriya Fizika, No.9, p.99. 1968.

834. KREKOV, G.M. and G.A. MIKHAILOV. – Izvestiya Vuzov SSSR, Seriya Fizika, No.10, p.145. 1968.

835. KREKOV, G.M., M.M. KREKOVA, and I.V. SAMOKHVALOV. – Izvestiya Vuzov SSSR, Seriya Fizika, No.5. 1969.

836. KREKOV, G.M. – Izvestiya Vuzov SSSR, Seriya Fizika, No.6, p.74. 1969.

837. KREKOV, G.M. Reports 9th All-Union Conf. on Radiowave Propagation. Khar'kov. 1969.

838. IVANOV, A.P. and A.Ya. KHAIRULLINA. – Izvestiya AN SSSR, Fizika Atmosfery i Okeana, Vol.3, No.5, p.537. 1967.

839. KHAIRULLINA, A.Ya. and A.P. IVANOV. – Zhurnal Prikladnoi Spektroskopii, Vol.7, No.2, p.255. 1967.

840. IVANOV, A.P. and I.L. KATSEV. – Izvestiya AN SSSR, Fizika Atmosfery i Okeana, Vol.3, No.7, p.714. 1967.

841. KATSEV, I.L. and A.P. IVANOV. – Zhurnal Prikladnoi Spektroskopii, Vol.7, No.5, p.714. 1967.

842. BELYANTSEV, A.N., L.S. DOLIN, and V.A. SAVEL'EV. – Izvestiya AN SSSR, Seriya Radiofizika, Vol.10, No.4, p.489. 1967.

843. KREKOV, G.M. – Izvestiya Vuzov SSSR, Seriya Fizika, No.7, p.7. 1969.

844. ARMAND, S.A. – Radiotekhnika i Elektronika, Vol.14, No.4, p.587. 1969.

845. SAIEDY, F., H. JACOBONRITZ, and D.Q. WARK. – J. Atmos. Sci., Vol.21, No.1. 1967.

846. MONIN, A.S. and A.M. OBUKHOV. – Trudy Geofizicheskogo Instituta, No.24, p.151. 1954.

847. TSVANG, L.R. – Izvestiya AN SSSR, Seriya Geofizicheskaya, No.8, p.1252. 1960.

848. GURVICH, A.S. and B.N. MELESHKIN. – Izvestiya AN SSSR, Fizika Atmosfery i Okeana, Vol.2, No.7, p.688. 1968.

849. ROGERS, C.B. – J. Opt. Soc. Amer., Vol.55, No.9, p.1151. 1965.

850. WARD, R.C. and G.V. BERRY. – Appl. Optics, Vol.6, No.6, p.1136. 1967.

851. TATARSKII, V.I. – Izvestiya Vuzov SSSR, Seriya Radiofizika, Vol.10, No.1, p.48. 1967.

852. RICHARD, T., and J. BROWN. – In: Proc. 12th Conf. Radar Meteorology, Norman, Okla. 1966.

853. KABANOV, M.V. and I.V. SAMOKHVALOV. – Izvestiya Vuzov SSSR, Seriya Fizika, No.3, p.80. 1969.

854. FRIED, D.L. – J. Opt. Soc. Amer., Vol.57, No.2, p.169. 1967.

855. CHU, T.S. – Appl. Optics, Vol.6, No.1, p.163. 1967.

856. BERAN, M.J. – J. Opt. Soc. Amer., Vol.56, No.11, p.1475. 1966.

857. BONCH-BRUEVICH, A.M. and V.A. KHODOVOI. – Uspekhi Fizicheskikh Nauk, Vol.85, No.1, p.3. 1965.

858. WALLACE, R. – Molec. Phys., Vol.11, p.457. 1966.

859. LUKASZEWSKI, M. – Postepy fiz., Vol.18, p.287. 1967.

860. RAIZER, Yu.P. – Uspekhi Fizicheskikh Nauk, Vol.87, No.1. 1965.

861. MORTON, V.M. – Proc. Phys. Soc., Vol.92, p.301. 1967.

862. PIERRE, N. and V. PIERRE. – C. r. Acad. Sci., Vol.262, p.1337. 1966.

863. BERAN, M.J. and G.B. PARRENT. Theory of Partial Coherence. Englewood Cliff, N.J., Prentice-Hall. 1964.

864. TAYLOR, L. – J. Opt. Soc. Amer., Vol.57, No.3, p.304. 1967.

865. GUSHCHIN, G.P. (editor). Atmospheric Ozone. Gidrometeoizdat. 1966.

866. VASILEVSKII, K.P., V.A. KAZBANOV, and V.A. BOLDYREV. Report 7th All-Union Conf. on Actinometry and Atmospheric Physics, Leningrad. 27–31 May 1968.

867. GOLUBITSKII, B.M. and N.I. MOSKALENKO. – Izvestiya AN SSSR, Fizika Atmosfery i Okeana, Vol.4, No.1, p.90. 1968.

868. GOLUBITSKII, B.M. and N.I. MOSKALENKO. – Izvestiya AN SSSR, Fizika Atmosfery i Okeana, Vol.4, No.3, p.346. 1968.

869. MOSKALENKO, N.I., B.M. GOLUBITSKII, and S.P. MIRUMYANTS. Report 7th All-Union Conf. on Actinometry and Atmospheric Physics, Leningrad. 27–31 May 1968.
870. GOLUBITSKII, B.M. and N.I. MOSKALENKO. – Izvestiya AN SSSR, Fizika Atmosfery i Okeana, Vol.4, No.3, p.360. 1968.
871. FARMER, C.B. and J.T. HOUGHTON. – Nature, Vol.209, No.5030, p.1341. 1966.
872. GOLUBITSKII, B.M., S.O. MIRUMYANTS, M.V. TANTASHEV, V.L. FILIPPOV, and L.M. ARTEM'EVA. Report 7th All-Union Conf. on Actinometry and Atmospheric Physics, Leningrad. 27–31 May 1968.
873. FILIPPOV, V.L. and S.O. MIRUMYANTS. – Izvestiya AN SSSR, Fizika Atmosfery i Okeana, Vol.5, No.6, p.638. 1969.
874. ASHCHEULOV, S.V., K.Ya. KONDRAT'EV, and D.B. STYRO. – In: Problems of Atmospheric Physics, No.5. Izdatel'stvo Leningradskogo Gosudarstvennogo Universiteta. 1967.
875. BOLLE, H.J. – Memoires de la Société Royale des Sciences de Liège, Vol.26. 1964.
876. DAVIS, P.A. and W. VIEZEE. – J. Geophys. Res., Vol.69, No.18. 1964.
877. KISELEVA, M.S., I.N. MILININA, and E.O. FEDOROVA. Report 7th All-Union Conf. on Actinometry and Atmospheric Optics, Leningrad. 27–31 May 1968.
878. WACKER, P.F. and M.R. PRATTO. Microwave Spectral Tables, Vol.II. Line Strengths of Asymmetric Rotors. – NBS Monograph, 70. National Bureau of Standards, Washington, D.C. 1964.
879. GATES, D.M., R.F. CALFEE, and W. HANSEN. – Appl. Optics, Vol.2, No.11, p.1117. 1963.
880. GATES, D.M., R.F. CALFEE, D.W. HANSEN, and W.S. BENEDICT. Line Parameters and Computed Spectra for Water Vapor Bands at 2.7 μ. – NBS Monograph, 71. National Bureau of Standards. Washington, D.C. 1964.
881. BENEDICT, W.S. and R.F. CALFEE. Line Parameters for the 1.9 and 6.3 Micron Water Vapor Bands. – ESSA Professional Paper 2. U.S. Department of Commerce, Washington, D.C. 1967.
882. POKROVSKII, A.G. – In: Problems of Atmospheric Physics, No.5. Izdatel'stvo Leningradskogo Gosudarstvennogo Universiteta. 1967.
883. TIMOFEEV, Yu.M. – In: Problems of Atmospheric Physics, No.5. Izdatel'stvo Leningradskogo Gosudarstvennogo Universiteta. 1967.
884. KONDRAT'EV, K.Ya. and Yu.M. TIMOFEEV. Report 7th All-Union Conf. on Actinometry and Atmospheric Optics, Leningrad. 27–31 May 1968.
885. BEN ARYEH, Y. – J. Quant. Spectry. Rad. Trans., Vol.7, No.1, p.211. 1967.
886. SAKAI, H. Line Strengths and Widths in the ν_2 Water Band.
887. PONGTONGCHAROEN, P. Intensities in Heated Water Vapor in the Region 4–5 μ. – M.S. Thesis, Catholic Univ., Washington, D.C. 1965.
888. KIRIENKO, G.A., T.P. TOROPOVA, G.A. KHARITONOVA, and N.M. IBRAGIMOV. Report 7th All-Union Conf. on Actinometry and Atmospheric Optics, Leningrad. 27–31 May 1968.
889. GAL'TSEV, A.P. and V.M. OSIPOV. Report 7th All-Union Conf. on Actinometry and Atmospheric Optics, Leningrad. 27–31 May 1968.
890. GAL'TSEV, A.P. – In: Problems of Atmospheric Physics, No.5. Izdatel'stvo Leningradskogo Gosudarstvennogo Universiteta. 1967.
891. KOZHEVNIKOV, N.I., A.P. SARYCHEV, G.F. SITNIK, and A.I. KHLYSTOV. Report 7th All-Union Conf. on Actinometry and Atmospheric Optics, Leningrad. 27–31 May 1968.
892. SIEDENTOPF, H. and U. MAYER. – Optik, Vol.22, No.9, p.626. 1965.
893. ZUEV, V.E., A.V. SOSNIN, and S.S. KHMELEVTSOV. Report 7th All-Union Conf. on Actinometry and Atmospheric Optics, Leningrad. 27–31 May 1968.
894. JUNGE, H. Air Chemistry and Radioactivity. New York, Academic Press. 1963.
895. ROZENBERG, G.V. – Uspekhi Fizicheskikh Nauk, Vol.95, No.1, p.157. 1968.
896. SHEDLOVSKY, J.P. and I.H. BLIFFORD, Jr. Report 14th General Assembly of the International Geophysical Union, Geneva. September 1967.

897. GORCHAKOV, G.I. and G.V.ROZENBERG. – Izvestiya AN SSSR, Fizika Atmosfery i Okeana, Vol.3, No.6, p.611. 1967.

898. ROZENBERG, G.V. and G.I.GORCHAKOV. – Izvestiya AN SSSR, Fizika Atmosfery i Okeana, Vol.3, No.7, p.699. 1967.

899. GEORGIEVSKII, Yu.G. Report All-Union Interdepartmental Conf. on Scattering of Light in the Atmosphere, Chernovtsy. June 1967.

900. SELEZNEVA, E.S. Atmospheric Aerosols. Gidrometeoizdat. 1966.

901. SANDOMIRSKII, A.B., N.P.AL'TOVSKAYA, and G.I.TRIFONOVA. – Izvestiya AN SSSR, Seriya Geofizicheskaya, No.7, p.121. 1964.

902. PYLDMAA, V.K. and G.V.ROZENBERG. – Izvestiya AN SSSR, Fizika Atmosfery i Okeana, Vol.2, No.8, p.820. 1966.

903. KOPROVA, L.I. Report All-Union Interdepartmental Conf. on Scattering of Light in the Atmosphere, Chernovtsy. June 1967.

904. ROZENBERG, G.V., A.B.SANDOMIRSKII, and V.K.PYLDMAA. – In: Noctilucent Clouds. Trudy Mezhdunarodnogo Simpoziuma (Tallin, 1966), p.94. Izdatel'stvo VINITI. 1967.

905. ROZENBERG, G.V. and A.B.SANDOMIRSKII. – Izvestiya AN SSSR, Fizika Atmosfery i Okeana, Vol.3, No.2, p.151. 1967.

906. ROZENBERG, G.V. and Yu.S.LYUBOVTSEVA. – Izvestiya AN SSSR, Fizika Atmosfery i Okeana, Vol.3, No.2, p.172. 1967.

907. ZUEV, V.E., I.I.IPPOLITOV, and Yu.S. MAKUSHKIN. Report 7th All-Union Conf. on Actinometry and Atmospheric Optics, Leningrad. 27–31 May 1968.

908. ZUEV, V.E., M.V.KABANOV, and B.A.SAVEL'EV.– Appl. Optics, Vol.8, No.1, p.137. 1969.

909. ZUEV, V.E. Report 7th All-Union Conf. on Actinometry and Atmospheric Optics, Leningrad. 27–31 May 1968.

910. ZUEV, V.E., Report 7th All-Union Conf. on Actinometry and Atmospheric Optics, Leningrad. 27–31 May 1968.

INDEX

Absorption coefficient 66, 70, 75, 118, 137–140, 142–148, 193, 194, 227, 335
- in a gaseous medium 66, 70, 75, 118, 137–140, 142–148, 335
- in a scattering medium 193, 194, 227

Absorption functions 66, 67
- calculation utilizing spectral models 74–81, 106, 107, 118
- carbon dioxide 117–120
- carbon monoxide 125, 126
- experimental determination 81, 82, 107–111, 119, 120, 159–164
- in homogenous atmosphere 146–150
- – allowance for the earth's curvature 156–159
- – approximate method 150–156
- methane 122–124
- methods of direct calculation 73, 104–106, 117
- nitrous oxide 124
- of a single line 71–73
- of groups of overlapping lines 73–82, 126–128
- oxygen 122
- ozone 120–122
- water vapor 104–115, 160–164

Absorption of overlapping lines and bands 73–82, 126–128

Absorption spectra models 74–81, 106, 107, 118
- combined model 78, 79
- Elsasser model 75, 76
- quasi-random model 79
- statistical model 76–78

Anharmonicity 36, 37

Asymmetric rotator molecules 33–35, 42

Atmosphere
- aerosol structure 168–187
- – cloud and fog 168–172, 173, 176
- – haze 177–187
- – precipitation 187, 188
- carbon dioxide distribution 22–24
- distribution of minor constituents 22–24
- gaseous constituents 2
- oxygen distribution 1
- ozone distribution 21, 22
- pressure 1, 2
- temperature 2, 4
- turbulence characteristics 348–351, 366
- – correlation functions 349, 354
- – correlation radices 362
- – structure constants 349
- – structure functions 349, 352, 354
- water vapor distribution 4–20

Backscattering 304–307, 316

Bouguer law 66, 82–86, 146, 227, 245, 254, 271–275, 335
- limits of applicability for an absorbing medium 82–86
- limits of applicability for a scattering medium 271–275, 301–304

Carbon dioxide 22–24, 56–59, 115–120, 144, 145
- absorption coefficients 144, 145
- absorption functions 117–120
- components of fine spectral structure 115–117
- distribution in the atmosphere 22–24
- general characterization of absorption spectrum 56–59

Carbon monoxide 22–24, 65, 125, 126
- absorption function 125, 126
- distribution in the atmosphere 22–24
- fine spectral structure 125, 126
- general characterization of spectrum 65

Clouds and fogs 168–177, 234, 237–245, 263–266, 307–310
- chemical composition of particles 176, 177
- extinction coefficient 234, 237–243
- frequency of meteorological visibility values 243–245
- microstructural parameters 171–173

Clouds and fogs (contd.)
- particle concentration 172, 244
- particle size spectra 168–170
- polarization of scattered radiation 307–310
- scattering diagrams 263–266
- water content 172, 173, 244

Complex refractive index 204, 235–237
Correlation radius 362
Creation and annihilation operators 323

Depth (steady) regime 299–301
Doppler effect 44, 48

Effective field 267
Extinction coefficient 194, 196, 199, 226, 230, 234, 237–243, 246–254, 262

Fluctuations of parameters of light beams in a turbulent atmosphere 348–367
- beam shift 354
- coherence 366, 367
- cross section fluctuations 363, 364
- fluctuations in the angle of arrival 365, 366
- intensity fluctuations 356–363

Forward scattering 286–301, 310–314
- experimental investigations 291–301
- Monte Carlo method 289, 311
- solution of transport equation 286

Hamiltonian 88–91, 101, 321–323, 332–334
Harmonic oscillator 35, 36
Haze 177–187, 246–266
- chemical composition of particles 186, 187
- extinction coefficient 246–254
- particle concentration 179–187
- particle size spectrum 177–179
- scattering diagrams 263–266
- transmittance along horizontal paths 250, 252
- transmittance along slant paths 254–260

Integral absorption 67
Integral strength of bands 67, 111, 112, 120

Lasers 276–285
- absorption of radiation in the atmosphere 114–145, 165–167
- coherence 279, 280
- divergence 279, 282
- emission spectra 277–285
- fluctuation of parameters in the atmosphere 356–367
- gas 280–282
- glass 279–280
- operating regime 277–285
- output and energy 277–285
- scattering of radiation in the atmosphere 266–268, 286–319

Linear molecules 27–30, 41

Matrix element 50, 96–98
Meteorological visibility 186, 243–245, 256
Methane 22–24, 64, 65, 122–124, 142, 144
- absorption function 122–124
- absorption of laser radiation emitted at $\lambda = 3.39\,\mu$ 142, 144
- distribution in the atmosphere 22–24
- fine structure of spectrum 122, 123
- general characterization of spectrum 64, 65

Molecular moments of inertia 27, 28, 31, 33, 51
Molecular scattering 224, 225
Molecular spectra 25–27
- carbon dioxide 56–59
- carbon monoxide 65
- electronic 42, 43
- methane 64, 65
- nitrous oxide 63, 64
- oxygen 61–63
- ozone 59, 61
- purely rotational 27–35
- vibration-rotation 35–42
- water vapor 50–56

Nitrous oxide 22–24, 63, 64, 124
- absorption by individual bands 125
- distribution in the atmosphere 22–24
- fine structure of spectrum 124
- general characterization of absorption spectrum 63, 64

Nonlinear effects 320–347
- evaporation of droplets 338–347
- multiple-photon processes 320–332
- saturation effect 332–336
- self-focusing of a beam 336–338

Optical thickness 66
Oxygen 1, 61–63, 122
- absorption coefficients and functions 122
- diffuse absorption due to complex molecules 62

Oxygen (contd.)
- distribution in the atmosphere 1
- fine structure of spectrum 122
- general characterization of spectrum 61-63

Ozone 21, 22, 59, 61, 120-122
- absorption functions 121
- distribution in the atmosphere 21, 22
- fine structure of spectrum 120, 121
- general characterization of spectrum 59, 61

Parametric scanning of laser emission frequency 283-285
Perturbation theory 88-91
Polarization of scattered radiation 307-310
Precipitation 176, 177, 187, 188, 262, 263
- chemical composition 176, 177
- extinction coefficient 262, 263
- intensity 188
- particle size spectra 187, 188
- scattering diagrams 263

Pulsed radiation 278, 280, 285, 310-319
- deformation on propagation through a medium 310-319

Quantum numbers 28-30, 36-40

Radiation scattering by a single particle 189-222
- efficiency factors for extinction, scattering and absorption 204-218
- - approximate expressions 218
- - large particles 205, 206
- - small particles 205
- - "soft" particles 206
- - tabulation 206-218
- inhomogeneous field 199-204
- nonspherical particles 221, 222
- plane monochromatic wave 191-193
- scalar approximation 195-199
- scattering, absorption and extinction coefficients 193-195
- scattering diagram 220, 221
- scattering matrix 192
- Stokes parameters 190, 191

Scattering coefficient 193, 194, 224, 225, 227
Scattering matrix 192
Schrödinger equation 88, 90, 321
Selective radiation absorption 128, 129
Spectral absorption 66, 146, 147, 165-167
- for an inhomogeneous path 146, 147, 165-167

Spectral line 43-50, 68-73, 86-94, 105, 111-115
- absorption function 72-74
- integral absorption 67, 71-73
- location of centers 86-94, 105
- shape 43-49, 68-71
- - collision broadening 45-48
- - Doppler broadening 44
- - far wings 48, 111-115
- - half widths 44, 46-48
- - natural broadening 43
- strength 49, 50, 94-99, 104

Spectral transmission 66, 146, 147, 165-167
- for an inhomogeneous path 146, 147, 165-167

Spherical rotator molecules 30, 31, 41
Structure constants 349, 367
Structure functions 349
Symmetric rotator molecules 31, 32, 41, 42

Unsteady scattering 311-319

Water vapor 4-19, 50-56, 86-115, 136-139
- absorption by far wings of lines 111-115
- absorption coefficients 137-140, 143-145
- absorption functions 104-111, 159-164
- distribution in the atmosphere 4-19
- general characterization of absorption spectra 50-56
- half widths of lines 99-104
- line strengths 94-99, 104
- location of line centers 86-94, 104

Wave function 88, 90, 96, 100, 321-323, 332, 333